Variation

Dedicated to Ernst Mayr (1904–2005) who recognized the fundamental importance of population thinking in evolutionary biology.

Variation

Edited by
Benedikt Hallgrímsson
Brian K. Hall

AMSTERDAM • BOSTON • HEIDELBERG • LONDON
NEW YORK • OXFORD • PARIS • SAN DIEGO
SAN FRANCISCO • SINGAPORE • SYDNEY • TOKYO

Acquisition Editor: Nancy Maragioglio
Project Manager: Justin Palmeiro/Brandy Lilly
Associate Editor: Kelly Sonnack
Marketing Manager: Philip Pritchard
Composition: CEPHA IMAGING PVT LTD
Cover Printer: Phoenix Color
Interior Printer: The Maple-Vail Book Manufacturing Group

Elsevier Academic Press
30 Corporate Drive, Suite 400, Burlington, MA 01803, USA
525 B Street, Suite 1900, San Diego, California 92101-4495, USA
84 Theobald's Road, London WC1X 8RR, UK

This book is printed on acid-free paper. ♾

Library of Congress Cataloging-in-Publication Data

Application Submitted

British Library Cataloguing in Publication Data
A catalogue record for this book is available from the British Library

ISBN: 0-12-088777-0

For all information on all Elsevier Academic Press Publications
visit our Web site at www.books.elsevier.com

Printed in the United States of America
05 06 07 08 09 10 9 8 7 6 5 4 3 2 1

CONTENTS

FOREWORD

It is amazing to what great extent variation in natural populations has been neglected in the study of evolution. Amazing because natural selection would be meaningless without variation. This conclusion gave me the idea to consider the production of variation as a step in the process of natural selection. Periods of high phenotypic variation (as in the Pre-Cambrian, Cambrian) have alternated at other geological periods with periods of relative stasis. As natural selection is active at all periods, it would seem that evolutionary change depends more on the availability of variation than that of selection.

The discovery by Monod and Jacob that there are regulatory genes in addition to the structural genes has shed important light on the activity of genes. What, in turn, regulates the regulatory genes? What environmental conditions favor high variability? What others favor stability? Why has no geological period in the last 400 million years been able to produce such a reckless invention of new structural types as the Pre-Cambrian, Cambrian times? Can high variability by itself be favored by selection because it increases the probability of the production of unusual character combinations? Why are alterations of rapid and stagnant evolution (punctuated equilibria) so frequent in many phyletic lineages? Only the study of variation will enable us to explain why there are so many extraordinary productions like the peacock tail or certain Pre-Cambrian genes, which on first sight, would seem difficult to explain as the product of natural selection.

In short, variation is an endless source of challenging questions.

Ernst Mayr

CONTRIBUTORS

ALEXANDER V. BADYAEV, Department of Ecology and Evolutionary Biology, University of Arizona, Tucson, Arizona, USA

AARON M. BAUER, Department of Biology, Villanova University, Villanova, Pennsylvania, USA

PETER J. BOWLER, School of Anthropological Studies, Queen's University, Belfast, Northern Ireland, United Kingdom

JEVON JAMES YARDLEY BROWN, Joint Injury and Arthritis Group, University of Calgary, Calgary, Alberta, Canada

THEODORE M. COLE III, Department of Basic Medical Science, School of Medicine, University of Missouri-Kansas City, Kansas City, Missouri, USA

IAN DWORKIN, Department of Genetics, North Carolina State University, Raleigh, North Carolina, USA

REBECCA Z. GERMAN, Department of Physical Medicine and Rehabilitation, Johns Hopkins Hospital, Baltimore, Maryland, USA

BRIAN K. HALL, Department of Biology, Dalhousie University, Halifax, Nova Scotia, Canada

BENEDIKT HALLGRÍMSSON, Department of Cell Biology and Anatomy, Joint Injury and Arthritis Research Group, The Bone and Joint Institute, University of Calgary, Calgary, Alberta, Canada

ARY A. HOFFMANN, Centre for Environmental Stress and Adaptation Research, Department of Genetics, University of Melbourne, Parkville, Victoria, Australia

DONNA CARLSON JONES, Department of Physical Medicine and Rehabilitation, Johns Hopkins Hospital, Baltimore, Maryland, USA

CHRISTIAN PETER KLINGENBERG, Faculty of Life Sciences, The University of Manchester, Manchester, United Kingdom

ELLEN LARSEN, Department of Zoology, University of Toronto, Toronto, Ontario, Canada

SUBHASH R. LELE, Department of Mathematical and Statistical Science, University of Alberta, Edmonton, Alberta, Canada

JOHN A. MCKENZIE, Centre for Environmental Stress and Adaptation Research, Department of Genetics, University of Melbourne, Parkville, Victoria, Australia

DANIEL W. MCSHEA, Biology Department, Duke University, Durham, North Carolina, USA

LAUREN ANCEL MEYERS, Section of Integrative Biology, University of Texas at Austin, Austin, Texas, USA

A. RICHARD PALMER, Systematics and Evolution Group, Department of Biological Sciences, University of Alberta, Edmonton, Alberta, Canada; Bamfield Marine Sciences Centre, Bamfield, British Columbia, Canada

DAVID M. PARICHY, Department of Biology, University of Washington, Seattle, Washington, USA

JOAN T. RICHTSMEIER, Department of Anthropology, The Pennsylvania State University, University Park, Pennsylvania, USA; Center for Craniofacial Development and Disorders, The Johns Hopkins University, Baltimore, Maryland, USA

DEREK A. ROFF, Department of Biology, University of California, Riverside, California, USA

V. LOUISE ROTH, Biology Department, Duke University, Durham, North Carolina, USA

ANTHONY P. RUSSELL, Department of Biological Sciences, University of Calgary, Calgary, Alberta, Canada

SONIA E. SULTAN, Biology Department, Wesleyan University, Middletown, Connecticut, USA

SAMUEL SHOLTIS, Department of Anthropology, Pennsylvania State University, University Park, Pennsylvania, USA

STEPHEN C. STEARNS, Department of Ecology and Evolutionary Biology, Yale University, New Haven, Connecticut, USA

LEIGH VAN VALEN, Department of Ecology and Evolution, University of Chicago, Chicago, Illinois, USA

KENNETH M. WEISS, Department of Anthropology, Pennsylvania State University, University Park, Pennsylvania, USA

KATHERINE E. WILLMORE, Department of Medical Sciences, University of Calgary, Health Sciences Centre, Calgary, Alberta, Canada

MIRIAM LEAH ZELDITCH, Museum of Paleontology, University of Michigan, Ann Arbor, Michigan, USA

ACKNOWLEDGMENTS

This book represents the collaborative effort of many people. The editors would like to thank the contributors for accepting the invitation to contribute this volume and for the work and intellectual effort that went into writing each contribution. We appreciate that the time spent on these chapters is time not spent on something else and we hope that the final product has made that investment worthwhile. We would also like to thank Chuck Crumly for soliciting the original book proposal from us and for his role in the early stages of this book. We are grateful to Suzanne Harrison for kindly giving us permission to publish her father's foreword to our book, which he so generously wrote, as it turned out, only shortly before his death. We thank Kelly Sonnack, Lisa Royse, and Nancy Maragioglio for all of their hard work and professionalism and for keeping the production on track. We are grateful for the excellent assistance provided by Courtney Rippin, Wendy Swales, and Mitzi Murray at different times during this process.

Finally, we would both like to thank our spouses Leigh (BH) and June (BKH) for their support and encouragement and for patiently tolerating all of the impositions and inconveniences of academic life and thus allowing us to collaborate on projects such as this.

Benedikt Hallgrímsson
Brian K. Hall

CHAPTER 1

Variation and Variability: Central Concepts in Biology

BENEDIKT HALLGRÍMSSON* AND BRIAN K. HALL†
*Department of Cell Biology and Anatomy, Joint Injury and Arthritis Research Group, University of Calgary, Calgary, Alberta, Canada
†Department of Biology, Dalhousie University, Halifax, Nova Scotia, Canada

Phenotypic variation is the raw material for natural selection, yet a century after Darwin, it is an almost unknown subject.

Leigh Van Valen, 1974

Van Valen's assessment of our understanding and appreciation of phenotypic variation is only slightly less true today than it was 30 years ago. Yet, variation is a central topic, both conceptually and historically in evolutionary biology. Phenotypic variation was Darwin's fundamental observation. Indeed, the first two chapters of *On the Origin of Species* deal explicitly with variation. Variation within and among species has certainly been as central to the thinking of Ernst Mayr (1963) as it was to the thinking of Sewall Wright (1968), two of the fathers of the modern synthesis. However, the study of variability or the propensity to vary, with few exceptions (Bateson, 1894; Schmalhausen, 1949; Waddington, 1957), has remained peripheral to study of the mechanisms of evolutionary change at any level of the biological hierarchy. Although implicit in virtually all research in the biological sciences, whether one is seeking understanding at the genetic, developmental, organismal, species, population, or ecologic/community levels, variation is seldom treated as a subject in and of itself. The perception of this major gap prompted the present volume.

Variation is an extremely broad topic, and a modern treatment of this subject is not possible without a thematic focus. In this volume, we focus on the determinants and constraints on the generation of variation. We address this theme through both a hierarchical treatment and integrative approaches that point toward new directions of research.

Although it is not overtly noted, the book is organized thematically, beginning with Peter Bowler's overview of the historical and conceptual foundations of the topic.

We chose to solicit an historical perspective for this chapter because of the deep roots of the conceptualization of variation, many of which precede Darwin's synthesis. Bowler also deals with alternate theories of variation, thus providing a broader historical perspective against which later studies can be viewed.

The second section (Chapters 3–5) deals with the analysis of variation. The analytical issues deal with ways to compare and separate size and shape in biological structures. The field is surprisingly complex, and there are several methodologic approaches. Leigh Van Valen's treatment of approaches to variation is a classic that must be read by all serious investigators in this area (Van Valen, 1978). Current approaches include the geometric morphometric school as well as Euclidean distance matrix analysis. The methodologic debates are well beyond the scope of this volume, but readers interested in exploring these issues will find pertinent chapters in a forthcoming volume by Slice (2005).

The chapter by Joan Richtsmeier *et al.* lays out a recently developed and particularly useful approach for the analysis of variation. Their approach has been successfully used for the analysis of variability in dysmorphology in both mice and humans (Richtsmeier *et al.*, 2000). The contribution by Jones and German deals with the conceptual and methodologic issues that relate to the ontogenetic analysis of biological variation. Important aspects of variation can only be seen through ontogenetic analyses, and this is often overlooked. Jones and German deal with issues such as appropriate units of analysis and the levels at which variation can be measured.

The third section (Chapters 6–12) deals with the genetic and developmental determinants of the propensity to vary. The chapter by Lauren Myers deals with constraints on the propensity to vary at the molecular and genomic level, particularly in RNA. Her treatment includes mutational constraints and their evolution, the evolution of epistatic constraints, and modularity at the molecular level. A particular focus of her approach is the interaction of constraint and modularity in limiting and enabling the generation of molecular level variation.

Ellen Larsen's chapter focuses on the role of intrinsic developmental variation, by which she means variation that originates inherently in developmental processes and translates to heritable phenotypic variation. Her chapter essentially takes a broad view of the origin and role of variation that arises from emergent properties of complex developmental systems.

Chapter 8 by Dworkin focuses on canalization and the limitation of variation by development. Dworkin outlines different frameworks for the study of canalization. He relates canalization to reaction norms, and he devotes considerable thought and insight to the issue of how canalizing mechanisms influence the translation of genetic into phenotypic variation.

Ary Hoffman and John McKenzie take a genetic approach to the analysis of canalization. They discuss the evidence for the existence of specific genetic factors, such as heat shock proteins, which modulate variability in phenotypic traits in natural populations. Their work adds another layer to the hierarchy, that of interaction between genetic and environmental factors.

In the tenth chapter, Willmore and Hallgrímsson review the developmental basis for phenotypic variation that derives from developmental instability. This chapter takes a hierarchical and integrative approach to the developmental mechanisms from which developmental instability may arise. In particular, this chapter devotes thought to the issue of specific mechanisms versus diffuse emergent properties as the basis for variation in developmental stability, a theme that emerges in several other chapters such as that of Sultan and Stearns.

Chris Klingenberg continues the developmental and integrative theme from the previous chapter with a discussion of modularity and evolvability and developmental constraints. His conceptual work builds on the foundation laid by Wagner and others (Wagner and Altenberg, 1996). He presents methods for discriminating modularity caused by direct and indirect developmental and genetic interaction and discusses the implications of the finding based on these methods for evolvability of developmental systems.

The final chapter in this section, by Miriam Zelditch, takes an epigenetic view of the developmental regulation of variability. Focusing on the role of the mechanical environment and particularly muscle–bone interactions, Zelditch uses morphometric techniques to test hypotheses about the mechanisms responsible for the ontogenetic patterning of phenotypic variation. Although morphometri-cally based, her perspective is overtly developmental in that the focus is on the developmental mechanisms that generate or remove variance throughout ontogeny.

Following our hierarchical theme, the fourth section (Chapters 13–15) deals with environmental determinants of variation as well as genotype–environment interactions. The theoretical thrust of this section is to place the generation of variation into ecologic contexts. Alex Badayev's chapter deals with the relationship between stress, developmental integration, and both phenotypic and genetic variability. Two major themes of his contribution are that stress-induced increases in variation are structured by developmental systems and that the resulting pattern can have evolutionary consequences.

In the next chapter, Sonia Sultan and Steve Stearns review the vast area of phenotypic plasticity and the norm of reaction and lay out new directions of research in this vibrant area. The norm of reaction concept is central to the understanding of the interaction of genetic and environmental factors in the generation of phenotypic variation. This is true both historically, through Schmalhausen's work, and conceptually in current work in the area. Despite the large amount of work in this area, the relationship between norms of reaction and canalization is underappreciated, a point that is developed in this chapter.

The discussion of phenotypic plasticity is further elaborated in the context of life history by the following chapter by Roff. This chapter emphasizes the role of predictable versus stochastic environments in affecting patterns of phenotypic plasticity over life history using a population-genetic framework. Roff reviews the current population genetic concepts related to the central issues of the maintenance and reduction of genetic variance and relates these to life-history evolution.

The fifth section (Chapters 16–19) deals with comparative and phylogenetic approaches to the study of variation. Obviously, this is a vast field encompassing many areas of study and perspectives. The chapters solicited address four important topics within this vast area. These are variation in relation to organismal symmetry, structure and function, the evolution of developmental and morphologic complexity, and macroevolution. This selection of topics is not intended to be comprehensive or exhaustive. Rather, these are areas of important intersection between the study of the processes that underlie variation and large-scale evolutionary patterns.

Rich Palmer's chapter entitled "Antisymmetry" deals with variation across planes of symmetry and evolution of asymmetric phenotypes. This is an interesting area because the evolution of directional asymmetry such as that seen in flounder heads or in the human heart involves the evolution of heritable variation across planes of symmetry from developmental systems in which, presumably, such variation does not exist. Palmer's argument, supported by hundreds of examples, is that antisymmetry, or the tendency for negative covariation across planes of symmetry, is a critical intermediary step in the evolution of asymmetry phenotypes.

Tony Russell and Aaron Bauer tackle the ambitious topic of how variation in structure relates to variation in function. They review different approaches to this issue, including the advantages and limitations of natural (*in situ*) versus laboratory (*ex situ*) studies. They advocate an approach that combines these two kinds of studies and point out that the degree to which structure–function relationships can ever be worked out depends critically on the level of detail specified by the theoretical model.

Dan McShea tackles the issue of how long-term evolutionary trends in developmental complexity, developmental buffering, and the generation of variability interact. He discusses the selective forces and trade-offs that may determine the overall evolutionary trend and argues that there is a long-term increase in internal variance that goes along with a tendency for complexity to increase. This chapter builds on his extensive work on complexity and evolution (McShea, 1996), adding an explicit consideration of the role of variability.

Having begun with a historical perspective and an analysis of current concepts, the sixth section (Chapters 20–22) looks to the future in dealing with new directions of research relating to variation and variability. This section explores important intersections and integrations between disciplines, principally evo-devo, phenogenetics, and evolutionary theory.

In the first chapter of this section, David Parichy argues that, despite the strong history of population thinking in evolutionary biology since Darwin, developmental biology has remained essentially typologic in approach. He argues that this will finally change in coming years as developmental biologists are forced to turn their attention to issues of variability by addressing the remaining large issues that face that field. This includes understanding the developmental genetic basis for variation among evolutionary lineages, among closely related species, and phenotypic variation within species. The latter topic, he argues, important for both evolutionary and biomedical research, will focus on how developmental systems translate genetic to phenotypic variation. In this area, canalization is a central concern.

In the second chapter, Sholtis and Wiess lay out the theoretical perspective of phenogenetics. Their perspective focuses on the complexities of the genotype to phenotype relationship. Arguing, like Minelli (2003) and others, that modern developmental biology is overly "gene centric," they point out that the complications introduced by gene regulation, epigenetic factors, and the environmental context of development create a many-to-many relationship between genotype and phenotype. The relationship is further complicated by the fact that the genetic basis for development can and does evolve without phenotypic change and by developmental stability. None of this is individually contentious or novel, but their overall perspective is in that they argue that the implications of these complications are that the gene-driven paradigm of modern developmental genetics is fundamentally flawed. They argue for a more phenotype-focused approach and explicit consideration of these complications in experimental investigation of the developmental basis for evolutionary change and the developmental-genetics of human disease.

A theme that emerges from the approaches that authors have taken is the distinction between variation and variability and an emphasis on the latter. Günter Wagner *et al.* (1997) have made this distinction explicitly, defining variation as the set of observed differences and variability as the tendency of a system to generate differences. The distinction, therefore, is one of pattern versus process. Although they were not explicitly asked to do so, the authors in this volume by and large chose to focus on process and, therefore, on variability. Current questions about variability deal with factors that enhance or limit the tendency for biological systems to exhibit variation at the different levels of the biological hierarchy. A common theme that emerges through the chapters by Larsen, Roth, McShea, and Russell and Bauer is the phylogenetic diversity in the processes that influence variability from variation in molecular, developmental, and functional constraints. Canalization is another central theme that emerges through many of the chapters. As a conceptual framework for understanding variability, Waddington's concept of canalization is clearly pivotal. Nearly all of the chapters touch on canalization in one form or another. In particular, the chapters by Larsen, Dworkin, Hoffman and McKenzie, Willmore, and Parichy as well as our own chapter deal explicitly with canalization, and this arose without instruction or encouragement from the editors despite our own obvious interest in the subject. In the final chapter, we attempt to pull together some of these themes to define a theoretical framework for the study of phenotypic variability at the level of the organismal phenotype. Understanding the mechanisms by which developmental systems buffer, augment, or structure phenotypic variation is central to understanding the relationship between genetic and phenotypic variation. The complexities of that relationship, after all, are principally what make the study of development so relevant to understanding evolutionary processes.

A central motivation behind much of the work in this book is the need for a coherent theory of phenotypic variation. The most fundamental task is a coherent framework for relating genetic to phenotypic variation. This would be a trivial task if genetic variation mapped directly onto phenotypic variation as is often assumed in population genetic models. We argue that developmental processes are the level at which one can understand how genetic variation is translated to phenotypic variation within and among species. Experimental developmental biology, complex systems modeling, and bioinformatics all have important roles to play for the theory. Morphometrics also have an important role by providing methods to quantify phenotypic variation in shape and size. A theory of phenotypic variation must also be firmly grounded in population genetics and must be able to relate selection, gene flow patterns, population structure and size, geographic range, and spatiotemporal environmental variation to the developmentally based expression of phenotypic variation. Finally, a theory of phenotypic variation must draw on both developmental biology and population genetics to provide a framework to understand how developmental systems influence the dynamics of macroevolutionary change. These tasks situate variation centrally within the evolutionary developmental biological paradigm and thus return the concept to the forefront of evolutionary thought.

This volume appears at a time when the synthesis of developmental and evolutionary biology (evo-devo) is reaching a mature phase. Indeed, the prospect for a new synthesis bridging genetics, development, ecologic, and evolutionary biology now seems more likely than at any time in the past. Understanding the significance of both variation and variability will be crucial to this emerging synthesis. One trend that may contribute to increased understanding and appreciation of variation and variability is the increased focus on systems (as opposed to gene or developmental process) level understanding of development, which is enabled by the ongoing growth and maturation of bioinformatics and computational biology. Recent work by Siegal and Bergman is an example of early contributions to this emerging area (Siegal and Bergman, 2002; Bergman and Siegal, 2003). This area is only tangentially treated here because the subject of modeling biological systems to understand the origins of variation would require a separate volume. The goal here is to bring together a diversity of treatments of variation and variability at multiple levels and thus situate the role of variability within this broad emerging synthesis.

Only three other volumes have attempted a broad treatment of phenotypic variation since Darwin: William Bateson's *Materials for the Study of Variation* (Bateson, 1894), Sewall Wright's *Variability Within and Among Natural Populations* (1984), and Yablokov's *Variability in Mammals* (1966). This volume aims to examine the concept of variation through the lenses created by different levels and areas of research. The processes related to phenotypic variation emerge from the complexities of interaction among and within different levels of organization. For this reason, a hierarchical approach is critical. We believe that the resulting treatment is unique in juxtaposing a series of perspectives with a transdisciplinary approach that seeks to contextualize variation as a foundational concept in biology.

REFERENCES

Bateson, W. (1894). *Materials for the Study of Variation Treated with Especial Regard to Discontinuity in the Origin of Species* (reprinted in 1992). Baltimore and London: Johns Hopkins University Press.

Bergman, A., and Siegal, M. L. (2003). Evolutionary capacitance as a general feature of complex gene networks. *Nature* 424(6948), 549–552.

Mayr, E. (1963). *Animal Species and Evolution*. Cambridge, MA: Belknap Press.

McShea, D. W. (1996). Metazoan complexity and evolution: Is there a trend? *Evolution* 477–492.

Minelli, A. (2003). *The Development of Animal Form: Ontogeny, Morphology, and Evolution*. Cambridge, England: Cambridge University Press.

Richtsmeier, J. T., Baxter, L. L., and Reeves, R. H. (2000). Parallels of craniofacial maldevelopment in Down syndrome and Ts65Dn mice. *Development Dynamics* 217(2), 137–145.

Schmalhausen, I. I. (1949). *Factors of Evolution*. Chicago: University of Chicago Press.

Siegal, M. L., and Bergman, A. (2002). Waddington's canalization revisited: Developmental stability and evolution. *Proceedings of the National Academy of Science USA* 99(16), 10528–10532.

Slice, D. E., ed. (2005). *Modern Morphometrics in Physical Anthropology*. New York: Kluwer.

Van Valen, L. M. (1978). The statistics of variation. *Evolutionary Theory* 433–443.

Waddington, C. H. (1957). *The Strategy of the Genes*. New York: MacMillan Company.

Wagner, G. P., and Altenberg, L. (1996). Complex adaptations and the evolution of evolvability. *Evolution* 50, 967–976.

Wagner, G. P., Booth, G., and Bagheri-Chaichian, H. (1997). A population genetic theory of canalization. *Evolution* 51(2), 329–347.

Wright, S. (1968). *Evolution and the Genetics of Populations: A Treatise*. Chicago: University of Chicago Press. Vol. 4.

Wright, S. (1984). Evolution and the Genetics of Populations, Volume 4: Variability within and among Natural Populations. Chicago: University of Chicago Press.

Yablokov, A. V. (1966). *Variability of Mammals*. New Delhi, India: Amerind.

CHAPTER 2

Variation from Darwin to the Modern Synthesis

PETER J. BOWLER
School of Anthropological Studies, Queen's University, Belfast, Northern Ireland, United Kingdom

INTRODUCTION

Darwin's theory of natural selection inevitably focused attention on the problem of variation within species. For selection to work, there must be significant differences between the individuals making up the population, which was what Darwin meant by the term *variation*. Understanding the nature of this variation would remain a major problem within evolution theory from Darwin's time through to the creation of the Modern Synthesis of genetics and natural selection—and beyond. Darwin did not think about heredity and variation in terms that would be familiar to a post-synthesis biologist, so the first efforts to create a history of the problem of variation were constructed around the assumption that it was necessary to clarify the issues about which Darwin was confused. However, historians of science no longer think in these terms: Darwin may have been "confused" by modern standards, but it is important to try to understand how he viewed the nature of variation and why he thought about it in those terms. Only then will we be able to understand properly the transitions that were necessary to establish the foundations of how we think about the issue today.

Consequently, despite the boldness of his theorizing in other respects, we have to accept that in this area he remained content to think within a conventional framework. He tended to accept without question the traditional view that variation and heredity were two antagonistic processes: Heredity tried to make the offspring an exact copy of its parents, while variation was a disturbing force that limited the exactness of the copying. Although modern biologists see variation as something, which, by definition, occurs within a population, Darwin still thought in terms of a "norm" for the species, a norm that varies in some individuals, perhaps because a changed environment interfered with the copying involved in heredity. He had no sense of the population as a reservoir of variant characters actually maintained by heredity. For Darwin, each variant individual was created by forces affecting that individual's conception and ontogeny. Because he thought the "interference" that created variation came from outside the organism, Darwin still saw characters acquired during the organism's life before reproducing as part of the variability relevant for evolution. Disturbances could be produced both before conception and afterward; there was no distinction between genetic and somatic characters, and Darwin could still retain a role for the Lamarckian effect of the inheritance of acquired characters.

Because Darwin still thought of variation as a collection of individual acts of "disturbance," he drew no sharp line between small or trivial individual variations and large-scale monstrosities or saltations, sometimes called "sports of nature." Although Darwin did not think that natural selection normally made use of sports, he saw the better-adapted individuals favored by selection as being just as *individual* as sports and just as rare except when the species was subjected to a major change in its environment. Many of his contemporaries did believe that evolution made use of sports; indeed it was a common assumption that only sports had the capacity to establish a new breeding population with a different basic character to the old, i.e., a new species. However, where Darwin assumed that individual acts of variation were in a sense "random"—because they represented disturbances of the copying process—many of his contemporaries thought that the forces generating sports or even smaller variations might be controlled from within the organism, by whatever drove the embryo to develop toward a mature organism. Even if the variant were triggered by an external factor disturbing heredity, the disturbance was likely to generate a change that was to some extent latent within the processes of ontogeny. It would thus be to some extent predetermined, and there would be only limited directions of variability available to the species. This assumption—the foundation of theories of orthogenetic evolution—has reemerged in a modern form in our current preoccupation with the possibility of developmental constraints. However, the old viewpoint tended to see the directed variations exploited by Lamarckian or orthogenetic evolution as *additions* to development, not disturbances of it. Here was a crucial difference between Darwin and his opponents: He accepted that variation was a disturbance of the copying process, while they often saw it as the cumulative addition of stages to the existing process of ontogeny. As Gould (1977) showed, this is why the recapitulation theory flourished in the late nineteenth century: If evolution is the addition of stages to individual development, old adult forms will be preserved as earlier stages in that development. There have been recent efforts to link Darwin himself to the recapitulation theory (Richards, 1991), but the position adopted in this article is that his willingness to see variation as relatively free from direction by the existing course of ontogeny allowed him to formulate a theory that could be translated into the modern language of genetics and mutation.

The transition to the model of variation that became incorporated into the genetic theory of natural selection thus involved far more than the clarification of old-fashioned confusions. It required the dismantling of key aspects of the old view of heredity and variation, which assumed that variation was a product of additions (internally or externally directed) to the process of ontogeny. The old viewpoint was, up to a point, a coherent theoretical model in which Lamarckism, orthogenesis, and the recapitulation theory all seemed perfectly natural and obvious, indeed more obvious than the kind of random variation that Darwin saw as the raw material of natural selection. Explaining why Darwin made his breakthrough to a theory based on undirected variation—even though he could not escape other aspects of the contemporary view—is vital for our understanding of the emergence of the selection theory and the model of variation it entails. Equally crucial is the process by which that insight was transformed by Mendelian genetics in the early twentieth century, a process I once dubbed the "Mendelian revolution" because I think it was as crucial as Darwin's original popularization of the basic idea of evolution (Bowler, 1989). The breakdown of the old developmental viewpoint, to which Darwin himself still subscribed in some areas, involved the following key transitions:

1. Recognition that acquired characters can have no genetic basis and hence are not to be confused with the kind of variation relevant to evolution.
2. Recognition that variation is best understood as a property of populations, not of individuals.
3. Recognition that heredity and variation are not antagonistic because heredity preserves the variant characters within a population.
4. Rejection of the view that variant characters are likely to be additions to development controlled by the existing process of development (which does not rule out the possibility that ontogeny might impose constraints on the range of variations).
5. Clarification of the relationship between large sports and what Darwin would have regarded as normal individual differences. In part this required a recognition that genetic mutations can have both large (often severely deleterious) and small effects.

I. VARIATION BEFORE DARWIN

Space forbids any detailed exploration of this topic, but we need to be aware of the extent to which pre-Darwinian naturalists and breeders allowed for variation within species. According to Mayr (1982), this period was dominated by a typologic view of species, in which the species was defined by an ideal type or pattern, perhaps existing in the mind of the Creator. There is now much disagreement over Mayr's view, but his position can be accepted at least to the extent that most naturalists believed that there were limits defining the range of characters members of the species could acquire (for a survey, see Bowler, 2003, Chapters 3 and 4).

Almost everyone accepted that there was some room for modification under changed conditions, because all experienced naturalists knew that there were well-marked local varieties or subspecies, which were almost certainly derived by natural processes from whatever the parent form had been. Mayr (1982: 340–341), for instance, points to the recognition of this point in Linnaeaus's *Philosophia Botanica* of 1750. There was, however, disagreement over the extent to which such natural variation could change the species. Some taxonomists (colloquially known nowadays as "splitters") were loath to admit significant variation within the species. For them, every local variant was actually an originally created species. The nineteenth-century Swiss-American naturalist Louis Agassiz is an example of this approach. Others ("lumpers") were inclined to include quite widely divergent local variations as belonging to the same species. The most extreme example of this is the eighteenth-century French naturalist Buffon, who eventually decided that all the members of a Linnean family (such as the lion, tiger, and leopard) are nothing more than strongly marked and potentially interfertile members of the same species (Bowler, 1973). This has often been seen as an anticipation of evolution, but Buffon still insisted that there could be no transition between the types—each was governed by its own "internal mold," that fixed its basic character for all time.

Recognizing the existence of naturally produced varieties within each species did not require serious study of the extent of individual variation within a population, nor did it imply any particular understanding of how individual variation from the hypothetical norm for the species was produced. On the question of causation, many naturalists assumed that organisms exposed to different conditions would be affected in some way, and if such changes were inherited, they would accumulate to generate well-marked local varieties. This was the position adopted by Buffon and the same view was adopted, at least in the case of plants, by one of the founders of transformism (what we now call evolutionism), J. B. Lamarck (see Hodge, 1971). Lamarck became better known for the alternative mechanism he suggested for animals: the inheritance of new characters acquired through bodily exercise in response to changed needs and habits. Both of these mechanisms blurred the distinction between what we would call somatic and genetic characters by simply assuming that a somatic change would be inherited to at least a small extent. What was truly revolutionary about Lamarck's theory was his claim that such a mechanism could produce an indefinite amount of change so as to produce new forms that naturalists would classify as distinct species. Here was a clear rejection of the notion of a fixed type to which the species would always revert in a few generations if the environment returned to its original state.

Lamarck, like most naturalists of his generation, made no effort to study variation beyond noting that the existence of varieties within species confirmed that individual variability was significant. To the extent that there was serious observation of individual variation, it was undertaken by animal and plant breeders and the few naturalists who took an interest in their work. They had always sought to "improve the breed" by allowing only their best specimens to reproduce, and at a practical level, they were adept at picking out the minute differences that guaranteed success. In horticulture especially, there were self-conscious efforts to identify individual "sports of nature" with an unusual character that might be fixed in the breed to produce a commercially useful variety. At a practical level, this work confirmed the extent of individual variability and showed that variant characters bred true at least some of the time, but it threw little light on the nature and origin of those characters. Many authorities saw a clear distinction between the trivial differences, which mark animals off as individuals, and rare, large-scale sports or monstrosities that seem to introduce new characters. Even allowing for such sports, hardly anyone thought that the changes could accumulate enough to transform the species, and most naturalists dismissed the whole effect as so artificial as to tell them nothing about the nature of species. The only exceptions were those such as Kölreuter and von Gaertner who were interested in the cross-breeding of varieties. Although hailed as "precursors of Mendel" (Roberts, 1929), they were more concerned to show that the range of artificially produced varieties did not upset the traditional doctrine of the fixity of the underlying specific types.

In the early nineteenth century, the French naturalist Geoffroy Saint-Hilaire undertook an extensive study of teratology and the production of monstrosities by disturbances of the developmental process (Appel, 1987). He joined Lamarck in challenging the fixity of species, but to him, the occasional monstrosity with a viable constitution founded a new species instantaneously. Geoffroy thus pioneered a theory of evolution by saltation, which was to have considerable influence in the later part of the century and seems to anticipate Richard Goldschmidt's (1940) concept of the "hopeful monster" as the founder of new species.

II. DARWIN AND VARIATION

One of Charles Darwin's most important contributions was to recognize that the breeders' understanding of variation offered an important insight into the nature of variability in wild species. The theory of natural selection, developed in the late

1830s after he had returned from the voyage of the *Beagle,* depended on the existence of a fund of individual variation within a wild population and on the assumption that such variants bred true to a significant extent. At a very early stage in his project, he turned to a study of the breeders' literature and practice in the hope of throwing light on the problem of how species might change under natural conditions. This is not the place for a discussion of the controverted question of whether Darwin based the idea of natural selection on a model provided by the breeders' method of artificial selection. However, Darwin's willingness to use their evidence of variability as a basis for understanding what happened within wild populations was a vital new initiative. Darwin undertook a lifelong program of study devoted to this topic and published his *Variation of Animals and Plants Under Domestication* in 1868 as a vital supplement to the argument of the *Origin of Species*. As part of this project, he developed a theory of heredity and variation, pangenesis, which he outlined in the 1868 book.

Darwin's earliest descriptions of natural selection, including the substantial essay he wrote in 1844 (eventually published in Darwin and Wallace, 1958), took artificial selection as the model for natural selection and hence implied that the variability of domesticated populations was comparable, in kind if not in magnitude, with that in the wild. Like Lamarck, he took the existence of well-marked geographic varieties as proof that natural variation could extend to produce significant modifications of the species' original character. However, Darwin shared the common preconception of his time that the amount of individual variation shown by a wild population was very small. It is so small, he admits, that we often do not notice it at all, but the fact that animals recognize one another as individuals shows that it exists. We do notice individual variants within domesticated populations, partly because breeders develop the skills necessary to pick out even the smallest individual difference, but also (he believed) because there is, in fact, more individual variability in domesticated animals and plants than found in the wild.

To understand why Darwin was convinced of this last point, we must turn to the theory of heredity and variation Darwin began to put together at the very start of his work, although it was not fully articulated until 1868 (Hodge, 1985; Gayon, 1998). Darwin visualized heredity as a process in which minute particles, which he called "gemmules," were budded off from the various tissues of the body and transmitted to the reproductive organs. Because they were "'buds," they carried the potential to generate exactly the same kind of tissue or organ as the parent in the offspring. Because sexual reproduction mixed gemmules from all parts of both parents' bodies however, the resulting offspring often appeared to blend their characters together. There were also complex processes by which ancestral characters might reappear after having lain dormant for some generations. At this level, the theory preserved the traditional notion of heredity as a process of exact copying of parental characters. For natural selection to work as a mechanism of genuine transmutation, there had to be a source of entirely new characters. To provide this,

Darwin hypothesized that the copying process can be disturbed to some extent by changes in the external environment. One aspect of this was the inheritance of characters acquired through use and disuse or through direct response to environmental change. If the parent's body changed, then the gemmules it produced would reflect that change, allowing Darwin to preserve a minor role for what became known as Lamarckism. However, the breeders had taught him that most of the variation they found was not directed or adaptive; it was "random" in the sense that it seemed to produce a wide range of useless characters (useless to the organism, but occasionally valuable to the breeder). This could be explained by supposing that the copying process of heredity was disturbed by the effect of changed conditions so that what were, in effect, copying errors were introduced. The fact that wild populations showed little variability when compared with domesticated ones now made perfect sense because domestication would expose the organisms to unnatural conditions and hence trigger the copying errors. Wild populations living in their normal environments would experience little disturbance and hence show little variation. If the surrounding conditions changed however, the situation would more closely approach that seen in domesticated populations, and at least a small amount of individual variation would appear.

It must be stressed that Darwin still saw variation as a force that disturbed the normal process of exact copying provided by heredity, although he accepted that the copying errors would be reproduced once they had appeared. Darwin did not make a distinction between what would later be called somatic and germinal (or genetic) variations, and so his explanation of these effects was very different from that developed by his twentieth-century followers. His was still a theory of "generation" in the old tradition, which did not distinguish between what we call heredity and embryologic development.

In addition, some historians insist that Darwin had still not achieved what Mayr (1982) calls a "populational" view of species. He did not think of a wild population as normally exhibiting a fund of variability, and he still tended to think of individual variants as unique and probably quite rare deviations from the normal character of the population. This became apparent in Darwin's reaction to the widely publicized critique of the selection theory by the engineer Fleeming Jenkin in 1867 (Vorzimmer, 1970; Bowler, 1974; Gayon, 1998).

Jenkin followed the common assumption that individual variation came in two forms: trivial everyday variations and large-scale sports or monstrosities. He accepted that natural selection could act on small variations to produce local varieties or subspecies, but defended the traditional view that there was a limit beyond which such changes could not go, in effect defining the limits of the species. Only saltations could break this barrier—but Jenkin argued that the "hopeful monster" (to use a later term) would be a single individual, which would have to breed with a member of the original species. On the prevailing view that heredity normally blended the characters of the two parents, this meant that the sport's offspring would have only half its new character, their offspring only a fourth, and so on. Whatever the advantage of the mutated character, it would soon be diluted beyond recognition and could not affect the whole population.

Darwin had never believed that natural selection made use of large-scale sports. His was always a gradualistic theory based on the accumulation of minute individual differences. Yet his letters show that he was deeply disturbed by Jenkin's argument. To understand why, we have to recognize that Darwin did not make the common distinction between everyday variations and sports. For him, all variations, small and large, were individual deviations from the normal pattern of development. Jenkin's swamping argument was thus valid for small variations too, because for Darwin even these were quite rare and would be subject to dilution through interbreeding with the unchanged mass of the population. At this late stage in his career, Darwin even gave up his original assumption that evolution occurred best in small, isolated populations, because he now feared that such small populations would not throw up enough individual variants for selection to be effective.

Alfred Russel Wallace, the codiscoverer of natural selection, pointed out the problem with the way Darwin had conceptualized variation. Wallace's original 1858 paper on selection had been ambiguous on the nature of the variation upon which the process acts. Some historians, the present author included, think that Wallace was originally thinking in terms of selection eliminating local varieties or subspecies, not individuals within a single population (Bowler, 1976). By the late 1860s, however, Wallace had fully assimilated the Darwinian concept of individual selection and had begun to study the variability of populations. He now told Darwin that he was wrong to think of small variants and single individuals within an otherwise uniform population. All populations, wild or domesticated, exhibit a range of individual variation for any measurable character. If one side of that range is favored by selection, then, by definition, half the population will be favored to some extent and the other half disadvantaged. Selection always has ample raw material to work upon because widespread individual differences are a natural feature of every population. In his *Darwinism* (1889), Wallace provided primitive distribution curves illustrating the range of variation he had measured within populations, showing the soon-to-be-familiar bell-shaped curve in which most individuals cluster around the mean, with smaller proportions varying to a greater extent in either direction. This was a true breakthrough into a populational view of species and is exactly what would be seized upon by Darwinism's twentieth-century supporters.

III. ALTERNATIVE THEORIES OF VARIATION AND EVOLUTION

Jenkin's criticism of the selection theory was one among many, and the late nineteenth century saw other mechanisms of evolution explored during what Julian Huxley called the "eclipse of Darwinism" (for further details of the theories outlined below see Bowler, 1983). These all had implications for concepts of heredity and variation and channeled biologists' interests in directions that, by hindsight, can be seen as blind alleys in the development of modern genetics and Darwinism.

The most obvious alternative for those who shared Darwin's interest in adaptation as the guiding theme of evolution was the Lamarckian theory of the inheritance of acquired characters. Many apparent supporters of Darwin, including Ernst Haeckel and Herbert Spencer, gave this mechanism a greater role than natural selection. In the last decades of the century, Lamarckism was frequently portrayed as an alternative to selectionism by biologists such as Theodor Eimer and thinkers such as Samuel Butler. A substantial school of neo-Lamarckism flourished in America, led by the entomologist Alpheus Packard and the paleontologists Edward Drinker Cope and Alpheus Hyatt.

Neo-Lamarckism was built on a deliberate refusal to contemplate any rigid distinction between what August Weismann would call somatic and germinal variation; indeed most Lamarckians held a holistic view in which it was inconceivable that changes could take place in the adult body and yet not be reflected in the process of heredity. If the body acquired new characters in response to a new environment, either directly or through changed habits, then the acquired character must to some slight extent be transmitted to future generations. Much effort was devoted to showing that there were indeed acquired characters (hardly a point anyone would dispute) coupled with an automatic assumption that these must be inherited and hence be the source of adaptations. In other words, acquired characters were "variations" in the sense required by heredity and evolution. Evolution was the summation of individual acts of self-development. Most Lamarckians visualized variation as a process by which new stages were added to the development of the individual organism; the new stage was then inherited by being incorporated as the last stage in ontogeny. Ontogeny was thus understood as an ascent through all the earlier adult stages, allowing Lamarckism to play a key role in the recapitulation theory promoted by Haeckel and the American neo-Lamarckians (Gould, 1977). Lamarckians were also inclined to think of heredity as a process analogous to memory—the developing organism was, in effect, remembering all the new characters acquired by its ancestors in the course of their evolution.

The more extreme anti-Darwinian naturalists rejected the role of adaptation altogether. They held that evolution was actively directed by the process that generated new individual variations, and they assumed that these variations appeared persistently in a single direction; which therefore became the direction of evolution. This was the theory of orthogenesis popularized by Eimer and (under various names) by paleontologists such as Cope, Hyatt, and Osborn. Variation was directed and nonadaptive and in extreme cases might produce evolution in a direction that was positively harmful to the species, leading to extinction. It was assumed that the direction was imposed by forces arising from within the process of individual development.

Like Lamarckism—and the two often went hand in hand—orthogenesis presented variation as an addition to ontogeny. The difference was that the new characters owed nothing to an adaptive response to the environment, but were entirely generated by internal pressures in ontogeny. It was as though individual development, and hence evolution, acquired a kind of momentum that kept on adding characters in a single direction whatever the cost to the species. Eimer saw regular patterns of nonadaptive variation in species of butterflies, and because he held that the same patterns could affect different species, the effect was supposed to produce the similarities that the Darwinians attributed to mimicry. Paleontologists supported orthogenesis by constructing patterns of parallel lines of development within the fossil record, in which whole collections of related species advanced through the same predetermined sequence of development ending in "racial senility" and extinction.

One problem with Lamarckism was that there was little demonstrable evidence that acquired characters really were inherited. Supporters of both this theory and orthogenesis frequently argued that the effects they postulated did not operate continuously; instead they were concentrated at what Cope called "expression points," when the pent-up pressure for change suddenly manifested itself in a quite abrupt transition to a new form. These theories thus made common cause with a third non-Darwinian tradition, the claim that the only form of variation relevant to evolution was that provided by sports or saltations. Even staunch Darwinians favored this alternative, including T. H. Huxley and Francis Galton. The theory was, in effect, a revival of Geoffroy Saint-Hilaire's idea that monstrosities could become the founding fathers (or mothers) of new species by instantaneous transition from one form to the next. Most saltationists assumed that the normal individual variation seen within a population was trivial in evolutionary terms; if inherited, it nevertheless would eventually reach a barrier beyond which it could not be pushed even by selection. Only a saltation could establish a totally new form, creating a new "normal" type around which trivial variation would center. This idea was expressed by Galton (1889) through a model that drew an analogy with a rolling polyhedron. If the polyhedron rests stably on one of its bases, pressure to move will at first merely rock it from side to side—this is the equivalent of normal individual variation. If the pressure builds up, however, eventually the polyhedron topples over onto another face, which then becomes a new center for rocking motion—this is the saltation. Significantly, this analogy implies that the transition is preordained by the structure of the polyhedron, and many saltationists held that the sudden variations they postulated were predetermined by existing developmental process. In this sense, saltationism and orthogenesis were related theories, although it was clear that the saltations observed in nature did not consistently push the species in a single direction. Perhaps ontogeny imposed limits on the kind of saltations that were possible, without actually constraining them into a single direction.

The theory was, of course, subject to the same objection as that raised against selectionism by Fleeming Jenkin: If there was only a single "hopeful monster," how could it breed so as to transmit its new character to new generations? Most supporters of saltationism seem to have assumed that the new character would be expressed in enough individuals to form a small, distinct breeding population from which the new species would be derived. This was certainly the view of a new generation of saltationists who emerged in the 1890s, basing their ideas on observations that, they claimed, undermined the Darwinian theory that random individual variation could have any evolutionary significance. Leading figures here were William Bateson, Hugo De Vries, and T. H. Morgan, all of whom (not coincidentally) played roles in the "rediscovery" and promulgation of Gregor Mendel's laws of the inheritance of discontinuous characters. Bateson's *Materials for the Study of Variation* (1894) promoted a strongly antiselectionist and antiadaptationist position. He claimed to show that many new characters must have appeared discontinuously because they represented merely changes in the numbers of existing elements. If a flower, for instance, changed from five to six petals, this could not happen by the additional petal beginning as a rudiment and then growing larger over successive generations. There would be an instantaneous rearrangement of the developmental process that simply repeated the petal structure an extra time. Such saltations were the real source of new varieties and ultimately of new species.

De Vries's "mutation theory" (translation, 1910) promoted the same view, backed up by evidence that seemed to show that discrete new varieties or subspecies were appearing in the evening primrose, *Oenothera lamarckiana*, which he was studying. Like Bateson, De Vries dismissed continuous variation as irrelevant and held that his mutations were the only source of new characters. He believed that all species underwent occasional periods in which they produced large numbers of new mutations. In one respect, however, De Vries threw off the orthogenetic associations of saltationism and claimed to provide new support for Darwin. He insisted that the mutations were undirected, generating many different, apparently purposeless new forms, and he accepted that a form of natural selection would ultimately determine which of the new subspecies would survive. This view was strongly resisted by the theory's leading American supporter, Thomas Hunt Morgan (1903), who insisted that selection and adaptation played no role in evolution: The direction of change was determined solely by what mutations were produced, although Morgan resisted the claim that this would produce orthogenetic change through a series of cumulative mutations.

IV. NEO-DARWINISM

In one important sense, the mutation theory that flourished at the turn of the century differed from earlier saltationist theories: Bateson, De Vries, and Morgan were all now working within a paradigm that made a clear distinction between heredity and individual development (Bowler, 1989). It was no longer possible to believe that a character acquired during ontogeny (whether adaptive or merely some "accident of growth") could be inherited so as to form the basis for evolutionary change. Significant variations were germinal, not somatic in origin: Mutations were spontaneous rearrangements of the material responsible for transmitting characters to the offspring—this is why it was no longer possible to believe that they could occur in a predetermined direction through an extension of ontogeny. Also, as De Vries especially insisted, the germinal rearrangements seem to occur in an undirected fashion, producing a range of different new characters, which he (but not Bateson and the younger Morgan) saw as the basis for a process of natural selection.

This new way of thinking came not from within the saltationist program but from one of the leading supporters of Darwinian selectionism in the late nineteenth century, August Weismann. Although Weismann began as a recapitulationist, his theory of the germ plasm (1893) undermined the developmental model in which heredity and ontogeny were inextricably linked (Churchill, 1999). The germ plasm was the material of heredity located in the chromosomes; it contained the information or program (as we would call it today) for building the new organism. However, Weismann insisted that it was completely isolated from the body or soma that surrounded it so that any changes taking place in the parents' bodies could not be reflected by corresponding changes in the germ plasm. This was the modern view in which there is a one-way flow of information from the germ plasm (the genes, as we would now say) to the developing organism, but no mechanism for feedback from development to germ. At a stroke, Weismann declared invalid the Lamarckian mechanism and any theory that supposed that variation could be directed by the process of ontogeny.

For Weismann himself, this position vindicated Darwinism. The germ plasm was composed of "determinants" for the various parental characters that obviously existed in numerous different forms corresponding to the range of individual variation for each character. New characters might occasionally be introduced by spontaneous changes within the determinants, but these would be undirected because there was no way in which the physical changes within could be monitored or directed by ontogeny or the needs of the adult organism. In effect, the germ plasm provided the source of the "random" or undirected variation postulated by Darwin as the raw material of natural selection, and the only way of imposing a direction on this random variation was through changing the proportion of determinants within the population through differential reproduction.

Weismann had little interest in large variations or saltations, favoring Darwin's view that most variant characters useful to selection lay within the range of small-scale variations seen from time to time within any population. However, already his theory was transforming the logic of the selection theory by suggesting that determinants for a wide range of characters persisted within any normal population. The old idea that variation was a force for change antagonistic to the copying process of heredity was breaking down. Variation was now part of the same phenomenon as heredity; it was the rigid inheritance of the various determinants within the population that maintained overall variability. When new characters did appear through germinal transformations, they were preserved by heredity.

Weismann's theory was incorporated into what became known as neo-Darwinism: the claim that natural selection was the sole mechanism of evolution. The same point was being made independently in Britain by Wallace, Galton, and the research school known as biometrics (Provine, 1971; Gayon, 1998). We have already seen how Wallace challenged Darwin to accept that variant characters were not produced as single, abnormal individuals, but were part of a range of variation within the population for any measurable characteristic. Darwin's cousin, Francis Galton, made the same point and began a program to describe the range of variation existing within the human population and that of many other species. For Galton, variation within a population would normally follow the same pattern as that of the shots fired by a marksman at a point target: a heavy concentration around the bull's-eye and a diminishing frequency of hits spreading outward. In graphic terms, Wallace's primitive bell-shaped distribution curves could be put on a firm foundation by detailed population surveys. Like Weismann, Galton sensed that variation and heredity were not antagonistic forces; heredity maintained the range of variant characters within the population. Galton had his own theory of heredity, the law of ancestral inheritance, which sought to explain how each individual's unique character was derived from its parents, grandparents, and so on, in diminishing proportions. The theory did not map onto the laws of inheritance promoted by the Mendelians, but it did encapsulate the view that the population is a collection of unique individuals, the ensemble of which constitutes the range of variation for the species.

In one crucial respect, Galton did not agree with Weismann: He thought that ancestral inheritance would tend to maintain the original norm for the population even when variation was temporarily skewed by selection. The production of genuinely new species would thus depend on saltations, which, in effect, would define a new norm, a new center of variation for the transformed population. Galton's disciple, the statistician Karl Pearson, showed that there was no need for this saltationism: Galton's own law of inheritance would allow selection to shift the mean of the range of variation for the population, in effect transforming the species. In collaboration with W. F. R. Weldon, Pearson undertook an extensive series of biometrical researches on wild populations of crabs and snails, building up detailed information on the range of natural variation for a number of characters within the wild populations (Weldon, 1894–1995, 1901). They were also able to show that when the environment changed, the range of variation shifted accordingly as selection favored individuals at one end of the range. In Plymouth harbor, where the water was muddied by an extensive dredging program, the crabs became slightly larger, apparently because larger individuals were better able to cope with the threat of their gills being clogged. The biometrical school thus demonstrated the existence of a range of natural variation within wild populations and the occurrence of microevolution by the action of natural selection on that variation.

V. THE EVOLUTIONARY SYNTHESIS

In 1900, De Vries and Carl Correns announced the "rediscovery" of Mendel's laws of particulate inheritance. The new science of genetics emerged with a strong bias toward the saltationism of the mutation theory, although De Vries himself soon lost interest in Mendelism. The discrete characters studied in domesticated species by Mendel and his twentieth-century followers seemed to cry out for a theory in which unit characters were produced by single mutations. In species subjected to artificial selection, distinct breeds had been produced, each with its own unique characters, and hybridization experiments showed that the characters bred true as units. The lack of apparent intermediates in the breeds favored by the Mendelians seemed to vindicate Bateson's and De Vries's assumption that such discrete characters were produced instantaneously by mutation (Bowler, 1989). How was this situation to be related to the range of continuous variation being studied in other species by the biometrical school, which they presumed to be the raw material upon which natural selection operated? At first it seemed that no reconciliation was possible, and the two rival schools of thought on the nature of variation and evolution engaged in a bitter and ultimately fruitless controversy (Provine, 1971). To the Mendelians, the continuous range of variation studied by Pearson and Weldon was irrelevant because it was probably the result of mere somatic variation produced by differences in nutrition, etc., affecting genetically identical individuals. It could have no relevance to the production of new characters in evolution, and the effects demonstrated by the biometricians were too small to provide convincing evidence of Darwinian selection. To the biometricians, the discrete characters traced through Mendelian hybridization experiments were the products of rare sports preserved by rigorous inbreeding, which offered no analogy with evolution under natural conditions.

The situation gradually became more fluid as the Mendelians began to realize that the genetic structure of large populations was much more complex than their experiments with artificial breeds had suggested. Wilhelm Johannsen had shown that the apparently continuous range of variation for size in the beans he studied was actually composed of several discrete "pure lines" defined by Mendelian factors: Selection could isolate the gene for one extreme of the range, but any remaining variation was somatic. Selection appeared to have reached a limit. Johannsen's work was thus seen as a vindication of the claim that a discrete mutation would be needed to produce a character beyond the original range.

The idea that an apparently continuous range could actually be produced by a number of interacting Mendelian factors had already been noted as a potential bridge between the Mendelian and biometrical schools by G. Udney Yule as early as 1902. When T. H. Morgan adopted the Mendelian position and began his classic studies of mutations in the fruit fly *Drosophila*, he effectively abandoned De Vries's theory in which new subspecies were produced instantaneously by mutated characters (see Allen, 1978). The mutated fruit flies bred with other members of the population and effectively fed their new character into the existing range of variation. In 1909, the Swedish biologist H. Nilsson-Ehle performed breeding experiments with grain that confirmed Yule's insight: He showed that an apparently continuous range of variation could be accounted for in terms of three or four Mendelian factors, each segregating independently, but their effects intertwining so completely in the whole population that the result was apparent continuity. The more Mendelian factors were involved, the greater the degree of continuity. The same point was noted in America by Edward East. It was becoming increasingly clear that mutation could produce both discrete new characters (although these were often deleterious and would not survive in the wild) and relatively small modifications of existing characters that would simply blend in with and extend the range of variation in a large population. By 1916, even T. H. Morgan—once a strong opponent of Darwinism—was admitting that selection by the environment would determine whether or not a mutated gene would spread into the population.

These insights were translated into the study of wild populations by the founders of population genetics and by the biologists who linked their work to the field naturalists studying geographic variation (Mayr and Provine, 1982). The most important figures were R. A. Fisher and J. B. S. Haldane in Britain and Sewell Wright and Theodosius Dobzhansky in America. Fisher was a statistician working within the tradition established by Pearson, but unlike Pearson, he realized that a Mendelian approach to heredity could be used to analyze the variation and evolution of large populations. Fisher's work culminated in the classic *Genetical Theory of Natural Selection* (1930), in which he showed how variation in a wild population was maintained by the circulation of genes through sexual recombination and how that variability could be shaped by natural selection when some genes reproduced more effectively than others because they conferred a slight adaptive benefit. Fisher modeled the effect on the behavior of molecules according to the kinetic theory of gases: Large-scale effects were produced by the summing up of myriads of individual acts too small to be studied except through their statistical effects on the whole. The population was seen as a pool of genetic variability retained by recombination, fueled by the production of new characters by mutation, and shaped by the gradual increase in the frequency of any gene that conferred an advantage in a changed environment. Fisher ignored sports, following the biometrical model of Darwinism in which change was always slow and gradual, exploiting only genes that survived in the population because their effect was relatively small. His model of the gene pool allowed for a range of variation to exist even where the extremes were disadvantaged by the environment, because new genes were constantly being produced by mutation and no gene, not even a potentially harmful one, could be eliminated altogether if it were recessive. When one extreme of the range became favored in a changed environment, the genes responsible for the favored character would slowly but steadily begin to increase their frequency.

J. B. S. Haldane (1932) worked with similar assumptions but allowed for the effect of larger genetic differences that could confer greater adaptive and hence selective advantage. As an example, he used the case of industrial melanism in the peppered moth, *Biston betularia*, in which there was a significant difference between the normal population and the dark or melanic form. The latter had appeared as a rarity in the mid-nineteenth century but had spread rapidly in industrial areas, apparently because its color offered some protection as camouflage against predators (see Kettlewell, 1955). This explanation has generated a good deal of controversy among historians (Hagen, 1999; Rudge, 1999), but whatever the actual advantage, here selection was, in effect, making use of what would once have been called a sport or saltation.

In America, different mathematical models were used by Sewell Wright, whose vision of evolution was shaped more by his early experiences with artificial selection (see Provine, 1986). Wright drew upon the evidence provided by William E. Castle, who had shown that small inbreeding populations could generate a much wider range of variation than was possible in a larger group, because of interactions between unusual combinations of genes. Wright saw evolution taking place most readily not in large, uniform populations, but in species that were geographically broken up into many partially-isolated subpopulations. Successful new genetic combinations produced within one subpopulation would gradually spread to others and thus affect the whole species. These conclusions were translated into nonmathematical terms by Theodosius Dobzhansky, who had come to America after working on the genetics of wild populations in Russia (Adams, 1994). Here S. S. Chetverikov and others had promoted the concept of the gene pool as a reservoir of variability, before their work was cut off by the rise of the Marxist-inspired Lamarckism of T. D. Lysenko. Dobzhansky now drew on Wright's work to write his *Genetics and the Origin of Species* (1937), and the two biologists subsequently collaborated in a long series of studies of the genetics of natural populations (collected in Dobzhansky, 1981). Dobzhansky's work established the foundations of the "balance" hypothesis of population structure, in which the population was seen as a reservoir of genetic variation, with many genes being maintained at a low frequency even though they conferred no advantage. On this model, if the environment changed, there would be ample genetic variability available for selection to act upon, without the need for immediate new mutations. This view was challenged by T. H. Morgan's student H. J. Müller, who preferred a model in which most individuals within a wild population had a single genetic structure, all unfavorable mutations being rapidly eliminated. Here new mutations would be needed for selection to work when the environment changed.

Dobzhansky's view of the genetics of natural populations played a vital role in the establishment of the so-called "Modern Synthesis" of Darwinism and genetics. It allowed field naturalists such as Ernst Mayr to relate the new population genetics to their studies of geographic variation in species that were spread over a fragmented habitat. Mayr drew on a long tradition of studying geographic variation, to which Darwin and Wallace had both contributed. Naturalists such as Karl Jordan, David Star Jordan, and Francis B. Sumner were convinced that such varieties were produced by adaptation to the local environment, but many accepted a Lamarckian rather than a selectionist explanation. As late as 1936, a major survey of variation by G. C. Robson and O. W. Richards threw doubt on the effectiveness of selection and favored Lamarckism. Sumner, however, had begun to realize that the differences between local varieties had a genetic basis (Provine, 1979). Like his mentor Bernhard Rensch, Mayr began from a Lamarckian position, but thanks to Dobzhansky's work, he and many others were able to see how the new genetic theory of natural selection could explain the geographic effects they had been studying. Mayr's *Systematics and the Origin of Species* (1942) built on Dobzhansky's model to become one of the founding documents of modern Darwinism. Significantly, Mayr and the other founders of the synthesis ridiculed Richard Goldschmidt's efforts (1940) to retain a role for saltations.

VI. CONCLUSIONS

The Modern Synthesis thus represented the culmination of several distinct stages of conceptual development that had established a new view of the nature of variation. Darwin had recognized the importance of hereditary variability and had drawn upon the breeders' understanding of this to throw light on the situation in nature, but although he studied the emergence of geographic varieties, he did not try to measure the variability of single populations. Indeed, he seems to have thought that such variability was very limited because variant individuals appeared only rarely even when the environment changed. Also, by treating these individual variants as the product of disturbances to individual development, he left intact a view of the relationship between heredity, variation, and ontogeny that was far better suited to rival theories such as Lamarckism, orthogenesis, and saltationism. To understand the origins of the modern view of variation and evolution, we have to look to the later revolutions pioneered by biologists such as Weismann and Galton, who established a distinction between germinal (genetic) and somatic variation, thus undermining the plausibility of those theories which presented variation and evolution as purposeful or directed modifications of the ontogenetic pathway.

Despite the controversy between the neo-Darwinians and the mutationists (including their successors, the early Mendelians), the elimination of the developmental model of heredity and variation created the framework within which a synthesis could be worked out, incorporating the idea that variation consisted of alternative genetic factors maintained within the population by heredity and potentially subject to selection. The eventual combination of population genetics with the study of geographic variation brought together two lines of interest pioneered by Darwin but held apart for half a century by conceptual and professional differences among the various schools of biological research.

REFERENCES

Adams, M., ed. (1994). *The Evolution of Theodosius Dobzhansky.* Princeton, NJ: Princeton University Press.

Allen, G. E. (1978). *Thomas Hunt Morgan: The Man and His Science.* Princeton, NJ: Princeton University Press.

Appel, T. A. (1987). *The Cuvier-Geoffroy Debate: French Biology in the Decades Before Darwin.* Oxford, England: Oxford University Press.

Bateson, W. (1894). *Materials for the Study of Variation Treated with Especial Regard to Discontinuity in the Origin of Species.* London: Macmillan.

Bowler, P. J. (1973). Bonnet and Buffon: Theories of Generation and the problem of Species. *Journal of the History of Biology* **6**, 259–281.

Bowler, P. J. (1974). Darwin's Concepts of Variation. *Journal of the History of Medicine* **29**, 196–212.

Bowler, P. J. (1976). Alfred Russel Wallace's Concepts of Variation. *Journal of the History of Medicine* **31**, 17–29.

Bowler, P. J. (1983). *The Eclipse of Darwinism: Anti-Darwinian Evolution Theories in the Decades Around 1900.* Baltimore: Johns Hopkins University Press.

Bowler, P. J. (1989). *The Mendelian Revolution: The Emergence of Hereditarian Concepts in Modern Science and Society.* London: Athlone.

Bowler, P. J. (2003). *Evolution: The History of an Idea.* 3rd ed. Berkeley, CA: University of California Press.

Churchill, F. B. (1999). August Weismann: A developmental evolutionist. In *August Weismann: Selected Letters and Documents* (F. Churchill and H. Risler, eds.), Vol. 2, pp. 749–798. Freiburg, Germany: Universitatsbibliothek, Freiburg i. Br.

Darwin, C. (1868). *The Variation of Animals and Plants Under Domestication.* London: John Murray, London.

Darwin, C. and Wallace, A. R. (1958). *Evolution by Natural Selection.* Cambridge, England: Cambridge University Press.

De Vries, H. (1910). *The Mutation Theory: Experiments and Observations on the Origin of Species in the Vegetable Kingdom.* London: Kegan Paul. 2 vols.

Dobzhansky. T. (1937). *Genetics and the Origin of Species.* New York: Columbia University Press.

Dobzhansky. T. (1981). *Genetics of Natural Populations, I–XLIII* (R. C. Lewontin, ed.). New York: Columbia University Press.

Fisher, R. A. (1930). *The Genetical Theory of Natural Selection.* Oxford, England: Clarendon Press.

Galton, F. (1889). *Natural Inheritance.* London: Macmillan.

Gayon, J. (1998). *Darwinism's Struggle for Survival: Heredity and the Hypothesis of Natural Selection.* Cambridge, England: Cambridge University Press.

Goldschmidt, R. (1940). *The Material Basis of Evolution.* Princeton, NJ: Princeton University Press.

Gould, S. J. (1977). *Ontogeny and Phylogeny.* Cambridge, MA: Harvard University Press.
Hagen, J. B. (1999). Retelling experiments: H. B. D. Kettlewell's studies of industrial melanism in peppered moths. *Biology and Philosophy* 14, 39–54.
Haldane, J. B. S. (1932). *The Causes of Evolution.* London: Longmans.
Hodge, M. J. S. (1971). Lamarck's science of living bodies. *British Journal for the History of Science* 5, 323–352.
Hodge, M. J. S. (1985). Darwin as a lifelong generation theorist. In *The Darwinian Heritage* (D. Kohn, ed.), pp. 207-234. Princeton, NJ: Princeton University Press.
Kettlewell, H. B. D. (1955). Selection experiments on industrial melanism in the *Lepidoptera. Heredity* 9, 323–342.
Mayr, E. (1942), *Systematics and the Origin of Species.* New York: Columbia University Press.
Mayr, E. (1982). *The Growth of Biological Thought: Evolution, Diversity and Inheritance.* Cambridge, MA: Harvard University Press.
Mayr, E., and Provine, W. B., eds. (1982). *The Evolutionary Synthesis: Perspectives on the Unification of Biology.* Cambridge, MA: Harvard University Press.
Morgan, T. H. (1903). *Evolution and Adaptation.* New York: Macmillan.
Morgan, T. H. (1916). *A Critique of the Theory of Evolution.* Princeton, NJ: Princeton University Press, Princeton.
Provine, W. B. (1971). *The Origins of Theoretical Population Genetics.* Chicago: University of Chicago Press.
Provine, W. B. (1979). Francis B. Sumner and the evolutionary synthesis. *Studies in the History of Biology* 3, 211–240.
Provine, W. B. (1986). *Sewall Wright and Evolutionary Biology.* Chicago: University of Chicago Press.
Richards, R. J. (1991). *The Meaning of Evolution: The Morphological Construction and Ideological Reconstruction of Darwin's Theory.* Chicago: University of Chicago Press.
Roberts, H. F. (1929). *Plant Hybridization Before Mendel.* Princeton, NJ: Princeton University Press.
Robson, G. C. and Richards, O. W. (1936). *The Variation of Animals in Nature.* London: Longmans.
Rudge, D. W. (1999). Taking the peppered moth with a pinch of salt. *Biology and Philosophy* 14, 9–37.
Vorzimmer, P. J. (1970). *Charles Darwin: The Years of Controversy.* Philadelphia: Temple University Press.
Wallace, A. R. (1889). *Darwinism: An Exposition of the Theory of Natural Selection.* London: Macmillan.
Weismann, A. (1893). *The Germ Plasm: A Theory of Heredity.* London: Walter Scott.
Weldon, W. F. R. (1894–1995). An attempt to measure the death rate due to the selective destruction of *Carcinas moenas. Proceedings of the Royal Society* 57, 360–379.
Weldon, W. F. R. (1901). A first study of natural selection in *Clausilia laminata. Biometrika* 1, 109–124.

CHAPTER 3

The Statistics of Variation

LEIGH VAN VALEN

Department of Ecology and Evolution, University of Chicago, Chicago, Illinois, USA

ABSTRACT

Univariate and multivariate measures of variation are reviewed, for both absolute and relative variation. Biologists almost never test variation appropriately; in particular, the *F* test should never be used for this purpose. There are, however, several useful and robust tests that apply, with modifications, to all useful measures of variation, including the multivariate coefficient of variation. Some of these applications are little known or new. A measure of the effective dimensionality of variation permits the relative variation of quantities such as volumes to be compared with that of linear measurements. Several measures exist for the tightness of a distribution in two or more dimensions, one of which generalizes this property of the correlation coefficient. Measurement error is part of observed variation and can be subtracted from it.

Variation

INTRODUCTION

We often want to estimate variation within populations and test for its equality among populations. Unfortunately there are serious difficulties with some of the most common methods, and these difficulties are not widely known. I therefore review these difficulties and ways to avoid them, as well as the main properties of the structural statistics. This chapter revises and extends an earlier paper (Van Valen, 1978). Some modifications and applications were new in that paper and have had relatively little use.

My domain will be continuous variation and that discrete variation that can (as most can) be approximated by continous distributions. Variation of more sharply discrete parameters, and of qualitative ones, is best measured by information-type statistics, which are beyond the scope of this review (but cf. Van Valen, 1974, for related suggestions). The variation can be of original measurements or of appropriate deviations from an axis of regression or coregression. (Coregression refers to trend lines, ordinarily the major axis or reduced major axis for a bivariate distribution, for mutually interdependent variables, where there is no independent variable, as in morphometrics.)

I. ABSOLUTE VARIATION: UNIVARIATE CASE

NEVER USE AN *F*-TEST TO TEST EQUALITY OF VARIANCES. The basis for this shocking statement is well known to statisticians (Box, 1953; Pearson and Please, 1975; Markowski and Markowski, 1990) but is almost never communicated to biologists.

The variance, σ^2 for population and s^2 for sample, is the most widely suitable measure of variation in one dimension. Independent variances are additive; for variances of interdependent characters, the amount of interdependence (the covariance) is subtracted from their sum. The standard deviation, σ or s, measures variation in the units in which measurements are taken and so is often heuristically preferable. Statistics such as the observed range, often excluding 10% or 25% at each end, are sometimes used. The full observed range is very sensitive to sample size and is rarely useful; truncated ranges can be useful when one is interested in only the central part of the distribution. Smith's test (see the following text) can be applied to ranges; see Stuart and Ord (1994, p. 364), together with Stuart, Ord, and Arnold (1999, pp. 152–154) for the variance of estimation.

Unfortunately, the *F* distribution (the distribution of the ratio of two sample variances) is exceptionally sensitive to deviations from normality of the measured distributions. When the *F* distribution is applied in the ordinary analysis of variance, this is unimportant except in extreme cases. The ordinary analysis of variance tests equality of means, and by the central limit theorem, the distribution of means of samples from a single population is asymptotically normal. However, there is no such asymptotic normality for the variation of the character itself. It has even been half seriously stated that the *F* test is a better test for normality than for equality of variances.

Also, one cannot just test one's distributions for normality and assume them to be normal if they pass the test. Such tests just are not sensitive enough except with very large samples. This is partly because of the difficulty in demonstrating any kind of small deviation from any distribution. It is also true because the tails of a distribution contribute disproportionately much to the variance and the tails are hardest to test.

It can easily happen that a distribution cannot be distinguished from normality but really is different enough to invalidate the *F* test. In practice, one can almost never have adequate reason to believe that a pair of distributions are each close enough to normality that the *F* test for equality of variances is suitable.

Almost every statistical comparison, in the biological literature, of the variances of two or more populations is wrong, often seriously so. Bartlett's test, being a generalization of the *F* test, is just as sensitive to nonnormality as is the *F* test itself (e.g., Rivest, 1986). There are, however, three useful tests that are robust to nonnormality and were developed for this reason. Each has its advantages, but I have rarely seen any of them applied in biology. Many other tests have been proposed, and I am sure that there are more than I have seen, but each of the others I know seems to be inferior to these in at least one respect (Balakrishnan and Ma, 1990; Baker, 1995). In particular, a commonly recommended (e.g., Chernick, 1999, p. 63 [but apparently unpublished]) method of bootstrapping has been found to be very nonrobust (Hall and Padmanabhan, 1997, p. 420; cf. also Hall and Wilson, 1991). The bootstrap itself can be inconsistent (give wrong but progressively better-supported results with increasing sample size) for heavy-tailed distributions (Wu, 1988).

A. Levene's Test

This is the simplest of the three tests. Its original description (Levene, 1960) is meant for statisticians, but the test itself is simple both conceptually and in execution. In formal terms, it directly tests the mean absolute deviation rather than the variance.

For each sample, find the mean or, better, the median (see the following text). For an even number of observations, the median is the mean of the two central values. Then calculate the absolute difference of each original datum from the median (or mean). This is a new variable; the more varying the sample, the higher the values will be. Then find the mean and variance of the new variable. The means of the new variable for each population are then tested for equality by a t test or an analysis of variance. Symbolically, $y_i = |x_i - x_{median}|$ and equality of the $\bar{y}$s are tested.

Note that Levene's test is not quite a test of equality of variances. It is, from this perspective, a joint test of all the even moments of the distributions: variance, kurtosis, etc. However, the effect of the variance predominates.

Levene's original version of his test can be improved by using the median of the distribution instead of the mean as the center from which to measure deviations (Brown and Forsythe, 1974). Although Brown and Forsythe say that the best procedure for symmetrical long-tailed distributions is to exclude the 10% of data at each end of the distribution and use the mean of the remaining 80%, I think that the elimination of such biologically relevant information is hardly ever worth the small increase in precision that they found for well-defined distributions (cf. Schultz, 1983). Keyes and Levy (1997) and Hines and O'Hara Hines (2000) give further incremental improvements for which I have not seen independent evaluations. Calibration by randomization is not warranted (Francis and Manly, 2001), but a bootstrap-like randomization does improve the test under some conditions, such as appreciably unequal sample sizes (Manly and Francis, 2002) and can be recommended when feasible. It reduces or eliminates the value of other modifications.

For the randomization, lump together all absolute deviations from each population's median to form a single set from all populations jointly. Then randomly reallocate all these deviations to samples, using the same sample sizes as before. Then test homogeneity of the means of the reconstituted samples by a t test or analysis of variance and retain the resulting values of t or F. Repeat the randomization and test many times, say 1000. This then gives a distribution of t or F, and the value from the unrandomized test can be compared to see whether it falls within the extreme 5% (or whatever) of the distribution.

B. Smith's Test

This is a very general but somewhat more complex test. It was described by Cedric A. B. Smith and has been published, as far as I know, only in Grüneberg *et al.*, 1996. It is a test directly of equality of variances. For a sample of size n, the variance of estimation of the variance (the square of the standard error of the variance) for the jth population is

$$s_{s_j^2}^2 = \frac{\sum_i (x_i - \bar{x})^4 - s_j^4\left(\frac{n-3}{n}\right)}{(n-2)(n-3)}, \tag{1}$$

correcting a misprint. For k samples, where k can be 2 or more, there is an approximately chi-square statistic with $(k-1)$ degrees of freedom:

$$\chi^2_{k-1} = \sum_j \frac{s_j^4}{s^2_{s_j^2}} - \frac{\left[\sum_j \frac{s_j^2}{s^2_{s_j^2}}\right]^2}{\sum_j \frac{1}{s^2_{s_j^2}}}, \tag{2}$$

correcting another misprint.

Smith's test can be used to test for homogeneity of any statistic for which the standard error is available, by replacing s_j^2, with the desired statistic. Again, normality is not required. Smith's test is the only one of the three that can be used with other people's summary statistics of variation.

C. Jackknifing

This is the most complex but most general test. It can be used for correcting biases of estimation and for tests on almost any statistic. However, the sample size n must be moderately large (at least 20 or so). Good references are Arveson and Schmitz (1970), Miller (1974), and Bissell and Ferguson (1975). My treatment and symbols are simplified from these references.

First, divide the data into n broadly overlapping groups of size $(n-1)$ by eliminating one datum for each group, a different datum each time. Then estimate the variance for each group (s_i^2) and for the total (s^2_{total}). Let

$$\hat{\sigma}_i^2 = ns^2_{total} - (n-1)s_j^2, \tag{3}$$

for the ith group. Then the bias-reducing estimate is

$$\hat{\sigma}^2 = \frac{1}{n}\sum \hat{\sigma}_i^2. \tag{4}$$

Because of the use of differences, at least two more significant figures than usual should be used in the calculations. The variance of estimation of $\hat{\sigma}^2$ is

$$s^2_{\hat{\sigma}^2} = \frac{\sum\left(\hat{\sigma}_i^2 - \hat{\sigma}^2\right)^2}{n(n-1)} \tag{5}$$

and is distributed as t with $(n-1)$ degrees of freedom. Thus ordinary t tests and analyses of variance can be performed on $\hat{\sigma}^2$. Bissell and Ferguson (1975) have given a straightforward (but necessarily even more complex) multivariate generalization. Jackknifing is a little less robust than Levene's test (Brown and Forsythe, 1974), being somewhat nonconservative under nonnormality. It is the only method that can be used to give confidence intervals for the variance.

II. ABSOLUTE VARIATION: MULTIVARIATE CASE

Beyond the mean, there are no unique multivariate generalizations of univariate statistics. We choose which ones we want to use on the basis of exactly what it is we want to measure and by convenience.

The usual scalar measure of variation in two or more dimensions is called the generalized variance, $|\mathbf{\Sigma}|$ for populations and $|\mathbf{S}|$ for samples. It is the determinant of the covariance matrix. Geometrically, it is roughly proportional to the square of the area, volume, or hypervolume of any of the family of equiprobability envelopes of the referent distribution, just as the univariate variance is roughly proportional to the square of the length of the line that corresponds to any proportion of points in a univariate distribution. The square root of the generalized variance is an analog of the standard deviation.

In two dimensions,

$$|\mathbf{\Sigma}| = \sigma_1^2\sigma_2^2(1-\rho^2), \tag{6}$$

where ρ is the correlation coefficient. In three dimensions,

$$|\mathbf{\Sigma}| = \sigma_1^2\sigma_2^2\sigma_3^2\left(1-\sum_{i>j}\rho_{ij}^2+2\prod_{i>j}\rho_{ij}\right). \tag{7}$$

The factor in parentheses is the determinant of the correlation matrix, and the same general form holds for higher dimensions.

The variance of a distribution can be measured in any direction. If we measure it in the direction of the longest axis of the distribution (Figure 3-1), it is called the first eigenvalue. However, it is still a variance. So are the other eigenvalues. They are variances in other directions perpendicular to the first and to each other. There are as many eigenvalues as there are real dimensions, although not always as many as the axes of original measurement.

A diagonal line in a plane has one real dimension and one eigenvalue but is located with respect to two original axes of measurement. An eigenvalue is

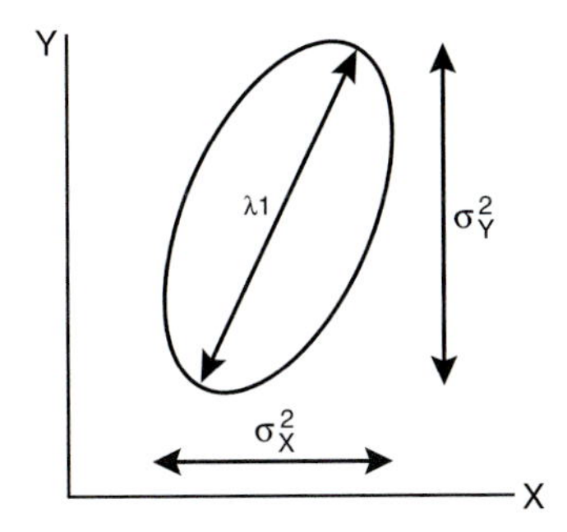

FIGURE 3-1. The ellipse represents an envelope of a bivariate distribution. The variances of X and Y are proportional to the squares of the lengths of their respective arrows: λ_1, the first eigenvalue, is also proportional to the square of the length of its arrow.

conventionally represented, by λ (not λ^2). The generalized variance can be viewed as the product of the eigenvalues:

$$|\Sigma| = \Pi \lambda_i \tag{8}$$

(Dempster, 1969, p. 137).

There are several serious problems with using $|\Sigma|$ to measure variation (Van Valen, 1974). For one thing, it vanishes when all the correlations are 1 and is small when most correlations are large. However, in such cases there may still be appreciable real variation left on the line or in the very elongate dimension of a line-like distribution. The volume of a line is 0, and that constitutes the difficulty.

The effect of the variances is worse. $|\Sigma|$ also vanishes when any one or more of the variances is 0, and the effect of some variances being much smaller than others is even more severe than the effect of large correlations. There can still be large variation in all the other dimensions. The geometry of this difficulty is that the volume of a plane is 0, and in general the hypervolume in p dimensions of a distribution with only $(p-1)$ dimensions is 0.

One may occasionally want to use the hypervolume of the distribution, but even then there is a problem, which is also perhaps the most severe one in using $|\Sigma|$ to measure variation. What, exactly, is the number of real dimensions, i.e., eigenvalues? In most real cases with more than three or four variables, there is enough correlation among the variables that the last eigenvalues are near 0 and do not differ significantly from 0. If they are included, $|\Sigma|$ will be near 0 for the same geometric reason as for the second difficulty. Moreover, small errors of estimation in small eigenvalues will cause proportionally large errors in $|\Sigma|$, especially when more than one eigenvalue is low. If we avoid this situation by including only those eigenvalues significantly different from 0, then the number we include will depend on the significance level we choose. Therefore our estimate of $|\Sigma|$ will also strongly depend on this quite arbitrary choice.

A better scalar measure of multivariate variation is the total variance (Van Valen, 1974). It generalizes the univariate variance in a different way. The univariate variance is the sum of the squares of the distances of each point in the population (or sample) from the mean, divided by the number of points. Exactly the same wording can be applied to any number of dimensions, using Euclidean distances. Figure 3-2 illustrates the concept.

Algebraically, for p dimensions and n points,

$$\sigma_p^2 = \frac{1}{n}\sum_{j=1}^{p}\sum_{i=1}^{n}(x_{ij} - \mu_j)^2, \tag{9}$$

where μ_j is the population mean for variable j. The denominator should be $(n-1)$ for samples, as in the univariate case. σ_p^2 would then be called s_p^2. It happens that the total variance also equals the sum of the univariate variances:

$$\sigma_p^2 = \sum \sigma_j^2. \tag{10}$$

This is quite a useful property, although a conceptually peripheral one. Because the sum of the variances in all independent directions must be the same whatever set of such directions is used, we see in particular that

$$\sum \lambda_j = \sum \sigma_j^2 \tag{11}$$

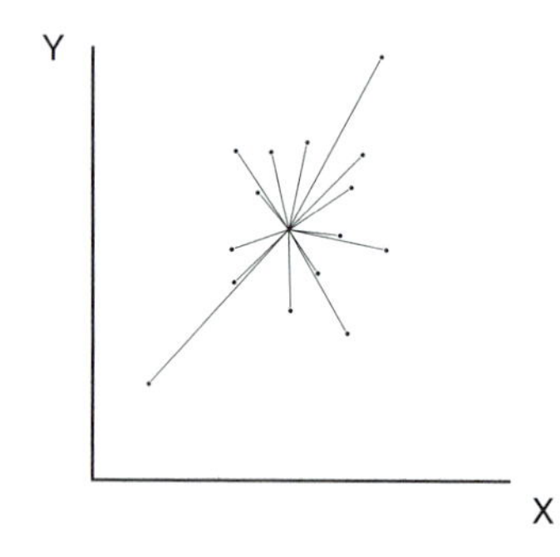

FIGURE 3-2. The disc at the convergence of the lines represents the bivariate mean of the distribution of dots. Each line is the distance from one point to the mean. These distances are squared in the calculation of the total variance.

(Dempster, 1969, p. 137): The total variance is also the sum of the eigenvalues. Use of the total variance requires that all variables be commensurable; e.g., lengths and angles cannot be combined. This is an entirely natural stipulation for a single multivariate distribution.

The univariate standard deviation, the square root of the variance, measures variation in the scale of the variable itself. Similarly, the square root of the total variance does so for several variables jointly. This follows directly from the definition, by an exact analogy with the univariate case. σ_p can thus be called the total standard deviation. σ_p can alternatively be thought of as the average of the univariate standard deviations measured in all possible directions from the joint mean.

The covariance matrix is itself a measure of multivariate variation, although one less easy to comprehend as a whole. It measures variation in shape and direction as well as in size. Two or more covariance matrices will differ if the total variances differ, but they may differ even if the latter do not. I know of no robust (and therefore genuinely useful) tests in the latter situation. Probably the best available test is the one (with two variants) that Zhang and Boos (1992, 1993) proposed, based on the bootstrap. Krzanowski (1993) has given a somewhat similar test for correlation matrices.

Statisticians commonly advise the standardizing of variables so that one variable does not dominate the results. Hotelling's T^2 test for homogeneity of multivariate means, current discrimination analysis, and Mahalanobis's distance measure D (or D^2) are examples of this perspective. Also, sometimes it is indeed appropriate for variables to be modified so as to contribute equally, but this is a question to be considered from a biologic perspective in each case, as it can distort biological meaning. Eigenvalues are not standardized, although they can be applied to standardized data, and the total variance would be meaningless if variables were adjusted to have the same variances, as is the usual recommendation. (Adjustment to a common mean avoids this difficulty and similar ones [if only in part when, as usual, the variance is related to the mean], if standardization is indeed desired.)

Intermediate degrees of unequal weighting are of course possible by intermediate adjustments. I once proposed a test for homogeneity of multivariate means that does not equalize the variables (Van Valen, 1980; useful especially when the populations differ in their eigenvectors), and a similar approach can be used for other problems of interest.

Levene's test, Smith's test, and jackknifing are each applicable to tests on the total variance and on eigenvalues, again without requiring normal distributions.

For Levene's test, one can either first estimate the multivariate mean (mean vector, centroid, joint mean) or estimate the deviation y_i of each point directly:

$$y_i = \sqrt{\sum_j (x_{ij} - \bar{x}_j)^2} \qquad (12)$$

for the jth dimension and the ith point. $\bar{y}$ is tested as before. Manly (1988) has called the version based on Equation (12) Van Valen's test and reasonably recommended that, as with the univariate case, the median be used for each dimension instead of the mean, so that

$$y_i = \sqrt{\sum_j (x_{ij} - x_{median_j})^2}. \tag{12a}$$

He found the test ordinarily appropriate, being robust and with good response to both type 1 and type 2 errors. Randomization, as with the univariate Levene's test, can of course be done if desired.

Two kinds of tests on eigenvalues can be appropriate. One is on whether different eigenvalues from the same population are really different, as would not, e.g., be true for a nearly (hyper)spherical distribution. The other kind of test is on corresponding eigenvalues from different statistical populations. In the latter case, there may be a concern as to the extent to which corresponding eigenvalues represent the same phenomena or measurements. The test itself is on the shape of the distribution, and if there is heterogeneity among populations in their eigenvectors (the mutually perpendicular axes of the distribution itself), the interpretation may be affected. Still, it is often of interest to compare variation of parts of the distributions per se, irrespective of the directions in which the distributions point.

For comparison of single eigenvalues of the same or different populations, it is best to find the median point along the corresponding eigenvector. This can be done for two eigenvectors jointly by using a standard bivariate plot of two eigenvalues and finding the median points for each axis. One then calculates the absolute distance of each point from the relevant medians, parallel to the respective axis. Figure 3-3 illustrates the geometry for the bivariate case in the original measurement space.

One now has two or more univariate distributions of deviations from their respective medians, and the univariate Levene's test is directly applicable. Because the eigenvalues are asymptotically independent, being measured in mutually perpendicular directions, the test can be used for eigenvalues of the same or different samples.

However, in samples the eigenvalues are not strictly independent, because the first eigenvector defines the direction in which the sample, not necessarily the population, is most variable. This produces a small upward bias for the first eigenvalue or so and a small downward bias for the last. My simulations indicate that the bias is not large, but some caution should be used when significance values as precise as possible are desired.

The preceding procedure must be modified if one is comparing sets of two or more eigenvalues from the same population, e.g., testing heterogeneity among populations in overall variation on the plane of two eigenvectors/eigenvalues. So far as I know, the median has not been defined for more than one dimension.

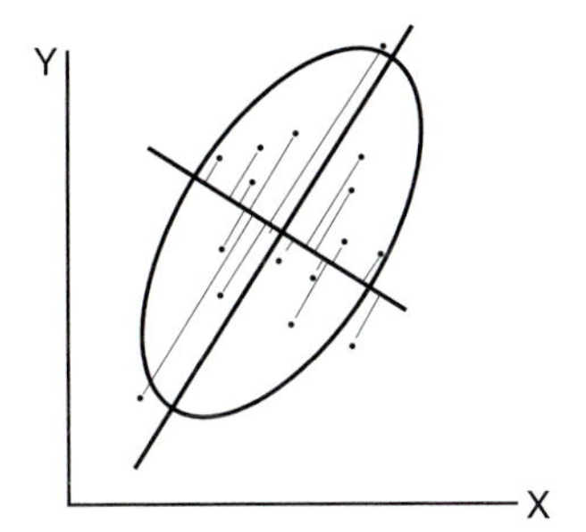

FIGURE 3-3. A distribution like that of Figure 3-2 now has an elliptical envelope, whose two axes are also represented. The distance of each point to the minor axis represents the distance of that point along the major axis. The distance shown is then squared in the calculation of the first eigenvalue. It is also the distance used as the deviation y_i in Levene's test for the first eigenvalue.

However, a bivariate/multivariate generalization seems straightforward. Find the median for each eigenvector of interest, as before. Its coordinate along the eigenvector can then be taken as one coordinate of the joint median. A joint median so defined will not in general be identical to one similarly defined on the basis of the axes of original measurement, but the eigenspace is defined by the distribution itself and is therefore preferable in inferring parameters dependent on the distribution.

Levene's test based on the joint median should behave comparably to the test based on the univariate median. Either method of calculating the distance from a point to the median is suitable.

Smith's test is best considered for eigenvalues (λ_i) first. Calculate the distance y_i as for Levene's test. Then

$$s_{\lambda_j}^2 = \frac{\sum y_i^4 - \lambda_j^2\left(\frac{n-3}{n}\right)}{(n-2)(n-3)}. \qquad (13)$$

Because an eigenvalue is a variance, the same formula applies with appropriate changes in notation. Then replace s_j^2 in Equation (2) by λ_j and solve. To prove the applicability of this and other tests to eigenvalues, we note that they are immediately applicable when the principal axes of the distribution are parallel to the referece axes. The eigenvalues are then the variances of the originally measured variables. Rotation of axes leaves the eigenvalues unchanged, so the tests are also unchanged and still apply.

Because the total variance $s_p^2 = \Sigma\lambda_i$ and all eigenvalues are approximately independent, the variance of estimation of s_p^2 is close to

$$s_{s_p^2}^2 = \sum s_{\lambda_j}^2. \tag{14}$$

Again, s_p^2 for each sample variance replaces s_j^2 in Equation (2).

Jackknifing is conceptually the same as before, but computation may be more complex than one can easily program.

III. RELATIVE VARIATION: UNIVARIATE CASE

Is a mouse as variable as an elephant, in a biologically meaningful way? Is rate of locomotion as variable as length of legs? To see, use the coefficient of variation

$$CV = 100\frac{s}{\bar{x}}. \tag{15}$$

Haldane's small-sample correction (1955) is of no practical use because it is always much smaller than the expected error. The coefficient of variation should almost always be used in comparing variation when means differ enough to matter.

There is an important assumption underlying use of the coefficient of variation, namely that absolute variation (σ, not σ^2) is proportional to the mean. This is so often true that we may tend to forget that there are cases where it is not.

The only real class of exceptional cases I know is of those in which variation starts from a more or less constant value above 0 (Mac Gillavry, 1965). These are cases where the structure is well canalized but may still vary a little. The number of presacral vertebrae in a population of a mammal is such a character. The number of lumbar vertebrae (presacrals posterior to the ribs) makes up most of this variation, and the number of neck vertebrae is almost always constant. *CV* for total presacral vertebrae is less than that for lumbars, but the structural variation is pretty much the same in each case. Threshold characters have the same problem (Lande, 1977).

Despite defeatist statements still prevalent in the literature, there are more suitable ways to test relative variation than there are for absolute variation. Each of the three methods given in the preceding text is appropriate directly for *CV*, and tests on a transformation are also often possible. The assumption that absolute variation is proportional to the mean is the same assumption one needs in order to use a logarithmic scale. Therefore the original data can be transformed to logarithms, and the tests for absolute variation can be performed on the transformed data (Wright, 1952; Lewontin, 1966). One cannot, however, do such tests on the logarithm of statistics such as the variance. The data themselves must be transformed.

The transformation is appropriate in both the univariate and the multivariate case. The scale of measurement (e.g., arithmetic or logarithmic) is important conceptually, although not with respect to the ease of making tests, and biological judgment is needed as to which scale is best.

However, logarithmic transformation is unnecessary for testing and can best be forgotten in this context except as a device to produce rather symmetric distributions. To adapt Levene's test to *CV*, divide each deviation by the mean:

$$y_i = \frac{|x_i - \bar{x}|}{\bar{x}}. \tag{16}$$

Then proceed as before. For Smith's test, it is merely necessary to know the variance of estimate of *CV*:

$$s_{CV}^2 \simeq \frac{CV^2}{2n}, \tag{17}$$

when the distribution is roughly normal; in other cases Stuart and Ord (1994) give the appropriate value. Substitute *CV* for s^2 and s_{CV}^2 for $s_{s^2}^2$ in Equation (2) and proceed as before. Similarly, jackknifing works by substituting the appropriate *CV* for s^2 in Equation (3).

Approximate confidence intervals on *CV* can be found by jackknifing, as for s^2 or s. However, there is a simpler way (Miller and Kahn, 1962) when skewness is relatively low and (as usual) *CV* is less than about 30:

$$\frac{nCV^2}{\chi^2_{\frac{\alpha}{2}}\left(1+CV^2\right)-nCV} < \text{true } CV^2 < \frac{nCV^2}{\chi^2_{1-\frac{\alpha}{2}}\left(1+CV^2\right)-nCV^2}, \tag{18}$$

where α is the significance level (1 − confidence level) and χ^2 has $(n-1)$ degrees of freedom. This method is of course also a direct test on whether the true *CV* has some specific value.

IV. RELATIVE VARIATION: MULTIVARIATE CASE

There is an immediate multivariate generalization of *CV* (Van Valen, 1974):

$$CV_p = 100\frac{\sigma_p}{|\mu|}, \tag{19}$$

where $|\mu|$ is the mean vector (multivariate joint mean).

$$|\mu| = \sqrt{\Sigma \mu_j^2}, \tag{20}$$

by Pythagoras's Theorem in p dimensions, so

$$CV_p = 100 \sqrt{\frac{\sum \sigma_j^2}{\sum \mu_j^2}}. \tag{21}$$

Like the univariate CV, CV_p is dimensionless in the sense that multiplying all measurements by the same constant leaves CV_p unchanged. CV_p is also independent of the number of variables and so is numerically comparable to CV.

Tests on CV_p are analogous to those on the univariate CV. For Levene's test, the deviation of each point i from the mean vector must be divided by the mean vector to produce y_i. Smith's test is identical to the univariate case. Similarly, jackknifing CV_p adds no new complexity to that for s_p^2.

V. DIMENSIONALITY OF VARIATION

As Lande (1977) has shown most clearly, the univariate CV gives too high values when applied to parameters such as volume that have dimensionality greater than one. An adequate correction for this effect cannot be derived from Lande's otherwise useful treatment because the latter is monotonically related to the correlations, yet even with perfect correlation CV for a volume can be almost the same as for one of its sides (as when a cylinder varies almost only in height.)

What is needed here is a measure of the effective dimensionality of variation. The equivalent number of independent variables, or amount of nonredundant information (Van Valen, 1974), is in one sense the true dimensionality of the shape of a distribution, being one for a line in any direction, near one for a linelike distribution, two for a circular disc, and three for a sphere. It measures precisely the dimensionality of variation. If variation of a volume is all, or mostly, in one dimension, the true variation is more nearly one-dimensional than three-dimensional. Biologic variation can be greater in three dimensions than it can in one, so we should have a way to adjust for differences in real dimensionality of variation.

In most cases, the dimensionality D of variation in the two-dimensional case is

$$D = (2 - r^2), \tag{22}$$

where r is the sample correlation coefficient. In the p-dimensional case,

$$D = 1 + (p - 1)(1 - r_p^2) \tag{23}$$

where r_p^2 is the total correlation.

$$r_p^2 = \frac{1}{p} \sum_{i=1}^{p} R_i^2, \tag{24}$$

R_i being the ith multiple correlation coefficient. In the three-dimensional case,

$$R_{i \cdot jk}^2 = 1 - (1 - r_{ij}^2)(1 - r_{ik \cdot j}^2). \tag{25}$$

When the measured variables differ appreciably in variance or other self-information, estimation of D must proceed sequentially (Van Valen, 1974, p. 242).

Once one has D, then dividing the directly estimated CV (for the multidimensional measure) by D gives a new CV that is numerically comparable to a univariate CV and CV_p. The same procedure can be applied to the standard deviation. However, estimating D in the case of something like mass may sometimes be difficult or impossible. In such a case, unless there is approximate similarity in shape or density (in which case D is close to three), no useful comparisons can be made using CV.

When the dimensions are perfectly correlated (no shape difference), variation in volume or area is greater than when they are uncorrelated but with the same variation as before. s_p^2 is the same in each case. I interpret this effect as a biologically real aspect of variation rather than as an artifact of dimensionality, even though it cannot occur in the one-dimensional case. It therefore needs no correction for dimensionality.

Usually the dimensions of variation will not be integers. Fractional dimensions are perhaps unfamiliar, but they occur naturally in other contexts, such as the dimensionality of a path of Brownian motion (Mandelbrot, 1977). This is something totally different from ontologic imprecision, or fuzziness, which can also be measured precisely (Van Valen, 1964; Zadeh, 1965).

VI. TIGHTNESS

The other important property of the correlation coefficient ρ, besides underlying the explanatory property of ρ or ρ^2, is that it measures the closeness of a distribution to it coregression. This is exactly true only at a coregression with a slope of 1, i.e., if the untransformed variances are equal, but with deviation from this slope, the error of using ρ to measure closeness is proportionally somewhat smaller than the deviation. This second property of ρ is not emphasized in recent work, perhaps because of the inaccuracy mentioned, but it is a major aspect of correlation, and in fact Galton (1888) considered it when he introduced "co-relation" into statistics.

A general treatment is straightforward and can be regarded as the variance of relative eigenvalues. If all the variance is explained by one principal component, the tightness of the distribution is maximal, which we may set equal to 1. If all the eigenvalues are equal, the tightness is minimal, or 0. Let the relative eigenvalue

$$t_i \equiv \frac{\lambda_i}{\sigma_p^2}. \tag{26}$$

Then the *tightness*

$$T = \sqrt{\frac{p}{p-1}\sum (t_i - \bar{t})^2} = \sqrt{\frac{p}{(p-1)\sigma_p^4}\sum (\lambda_i - \bar{\lambda})^2}, \tag{27}$$

for p dimensions. It is gratifying that in two dimensions with equal variances, $T = \rho$. This follows from the fact that $\lambda_1 \lambda_2 = \sigma_1^2 \sigma_2^2 (1-\rho)^2$.

In the general bivariate case,

$$T = \sqrt{1 - \frac{4\sigma_1^2\sigma_2^2(1-\rho^2)}{(\sigma_1^2 + \sigma_2^2)^2}} = \frac{\lambda_1 - \lambda_2}{\sigma_p^2}. \tag{28}$$

Another measure of tightness does not reduce to ρ. If all eigenvalues are equal, any equiprobability surface is a hypersphere with diameter proportional to the square root of an eigenvalue. With unequal eigenvalues, the axes of the hyperellipsoid have the same property. Then, with scaling, we get a measure of departure from sphericity:

$$T_1 = 1 - \frac{\sum \sqrt{\lambda_i}}{n\sqrt{\lambda_1}}. \tag{29}$$

A third measure of tightness, which also does not reduce to ρ, is the proportion of the maximum possible variance ($p\lambda_1$) that is not actually realized:

$$T_2 = 1 - \frac{\sum \lambda_i}{p\lambda_1} \tag{30}$$

These measures of tightness are taken from Van Valen (1974), rescaled for T and with a misprint corrected in (28).

VII. MEASUREMENT ERROR AND SINGLE SPECIMENS

Any measurement or count carries with it the possibility of error. Sometimes the error is small enough to be neglected, but in other cases, such as sometimes in studies of fluctuating asymmetry, where the variable of interest is a difference between two closely similar measurements, it can even overwhelm any real signal that may be present.

Measurement error is detected and estimated by repeated and independent measurements of the same specimens, or of a randomly chosen proper subset of them, when this is relevant; other methods are appropriate, if usually less adequate, in other cases.

First, test the means of the (preferably) several sets of repetitions to see whether they are detectably different. If they are not, as is usual, then the sets of repetitions can be treated as independent. (If there is heterogeneity among the means, and the specimens were measured together for a repetition, adjust each set of replicates to a common mean for the sequel, because the focus of real interest is presumably on variation among the specimens.)

For each measured structure or whatever ("character"), calculate the variance among the several repetitions for each specimen. (Yes, a variance can be calculated with a sample size of 2, if necessary.) Because the acts of measurement are presumably independent for different specimens, the measurement variances are additive. Find the mean of the measurement variances for the specimens. This mean is the measurement variance for the character, and it can be subtracted from the originally estimated variance, among specimens, for the character. Thus

$$s^2 = s_o^2 - \frac{\sum_{i=1}^{n} s_i^2}{n}, \tag{31}$$

s^2 being the estimate of the real variance among specimens, s_o^2 being the originally measured variance among specimens, and s_i^2 being the variance among replicates for the *i*th of *n* specimens.

When there is only one specimen or observation available from a population, standard theory says that it is quite impossible to test whether the mean of this population differs from the mean of another population. What is ordinarily done is to test whether the specimen itself could reasonably have come from the other population's distribution, by noting whether it falls into the central 95% or whatever of that distribution. This is, however, actually a different question and often it is not what we really want to know. Instead, our interest may be in whether the two sampled populations themselves differ from each other.

In most real cases, it is reasonable to assume that the variances of the populations are about the same. This then permits a test of the population means in the usual way, the standard error of the mean for a single specimen being the assumed standard deviation of the population. The significance levels will have to be loosened to the extent that the assumption of equal variances may actually be incorrect. The extent of this latter modification is a purely biologic judgment, and analogous procedures are possible in some other formally intractable situations.

REFERENCES

Arveson, J. N., and Schmitz, T. H. (1970). Robust procedures for variance component problems using the jackknife. *Biometrics* 26, 677–686.

Baker, R. D. (1995). Two permutation tests on equality of variances. *Statistics and Computing* 5, 289–296.

Balakrishnan, N., and Ma, C. W. (1990). A comparative study of various tests for the equality of two population variances. *Journal of Statistical Computation and Simulation* 35, 41–89.

Bissell, A. F., and Ferguson, R. A. (1975). The jackknife—toy, tool or two-edged weapon? *The Statistician* 24, 79–100.

Box, G. E. P. (1953). Non-normality and tests on variances. *Biometrika* 40, 318–335.

Brown, M. B., and Forsythe, A. B. (1974). Robust tests for the equality of variances. *Journal of the American Statistical Association* 69, 364–367.

Chernick, M. R. (1999). *Bootstrap Methods: A Practitioner's Guide*. New York: Wiley.

Dempster, A. P. (1969). *Elements of Continous Multivariate Analysis*. Reading, MA: Addison-Wesley.

Francis, R. I. C. C., and Manly, B. F. (2001). Bootstrap calibration to improve the reliability of tests to compare sample means and variances. *Environmetrics* 12, 713–729.

Grüneberg, H., Bains, G. S., Berry, R. J., Riles, L., Smith, C. A. B., and Weiss, R. A. (1966). *A Search for Genetic Effects of High Natural Radioactivity in South India*. London: Her Majesty's Stationery Office.

Haldane, J. B. S. (1955). The measurement of variation. *Evolution* 9, 484.

Hall, P., and Padmanabhan, A. R. (1997). Adaptive inference for the two-sample scale problem. *Technometrics* 39, 412–422.

Hall, P., and Wilson, S. R. (1991). Two guidelines for bootstrap hypothesis testing. *Biometrics* 47, 757–762.

Hines, W. G. S., and O'Hara Hines, R. J. (2000). Increased power with modified forms of the Levene (med) test for homogeneity of variance. *Biometrics* 56, 451–454.

Kendall, M. (1980). *Multivariate Analysis*. 2nd ed. London: Charles Griffin.

Keyes, T. K., and Levy, M. S. (1997). Analysis of Levene's test under design imbalance. *Journal of Educational and Behavioral Statistics* **22**, 227–236.

Krzanowski, W. J. (1993). Permutation tests for correlation matrices. *Statistics and Computing* **3**, 37–44.

Lande, R. (1977). On comparing coefficients of variation. *Systematic Zoology* **26**, 214–217.

Levene, H. (1960). Robust tests for equality of variances. In *Contributions to Probability and Statistics* (I. Olkin, S. G. Ghurye, W. Hoeffding, W. G. Madow, and H. B. Mann, eds.), pp. 278–292. Stanford, CA: Stanford University Press.

Lewontin, R. C. (1966). On the measurement of relative variability. *Systematic Zoology* **15**, 141–142.

Mac Gillavry, H. J. (1965) Variability of larger Foraminifera. Part I: Natural position of zero values. *Koninklijke Nederlandse Akademie van Wetenschappen (Amsterdam). Proceedings, Series B.* **68**, 335–355.

Mandelbrot, B. B. (1977). *Fractals*. San Francisco: W. H. Freeman.

Manly, B. F. J. (1988). Van Valen's test. In *Encyclopedia of Statistical Sciences* (S. Kotz and N. L. Johnson, eds.), Vol. 9, pp. 462–465. New York: Wiley.

Manly, B. F. J (1997). *Randomization, Bootstrap and Monte Carlo Methods in Biology*. 2nd ed. London: Chapman & Hall.

Manly, B. F. J., and Francis, R. I. C. C. (2002). Testing for mean and variance differences with samples from distributions that may be non-normal with unequal variances. *Journal of Statistical Computation and Simulation* **72**, 633–646.

Markowski, C. A., and Markowski, E. P. (1990). Conditions for the effectiveness of a preliminary test of variance. *American Statistician* **44**, 322–326.

Miller, R. G. (1974). The jackknife—a review. *Biometrika* **61**, 1–15.

Miller, R. L., and Kahn, J. S. (1962). *Statistical Analysis in the Geological Scienes*. New York: Wiley.

Pearson, E. S., and Please, N. W. (1975). Relation between the shape of population distribution and the robustness of four simple test statistics. *Biometrika* **62**, 223–241.

Rivest, L. P. (1986). Barlett's Cochran's and Hartley's tests on variances are liberal when the underlying distribution is long-tailed. *Journal of the American Statistical Association* **81**, 124–128.

Schultz, B. (1983). On Levene's test and other statistics of variation. *Evolutionary Theory* **6**, 197–203.

Stuart, A., and Ord, J. K., (1994). *Kendall's Advanced Theory of Statistics*, Vol.1. 6th ed. London: Edward Arnold.

Stuart, A., Ord, J. K. and Arnold, S. (1999). *Kendall's Advanced Theory of Statistics*, Vol. 2A. 6th ed. London: Edward Arnold.

Van Valen, L. (1964). An analysis of some taxonomic concepts. In *Form and Strategy in Science* (J. R. Gregg and F. T. C. Harris, eds.), pp. 402–415. Dordrecht, The Netherlands: D. Reidel.

Van Valen, L. (1974). Multivariate structural statistics in natural history. *Journal of Theoretical Biology* **45**, 235–247.

Van Valen, L. (1978). The statistics of variation. *Evolutionary Theory* **4**, 33–43; 122.

Van Valen, L. (1980). A test for the identity of multivariate means. *Evolutionary Theory* **5**, 135–138; 188.

Wright, S. (1952). The genetics of quantitative variability. In *Quantitative Inheritance* (E. C. R. Reeve and C. Waddington, eds.), pp. 5–41. London: Her Majesty's Stationery Office.

Wu, C. F. J. (1988). Discussion of the papers by Hinkley and DiCiccio and Romano. *Journal of the Royal Statistical Society, Series B* **50**, 364–365.

Zadeh, L. A. (1965). Fuzzy sets. *Information and Control* **8**, 338–353.

Zhang, J., and Boos, D. D. (1992). Bootstrap critical values for testing homogeneity of covariance matrices. *Journal of the American Statistical Association* **87**, 425–429.

Zhang, J., and Boos, D. D. (1993). Testing hypotheses about covariance matrices using bootstrap methods. *Communications in Statistics—Theory and Methods* **22**, 723–739.

CHAPTER 4

Landmark Morphometrics and the Analysis of Variation

JOAN T. RICHTSMEIER[*], SUBHASH R. LELE[†], AND THEODORE M. COLE III[‡]

[*]Department of Anthropology, The Pennsylvania State University, University Park, Pennsylvania, USA; Center for Craniofacial Development and Disorders, The Johns Hopkins University, Baltimore, Maryland, USA

[†]Department of Mathematical and Statistical Science, University of Alberta, Edmonton, Alberta, Canada

[‡]Department of Basic Medical Science, School of Medicine, University of Missouri—Kansas City, Kansas City, Missouri, USA

INTRODUCTION

But the problems of variability, though they are intimately related to the general problem of growth, carry us very soon beyond our limitations.

Sir D'Arcy Thompson (1917)

Landmark-based morphometrics consists of a series of approaches to the analysis of form and form difference in two or three dimensions. Landmarks represent the location of biologically relevant features that can be recorded from a form with an acceptable degree of accuracy and precision in two or three dimensions (Richtsmeier *et al.*, 1998). Morphometric methods provide measures of form and shape differences based on the relative location of landmarks. Modern morphometric approaches were inspired by the early work of Sir D'Arcy Thompson (1992) and matured with the belief that geometry of the whole, rather than analysis of disjointed measurements, might provide more complete answers to questions pertaining to morphology, phylogeny, functional morphology, and development. There are certain limitations associated with using the locations of landmarks to study morphology, however. For example, no information about surface features between the landmark locations is included in landmark data. But, other considerations (e.g., ease of data collection; developmental, evolutionary, or biomechanical correspondence of landmarks within, and potentially across, species; degree of precision; and the fact that coordinate data can be used to calculate linear distances or angles among landmarks) make them an appropriate choice for analysis in many biological research designs (see Lele and Richtsmeier, 2001).

Morphometrics has grown from a novel tool adopted by a few authors of morphological studies of the 1980s to a fairly standard methodology in biological inquiry. Still, problems persist in the estimation of certain parameters from landmark data, especially the parameters associated with the mean and variance–covariance structure (see Richtsmeier *et al.*, 1992; Lele, 1993; Lele and McCulloch, 2002). Measures of phenotypic variation estimated from linear measures have proven incredibly useful for understanding the relative roles of genes and environment in the production of the phenotype, in part because the genetic properties of a population can be characterized by partitioning phenotypic variation into components ascribed to various causative sources using methods from quantitative genetics (Falconer, 1981; Falconer and Mackay, 1996). These more traditional measures of variation have proven useful in statistical analysis of the phenotype and in quantitative genetics studies, but the ability to properly measure variation local to biologically significant landmarks could provide additional advantages. First, the ability to localize measured variation to a landmark, rather than to linear distances that traverse developmental and anatomical boundaries, could improve the delineation of the sources of variation within an organism. Given that landmark coordinate-based morphometric techniques enable the researcher to localize observations of morphological differences to particular points in space, landmark-based measures of variation should enable a fine-tuning of local measures of variation without the unavoidable overlap among linear distances that share a common endpoint. Second, using morphometric methods that enable comparison of growth patterns, estimation of anatomically localized measures of change in form from one generation to the next (Richtsmeier and Lele, 1993) should enable subsequent partitioning of local variance components, each attributable to a different cause, including localized selection differentials. Third, and more practically, if we can provide valid estimates of the variation around landmarks, rather than local to all unique distances between landmarks, there will be fewer parameters to estimate and to study. These advantages suggest that estimation of variation from landmark data may be useful. However, obtaining valid estimates of variation from landmark data has proven to be more difficult than originally thought.

In this chapter, we discuss the standard approaches that have been used to estimate variation in landmark data and explain why these methods do not properly estimate variation in biological forms. Though some of these approaches have become rather mathematically involved (Dryden and Mardia, 1998), their suitability to the realities of biological data has not improved in parallel. We present a generalized model for variation in landmark data. It has been shown that only certain features of this model can be consistently estimated (Lele, 1993; Lele and Richtsmeier, 2001; Lele and McCulloch, 2002). This model and the estimators are used in morphometric analysis as the basis for parametric bootstrapping procedures to test for differences in form using landmark data (Lele and Richtsmeier, 2001). Recognizing the limitation of the ability to estimate only certain features of this general model of variation, we discuss the need for the development of less general models that may reasonably characterize variation in landmark data. We show how biological knowledge pertaining to the organisms under study can be used to impose certain constraints on the models of variance–covariance structure. We suggest ways to integrate such constraints into the proposal of several less generalized models, some statistically convenient but biologically improbable, others less streamlined statistically but more biologically reasonable.

I. COORDINATE DATA AND THE COORDINATE SYSTEM

Most investigators credit Sir D'Arcy Thompson (1992) with the stimulus that initiated the field of morphometrics. Early geometry-based methodologies were proposed by Boas (1905) and Sneath (1967) (see Cole [1996] for the specifics of Boas's contribution), but were largely ignored until much later. The modern field of *geometric morphometrics*, defined as the fusion of geometry and biology (Bookstein, 1982), flourished from the 1980s onward and originated from a desire to analyze biological forms in ways that preserve the physical integrity of form in two or three dimensions. The nature of landmark data was particularly appealing to biologists. The expression of the relative location of landmarks in a coordinate system that displays geometric relationships among parts of the whole stood in stark contrast to the usual expression of a form as disarticulated series of linear and angular measures. Although the use of landmark data in the estimation of an average and variation around that average appears straightforward, the innate nature of landmark data, in particular the presence of the nuisance parameters of translation and rotation, makes estimation of the true variation complicated.

The researcher starts with coordinate data collected from a group of organisms. Biological organisms that constitute a group resemble each other to such a degree that we have an intuitive idea of the appearance of the typical or "average" form that is representative of all members of the group. Using statistics, an average or mean form can be estimated from the observations. When coordinate data are used, this mean form provides the average relative locations of landmarks. Genetic and environmental influences combine to affect structures so that all forms differ from each other in diverse ways. This variation in the sample is estimated as divergences in the relative location of landmarks from their configuration in the mean form.

When the researcher accumulates landmark data from a sample of forms, both the mean of the sample and the relationship of individual observations to the mean (the perturbation structure) are unknown. To estimate the mean and variance, it is necessary to specify a mathematical construct or *model*, that attempts to characterize certain aspects of the properties of the population. A *model* is used here as a mathematical construct that attempts to characterize certain aspects of the underlying phenomena (e.g., dimensions, dynamics, properties, interactions). Models include quantities called parameters, and these parameters are estimated according to the specified model using the sample data. Models and methods used to estimate mean and variance of a sample of coordinate data sets require that attention be paid to the unique coordinate system of each data set. Because of the need to address the varied coordinate systems within a sample of data sets, models for landmark data typically define a group of parameters that are unique to the properties of coordinate data. These parameters include rotation, translation, and in some cases, reflection and are required to place coordinate data sets into a common coordinate system so that the mean and variance can be estimated. This seems simple enough, and intuitively it makes sense: If we are going to build a mean form and determine phenotypic variation for a sample, the forms must all be expressed within the same coordinate system. The question is, of the infinite coordinate systems available, which one should be chosen for analysis?

From a biological perspective, one might argue that an orientation scheme should be adopted with reference to the primitive state of the organism or with reference to whatever is defined as standard anatomical position of the adult for a given species. Compelling arguments for the adoption of a particular coordinate system on mathematical grounds (i.e., ease of computation) could also be made. If adoption of varying coordinate systems gave the same sample estimates, this choice would be trivial. Unfortunately it has been shown that estimates of the mean and variance change with the coordinate system adopted (Richtsmeier *et al.*, 1992; Lele, 1993; Lele and McCulloch, 2002).

Because estimation of the mean and variance–covariance structure from landmark data requires that all forms be placed into a common coordinate system and the estimates of these parameters change according to the chosen coordinate system, that choice must be critically evaluated. How can a scientist know that the parameters estimated for a sample are valid if there is no information regarding the relative validity of any particular coordinate system? This question is the source of the complexity of estimating sample parameters for coordinate data sets. We detail the problems involved in estimation of mean and variance–covariance structure of coordinate data sets in the following text.

II. THE GENERAL PERTURBATION MODEL FOR LANDMARK VARIATION

Suppose we are interested in a sample of n organisms and we measure them using a series of K landmarks in D (=2 or 3) dimensions. A $K \times D$ matrix that we designate M describes the mean for the population, where each row represents the D dimensional coordinates of a single landmark. Although M is a mathematical construct and is not directly observable from a sample of specimens (even an infinite one), we can imagine the mean configuration of landmarks in a coordinate system that we will refer to as "natural space," where within-sample variation arises. No single individual is likely to be identical in form to the mean, and no two individuals are likely to be identical.

In the natural space, phenotypic variation is manifested as perturbations around the mean landmark configuration, M (Figure 4-1). Note that the dispersion patterns of these perturbations can vary in size and shape from landmark to landmark. Some landmarks are more variable than others, as indicated by the relative sizes of their dispersions. The perturbation scatters for landmarks can also vary in shape, with some being round and others being elliptical. Finally, there may be covariances between the landmarks because of developmental, biomechanical, or physical constraints, so that the relative positions of the observations at one landmark may be correlated with the positions of other landmarks. If we want to understand the biological processes and mechanisms that produce such a pattern of variation, we need first an objective way of describing it quantitatively. In statistical analysis, a model must be specified before data are analyzed. Parameters are then estimated from each sample using the specified model. Knowledge of particular properties of the phenomena that the data represent can and should be included in proposing a model. Whatever characteristics the biologists deem important or explanatory should be included as assumptions of the model. If a model appears appropriate and correct, methods for checking the assumptions can be developed (if they do not already exist) and applied.

In morphometrics, phenotypic variation as described by landmark data has been modeled routinely using a *general perturbation model*, sometimes called the *Gaussian* perturbation model (see Goodall, 1991; Lele, 1993). Goodall (1991)

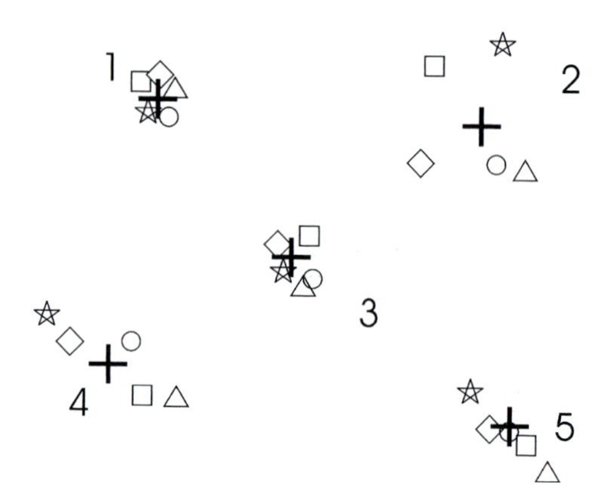

FIGURE 4-1. The "natural space" where individual differences in form originate. Plus signs (+) represent the parametric mean configuration for a hypothetical organism with five landmarks. The symbols represent the landmark locations of different specimens (where like symbols belong to the same specimen). These locations are phenotypic perturbations of the mean, which reflect underlying genetic and environmental variability. Note that the dispersion patterns differ from one landmark to the next. Some landmarks have roughly circular distributions (1, 2, and 3), while others are elliptical. Landmarks 1 and 3 have relatively small dispersions, while the dispersion around landmark 2 is large. In addition, some of the perturbations may be correlated—note the similarity in the rank order of perturbations (from upper left to lower right) for landmarks 4 and 5. As detailed in the text, the positions of the perturbations in natural space cannot be reconstructed; however, some descriptors of the dispersion patterns can be estimated. Adapted from Cole *et al.*, 2002. (From Morphology, Shape and Phylogenetics, [2002] pp. 194–219. Reprinted with the permission of Taylor and Francis.)

used this model in the development of morphometric methods based on Procrustes superimposition, and Lele and colleagues (e.g., Lele and Richtsmeier, 1990, 1991; Lele and McCulloch, 2002) used it in the development of Euclidean distance matrix analysis, or EDMA (pronounced ĕd·mă). Using the general perturbation model, the set of landmark data for each observation in a sample is described by a $K \times D$ matrix called X_i. Each X_i is related to the population mean, M, as follows:

$$X_i = (M + E_i)\Gamma_i + t_i$$

where E_i is a $K \times D$ matrix of perturbations that describe how X_i differs from M in the natural space (the *true* difference between X_i and M). For the population, these perturbations are assumed to have a multivariate normal distribution with a $K \times D$ mean matrix **0** and a covariance structure $\mathbf{\Sigma}_K \otimes \mathbf{\Sigma}_D$, where $\otimes$ denotes a Kronecker product. $\mathbf{\Sigma}_K$ is a $K \times K$ matrix that describes the variances and covariances of the landmarks, while $\mathbf{\Sigma}_D$ is a $D \times D$ matrix that describes the variances and covariances of the perturbations with respect to the natural space's coordinate-system axes (i.e., $\mathbf{\Sigma}_K$ and $\mathbf{\Sigma}_D$ describe the sizes, shapes, and orientations of the perturbation scatters). Γ_i is an orthogonal $K \times K$ matrix that describes the rotation of X_i (in the coordinate system where the data have been collected) relative to M (as it lies in the natural space), while t_i is a $K \times D$ matrix that describes the translation of X_i relative to the position of M in the natural space. Realize that both the position and orientation of M are arbitrary.

The mean, M, and the variance–covariance matrix of the perturbations, $\mathbf{\Sigma}_K \otimes \mathbf{\Sigma}_D$, are obviously of great biological interest. Unfortunately, they are not estimable given the general perturbation model because of the presence of the other terms in the equation. These entirely arbitrary parameters, Γ_i and t_i, are "nuisance parameters" (*sensu* Neyman and Scott, 1948). Although they have no biological meaning, these nuisance parameters are essential in describing the statistical nature of the observed data. Consequently, Γ_i and t_i are nuisance parameters from the scientific, but not the statistical perspective. This is because identification of the orientation of an object with respect to the mean is not needed to characterize the biological nature of phenotypic variation, but is essential in describing the statistical nature of the observed landmark data. Unfortunately, Γ_i and t_i are unobservable and cannot be estimated, for reasons explained in the following text. This means that reconstruction of the natural space from empirical data is impossible (Lele and Richtsmeier, 1990; Lele, 1993; Lele and McCulloch, 2002; Richtsmeier *et al.*, 2002). Lele and colleagues provide mathematical proof of the inestimability of M, $\mathbf{\Sigma}_K$, and $\mathbf{\Sigma}_D$ (Lele, 1993; Lele and McCulloch, 2002).

First, let us consider the number of unknowns in the general perturbation model, $X_i = (M + E_i)\Gamma_i + t_i$. Only X_i is known from the data. All parameters, including the mean form, M, the variance–covariance structure of the errors, E_i, and the rotation and translation parameters, Γ_i and t_i, are unknown. The parameters M, $\mathbf{\Sigma}_K$, and $\mathbf{\Sigma}_D$ are fixed and common to all the specimens. However, the parameters Γ_i and t_i are different for *every specimen* because each individual has a unique location and orientation with reference to its position in the natural space. This means that we have a total of $(2+2n)$ unknowns for the given equation, the first term referring to M and the variance–covariance structure of E_i for the sample, the second term referring to Γ_i and t_i for *each individual* in the sample of size n. The number of unknowns $(2+2n)$ is therefore larger than the sample size (n). A basic tenet of inferential statistics is that one cannot estimate more parameters than the number of observations. Therefore the parameters cannot be estimated and must remain unknown. Although, M, Γ_i, t_i, etc., are matrices, for the sake of exposition, we consider them as single entities. Strictly speaking and in mathematical terms, the number of parameters is of $O(n)$, which is the same order of magnitude as the sample size (Richtsmeier *et al.*, 2002; see Lele and Richtsmeier, 2001; pp. 63–66).

It is the nature of landmark data—that they require specification of a coordinate system for their expression—that makes them so attractive to biological inquiry, but also makes them difficult to analyze. No information is available from landmark data *per se* regarding how the coordinate systems for each specimen relate to the natural space. In morphometrics, values for translation and rotation are often chosen based upon a "rule" or goal. For example, in Generalized Procrustes

Analysis (GPA), rotation and translation are estimated so that the sums of the squared distances between corresponding landmarks among individuals in a sample are minimized. An alternate rule is used in the robust or resistant fitting Procrustes analysis (Siegel and Benson, 1982; Chapman, 1990; Rohlf and Slice, 1990), and the application of this rule results in a different superimposition and thus a different mean and variance–covariance structure. In fact, because no information is available from the landmark data about how the coordinate system of each individual and the natural space fit to one another, there are infinite ways to place all individuals from a sample into a common coordinate system, and therefore there are infinite choices for the possible values of Γ_i and t_i. If all these choices were equivalent, the issue of choice would be irrelevant. However, as proven elsewhere (Lele, 1993; Lele and McCulloch, 2002) and shown graphically later in this chapter, this choice directly affects the estimation of the mean and especially the variance. If the estimate of the variance–covariance structure is invalid, it certainly cannot be used to partition variance components.

It is often stated in analyses using Procrustes methods that the nuisance parameters of rotation and translation are "effectively removed" from the analysis once the specimens have been oriented into a consensus coordinate system. In reality, the nuisance parameters are not removed, effectively or otherwise, and this statement is a misrepresentation of what occurs when data sets are analyzed using Procrustes analyses. Even after the data have undergone rotation, translation, and in some cases reflection, according to whatever rule the Procrustes method specifies, those parameters are still embedded (although further obfuscated) within the data. As a consequence, they affect the estimates of the mean and variance–covariance structure.

In short, nuisance parameters are not just irritants that can be eliminated by distributing their effects among individuals. Nuisance parameters must be dealt with in ways that do not cause them to affect the estimates of the very parameters that we are trying to estimate. Fortunately, the nuisance parameters, Γ_i and t_i, are of no real interest (except for some specific research questions) because knowledge of the orientation of an object with respect to its original position or to the mean is unimportant. We are really only interested in estimating the mean form, M, and the variance–covariance structure of the errors, E_i. In the following section, we show how these nuisance parameters can be properly eliminated.

III. PROPER ELIMINATION OF NUISANCE PARAMETERS USING A COORDINATE SYSTEM INVARIANT METHOD OF ESTIMATION

An alternative general method for estimating the mean and variance–covariance structure does not require an attempt to estimate Γ_i and t_i (Lele and Richtsmeier, 1991; Lele, 1993). If we limit the focus to the mean, M, and the variance–covariance of E_i, then following the logic presented in the preceding text, the *number* of unknowns is fixed and does not change with sample size.

These biologically interesting components of the model that are identifiable can be estimated in several ways, one of which is by using method-of-moments techniques developed by Lele (1993) and Lele and McCulloch (2002). Although the coordinate locations of the landmarks of the population mean form, M, cannot be observed directly, we can compute the coordinates of a consistent sample estimate of the mean, called $\hat{M}$, up to translation, rotation, and reflection. The sample estimate, $\hat{M}$, consists of all linear distances measured between unique pairs of landmarks and is equivalent to a coordinate data set, except that it does not require a coordinate system for its expression. Similarly, although one cannot estimate the among-landmarks variance–covariance matrix ($\mathbf{\Sigma}_K$), one can obtain a consistent estimate of a singular version of it, called $\mathbf{\Sigma}_K^*$ (see Lele, 1993). Although neither the among-axes variance–covariance matrix ($\mathbf{\Sigma}_D$) nor its eigenvectors is estimable, the eigenvalues of this matrix can be estimated (Lele and McCulloch, 2002), describing the overall eccentricity of the perturbation scatters at each landmark. In summary, the following features of the mean and the variance parameters can be identified and estimated:

1. All pairwise distances in the mean form M
2. A singular version of $\mathbf{\Sigma}_K$
3. Eigenvalues of $\mathbf{\Sigma}_D$

We emphasize that these results hold true irrespective of the particular method of estimation, whether method of moments or maximum likelihood. Moreover, these quantities and their estimators are *coordinate-system invariant*, meaning that they are not affected by the positions and orientations of the observations in any arbitrary coordinate system. However, the centering that is required in the calculation of $\mathbf{\Sigma}_K^*$ is by its nature not coordinate system invariant and, therefore, $\mathbf{\Sigma}_K^*$ can only be used as a suitable measure of the variation in landmark coordinate data for specific purposes as described in the following text. Details of the calculations of $\hat{M}$ and $\mathbf{\Sigma}_K^*$ are provided by Lele (1993), Lele and Cole (1996), and Lele and Richtsmeier (2001).

At this point, it is critical to understand the relationship between $\mathbf{\Sigma}_K^*$, which we can estimate, and $\mathbf{\Sigma}_K$, which we cannot. $\mathbf{\Sigma}_K^*$ is a singular version of $\mathbf{\Sigma}_K$, and the two are related as follows:

$$\mathbf{\Sigma}_K^* = \mathbf{L}\mathbf{\Sigma}_K\mathbf{L}^T$$

where $\mathbf{L}$ is a *centering matrix*. In both superimposition methods and in EDMA, the centering of all observations into a common coordinate system (that is, the translation of all observations to the origin) is a necessary first step. However, the purpose of the centering is different under the two methods. For superimposition methods, the centering is an initial step in attempting to reconstruct the natural space and, as a result, the true variance–covariance structure $\boldsymbol{\Sigma}_K$. However, we know that the natural space is actually impossible to reconstruct and that $\boldsymbol{\Sigma}_K$ is not identifiable, so whatever is reconstructed using superimposition methods, it cannot be the true variance–covariance structure. Using EDMA, the purpose of centering is different: The observations are centered to get a *version* of $\boldsymbol{\Sigma}_K$, called $\boldsymbol{\Sigma}_K^*$, that is identifiable and consistently estimable. We emphasize again that in estimating $\boldsymbol{\Sigma}_K^*$ in EDMA, we do *not* intend to reconstruct the natural space. The preceding equation may lead the reader to ask why we cannot obtain $\boldsymbol{\Sigma}_K$, given both $\boldsymbol{\Sigma}_K^*$, which we can estimate from data, and the known matrix, $\mathbf{L}$. The answer is that the system of equations $\boldsymbol{\Sigma}_K^* = \mathbf{L}\boldsymbol{\Sigma}_K\mathbf{L}^T$ does not possess a unique solution. There are more unknowns than the number of equations in the system. So even if $\mathbf{L}$ and $\boldsymbol{\Sigma}_K^*$ are known completely, one cannot solve for $\boldsymbol{\Sigma}_K$ in a unique fashion. In estimating $\boldsymbol{\Sigma}_K^*$ from real data, we must decide on the form of $\mathbf{L}$, and we must realize that this decision is ultimately arbitrary. We may choose to center all of the observations on a particular landmark, or we may choose to center them on some function of two or more landmarks, such as the centroid. If we experiment with different centering criteria, we will see that different decisions will lead to different pictures of variability.

To demonstrate this, we have randomly generated a sample of 100 observations (each with three landmarks in two dimensions) in the natural space, using the mean form ($\mathbf{M}$) and variance–covariance matrix ($\boldsymbol{\Sigma}_K$) shown in Table 4-1. The data are illustrated in the upper left-hand corner of Figure 4-2. Because these are artificial data that we generated using known parameters, we know that this is a true picture of the sample variation. To illustrate the effect of different choices of centering, we centered $\boldsymbol{\Sigma}_K$ several different ways to get $\boldsymbol{\Sigma}_K^*$, using the relationship $\boldsymbol{\Sigma}_K^* = \mathbf{L}\boldsymbol{\Sigma}_K\mathbf{L}^T$, where $\mathbf{L}$ is the centering matrix (Table 4-1). If we center, in turn, on the centroid (Figure 4-2A), the first landmark (Figure 4-2B), and the second landmark (Figure 4-2C), we see that we get $\boldsymbol{\Sigma}_K^*$ matrices with structures that look very different (Table 4-1). When we generate random data using these matrices (see Cole *et al.*, 2002), we also get very different visual impressions of the variability in the samples. Which, if any of these is the most accurate depiction of $\boldsymbol{\Sigma}_K$? We simply have no objective way of knowing.

Because the centering criterion is arbitrary, we must resist the temptation to examine a particular estimate of $\boldsymbol{\Sigma}_K^*$ as if it were a picture of the natural space. Had we chosen a different centering criterion, we may have obtained a very different picture. Given that fact, we are led to an obvious question about $\boldsymbol{\Sigma}_K^*$: If $\boldsymbol{\Sigma}_K^*$ does not give us an accurate picture of the true variation in landmarks ($\boldsymbol{\Sigma}_K$), how is it useful? In practical applications, estimates of $\boldsymbol{\Sigma}_K^*$ should not be seen as ends in themselves but as aids in asking other interesting questions about variation in form, shape, or growth. For example, we have used $\boldsymbol{\Sigma}_K^*$ to generate random data sets under the general perturbation model (given sample estimates of

TABLE 4-1. Example Illustrating the Effects of Different Centering Criteria on $\mathbf{\Sigma}_K^*$ Matrices, where $\mathbf{\Sigma}_K^* = \mathbf{\Sigma}_K L^T$.[a]

True parameters:

$$M = \begin{bmatrix} 0 & 0 \\ 10 & 0 \\ 0 & 10 \end{bmatrix} \qquad \Sigma_K = \begin{bmatrix} 1.0 & 0.3 & 0.6 \\ 0.3 & 3.0 & 0.8 \\ 0.6 & 0.8 & 2.0 \end{bmatrix}$$

Centering on the centroid:

$$\Sigma_K = \begin{bmatrix} 0.778 & -0.656 & -0.122 \\ -0.656 & 1.311 & -0.656 \\ -0.112 & -0.656 & 0.778 \end{bmatrix}$$

Centering on landmark 1:

$$\Sigma_K = \begin{bmatrix} 0 & 0 & 0 \\ 0 & 3.4 & 0.9 \\ 0 & 0.9 & 1.8 \end{bmatrix}$$

Centering on landmark 2:

$$\Sigma_K = \begin{bmatrix} 3.4 & 0 & 2.5 \\ 0 & 0 & 0 \\ 2.5 & 0 & 3.4 \end{bmatrix}$$

[a]The mean form **M** is used to generate the random data shown in Figure 4-2, using both $\mathbf{\Sigma}_K$ and the different versions of $\mathbf{\Sigma}_K^*$.

M and $\mathbf{\Sigma}_K{}^*$ and assuming multivariate normality) (Richtsmeier *et al.*, 2000). We have used these random data as part of a parametric bootstrapping procedure in testing for differences between the mean shapes of two samples (Lele and Cole, 1996). We used the same random data to generate confidence intervals for sample differences in the relative sizes of specific linear distances. In another type of application, we used $\mathbf{\Sigma}_K^*$ estimates as part of tests for phylogenetic signals in landmark data (Cole *et al.*, 2002). Importantly, Lele and McCulloch (2002: Theorem 2) have demonstrated that the choice of centering does not have any effect on the results of these applications. In addition, as noted by Lele and Cole (1996: Appendix),

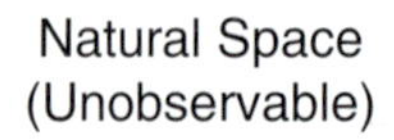

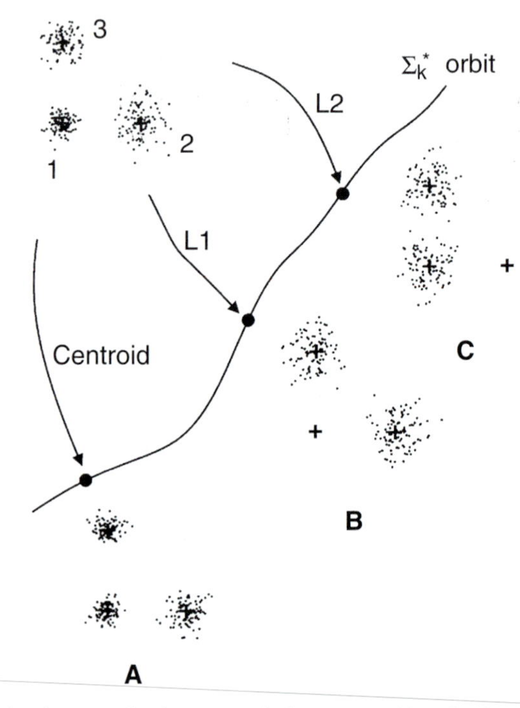

FIGURE 4-2. Sample of 100 randomly generated observations (three landmarks in two dimensions) in the natural space, based on the mean from M and variance–covariance matrix $\mathbf{\Sigma}_K$ shown in Table 4-1. The mean locations of the landmarks are indicated by the plus signs (+). The plot shows the true dispersion pattern of the sample in the natural space (top left). Also shown are random samples generated under different singular versions of the variance–covariance structure, calculated using different centering criteria, L, that used (A) the centroid; (B) landmark 1 (L1); and (C) landmark 2 (L2). $\mathbf{\Sigma}_K^*$ for each of these centering patterns are given in Table 4-1. The mean locations and the scales of the plots are the same in every case.

confidence intervals and the distributions of test statistics will be the same when calculated with any $\mathbf{\Sigma}_K^*$ as they would be if we could use $\mathbf{\Sigma}_K$ itself.

IV. ADDING ASSUMPTIONS TO THE PERTURBATION MODEL

Estimation of a $\mathbf{\Sigma}_K^*$ matrix, which is a nonunique, singular version of $\mathbf{\Sigma}_K$, is the best that we can do with the general perturbation model that we have specified. However, if we can add some additional assumptions to the model, and we are confident that these assumptions have a reasonable theoretical base, we can move another step closer to seeing a picture of the natural space, which is a picture of variation that we can interpret in biological terms. In essence, by imposing a more restricted structure, we make it possible for the system of equations, $\mathbf{\Sigma}_K^* = \mathbf{L}\mathbf{\Sigma}_K\mathbf{L}^T$, to have a unique solution. We impose the constraints in such a fashion that the number of unknowns is *at most* equal, and hopefully smaller, than the number of equations.

Remember that we are trying to estimate the mean form, M, and the variance–covariance structure of the perturbations, $\mathbf{\Sigma}_K \otimes \mathbf{\Sigma}_D$. E_i is a $K \times D$ matrix of perturbations that describe how the coordinates for each individual within a sample, X_i, differs from M in a space that will be specified by the assumptions of our model. The perturbations are assumed to have a multivariate normal distribution with a covariance structure $\mathbf{\Sigma}_K \otimes \mathbf{\Sigma}_D$. $\mathbf{\Sigma}_K$ is a $K \times K$ matrix that describes the variances and covariances of the landmarks, while $\mathbf{\Sigma}_D$ is a $D \times D$ matrix that describes the variances and covariances of the perturbations with respect to the coordinate-system axes.

A. Model 0: Isotropic Error Model

A simplifying assumption that often has been added to the general perturbation model to make it estimable is that $\mathbf{\Sigma}_K = \sigma^2 \mathrm{I}$ and $\mathbf{\Sigma}_D = \mathrm{I}$, I being the identity matrix (Bookstein, 1986). In fact, this assumption is *necessarily* made by generalized Procrustes analysis in order for the estimate of the mean to be consistent (Lele, 1993; Kent and Mardia, 1997; Lele and Richtsmeier, 2001). What this assumption means is that the variances of all landmarks (that is, the amount of dispersion) are expected to be the same. It also means that the patterns of dispersion across landmarks are expected to be uncorrelated, so that different landmarks vary independently. Biologically, the idea of equal variances at each landmark seems unlikely. Lele and Richtsmeier (1990) demonstrated strong rejection of the hypothesis of equality of variances local to landmarks in an analysis of three biological data sets. Sometimes, especially in Procrustes analysis (Dryden and Mardia, 1998), it is assumed that the variation around the landmarks is small enough that the equal variation model is a reasonable approximation of reality. Assuming small variance local to landmarks seems biologically ill-advised as well, especially when this assumption cannot be validated using the real data. Moreover, an assumption of landmark independence seems extremely improbable, and it logically precludes the use of these models for studies of morphological integration or modularity, which focus on and interpret the covariances and correlations among landmarks. If we assume the covariances are all zero, there is, by definition, nothing for us to study. Biology has presented us with a number of constraints on the description of form by landmark data based solely on anatomy and morphology. It seems unwise to place additional constraints on analyses solely based on statistical considerations (e.g., the assumptions of Procrustes analyses) that may not reflect biological reality.

We now propose some models for the variance–covariance matrix that are more restricted than the general perturbation model, but are not as restricted as the isotropic error model. To demonstrate our models, we use the mouse dentary as the mean form on which we have located 11 landmarks, each landmark associated with one of six morphogenetic units, as defined by Atchley and Hall (1991) (Figure 4-3). As we progress through the development of models, we add increasingly complex assumptions in an effort to increase their biological validity and realism.

B. Model 1: Σ_K = Independent Local Variation

In this model, we introduce heterogeneity of phenotypic variation at landmarks by allowing different magnitudes of variation local to each landmark. Solely, for exposition of our models, we adopt the convention that variation can take any value

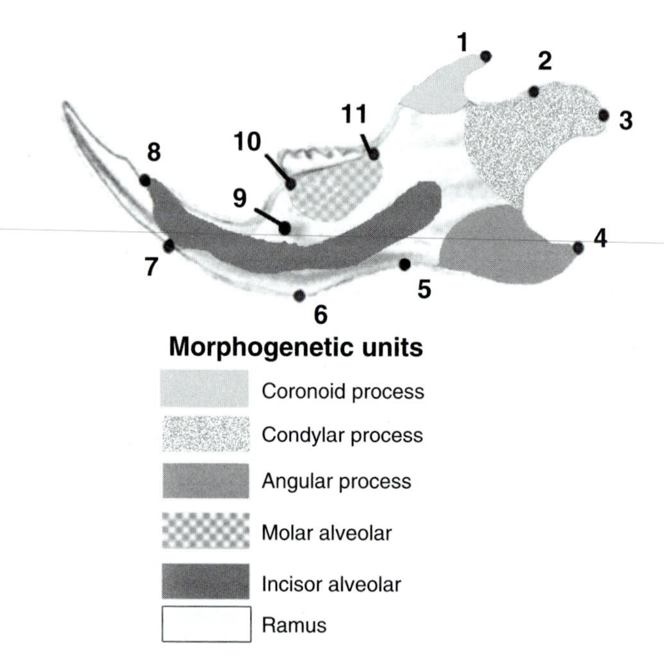

FIGURE 4-3. The adult mouse hemimandible (dentary) subdivided into the mandibular morphogenetic units (modules) proposed by Atchley and Hall (1991). (From Biological Reviews of the Cambridge Philosophical Society, Vol. 66 [1991], pp. 101–157. Reprinted with the permission of Cambridge University Press.)

from 0 to 6, 0 representing a lack of variation (or covariation), and 6 representing the maximum phenotypic variation possible. In this particular model, correlations among landmarks are not permissible, so off-diagonal elements remain constrained to zero. This is an extremely simple model that is relatively easy to estimate. However, it is biologically unrealistic since studies of morphological integration and allometry have demonstrated that local measures of variability are correlated within an organism because of relationships that have a developmental or functional origin. Because this model does not allow the most basic relationships among neighboring structures, further assumptions are required to build a more realistic model.

C. Model 2: Σ_K = Independent Modules

In this model, landmarks are grouped into modules that exhibit common properties of phenotypic variation based on developmental or functional considerations. The developmental organization of the mammalian dentary provided by Atchley and Hall (1991) divides the mandible into six morphogenetic units, each being derived from a separate cell condensation (Figure 4-3). Studies of modularity and morphological integration suggest that landmarks within the same module may share aspects of the perturbation structure.

In the independent modules model, variation local to landmarks (values of diagonal elements) that coexist within a module are similar, as are the values of the off–diagonal elements within a module (Figure 4-4). Correlation patterns exist for the perturbation pattern of landmarks within a module, but there is no correlation in patterns between modules. For example, landmarks 2 and 3 are both located within the condylar process morphogenetic unit. We set our expected value of variation to be similar local to landmarks 2 and 3, and we set an expected level of covariation within this module that reflects our knowledge (or hypotheses pertaining to identification) of the relative potential variation within this part of the mandible. Landmarks that represent the ramus (5, 6, 9) are expected to show a lesser magnitude of variation and covariation because formation of the ramus has proven to be relatively stable even in the case of knock-out experiments (see codependent module model, in the following text).

In our example, modules are defined and values are assigned using developmental information, but can be based on other relevant biological considerations. There are many other sources of information for modeling the variance–covariance structure (e.g., phylogenetic considerations, trajectories of disease processes, fate maps, teratogen targets, functional anatomy). Depending on the factors that define the modules, landmarks within a module may or may not be physically close to one another. Importantly, in our example, we are assigning values to demonstrate

LANDMARKS	1	2	3	4	5	6	7	8	9	10	11
1	**6**	*0*	*0*	*0*	*0*	*0*	*0*	*0*	*0*	*0*	*0*
2	*0*	**3**	*4.2*	*0*	*0*	*0*	*0*	*0*	*0*	*0*	*0*
3	*0*	*4.2*	**3**	*0*	*0*	*0*	*0*	*0*	*0*	*0*	*0*
4	*0*	*0*	*0*	**4**	*0*	*0*	*0*	*0*	*0*	*0*	*0*
5	*0*	*0*	*0*	*0*	**2**	*2.1*	*0*	*0*	*2.1*	*0*	*0*
6	*0*	*0*	*0*	*0*	*2.1*	**2**	*0*	*0*	*2.1*	*0*	*0*
7	*0*	*0*	*0*	*0*	*0*	*0*	**3**	*2.9*	*0*	*0*	*0*
8	*0*	*0*	*0*	*0*	*0*	*0*	*2.9*	**3**	*0*	*0*	*0*
9	*0*	*0*	*0*	*0*	*2.1*	*2.1*	*0*	*0*	**2**	*0*	*0*
10	*0*	*0*	*0*	*0*	*0*	*0*	*0*	*0*	*0*	**3.5**	*2.6*
11	*0*	*0*	*0*	*0*	*0*	*0*	*0*	*0*	*0*	*2.6*	**3.5**

FIGURE 4-4. Variance–covariance structure for independent modules. In this model, we constrain measures of variation at a landmark (diagonal elements of the matrix) and covariation between landmarks (off-diagonal elements of the matrix) to values between 0 and 6, 0 indicating a lack of variation or covariation and 6 indicating the maximum amount of variation or covariation possible. In this model, variation local to landmarks within a morphogenetic unit or module are set to similar values and covariation between landmarks within a module is allowed. No covariation is allowed between landmarks that occupy different morphogenetic units.

the expected patterns based on our knowledge of developmental relationships. When estimating the variances and covariances from data, we only specify the pattern (constraints within and between modules). The specified constraints comprise the pattern. Variances and covariances are estimated using the data.

D. Model 3: Σ_K = Co-dependent Modules

In this model, correlation in the perturbation pattern for landmarks within and between modules is allowed. One potential way to model such relationships is to predict a correlation structure that reflects what is known about the effects of particular genes, functions, selection differentials, or teratogens on different modules. An alternate way to model the relationships is to predict a correlation structure that is a function of the dissimilarity between landmarks within modules. If a dissimilarity metric is used, we might predict that all landmarks will be correlated but that the *degree* of correlation will be a function of the dissimilarity of the landmarks. The measure of dissimilarity does not need to be based on the physical proximity of landmarks or modules but can be based on an alternate measure of "adjacency" (*sensu* Chernoff and Magwene, 1999) that indicates whether two traits can be considered "neighbors" with respect to a specific criterion. The criterion for specifying adjacency may be functional, developmental, spatial, or based on any sensible biological data (Cheverud, 1982). For example, if development proves to be a driving force in patterns of variability, we might construct a dissimilarity measure based on the embryonic source of particular structures. With the appropriate data sets, we might expect that landmarks representing features derived from a specific population of neural crest cells might share a dissimilarity metric of lower magnitude (indicating a stronger relationship) even though the structures are anatomically remote.

As a simplified example of the codependent modules model using the mouse dentary, we assign expected values of variation and covariation based on information gained from development and from knock-out experiments in mice as summarized by Hall (2003). Normal dentary bones develop in mice in which Msx-1 has been knocked out, but the teeth and the associated alveolar bone fails to form (Satokata and Maas, 1994). When working with a developmental system in which we know that some aspect of the molecular pathway involving Msx-1 is affected, we might propose that the expressed phenotypic variation in the incisor and molar alveolar morphogenetic units would be similar and amplified, compared with phenotypic variation expressed local to the other morphogenetic units (ramus, coronoid process, condylar process, and angular process). Because the Msx-1 pathway has been manipulated in some way, it is expected that the proposed experiment will yield mice that display increased variation local to the molar and incisive alveolar modules. Moreover, since the same genetic pathway disrupts development in these two modules, we expect covariation between the incisive and molar alveolar modules. To accommodate these observations, we adjust our model by maintaining the expectations of the perturbation pattern for landmarks within those modules, increase the expectation of variation within the molar and incisive alveolar modules, and introduce covariation between modules (Figure 4-5). Similar magnitudes of perturbations are expected for the incisive and molar alveolar modules, while a lower magnitude of variation is proposed between the two alveolar modules and the ramus, coronoid, condylar, and angular morphogenetic units.

How can estimation of $\mathbf{\Sigma}_K$ be achieved under any given model? What we seek is the model of $\mathbf{\Sigma}_K$ that is the best "fit" to $\mathbf{\Sigma}_K^*$, which we compute from real data. To measure the fit of $\mathbf{\Sigma}_K^*$ (which we observe) to the $\mathbf{\Sigma}_K$ model (which we propose based on theory), we consider the following statistic:

$$\gamma = trace\,[(\mathbf{\Sigma}_K^* - \mathrm{L}\mathbf{\Sigma}_K\mathrm{L}^T)(\mathbf{\Sigma}_K^* - \mathrm{L}\mathbf{\Sigma}_K\mathrm{L}^T)^T]$$

LANDMARKS	1	2	3	4	5	6	7	8	9	10	11
1	**6**	0	0	0	0	0	0	0	0	0	0
2	0	**3**	4.2	0	0	0	0	0	0	0	0
3	0	4.2	**3**	0	0	0	0	0	0	0	0
4	0	0	0	**4**	0	0	0	0	0	0	0
5	0	0	0	0	**2**	2.1	0	0	2.1	0	0
6	0	0	0	0	2.1	**2**	0	0	2.1	0	0
7	0	0	0	0	0	0	**5**	4.2	0	3.7	3.7
8	0	0	0	0	0	0	4.2	**5**	0	3.7	3.7
9	0	0	0	0	2.1	2.1	0	0	**2**	0	0
10	0	0	0	0	0	0	3.7	3.7	0	**5**	4.2
11	0	0	0	0	0	0	3.7	3.7	0	4.2	**5**

FIGURE 4-5. Variance–covariance structure for codependent modules. In this model, we constrain measures of variation at a landmark (diagonal elements of the matrix) and covariation between landmarks (off-diagonal elements of the matrix) to values between 0 and 6, 0 indicating a lack of variation or covariation possible. In this model, variation local to landmarks within a morphogenetic unit or module are set to similar values and covariation between landmarks within a module is allowed. In addition, covariation is allowed between landmarks that occupy different morphogenetic units when the scientist has knowledge of an influence that affects more than one module simultaneously. In our example, we propose a murine model in which the Msx-1 pathway has been manipulated. Because knock-out experiments have demonstrated Msx-1 to be necessary to the development of the incisive and molar alveolar morphogenetic units, we expect phenotypic variation to be similarly elevated in these two modules and coviaration to exist among landmarks in these two modules.

Given $\mathbf{\Sigma}_K^*$, the best-fitting estimate of $\mathbf{\Sigma}_K$ (subject to the constraints of the model) is the one that minimizes γ. The estimate must be found by solving a system of nonlinear, simultaneous equations, which can be a formidable computational task when the matrices are large. If $\mathbf{\Sigma}_K^*$ were simply a centered version of our theoretical model, then γ would equal zero. While this exact relationship is unlikely with real data, we recognize values of γ that are closer to zero as indicative of better-fitting models.

In comparing the fits of two different models, we could use their respective γ values to develop a test statistic that would determine which model was the better fit to the data. The bootstrap could be then used to measure uncertainty in the test statistic, allowing us to decide whether one model was significantly better supported than another. We should reemphasize that estimation of Σ_K using this method requires us to specify a constrained model of variation *a priori;* this specification is far from being an automated process and can be very labor intensive. In theory, there are infinite numbers of potential models, and our most imposing challenge is to use our biological knowledge and experience to specify models that will be as realistic as possible.

V. CONCLUSIONS

The translation of genetic information into phenotype is central to biology. We are, however, a long way from understanding the connection between the underlying genetic architecture and expression of phenotypic variation. Because the entire range of genetic effects varies relative to the genetic variants present in any given genome within any given environment (Rutherford, 2000), evaluation of phenotypic variation and measures of the effects of various factors on the production of the phenotype remain critical to our understanding of many fundamental processes, among them development and evolution. To be of any use to science, however, these quantitative observations need to be precise, and sample parameters estimated from observations must be valid.

Our discussion has focused on the nature of landmark coordinate data and how their dependence upon a coordinate system limits what we can know about phenotypic variation within a sample. The position of a set of landmarks on a form provides a geometric representation of the relative locations of a set of salient features. When landmarks are collected from a sample of forms, each form is captured in its own coordinate system. We have shown that the true correspondence among the various coordinate systems of individuals within a sample can never be known. Models proposed to characterize the perturbation structure of landmark data sets contain nuisance parameters (Neyman and Scott, 1948) that include rotation, translation, and sometimes, reflection. These nuisance parameters make the determination of a valid common coordinate system impossible and prevent the estimation of the variance–covariance structure from empirical data (Lele and McCulloch, 2002; Lele and Richtsmeier, 1990). This limitation should be regarded as a challenge, forcing us to be clever in using what we *can* know to probe the data further.

We have presented alternative, coordinate-system-invariant features of the mean and the variance parameters that can be identified and estimated, and we have suggested ways that they might be used in biological contexts. Importantly, we have proposed how the addition of assumptions based on scientific knowledge might move our model closer to a characterization of variation that can be interpreted in biological terms. As always, the investigator's knowledge of the biology of the forms under study and proper incorporation of this knowledge into the model are key to its correctness and suitability.

ACKNOWLEDGMENTS

We thank Benedikt Hallgrímsson and Brian Hall for inviting us to contribute to this fine volume on the critical issue of variation. The ideas presented in this chapter are the result of years of collaborative effort to which all three authors contributed equally, but with varying points of emphasis. This work was supported in part by Public Health Service awards, 1F33DE/HD05706, 1 P60 DE13078 (Project VI), HD24605, and an NSERC grant.

REFERENCES

Atchley, W., and Hall, B. (1991). A model for development and evolution of complex morphological structures. *Biological Reviews* **66**, 101–157.

Boas, F. (1905). The horizontal plane of the skull and the general problem of the comparison of variable forms. *Science* **21**, 862–863.

Bookstein, F. L. (1982). Foundations of morphmetrics. *Annual Review of Ecology Systematics* **13**, 451–470.

Bookstein, F. (1986). Size and shape spaces for landmark data in two dimensions. *Statistical Science* **1**, 181–242.

Chapman, R. (1990). Conventional Procrustes approaches. In *Proceedings of the Michigan Morphometrics Workshop* (F. Rohlf and F. Bookstein, eds.), pp. 251–267. Ann Arbor, MI: University of Michigan Museum of Zoology.

Chernoff, B., and Magwene, P. (1999). Afterword. In *Morphological Integration* (E. Olson and R. Miller, eds.), pp. 319–348. Chicago: University of Chicago Press.

Cheverud, J. (1982). Phenotypic, genetic, and environmental morphological integration in the cranium. *Evolution* **36**, 1737–1747.

Cole, T., III (1996). Historical note: Early anthropological contributions to "geometric morphometrics." *American Journal of Physical Anthropology* **101**, 291–296.

Cole, T. M., III, Lele, S., and Richtsmeier, J. T. (2002). A parametric bootstrap approach to the detection of phylogenetic signals in landmark data. In *Morphology, Shape, and Phylogeny* (N. MacLeod and P. Forey, eds.), pp. 194–219. London: Taylor and Francis.

Dryden, I., and Mardia, K. (1998). *Statistical Shape Analysis*. Chichester, England: Wiley.

Falconer, D. (1981). *Introduction to Quantitative Genetics*. London: Longman.

Falconer, D., and Mackay, T. (1996). *Introduction to Quantitative Genetics*. Essex, England: Longman.

Goodall, C. (1991). Procrustes methods in the statistical analysis of shape. *Journal of the Royal Statistics Society Series B* **53**, 285–339.

Hall, B. (2003). Unlocking the black box between genotype and phenotype: Cell condensations as morphogenetic (modular) units. *Biology and Philosophy* **18**, 219–247.

Kent, J. T., and Mardia, K. (1997). Consistency of Procrustes estimators. J. R. *Statist. Soc. Ser. B* **59**, 281–290.

Lele, S. (1991). Some comments on coordinate free and scale invariant methods in morphometrics. *American Journal of Physical Anthropology* **85**, 407–418.

Lele, S. (1993). Euclidean distance matrix analysis (EDMA) of landmarks data: Estimation of mean form and mean form difference. *Mathematical Geology* **25**, 573–602.

Lele, S., and Cole, T. M. III. (1996). A new test for shape differences when variance–covariance matrices are unequal. *Journal of Human Evolution* **31**, 193–212.

Lele, S., and McCulloch, C. (2002). Invariance, identifiability, and morphometrics. *Journal of American Statistical Association* **971**, 796–806.

Lele, S., and Richtsmeier, J. T. (1990). Statistical models in morphometrics: Are they realistic? *Systematic Zoology* **39**, 60–69.

Lele, S., and Richtsmeier, J. T. (1991). Euclidean distance matrix analysis: A coordinate-free approach for comparing biological shapes using landmark data. *American Journal of Physical Anthropology* **86**, 415–427.

Lele, S., and Richtsmeier, J. (2001). *An Invariant Approach to the Statistical Analysis of Shapes*. London: Chapman and Hall/CRC Press.

Neyman, J., and Scott, E. (1948). Consistent estimates based on partially consistent observations. *Econometrika* **16**, 1–32.

Richtsmeier, J. T., and Lele, S. (1993). A coordinate-free approach to the analysis of growth patterns: Models and theoretical considerations. *Biological Reviews of the Cambridge Philosophical Society* **68**, 381–411.

Richtsmeier, J., Cheverud, J., and Lele, S. (1992). Advances in anthropological morphometrics. *Annual Review of Anthropology* **21**, 231–253.

Richtsmeier, J. T., Cole, T. M., 3rd, Krovitz, G., Valeri, C. J., and Lele, S. (1998). Preoperative morphology and development in sagittal synostosis. *Journal of Craniofacial Genetics and Developmental Biology* **18**, 64–78.

Richtsmeier, J. T., Baxter, L. L., and Reeves, R. H. (2000). Parallels of craniofacial maldevelopment in Down syndrome and Ts65Dn mice. *Developmental Dynamics* **217**, 137–145.

Richtsmeier, J., DeLeon, V., and Lele, S. (2002). The promise of geometric morphometrics. *Yearbook of Physical Anthropology* **45**, 63–91.

Rohlf, F., and Slice, D. (1990). Extensions of the Procrustes method for the optimal superimposition of landmarks. *Systematic Zoology* **39**, 40–59.

Rutherford, S. L. (2000). From genotype to phenotype: Buffering mechanisms and the storage of genetic information. *Bioessays* **22**, 1095–1105.

Satokata, I., and Maas, R. (1994). Msx1 deficient mice exhibit cleft palate and abnormalities of craniofacial and tooth development. *Nature Genetics* **6**, 348–356.

Siegel, A. F., and Benson, R. H. (1982). A robust comparison of biological shapes. *Biometrics* **38**, 341–350.

Sneath, P. H. A. (1967). Trend-surface analysis of transformation grids. *Journal of Zoology London* **151**, 65–122.

Thompson, D. (1917). *On Growth and Form*. Cambridge, England: Cambridge University Press.

Thompson, D. (1992). *On Growth and Form*. The complete revised edition. New York: Dover.

CHAPTER 5

Variation in Ontogeny

DONNA CARLSON JONES AND REBECCA Z. GERMAN
Department of Physical Medicine and Rehabilitation, Johns Hopkins Hospital, Baltimore, Maryland, USA

INTRODUCTION

The extensive variation that exists in morphometric form, produced by a wide array of mechanisms, is largely enacted during growth. Historically, morphometric studies of plant and animal ontogenetic series focus on the central values of growth parameters: rates, durations, time of onset and offset. Despite the obvious existence of variation in these parameters, even recent examinations of plants (e.g., Niklas, 1997; Doust, 2000; Wright and McConnaughay, 2002; Jordy, 2004) and animals (e.g., Kobyliansky and Livshits, 1989; Burns, 1993; Fischer and Fiedler, 2002;

Jabbour *et al.*, 2002; Smith and Kamiya, 2002; Line, 2003) ignore the changes in variability inherent within these systems, although some note variation (e.g., Kobyliansky and Livshits, 1989; Riddle *et al.*, 1992; Burns, 1993; Riddle and Purves, 1995; Gedroc *et al.*, 1996; Doust, 2000; Line, 2003; Jordy, 2004) and occasionally study it (e.g., Klingenberg, 1996; Zelditch and Fink, 1996). A few studies use variation among developmental trajectories as evidence of species definitions (as reviewed by True and Haag, 2001) or as the basis for delimitating evolutionary sequences in higher taxa, particularly heterochrony (e.g., Alberch *et al.*, 1979; McNamara, 1995; Klingenberg, 1998) and heterotopy (e.g., Zelditch and Fink, 1996; Guralnick and Lindberg, 1999; Zelditch *et al.*, 2000; Rudall and Bateman, 2002). In such studies, the life history of the organism is most often the unit of analysis, with comparisons of growth parameters among groups defined by sex, treatment, or species.

The hierarchy that exists from organs to populations is critical to studies of variation. Although growth is a property of an individual, measurements of growth, and the variation in those measurements, exist at several obvious levels of organization: within individual and among individuals, as well as between/among whatever biological groups constitute the hypothesis being tested (Figure 5-1). This hierarchy is ubiquitous and complicating to any analysis of variation. Furthermore, because variation is a property of a sample (i.e., a set of points), valid scientific questions compare *groups* that provide more than one sample of variation. In addition, any analysis of variability must include multiple levels of hierarchy, or find an appropriate estimate or calculation that accounts for the missing levels.

The identification of the components of this hierarchical relationship can be simple or complex, but certain elements are consistent. One element, the *unit of analysis*, is the component for which measurements are collected. In most studies, this is a single, biological individual. Another element is the *hierarchical level of the question* being asked or the hypothesis being tested. In a standard study, comparing mean values between two groups, such as males and females, two levels of hierarchy suffice: 1) a sample of individuals, which are 2) organized into sexes. However, to test if the *variation* of males differs from females, an additional level is necessary. A single set of males and a single set of females produce only one measure of variation for each sex, with no replication, and therefore no possibility of testing differences in variation between the sexes. The unit of analysis in this case becomes the group (males or females) that generates a value of variation among them (Figure 5-1C). Statistical testing requires replicates of the unit of analysis. In either study (mean values or variation), the level of interest is the difference between sexes, but the two studies ask and can answer different questions because the unit of analysis differs.

A further complication arises in the study of ontogenetic variability. Although the hierarchical levels in both genetic and phylogenetic studies are relatively well defined, the levels of variation in growth studies are not as obvious. For instance, potential variation exists as a consequence of development. This is analogous to the concept of a sample in inferential statistics. It is the population (in the statistical

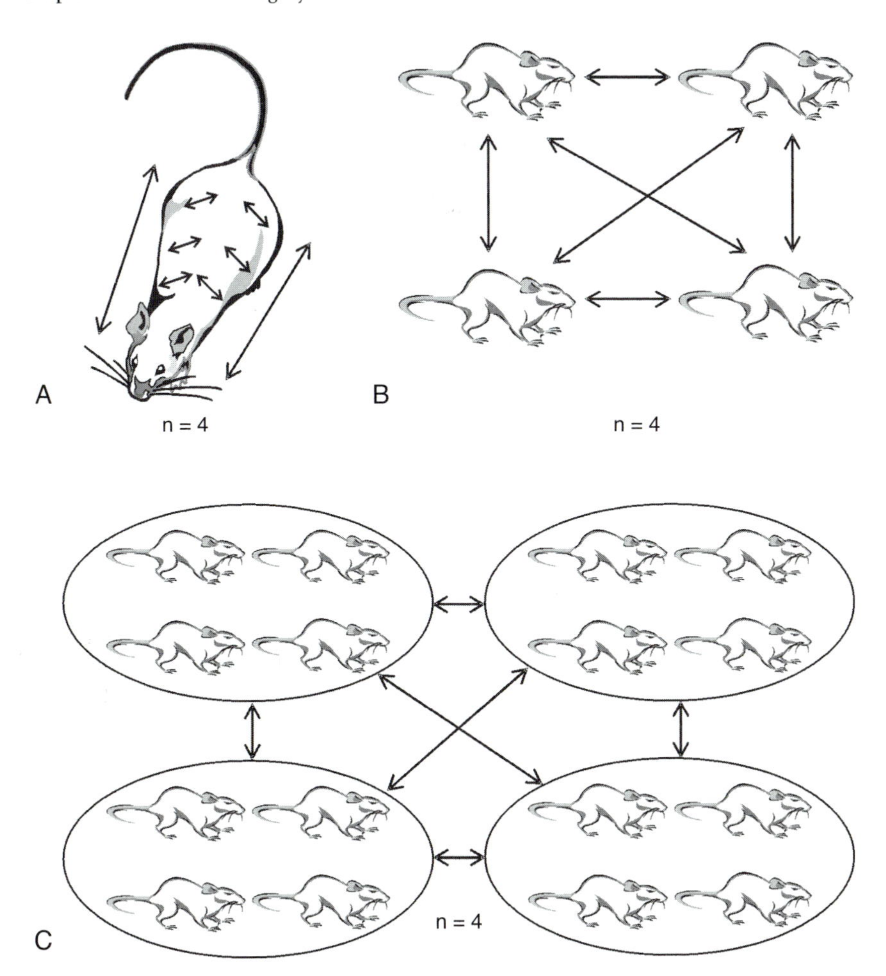

FIGURE 5-1. Hierarchical levels of variation. (A) Within individual variation, represented by left–right asymmetry. (B) Among individual variation. (C) Among groups variation. Because variation is produced only with replicate measurements, it is essential that at least two of the hierarchical levels of data be examined—one to produce the variation and one to compare the levels of variation of interest.

sense), including all individuals, existing or in the future, that is of interest; however, only a small sample of the population is available for analysis. Likewise, for any individual, there is a landscape of developmental possibilities that exist for a given genotype, equivalent to a hypothetical population. The extent or range of these possibilities, termed "developmental stability," is within-individual variation. Yet only a single pathway is realized; this is the sample available for analysis and presents difficulties for measurement of this variation.

Several alternative strategies for examining developmental variation exist. Within individual variation can be measured at a single point in time, e.g., by left–right asymmetry (e.g., Leamy, 1993; Polak *et al.*, 2003). A sample of these values can be compared over the period of growth or within a homogenous sample of individuals. Another way of examining variation within an individual is to measure the departure of an individual from an externally determined baseline, or *a priori* model of growth.

A carefully selected sample with a well defined factor of interest (i.e., level of the question), avoids some of the problems of obtaining data across multiple levels of hierarchy. If the unit of analysis is the realized developmental pathway of a single organism, out of the population of all potential pathways, it is not possible to collect replicates of that individual. If the question of interest is how an additional factor, such as an environment perturbation or a specific genetic background, impacts on that pathway, then it is possible to construct a set of replicates. Variation in a sample of different individuals, if homogeneous with respect to the factor of interest, will estimate the variation among the individuals, even though the unit of analysis is within individual.

The goal of this chapter is to elucidate the significance of variation during growth at these different hierarchical levels. We present an example of an analysis of growth data that includes partitioning within individual and among individual variation to compare differences in variation among externally or experimentally defined groups. In doing so, we will show that a single mechanism affects variation differently at these hierarchical levels.

I. MEASURING VARIATION: A CASE STUDY

A. Data Analyzed

To examine all hierarchical levels of variation through an ontogenetic series, we reanalyzed a set of longitudinal data from Miller and German (1999). This work examines the impact of two diets, a low-protein diet (4% protein) and a control, isocaloric diet (24% protein). The raw data are serial radiographs of individuals taken for each individual from 22 days (weaning) until adult size, approximately 400 days for the slowest growing group. The total sample size was 37 individuals, and each individual had an average of 38 radiographs, concentrated during the times of most rapid growth (see Miller and German, 1999, for complete details).

On each radiograph, 19 measurements (Figure 5-2) of the craniofacial skeleton are calculated as the distance between Cartesian points digitized from landmarks on the skull. To compare the rates and durations of growth for all measurements, each of the 19 measurements, in addition to weight over time, was fit to a Gompertz model for each individual. This model is widely accepted because it provides an

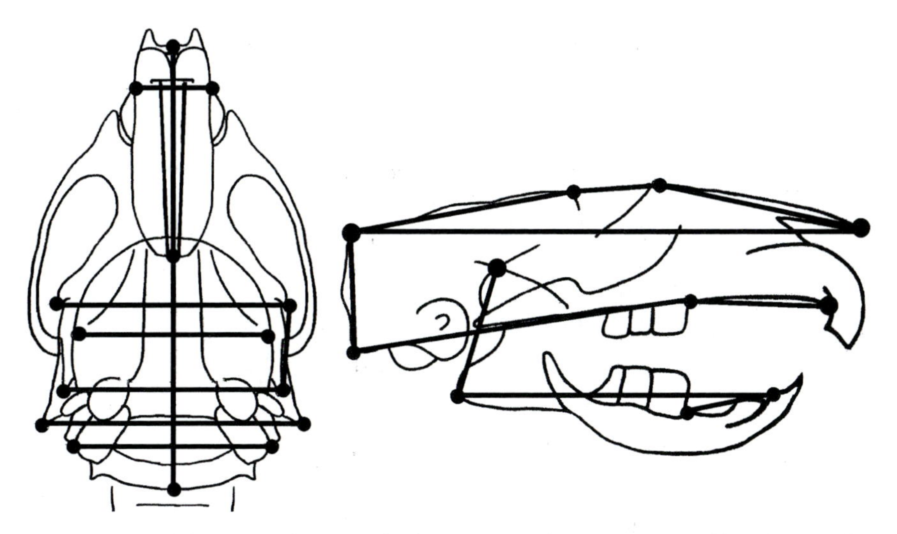

FIGURE 5-2. Adult rat skull with 19 craniofacial measurements from dorsoventral and lateral radiographs. Details of specific measurements are available in Miller and German (1999). (From J. Nutr.: [Vol. 129–11, pp. 2061–2069], American Society for Nutritional Sciences.)

excellent fit to the sigmoidal nature of mammalian growth (reviewed in Miller and German, 1999) and its parameters have associated biological meanings (Gille *et al.*, 1996). The resultant values of these parameters describe the timing and rate of growth (see also German, 2004, for a review).

The original work addressed the question: Are there any differences between the central values of the growth parameters between treatments and between sexes? That is, does diet change the rates and duration of growth, and does diet impact identically on the two sexes? A low protein diet has a significant impact on the rates, timing and durations of growth (Miller and German, 1999, Tables 1 and 2), relative to control levels of protein. The original analyses document the mean size (through distance measures between landmarks and weight) and mean growth trajectories (via Gompertz curves) of sex or diet. But, other than to assist with the determination of statistical significance, this study did not address variation during growth.

B. Levels of Variation in Data on Growth and Protein Malnutrition

These data permitted us to examine ontogenetic variation between groups defined by diet and sex at two levels: variation among individuals, and variation within an individual. These two comparisons suggest biologically different questions. The simplest questions are: (1) How much variation exists *within individuals* of each treatment, (2) How much variation exists *among individuals* of each treatment, (3) How does *within individual* variation differ among experimental treatment groups, and (4) How does *among individual* variation differ among experimental treatment groups? Both the data and the questions suggest that the appropriate analyses will differ between the studies defined by these questions. Finally, beyond understanding the role variation plays in this particular case, we wish to demonstrate that the hierarchical nature of biological variation requires a well defined experimental design, and one that can lead to different results depending on the level of organization being examined. It is the hierarchical nature of these data, which included multiple measurements on each individual at several points in time, that made valid statistical testing of these different levels possible.

C. Measuring within Individual Variation

One measure of within individual variation is how closely an individual's growth tracks the model trajectory for the Gompertz curve (Figure 5-3). For this study, we calculated the residuals from the fit of the data to the Gompertz model for each craniofacial measurement. Each individual had multiple residuals for each measurement at different points in time. These are equivalent to repeated measures on each individual animal. Increases or decreases in the absolute values of the residuals throughout growth would indicate an increase or decrease in the amount of variation within the individual. This produced a measure of within individual variation for 37 individuals. We then tested whether the amount of within

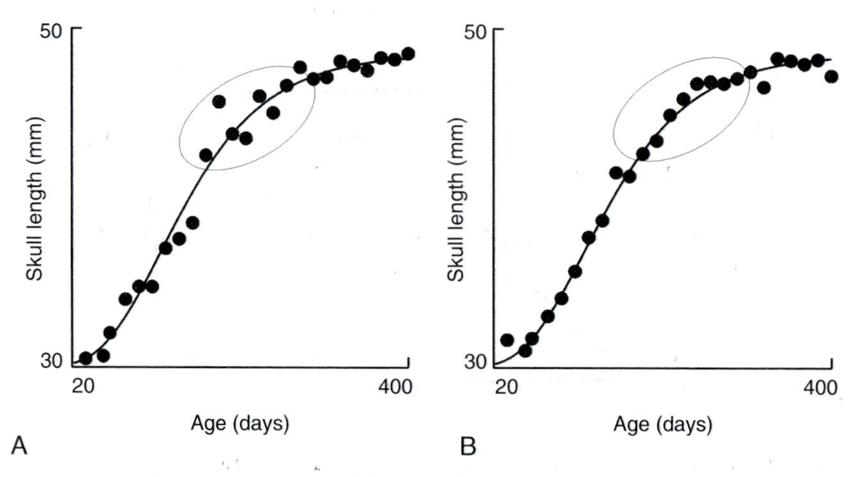

FIGURE 5-3. The fitted Gompertz model for skull length for two specimens. Specimens A and B, which are representative data to illustrate this point, do not fit their respective curves equally; A has more variation than B.

individual variation differed among groups defined by sex and diet. We would have used fluctuating asymmetry for this analysis but the sample size of our data set did not have sufficient power for such an analysis.

D. Measuring among Individual Variation

Testing variation among individuals requires careful definition of the unit of analysis. To test whether a central value such as size, or other growth parameter, differed among individuals, one would take measurements of that variable for each individual and then test for differences among individuals. The unit of analysis would be the individual. However, to measure the variation among individuals and test whether that variation is consistent across groups, the individual can no longer be the unit of analysis. Variation, which has become the parameter of interest, is the characteristic of a sample, and the design requires hierarchical organization of the data (Figure 5-4). Thus to ask whether variation in duration of growth (not the magnitude of that duration of growth) differs with sex or diet, a single measurement of variation is inadequate for testing. Multiple replications of variation in the duration of growth are required to test whether *variation* among individuals differs among groups. In this case, we used 19 different skull measurements (generating a sample of variations) for each growth parameter (duration, rate, etc.). Thus our replicate samples of variation among individuals were different across the morphological measurements. We recognize that variation may not be consistent across different morphological areas of the same individual. However, given the limitations of these data, we felt this was the best compromise for this analysis. Using different measurements as an aggregate will not bias our results, but add noise, and thus reduce our ability to discern differences that might exist. The unit of analysis remains the individual, but the multiple measurements necessary for statistical analysis were of different morphologic measurements taken on each individual.

To measure variation, we used the coefficient of variation, because it is scaled for the absolute size of the measurement. Our parameters, which include duration of growth and rate of decay of growth, differ by several orders of magnitude, and this scaling facilitated our quantitative comparison.

E. Testing for Variation between Treatment Groups

Once the lower levels of hierarchy and sampling are defined, and the amount of variation measured at the two levels, testing for larger group differences is a consequence of the units defined. We tested whether variation, either within or among individuals, differed between groups as defined by the two treatment factors, sex and diet. These fixed factors, each with two levels, were crossed so that we also tested for

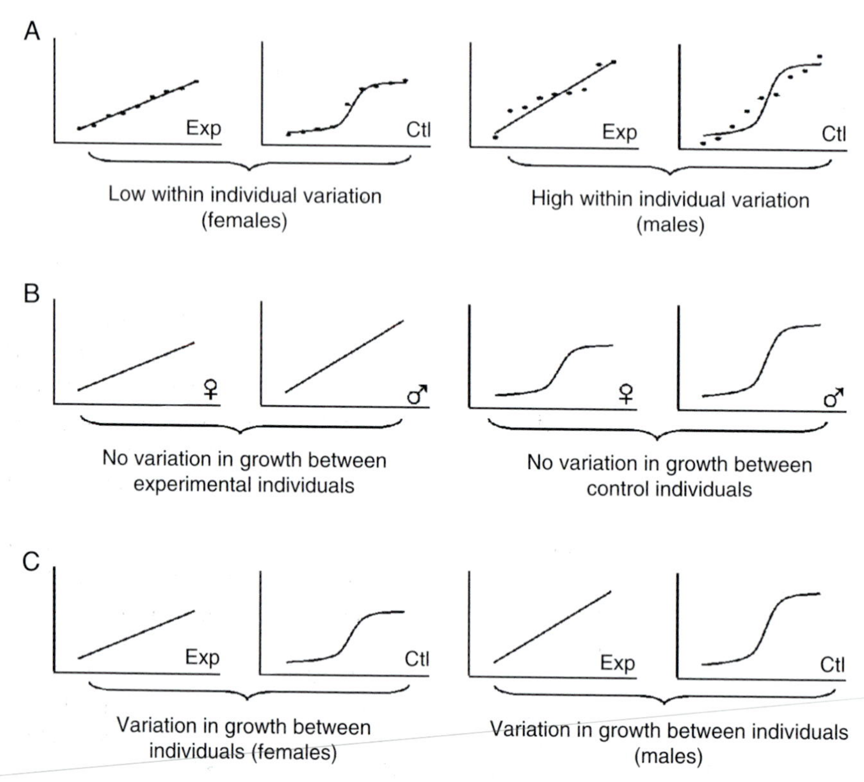

FIGURE 5-4. Graphical representation of hierarchical levels of variation, A) within individuals, B) between individuals as grouped by (exp, experimental; Ctl, control) treatment and C) between individuals as grouped by sex. In this case, within individual variation (as measured by model fit residuals) is lower for females than for males, regardless of experimental treatment (A). There is no between individual variation with experimental groups, independent of sex (B), but significant between individual variation within sexes, based on experimental treatment (C). Both within and between individual variation is tested in the context of questions about group or factor differences.

a significant interaction between the two. There are numerous outcomes of such an analysis, each with significant implications for biological interpretation (Farmer and German, 2004). These analyses were done in one complete step, i.e., testing for individual variation, as a repeated measure, which accounts for the random factor, in a model with the two fixed factors, sex and diet (Wilkenson, 2000).

F. Factor Differences for within Individual Variation

Patterns of within individual variation were not consistent across the different craniofacial measurements in this study, although a pattern did emerge. For 15 of 19 measurements, variation within individuals was not significantly different between

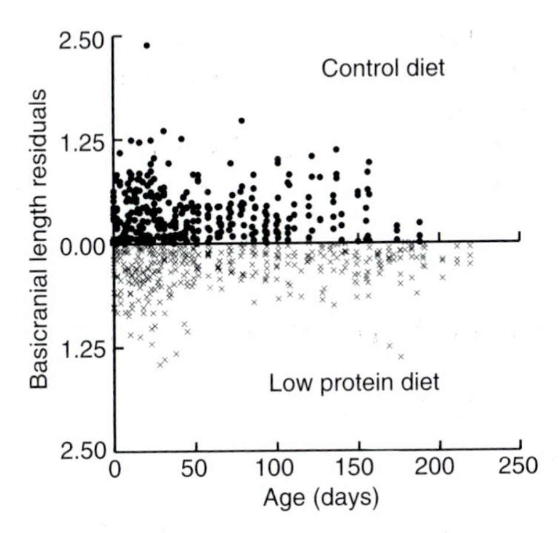

FIGURE 5-5. A mirror plot of the residuals from fit of Gompertz equation to longitudinal data of basicranial length for males of both diet groups, showing equivalent amounts of variation in the two treatments. Control diet is figured to the top of the plot, low protein to the bottom.

the sexes or between the two diets. Figure 5-5 is a plot of the residuals from a typical craniofacial measurement, basicranial length, over developmental time, for males of both diet groups. This illustrates no difference in variation between the treatments. Of the four exceptions, two were significantly different in residual variation by sex and two showed both diet and sex to be significant factors. Three of these four significant measurements were of the mandible: mandibular height, mandibular length, and length of the mandibular notch (between the coronoid and angular processes; see Figure 5-2). The fourth, the only neurocranial measurement with significant within individual variation, was the width of the neurocranium, which was significant for sex and diet.

G. Factor Differences for among Individual Variation

In no case was among individual variation different between sexes ($p > 0.10$). That is, variation in growth among females was equal to the variation among males. However, with the exception of *w*, initial relative size, all parameters describing rates and duration of growth were different between diets ($p < 0.01$). Variation among individuals on the low protein diet was significantly higher than variation among individuals on the control diet (Figure 5-6). There were also significant interactions for the duration of growth, the maximum rate of growth, and the decay of growth ($p < 0.01$). These patterns are shown in Figure 5-7 (note the order of magnitude difference in the scales of Figure 5-7A and Figure 5-7B). The amount of variation in initial size is not different among groups. There was, however, a significant interaction between sex and diet for variation in the duration of growth. In this case, not only was variation among individuals on the low protein diet higher than those on the control diet, but that males on the low protein diet had the highest amount of variation, and males on the control diet the least. This pattern of sex and diet interaction among groups was true for all variables with an interaction.

II. IMPLICATIONS FOR STUDIES OF VARIATION

A. Utility of Standard Laboratory Animal Models

The model species we used, *Rattus norvegicus*, has been purposefully inbred for reduced variation as a way of increasing the power of biomedical studies (although variation does still exist, see Riddle *et al.*, 1992; Riddle and Purves, 1995). Our results suggest that variation, even within this restricted setting, is complex. In response to the developmental perturbation generated by diet, variation was not consistent across the different hierarchical levels or between treatment groups. The answer to the question "does diet or sex impact on the amount of variation?" depends on the level being analyzed. Within individuals, variation in growth was for the most part constant despite dietary differences and, in fact, despite significant differences in growth parameters described by Miller and German (1999). In contrast, there was significant variation among individuals of the different diets. The overall pattern of these results suggests that, irrespective of perturbation, an animal holds to a standard growth trajectory, which may be a species or higher taxon characteristic (Farmer and German, 2004). However, external perturbations do impact differences among individuals. One goal of this chapter was to elucidate the significant existence of variation at different hierarchical levels during growth. This raises more questions, at each level, as well as questions that pertain to the relationships among levels.

B. Within Individual Variation

The relative lack of differences in within-individual variation between our treatment groups bears on the current debates concerning integration and modularity of growth in the craniofacial skeleton (Zelditch, 1988; Zelditch and Carmichael, 1989; Zelditch *et al.*, 1990, 1993; Raff, 1996; Klingenberg, 2002; Klingenberg *et al.*, 2003). Modules are distinct units that are more susceptible to forces of selection than an entire system, because the module is a largely separate biological entity, and changes in it do not necessarily require system-wide alterations to be functional (Cheverud, 1982; Cheverud *et al.*, 1989; Leamy, 1993; Klingenberg *et al.*, 2001, 2003). One way to quantitatively document the existence of modules is by studying within individual variation (e.g., among groups of structures, Cheverud *et al.*, 1989; Leamy, 1993). Our results suggest that individual growth trajectories are robust to perturbation. Although our measurements, by historical default (Miller and German, 1999), are not distinct modules, they do measure growth in structures that behave as would be predicted by a theory of craniofacial modules.

C. Between or Among Individual Variation

The significant variation we found among individuals, especially in response to treatment differences, suggests that even when animals have been selected for reduced variation in growth trajectories under a standard external environment, the capacity to respond to environmental differences remains. The variation in growth among control individuals is expected, given the artificial selection regimen that generated this breed of animal. Artificial selection clearly can limit variation among individuals in a constant environment. Nevertheless, when the full sibs of these individuals were challenged during growth, the response was not as predictable. Significant external changes can produce a significant amount of variation in response to those changes and produce the variation required for natural selection to operate (Figures 5-6 and 5-7).

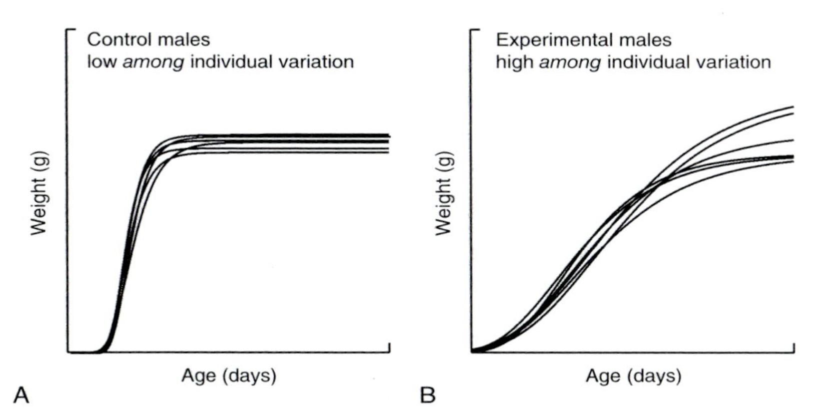

FIGURE 5-6. Variation in the Gompertz models for A) low protein and B) control diet males for weight over time. The experimental males have more among individual variation than the control males.

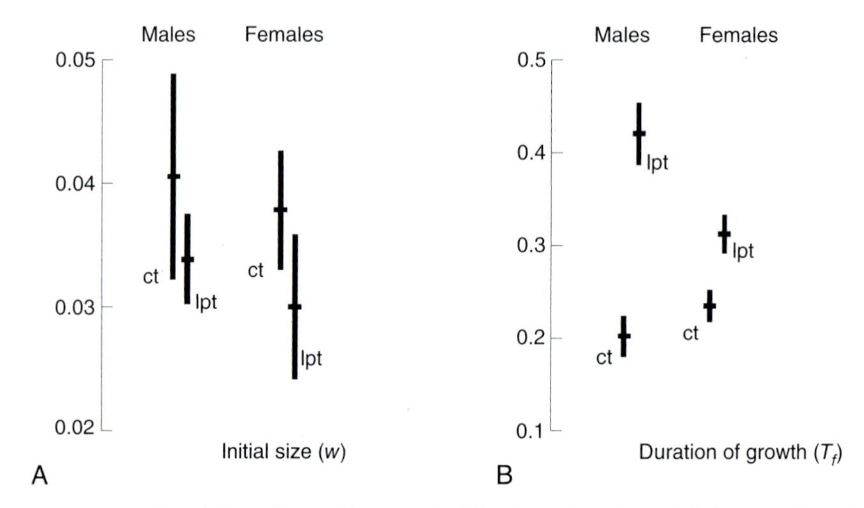

FIGURE 5-7. Plots of the coefficient of variation (CV) for A) initial size (w) and B) duration of growth (T_f) for groups defined by sex and diet. The CV was calculated over 17 measurements of craniofacial growth, to measure among individual variation. ct = control treatment, lpt = low protein treatment. Variation in initial size was not different for the sex or the diet factor. There was a significant interaction between sex and diet in variation in the duration of growth. Error bars are standard errors. Note the scale differences between the two graphs.

D. Variation across Hierarchical Levels

As Hallgrímsson *et al.* have pointed out (2003), two opposing concepts influence the consequences of morphologic change with growth: canalization and developmental plasticity. With canalization, any variation within a population that is present early in ontogeny is removed as all individuals converge onto the targeted adult morphology. Developmental plasticity, where larger amounts of morphologic variability in adult forms are produced from a population of relatively less variable offspring, explains an increase in variability through ontogeny (Gedroc *et al.*, 1996). Although these are processes or patterns of the individual level, both our within and between individual results indicate that canalization and plasticity may work simultaneously in the same population (or sample) at different levels.

It is also possible that plasticity and/or canalization alter during ontogeny. As Wright and McConnaughay (2002) point out, "there may be a 'window' in a plant's ontogenetic program during which plasticity is possible, and after which development is fixed such that allocational plasticity is constrained" (p. 126). The lack of within individual variation strongly supports such a canalized nature of growth. Yet, the varied response of individuals within a sex to the low-protein diet suggests that early in an organism's development a great deal of plasticity exists in setting the trajectory that will be followed without variation. Absent an explicit and accurate assessment of the unit of analysis in a study, valid and significant results may be overlooked.

III. CONCLUSIONS

Why is the study of variation important, and why is the hierarchical level significant? Riddle and Purves (1995) examined studies of rat brain morphology that reported significant variation in right–left differences, despite the fact that these animals are inbred to reduce variation. Each paper Riddle and Purves reviewed that found variation dismissed as either an artifact, a resultant from inconsistent collection techniques, or made no notice of it. Further, a recent review of the evolution of developmental mechanisms (Rudel and Sommer, 2003) listed 10 key concepts, none of which included any measure of ontogenetic changes in variation. Variation through ontogeny clearly exists, and yet most biologists systematically ignore it.

Wright and McConnaughay's (2002) study of ontogenetic variability in plants highlights this problem. They cite many authors who understand the significance of variability and that variability changes over the course of development. As they point out, "plasticity in any trait may be a transient phenomenon, expressed for a limited time during growth and development, [and] careful examination of phenotypic expression throughout ontogeny may reveal a more complete picture of the organism's ability to respond to environmental heterogeneity" (p. 124). Yet none of their reviewed studies differentiated between within-unit and among-unit variation.

One of the most explicit statements on the importance of hierarchy to understanding variation comes from Klingenberg *et al.*'s (2003) description of morphologic variation in mice: "genetic and environmental factors causing variation among individuals produce patterns of variation that do not have a direct equivalent in the within-individual processes that generate differences between body sides" (p. 528). This contrasts with Klingenberg and coauthors' (2001) work that suggests clear correlations between among-individual and within-individual variation for insect wings, albeit these are for simpler structures with more similar developmental pathways. Thus variation, and the causes and consequences of that variation, may differ across hierarchical levels, or not, depending on the system. A more complete understanding of these levels and interactions awaits more detailed analyses.

ACKNOWLEDGMENTS

An early version of this chapter was originally presented at a symposium organized by Benedikt Hallgrímsson and Christopher Klingenberg for the European Society of Evolutionary Biology meetings in 2003. The authors are indebted to B. Hallgrímsson and C. Klingenberg, as well as other participants of that symposium (B. K. Hall, J. M. Cheverud, L. J. Leamy, J. T. Richtsmeier, and S. Renaud) for their input, comments, and feedback.

REFERENCES

Alberch, P., Gould, S. J., Oster, G. F., and Wake, D. B. (1979). Size and shape in ontogeny and phylogeny. *Paleobiology* 5(3), 296–317.

Burns, K. J. (1993). Geographic variation in ontogeny of the fox sparrow. *The Condor* **95**, 652–661.

Cheverud, J. M. (1982). Phenotypic, genetic, and environmental morphological integration in the cranium. *Evolution* **36**(3), 499–516.

Cheverud, J. M., Wagner, G. P., and Dow, M. M. (1989). Methods for the comparative-analysis of variation patterns. *Systematic Zoology* **38**(3), 201–213.

Doust, A. N. (2000). Comparative floral ontogeny in Winteraceae. *Annals of the Missouri Botanical Garden* **87**(3), 366–379.

Farmer, M. A., and German, R. Z. (2004). Sexual dimorphism in the craniofacial growth of the guinea pig (*Cavia porcellus*). *Journal of Morphology* **259**(2), 172–181.

Fischer, K., and Fiedler, K. (2002). Reaction norms for age and size at maturity in response to temperature: A test of the compound interest hypothesis. *Evolutionary Ecology* **16**(4), 333–349.

Gedroc, J. J., McConnaughay, K. D. M., and Coleman, J. S. (1996). Plasticity in root/shoot partitioning: Optimal, ontogenetic, or both? *Functional Ecology* **10**(1), 44–50.

German, R. Z. (2004). The Ontogeny of Sexual Dimorphism: the implications of longitudinal vs. cross-sectional data for studying heterochrony in mammals. In *Shaping Primate Evolution* (F. Anapol, R. Z. German, and N. G. Jablonski eds.), pp. 11–23. Cambridge: Cambridge University Press.

Gille, U., Salomon, F., Rieck, T., Gerickle, A., and Ludwig, B. (1996). Growth in rats (*Rattus norvegicus* Berkenhout): 1. Growth of body mass: A comparison of different models. *Journal of Experimental Animal Science* **37**(4), 190–199.

Guralnick, R. P., and Lindberg, D. R. (1999). Integrating developmental evolutionary patterns and mechanisms: A case study using the gastropod radula. *Evolution* **53**(2), 447–459.

Hallgrimsson, B., Miyake, T., Willmore, K., and Hall, B. K. (2003). Embryological origins of developmental stability: Size, shape, and fluctuating asymmetry in prenatal random bred mice. *Journal of Experimental Zoology Part B—Molecular and Developmental Evolution* **296B**(1), 40–57.

Jabbour, R. S., Richards, G. D., and Anderson, J. Y. (2002). Mandibular condyle traits in Neanderthals and other Homo: A comparative, correlative, and ontogenetic study. *American Journal of Physical Anthropology* **119**, 144–155.

Jordy, M.-N. (2004). Seasonal variation of organogenetic activity and reserves allocation in the shoot apex of *Pinus pinaster* Ait. *Annals of Botany* **93**(1), 25–37.

Klingenberg, C. P. (1996). Individual variation of ontogenies: A longitudinal study of growth and timing. *Evolution* **50**(6), 2412–2428.

Klingenberg, C. P. (1998). Heterochrony and allometry: The analysis of evolutionary change in ontogeny. *Biological Reviews of the Cambridge Philosophical Society* **73**(1), 79–123.

Klingenberg, C. P. (2002). Morphometrics and the role of the phenotype in studies of the evolution of developmental mechanisms. *Gene* **287**(1–2), 3–10.

Klingenberg, C. P., Badyaev, A. V., Sowry, S. M., and Beckwith, N. J. (2001). Inferring developmental modularity from morphological integration: Analysis of individual variation and asymmetry in bumblebee wings. *American Naturalist* **157**(1), 11–23.

Klingenberg, C. P., Mebus, K., and Auffray, J. C. (2003). Developmental integration in a complex morphological structure: How distinct are the modules in the mouse mandible? *Evolution and Development* **5**(5), 522–531.

Kobyliansky, E., and Livshits, G. (1989). Age-dependent changes in morphometric and biochemical traits. *Annals of Human Biology* **16**(3), 237–248.

Leamy, L. (1993). Morphological integration of fluctuating asymmetry in the mouse mandible. *Genetica* **89**(1–3), 139–153.

Line, S. R. P. (2003). Variation of tooth number in mammalian dentition: Connecting genetics, development, and evolution. *Evolution and Development* 5(3), 295–304.

McNamara, K. J. (1995). *Evolutionary Change and Heterochrony.* Chichester, England: Wiley.

Miller, J. P., and German, R. Z. (1999). Protein malnutrition affects the growth trajectories of the craniofacial skeleton in rats. *Journal of Nutrition* **129**(11), 2061–2069.

Niklas, K. J. (1997). Mechanical properties of black locust (*Robinia pseudoacacia* L.) wood: Size- and age-dependent variations in sap- and heartwood. *Annals of Botany (London)* **79**(3), 265–272.

Oster, G. F., Shubin, N., Murray, J. D., and Alberch, P. (1988). Evolution and morphogenetic rules: The shape of the vertebrate limb in ontogeny and phylogeny. *Evolution* **42**(5), 862–884.

Polak, M., Moller, A. P., *et al.* (2003). Does an individual asymmetry parameter exist? A meta-analysis. In *Developmental Instability: Causes and Consequences* (M. Polak ed.), pp. 81–96. Oxford: Oxford University Press.

Raff, R. A. (1996). *The shape of life: Genes, development, and the evolution of animal form*. Chicago: The University of Chicago Press.

Riddle, D., and Purves, D. (1995). Individual variation and lateral asymmetry of the rat primary somatosensory cortex. *Journal of Neuroscience* **15**(6), 4184–4195.

Riddle, D., Richards, A., Zsuppan, F., and Purves, D. (1992). Growth of the rat somatic sensory cortex and its constituent parts during postnatal development. *Journal of Neuroscience* **12**(9), 3509–3524.

Rudall, P. J., and Bateman, R. M. (2002). Roles of synorganisation, zygomorphy, and heterotopy in floral evolution: The gynostemium and labellum of orchids and other lilioid monocots. *Biological Reviews (Cambridge)* 77(3), 403–441.

Rudel, D., and Sommer, R. J. (2003). The evolution of developmental mechanisms. *Developmental Biology* **264**(1), 15–37.

Smith, R. J., and Kamiya, T. (2002). The ontogeny of *Neonesidea oligodentata* (Bairdioidea, Ostracoda, Crustacea). *Hydrobiologia* **489**(1–3), 245–275.

True, J. R., and Haag, E. S. (2001). Developmental system drift and flexibility in evolutionary trajectories. *Evolution and Development* 3(2), 109–119.

Wilkinson, L. (2000). SYSTAT 10. Chicago, SPSS Inc.

Wright, S. D., and McConnaughay, K. D. M. (2002). Interpreting phenotypic plasticity: The importance of ontogeny. *Plant Species Biology* **17**(2–3), 119–131.

Zelditch, M. L. (1988). Ontogenetic variation in patterns of phenotypic integration in the laboratory rat. *Evolution* **42**(1), 28–41.

Zelditch, M. L., and Carmichael, A. C. (1989). Ontogenetic variation in patterns of developmental and functional-integration in skulls of *Sigmodon-fulviventer. Evolution* **43**(4), 814–824.

Zelditch, M. L., and Fink, W. L. (1996). Heterochrony and heterotopy: Stability and innovation in the evolution of form. *Paleobiology* **22**(2), 241–254.

Zelditch, M. L., Bookstein, E. L., and Lundrigan, B. L. (1993). The ontogenic complexity of developmental constraints. *Journal of Evolutionary Biology* **6**(5), 621–641.

Zelditch, M. L., Sheets, H. D., and Fink W. L. (2000). Spatiotemporal reorganization of growth rates in the evolution of ontogeny. *Evolution* **54**(4), 1363–1371.

Zelditch, M. L., Straney, D. O., Swiderski, D. L., and Carmichael, A. C. (1990). Variation in developmental constraints in *Sigmodon. Evolution* **44**(7), 1738–1747.

CHAPTER 6

Constraints on Variation from Genotype through Phenotype to Fitness

LAUREN ANCEL MEYERS
Section of Integrative Biology, University of Texas at Austin, Austin, Texas, USA

INTRODUCTION

There exists a great diversity of genetic mechanisms for introducing novelty into the genome. Mutation, recombination, transposition, polyploidy, to name a few, produce the variation that is essential for character evolution. Although such mechanisms are numerous and widespread, the production of nondeleterious phenotypic variation is constrained on multiple levels. A phenotypic variant will arise if a genetic modification occurs in the first place, if that modification produces a change in phenotype, and if that change in phenotype is neither lethal itself nor pleiotropically correlated with a lethal change in another phenotype.

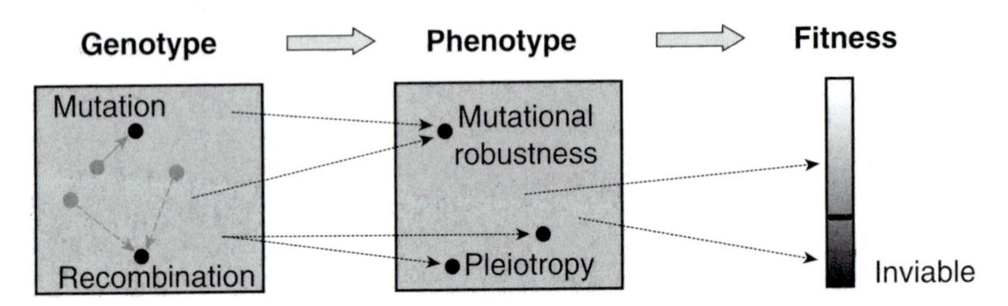

FIGURE 6-1. The genotype–phenotype–fitness relationship. Variation is constrained at the genotypic level by the mechanisms and rates of mutation and recombination, by the map from genotype to phenotype through redundancy (mutational robustness), and by the ultimate direct and pleiotropic contributions of the phenotype(s) to fitness.

The fitness landscape is a metaphor first described by Sewall Wright to conceptualize the relationship between genotypes and their fitnesses (Wright, 1932). As depicted in Figure 6-1, the relationship between genotype and fitness goes through the intermediate step of phenotype, that is, genes underlie the traits that in turn impact the fitness of an organism.[1] The map from genotype to phenotype can be quite complex. For example, Hsp90 is a gene that encodes a heat shock protein. This protein is implicated in regulating multiple complex signaling pathways, transcription factors, proteins participating in intracellular signaling pathways, and proteins that regulate cell division (Aligue *et al.*, 1994; Rutherford and Lindquist, 1998; Mayer and Bukau, 1999). At the same time, *Hsp90* has great deal of natural sequence variation (Gupta, 1995; Palmisano *et al.*, 1999; Passarino *et al.*, 2003). Thus the map from genotype to trait involves both pleiotropy—the mapping of a single gene to multiple phenotypes—and redundancy—the mapping of multiple genetic sequences to an identical phenotype. The map from phenotype to fitness can be similarly complex, depending on interactions among phenotypes within an organism and interactions between an organism and its population and environment. Despite these complexities, the overarching genotype–phenotype–fitness framework is generic and can aid in a general theory of variational constraints.

Constraints on variation take place on multiple levels (Figure 6-1). At the source of variation, the intrinsic rates, biases, and genetic requirements of mutational

[1]In mathematical terms, genetic operators (such as mutation) define a metric on the space of all possible genotypes (where distance is the number of mutations separating two sequences). The transformation from genes into traits is like a map from this metric space into a space representing traits (where there are metrics that reflect physiologic, biochemical, or behavioral similarity). Traits determine fitness, giving rise to a real-valued function on the space of phenotypes. Hence, the fitness landscape is like a graph of the real-valued function obtained by composing the function from genotype to phenotype with the function from phenotype to fitness.

mechanisms limit the range of possible genetic modifications. Moving up, the map from genes to traits is not perfect. Because of the influence of the environment, phenotypic variation can arise without genotypic variation; because of buffering within and among genes, genotypic variation can exist without corresponding phenotypic variation. This latter phenomenon—genetic robustness—significantly constrains the production of phenotypic novelty. Integrating across all three levels, natural selection itself can limit variation. A mutation may simply be deleterious and therefore be eliminated rapidly by natural selection. Furthermore, if a gene is involved in the production of multiple phenotypes, then a mutation may produce innocuous or beneficial variation in the trait of interest while damaging another trait. Natural selection may eliminate the mutation by virtue of the pleiotropic deleterious effects and thereby constrain the variation possible in the trait of interest.

In this chapter, we explore these three levels of constraints on variation and refer to them as mechanistic, epistatic, and viability constraints. We consider the large bodies of theoretical and experimental research on these topics and generate further intuition using a realistic genotype–phenotype model that has recently provided profound insights into the nature of phenotypic variation.

I. RNA EVOLUTIONARY MODEL

Throughout this chapter, we use a model of RNA structural evolution to illustrate various constraints on variation. For decades, RNA researchers have been developing algorithms for predicting secondary structure (Tinoco *et al.*, 1971, 1973; Waterman, 1978; Nussinov and Jacobson, 1980; Zuker and Stiegler, 1981; Turner and Sugimoto, 1989; Zuker, 1989, 1994; Cannone *et al.*, 2002). Thanks to these algorithms, evolutionary biologists have been able to simulate the evolution of novel structures, and in doing so, discover principles of evolution that were not apparent in simpler population genetic models. In particular, they computationally evolve populations of molecules, assuming that fitness is based on the similarity of a molecule's minimum free energy structure (henceforth ground state) to a predetermined target shape. One important insight gleaned from such work is that populations tend to evolve in a punctuated trajectory, experiencing long periods of stasis, during which mutations change the underlying sequence without changing the ground state (Fontana and Schuster, 1998; Reidys *et al.*, 2001). For a typical RNA shape, there is a vast network of sequences that assume that shape as their ground state and are connected by mutations. The size, breadth, and proximity of these *neutral networks* determine the rate and course of evolution (Fontana *et al.*, 1993; Reidys *et al.*, 1997, 2001; Fontana and Schuster, 1998; Stadler *et al.*, 2001).

Under natural conditions, RNA molecules do not simply freeze in their ground state, but exhibit a form of structural plasticity. Thermal fluctuations cause molecules to equilibrate among alternative low-energy shapes. We have extended the study of

RNA structural evolution to take into account the effects of thermodynamic noise (Ancel and Fontana, 2000). We can estimate the set of lowest free energy secondary structures of an RNA molecule using an extension (Wuchty *et al.*, 1999) of standard algorithms (Waterman, 1978; Nussinov and Jacobson, 1980; Zuker and Stiegler, 1981) that estimate RNA secondary structure. This suboptimal folding algorithm rapidly predicts the set of lowest energy secondary structures attainable for a given sequence. We ignore energy barriers and assume that a sequence equilibrates among all structures whose free energy is within $5kT$ of its ground state, which is approximately 3 kcal at 37° C, and corresponds to the loss of two C·G/G·C stacking interactions. We call the set of all possible shapes within 3 kcal/mol of the ground state (at 37° C) the "suboptimal repertoire" of an RNA sequence. The partition function (McCaskill, 1990) of a sequence is

$$Z = \sum_{\sigma} \exp(-\Delta G_{\sigma} / kT)$$

where ΔG_{σ} is the free energy of shape s and the sum runs over all possible shapes in the suboptimal repertoire. For any shape s in the repertoire of a sequence, the Boltzmann probability of s, $p_{\sigma} = \exp(-\Delta G_{\sigma}/KT)/Z$, measures the relative stability of σ with respect to the entire repertoire. Assuming equilibration, p_s estimates the amount of time the RNA molecule resides in shape s. The molecule spends the largest fraction of time in the ground state. For example, the molecule depicted in Figure 6-2 spends 4.1% of its time in its minimum free energy shape, another 4.1% of its time in its next lowest free energy shape, and the rest of its time in the 362 other lowest free energy shapes. The dots stand in for hundreds of shapes in the repertoire. This is a thermodynamically unstable molecule.

Using these methods, we can rapidly compute the shape repertoire of a sequence and approximate the amount of time spent in each shape. This constitutes a map from genotype (sequence) to phenotype (shapes). Equipped with this computational model of the RNA genotype–phenotype relation, we can simulate, for example, an experimental protocol that evolves molecules to optimally bind a ligand (Ellington, 1994). We select molecules according to their similarity to a prespecified target shape (Fontana and Schuster, 1998). In nature, the structural stability of an RNA sequence will presumably influence the overall binding constant of the molecule. At equilibrium, a fraction p_s of a large number of identical sequences assumes shape s and binds to a ligand with the corresponding constant. Our model calculates, for each shape s in the repertoire, a selective value $f(s)$ based on how well s matches the target shape. The overall fitness, r, of the sequence is the average of the selective values over the shapes in its repertoire, each weighted by its occupancy time, $r = \sum_{\sigma} f(\sigma) p_{\sigma}$.

We replicate molecules in a simulated flow reactor in proportion to their fitness. Genetic novelty is produced each generation through some combination of point

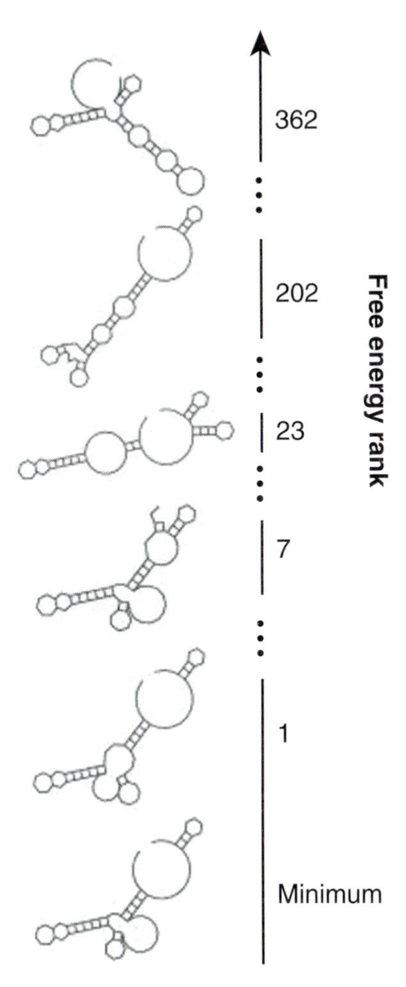

FIGURE 6-2. Thermodynamic repertoire of an RNA molecule.

mutation, deletions, insertions, recombination, and duplications to the underlying sequence. For further details of our simulation methodology, please see Ancel and Fontana (2000).

II. EVOLVING CONSTRAINTS ON VARIATION IN RNA

We consider a striking example of variation in RNA. Figure 6-3 depicts three typical evolutionary trajectories. In the top line, a population wanders neutrally

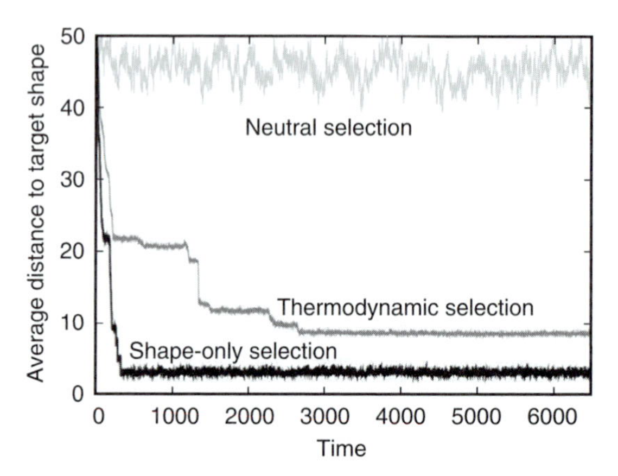

FIGURE 6-3. RNA evolutionary trajectories. Simulated populations randomly explore shape space via neutral evolution (light gray), selection for similarity to a target shape (black), and simultaneous selection for thermostability and similarity to a target shape (medium gray).

through genotype space with no selection whatsoever. In the bottom line, a population is selected for similarity between the ground state and a prespecified target structure. In the middle line, a population is simultaneously selected for thermodynamic stability and structural similarity to the target. As we describe in Ancel and Fontana (2000), natural selection for thermodynamic stability drives the third population into an evolutionary dead-end, from which it is unable to progress further toward the target. This is caused by a property of RNA folding that we termed *plastogenetic congruence*. In RNA, thermodynamic stability is correlated with mutational robustness. Molecules that fold rigidly into their ground state are little perturbed by point mutation, and conversely, molecules that wiggle among various alternative structures are easily perturbed by mutation. Thus selection for thermostability produces increasing mutational robustness as a by-product. Eventually a population can become so buffered against the potential affects of mutation that the variation necessary for further evolution never arises.

In Figure 6-4, we examine the evolution of accessible variation in these three evolutionary scenarios. We use two metrics to assay the potential for mutation to produce phenotypic (shape) variation in the populations. First, we measure the *neutrality* of the sequences in the populations. This is the fraction of all point mutations that leave the ground state unperturbed. Second, we can characterize the spectrum of mutational effects of a sequence or population by measuring the *dimensionality* of the set of shapes accessible through point mutation (Wagner, 1988; Hartl and Taubes, 1998; Haubold *et al.*, 1998). Loosely speaking, if we imagine that all shapes accessible through mutation are points in space, positioned at distances that reflect their structural differences, then the dimensions of the resulting

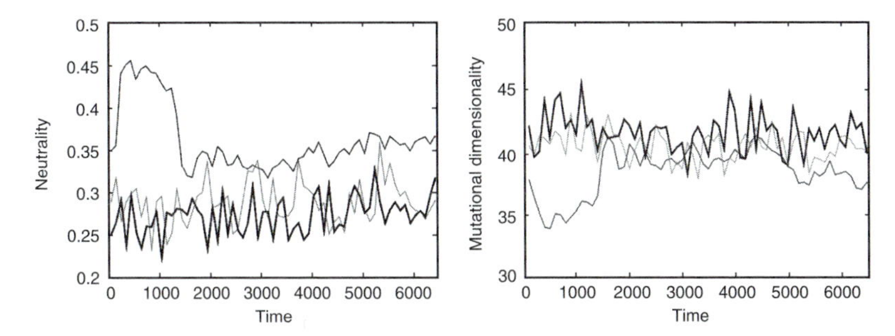

FIGURE 6-4. Evolving mutational potential. The neutral (light gray), shape-only selection (black), and thermodynamic selection (medium gray) populations vary in their ability to produce novel phenotypes (shapes) through mutation. Neutrality (left) is the fraction of all point mutations that leave the minimum free energy structure unperturbed and thus is inversely related to mutational potential. Dimensionality (right) is a measure of the overall shape diversity produced by point mutation.

space indicate the degrees of evolutionary freedom. Dimensionality indicates not only mutational robustness (as does mutational neutrality), but also the extent of innovation possible through mutation (even if it is not realized). To compute dimensionality, we embed the set of shapes mutationally accessible to a sequence in Euclidean space, such that structural distances between shapes are preserved. The dimension of the resulting space indicates the phenotypic diversity in the genetic neighborhood and hence the potential for further evolution. In general, such methods are called metric multidimensional scaling and are quite useful for exploration and visualization of high-dimensional data (Kruskal and Wish, 1978; Cox and Cox, 1994).

The potential for mutation to produce novel phenotypes will depend very much on the genetic makeup of a population, which is influenced by its evolutionary history. Populations wandering neutrally through genotype space generally maintain the ability to produce phenotypic novelty through mutation. Yet there are significant fluctuations in this potential as the population evolves. The other two populations are noticeably more constrained, with thermodynamic selection pushing populations into regions of genotype space in which sequences are largely impervious to mutation.

Not all variation is beneficial. In fact, seriously deleterious mutants may be readily eliminated and thus never contribute to standing genetic variation within these populations. In Figure 6-5, we analyze the mutational spectrum of a single typical sequence taken from each of the three populations toward the end of their evolutionary simulations. For each of the three sequences of length 76, we consider all (76)(3) = 228 possible point mutations. (Each base can be replaced by one of three alternative bases.) The left graphs illustrate the diversity of mutant ground states as a function of their structural distance from the original parental

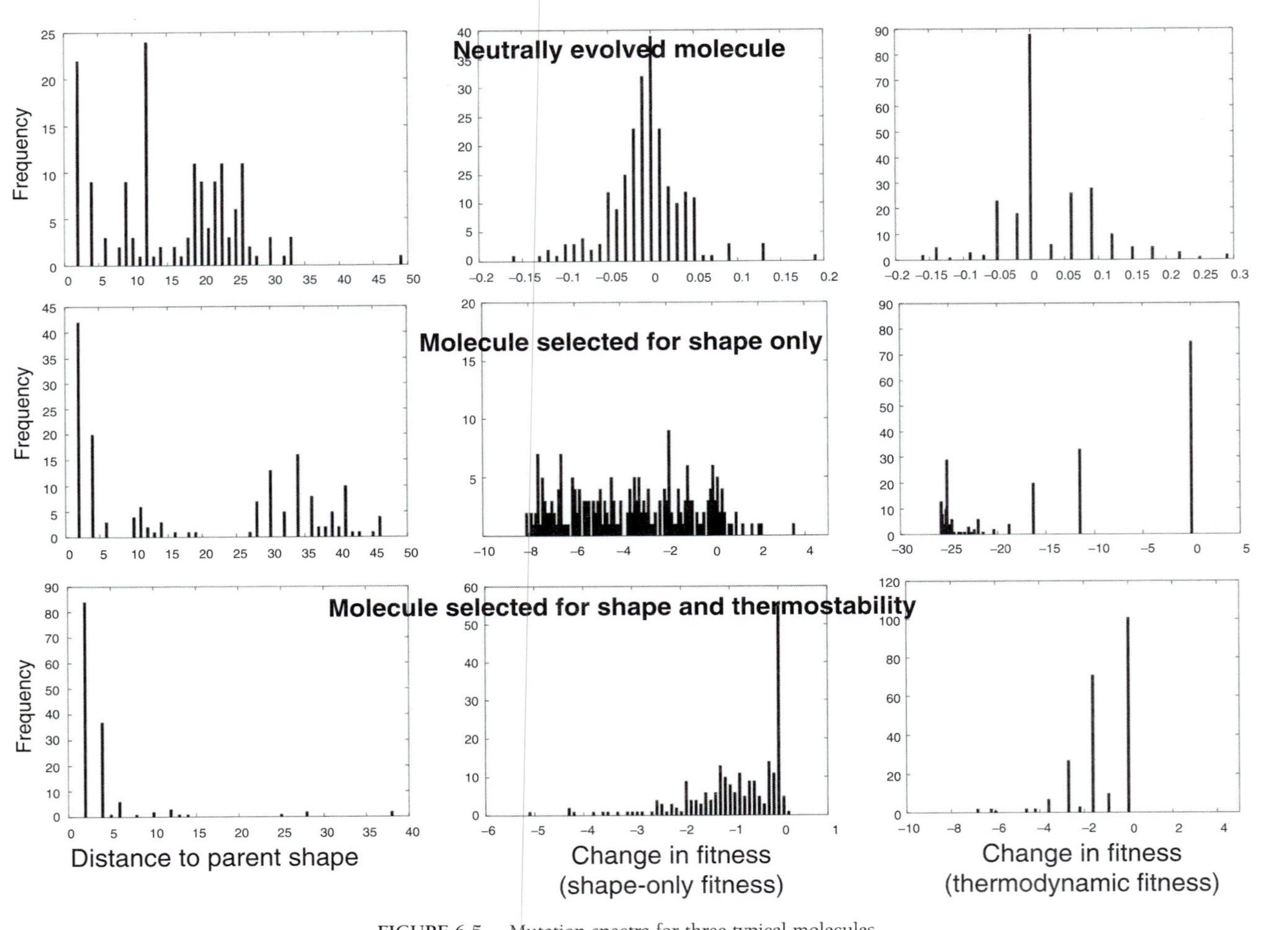

FIGURE 6-5. Mutation spectra for three typical molecules.

ground state. The middle and left graphs show the distribution in fitness effects of these mutations for the shape-only and thermodynamic fitness functions respectively. Neutrally evolved molecules have significantly more evolutionary potential than the other two classes of molecules, with ample probability of a beneficial point mutation. In contrast, the molecules that have undergone long-term shape-only selection tend to have mildly to strongly deleterious mutations, and the molecules that have undergone thermodynamic selection tend to have only neutral or mildly deleterious mutations.

Recall that a neutral network is a mutationally connected set of sequences that fold into the same ground state (Reidys *et al.*, 1997). In several respects, neutral networks are not truly neutral. For example, sequences that share the same ground state may have very different thermodynamic properties, ranging from those that fold stably and almost exclusively into the ground state, to those that unstably wiggle between the ground state and similarly stable alternative structures. Sequences within a neutral network can also be quite diverse with respect to their mutational potential. Figure 6-6 graphically illustrates the mutational dimensionality of three sequences that have the same ground state and thus lie in the same neutral network. The most stable (left) spends approximately 75% of its time in the ground state compared with 34% for the moderately stable molecule (middle) and 9% for the unstable molecule (right). The dots in these two-dimensional projections of the mutational neighborhoods represent each novel mutant ground state; the distances between the dots reflect structural distances; and a single line connects each dot, and the dot represents the most similar shape. Thus we see that evolutionary potential can vary substantially within a single neutral network. Numerical explorations of the RNA shape space suggest that neutral networks are made up of edge regions, from which other neutral networks (phenotypic variation) are easily accessed through mutation and interior regions in which mutations are largely internal to the network (neutral mutations).

Let us consider this analysis in light of the three levels of constraints presented in the introduction. At the lowest level, the only source of variations in the simulations is a modest rate of point mutation, approximately 0.001 mutations per replication per site. If we allowed for other mutational mechanisms such as deletions, insertions, or recombination, the potential genetic variation accessible to populations would expand significantly. At the genotype to phenotype level, these examples show that sequences can be quite diverse in their sensitivity to mutations and that the evolutionary history of a population influences the extent to which mutations have the potential to produce novel phenotypes. At the fitness level, we observe that a significant fraction of potential variation is deleterious and thus inconsequential. Phenotypes that require multiple mutations may be unlikely if the intermediate mutations have deleterious effects. As environments change, so do fitnesses. Figure 6-5 serves as a reminder that the fate of a mutation may change as the selective environment changes.

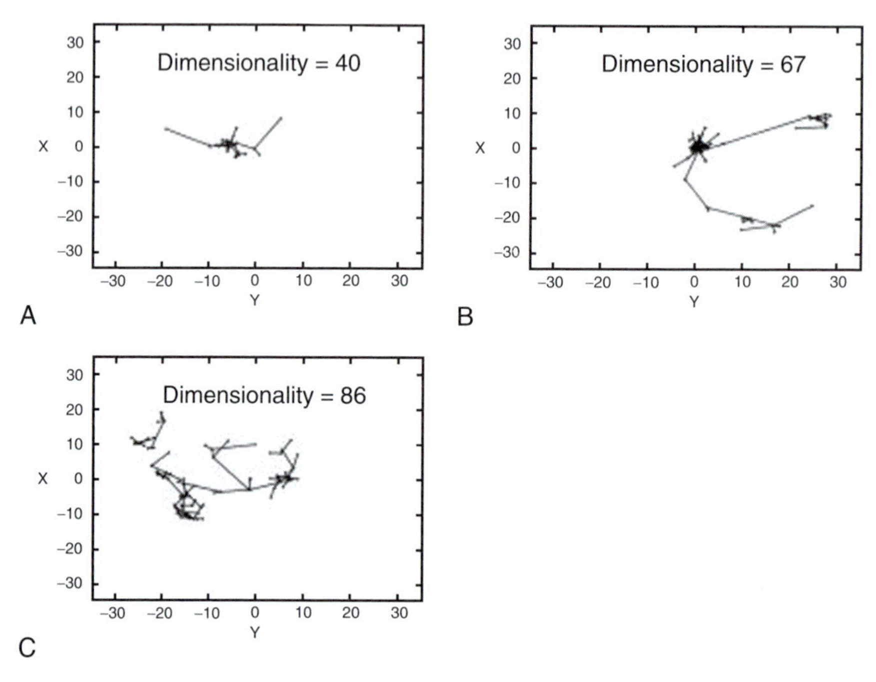

FIGURE 6-6. The mutational dimensionality of a very thermostable molecule (A), a moderately thermostable molecule (B), and an unstable molecule (C).

III. MECHANISTIC CONSTRAINTS

We now turn from the specific example of RNA to a more general discussion of variational constraints. Beginning at the source of novelty, mutational mechanisms vary in their presence and impact across taxa and even within single genomes. In the metaphor of Figure 6-1, mutational mechanisms determine the rapidity at which population can explore genotype space and the directions in which it can move. Rare point mutation may allow a population occasionally to take baby steps, whereas horizontal exchange of genetic material between different species may facilitate large leaps.

A. The Spectrum of Mutational Constraints

Both the tempo and direction of genetic change vary extensively within and between species. For example, mutation rates have been estimated to range from 10^{-10} and 10^{-11} in humans, mice, *Caenorhabditis elegans*, *Escherichia coli,* and *Saccharomyces cerevisiae* to as high as 10^{-4} in RNA viruses as measured in per base pair per replication mutation rates (Drake, 1999). The corresponding mutations per genome per generation rates are very high for both humans (over 0.1) and RNA viruses (as high as 0.76) while less than 0.01 for bacteria, bacteriophages, Archaea, and eukaryotic microbes (Drake *et al.*, 1998; Drake and Holland, 1999).

Mutation rate is determined by the underlying error rate—that is, the frequency of replication mistakes—and the stringency of error correction. DNA polymerase does a remarkably accurate job at replication, making less than one error per billion bases. When mistakes arise, they are often caused by exposure to a variety of mutagens. Some chemical mutagens, base analogues, so closely resemble purines and pyrimidines that they can be incorporated into DNA in place of normal nucleotides. Other chemical mutagens directly alter the structure or pairing properties of nucleotides, while others still cause single base insertions that can lead to codon frame shifts. Ionizing and ultraviolet radiation can also lead to a diversity of mutations and chromosomal rearrangements. Thus the environmental milieu of an organism's reproductive cells will constrain the underlying rate of initial replication mistakes.

There are also special polymerases, called mutases, which appear to intentionally introduce replication errors (Nelson *et al.*, 1996a,b; Reuven *et al.*, 1999; Tang *et al.*, 1999; Wagner *et al.*, 1999). Mutases have been identified in yeast, bacteria, and viruses, and some are components of stress-response systems. For example, mutases have also been implicated in SOS repair, which is a system that allows polymerase to bypass errors in situations of extensive DNA damage rather than grind to a halt (Goodman, 2002).

Many initial replication mistakes are caught and corrected by the polymerases and other cellular players in replication. Errors are corrected through either direct reversal mechanisms that restore a damaged base to its correct state, excision repair mechanisms that lead to a deletion of the damaged section, and postreplication mismatch repair mechanisms switch the newer member of a nonmatching pair of bases so that it matches the original base. Thus the rate and nature of mutations are further constrained by the assemblage of functional polymerases and other error-correction molecules in the organism. There is evidence that polymerases have evolved dramatically throughout the tree of life through a multitude of mutational mechanisms (Edgell *et al.*, 1998; Filée *et al.*, 2002). Of the six major polymerase families (A, B, C, D, X, and Y), none is conserved among all of three of the major domains: Bacteria, Archaea, and Eukaryota. Furthermore, the assortments of polymerases found on viruses and plasmids suggest an extensive and ongoing horizontal exchange (Filée *et al.*, 2002). The spectrum of mismatch repair genes and helix-hairpin-helix type DNA repair glycosylases likewise varies from genome to genome (Eisen, 1998; Denver *et al.*, 2003). For example, *E. coli* has a single MutS (mismatch repair gene) homologue while S. *cerevisiae* bears six such homologues in its genome (Eisen, 1998).

Mutation rates can vary not only between taxa, but also on a cell-by-cell basis within multicellular organisms. For example, mammalian somatic cells and germ-line cells exhibit different mutation rates (Mornet *et al.*, 1996; Bois *et al.*, 1997; Drake *et al.*, 1998). Some human cancers, including a hereditary form of colorectal cancer, have been linked to cell-specific deficiencies in the mismatch repair system (Loeb, 1991; Liu *et al.*, 1995). Colorectal cancer sometimes occurs in people with highly mutable microsatellites—tandemly repeated units of DNA—within a mismatch repair gene. Slip-stranded mispairing combined with oversight in error correction allows microsatellites to easily lose or gain a copy of the repeated unit. If and when a microsatellite mutation occurs within a mismatch repair gene, it can lead to genetic instability in the particular cells housing the mutation. Many other cancers have been linked to mutations in the *p53* gene that also leads to increased genetic instability through impairment of the system that detects and responds to the threat of mutations through programmed cell death (Lane, 1992).

These examples suggest that mutation rates also vary across the genome. Regions of the genome with long microsatellites are more prone to mutation. They are thought to be responsible for a number of other hereditary human diseases, including Huntington's disease and fragile X syndrome. They are also a possible mechanism for adaptive variation and occasional virulence in *Neisseria meningitis*, the bacteria responsible for epidemic meningitis, and other pathogenic bacteria (Moxon *et al.*, 1994; Meyers *et al.*, 2003). Microsatellites, however, are only one of many kinds of mutational hot spots found within genomes. For example, when a cytosine base is linked to a guanine base (CG), it is more susceptible to point mutation than in any other context (Jones *et al.*, 1987). In this combination, it is often methylated into 5-methyl-cytosine, which is unstable, and when deaminated, becomes thymine (T). Thus an unusually common point mutation is the transition-type substitution between pyrimidines C and T.

Mutational mechanisms not only vary in their level of activity within and among genomes, they also vary in the genetic modifications they produce. In the simplest case, point mutations are biased toward transitions, mutations that exchange a purine for a purine or a pyrimidine for a pyrimidine. Such mutations can be caused by factors such as nitrous acid, base impairing, or mutagenic base analogs such as bromouracil. Transversions between different classes of nucleotides are significantly less common. Insertions, often caused by transposable elements, like point mutations can be reversed. Deletions, on the other hand, are irreversible.

Recombination likewise varies in its rates within and among genomes. Mating, breeding, and other mechanisms for horizontal exchange of DNA vary widely among taxa, and therefore so do opportunities for genetic recombination. In plants alone, the diversity of mating strategies and the resulting ratio of selfing to outcrossing covers the spectrum of possibilities. Recent surveys show that approximately half of animal-pollinated species have intermediate rates of outcrossing, while the remaining have rates of outcrossing lower than 20% or higher than

80% (Vogler and Kalisz, 2001). As with mutation, there are recombination hot spots within genomes. For example, Chi motifs (5′GCTGGTGG) are known hot spots for homologous recombination by the *E. coli* RecBC pathway (Ponticelli *et al.*, 1985). The human immune system provides a striking example of a mutagenic hot spot within somatic cells. Billions of different antibodies are assembled during B-cell development via random recombination of a short stretch of DNA. A vast repertoire of antibodies emerges from this process, capable of interacting with almost any natural chemical structure, including human proteins.

Of course, point mutation and recombination are not the only mechanisms that generate genotypic variation. Inversions, duplications, deletions, and gene conversion can cause sweeping replacements and rearrangements within chromosomes. Insertions of transposable elements have been shown to cause frequent and large-scale interruptions throughout the genome. Each of these mechanisms likewise varies in its presence within a genome and across the tree of life.

Let us return to the context of Figure 6-1. The production of variation at the lowest level is constrained by the available combination of mutational mechanisms and error correction. In the case of a single RNA sequence, point mutation yields a mutational neighborhood in the immediate vicinity of the original sequence that totals $3L$ possible sequences, where L is the length of the original sequence. High rates of point mutation may allow for multiple simultaneous hits and thereby widen this variational horizon. A simple single-crossover recombination event with another sequence of the same length may not add that many new possibilities $[2(L - 1)]$, but it can potentially open up the space of possible variants to distant regions of the genetic landscape. Suppose we have two sequences that differ at every site. The resulting recombinational neighborhood spans a diverse region of the space. Let us consider total recombination, a complete mixing of two genotypes in which each base in the offspring sequence is chosen at random from the existing base in one of the two parental sequences. Even with this very liberal form of recombination in combination with mutation, two completely different sequences produce variants that occupy a miniscule portion of all of genotype space. While insertions, deletions, and other mechanisms increase the diversity of potential offspring, they also increase the size of the space of all possible genotypes. Thus the rate and nature of mutation combined with the standing variation significantly constrain future possible variation.

B. The Evolution of Mutational Constraints

The natural diversity of mutational mechanisms and error-correction machinery speaks to the evolvability of such mechanisms. Recently, scientists have observed the rapid evolution of mutation rates in bacteria found in the laboratory and in natural settings (Moxon *et al.*, 1994; Miller, 1998, 1999; Bucci *et al.*, 1999;

Tenaillon *et al.*, 1999; Oliver *et al.*, 2000; Richardson and Stojiljkovic, 2001; Meyers and Bull, 2002; Meyers *et al.*, 2003). Increases in mutation rates can arise through mutations that knock-out mismatch repair genes. In most cases, mutator strains with accelerated genome-wide mutation rates seem to evolve under heterogeneous conditions where beneficial mutations are possible. For example, mutator strains of *Pseudomonas aeruginosa* are observed much more in the lungs of patients with cystic fibrosis than in healthy lungs. In this environment, the bacteria face the deterioration of their physical environment from disease as well as a changing onslaught of immune defenses and antibiotic therapy (Oliver *et al.*, 2000). Although mutation rate has been observed to increase from the wild-type condition, studies that show an evolutionary decrease in mutation rate from the wild-type state are rare or absent.

How and why do mutational constraints evolve? Mutation and recombination are the primary sources of genetic variation and are therefore essential to biologic evolution. Yet these mechanisms can easily destroy essential functions and characters, and these mechanisms may exact a physiologic cost on the organism possessing them. Is there a *mutational happy medium*, that is, a mutation rate (or recombination rate) that optimally balances the production of useful novelty and deleterious phenotypes and thereby maximizes the evolutionary potential of a population?

Let us consider an evolving population that has not yet achieved the maximum fitness possible, perhaps because environmental or demographic conditions have recently changed. Intuitively, we conceive that mutation rates constrain the amount and nature of phenotypic variation in a population. One might be tempted to speculate that the rate of evolution will simply be proportional to the rate at which variation is introduced into a population through mutation. This would be incorrect. Figure 6-7 illustrates the pitfalls of both low and high rates of point mutation. In the simulations, RNA molecules are selected based on similarity to a target shape. Point mutations serve as the sole source of structural novelty. Evolutionary progress clearly depends on the mutation rate. Very high and very low mutation rates impede evolution, while intermediate mutation rates enable a population to successfully evolve toward the target shape.

On the low end, infrequent mutations simply mean that the novelty necessary for evolutionary progress is not produced. On the high end, the populations experiencing extremely high mutation rates are being swamped by deleterious genotypes. The resulting catastrophic loss of evolvability has been formally addressed by Eigen and others and is called an "error catastrophe" (Eigen *et al.*, 1989). Using a simple genotype–phenotype model in which there is a single optimal wild-type sequence, and all other sequences have equal and much lower fitness than the wild type, Eigen showed that there is a threshold mutation rate above which a population is unable to retain the high-fitness wild type or its close relatives. The error threshold is a function of the length of the sequence and the strength of selection for the wild type. Error catastrophes have been seen in both evolutionary simulation and laboratory experiments with viruses (Crotty *et al.*, 2001; Domingo, 2003).

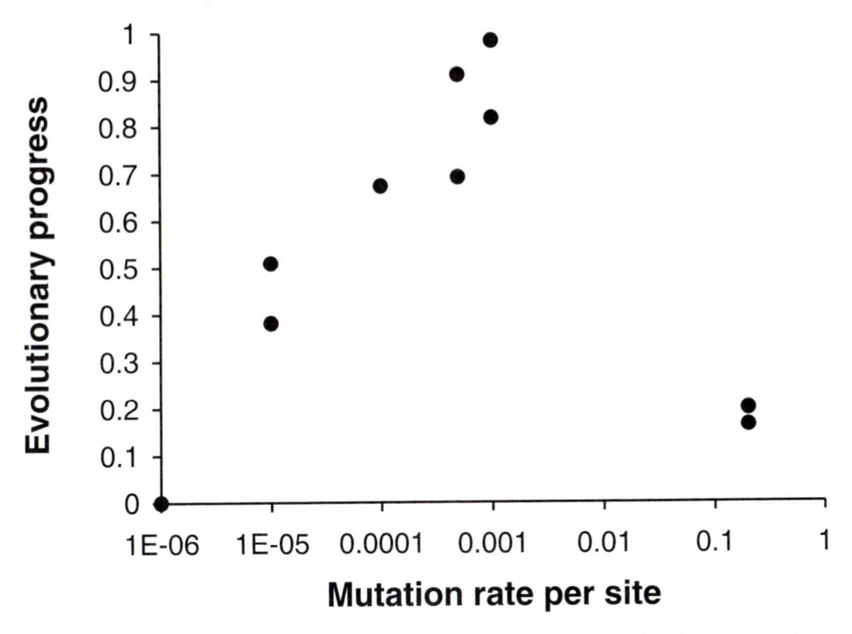

FIGURE 6-7. Mutation rate and evolvability. These are the results of pairs of evolutionary simulations for six different mutation rates (10^{-6}, 10^{-5}, 10^{-4}, $5 \cdot 10^{-4}$, 10^{-3}, 0.2). After 10,000 time steps, we recorded the evolutionary progress made by the population. Zero progress indicates that the population remained completely random with respect to the target structure, and one indicates that the population evolved all the way to the target structure. The pairs of runs at mutation rates 10^{-6} and 10^{-4} had identical equilibria and are thus indistinguishable in this graph.

This simple RNA example highlights some of the problems associated with extreme mutation rates on either the high or the low side. Evolutionary biologists have developed models to address the evolution of mutation rates. Most of this theory considers either the deleterious implications, the evolutionary benefits, or the biophysical costs of various mutation rates, but not all three at once. Models of evolution in a constant environment predict that natural selection should minimize the rate of mutation (Kimura, 1967; Liberman and Feldman, 1986; Kondrashov, 1995). The intuition behind this theory is that after a population has reached the maximum possible fitness, mutations can only lower the fitness of offspring. Models of evolution in a changing environment predict that mutation rates should remain relatively high, as long as the genes responsible for the high rates are linked to the genes that improve fitness upon mutation (Leigh, 1970, 1973; Ishii *et al.*, 1989). Recent theory suggests that the maintenance of mutator alleles that augment genome-wide mutation rates will depend on the distribution of effects of mutation, the ease with which mutator alleles are gained and lost through mutation and recombination, and population demographics (Taddei *et al.*, 1997; Tanaka and Feldman, 1999; Tenaillon *et al.*, 1999). Some of the predictions of these models are borne out in experimental populations of bacteria. In particular, mutators are seen to increase when frequent and decrease when rare (Chao and Cox, 1983; Trobner and Piechocki, 1985).

While the mechanisms that directly control the production of mutations may evolve for these reasons, evolutionary biologists have posited other important mechanisms for altering the evolutionary fate of mutations that do arise. For example, sexual reproduction may allow populations to purge themselves more easily of deleterious mutations and has been posited as an explanation for the evolution of sex (Feldman *et al.*, 1980; Kondrashov, 1982, 1983). This purging is enhanced if double deleterious mutants are less fit than predicted from multiplying the fitnesses of the single mutants. Another mechanism for altering the fate of deleterious mutations is inbreeding avoidance, where populations evolve to avoiding matings that bring together two copies of deleterious recessive alleles (Charlesworth and Charlesworth, 1987, 1999).

Recombination, like mutation, has potential pros and cons. Although it brings together novel, potentially beneficial combinations of genes, it can destroy existing combinations that already confer high fitness to the organism. In the early 1930s, Fisher and Muller asserted that recombination is critical to bringing beneficial mutations to fixation by bringing them together in the same genotype before one has the opportunity to exclude others (Fisher, 1930; Muller, 1932). Several decades later, this theory was formalized using two-locus mathematical models (Maynard Smith, 1968; Eshel and Feldman, 1970). Since then, a diverse body of theory on the evolution of recombination has emerged. Other theories consider the ability of recombination to reestablish relatively healthy individuals in populations suffering Muller's ratchet (Feldman *et al.*, 1980; Kondrashov, 1982); the ability of recombination to facilitate rapid adaptation under rapidly changing environments, and in particular, species undergoing "Red Queen" arms races with parasites or predators (Van Valen, 1973); and the possibility that the molecular underpinnings of recombination have been selected for the purposes of repairing DNA damage from uncorrupted templates or stabilizing homologous chromosomes during meiosis (Otto *et al.*, 1997).

IV. EPISTATIC CONSTRAINTS

In RNA, potential variation is not only constrained by the rate and nature of mutational events, but also by the particular sequence in which a mutation arises. The phenotypic effect of a mutation is almost always dependent on the bases present at many other positions in the sequence. Figure 6-6 shows the dimensionality of three sequences all of which fold into the same ground state. All three molecules have the same phenotype but are very differently impacted by mutations, illustrating the context-dependence of mutations. As a population evolves, the genetic milieu of a potential mutation changes, and thus the probability that a mutation will have a phenotypic impact and the magnitude of the impact changes (Figure 6-4).

Epistasis, or interactions among genes within a genome that affect a single phenotype, seems to be the rule rather than the exception in nature. In 1906, Bateson and Punnett were the first to experimentally measure epistasis (Punnett, 1923). They showed that two genes interact to produce the chicken comb phenotype. Thus the effect of a mutation at one locus will depend on which gene is present at the other locus. In the context of deleterious mutations, population geneticists have defined two categories of epistasis. "Synergistic epistasis" refers to compounding interactions among loci; that is, multiple deleterious mutations tend to be worse than expected from the effects of each mutation alone. "Antagonistic epistasis" is the opposite phenomenon, in which multiple deleterious mutations have a milder than expected impact on fitness. Our interest here is not in the ultimate impact of mutations on fitness, but in the immediate impact of mutation on phenotype. Rather than compare synergistic and antagonistic forms of epistasis, we address genetic interactions that buffer the phenotypic effects of mutation.

A. The Spectrum of Epistatic Constraints

The best-studied pattern of genetic buffering is dominance. The effect of a mutation of one allele depends on the companion allele. Theories for the evolution of dominance often pertain to epistasis in general. Fisher thought dominance evolves to buffer the effects of deleterious mutations. He wrote, "there is a tendency always at work in nature which modifies the response of the organism to each mutant gene in such a way that the wild type tends to become dominant" (Fisher, 1930). In contrast to this Fisherian population-level perspective, Wright's conception of gene interactions was much more mechanistic and individual centered. Wright argued that the selective pressure necessary for Fisher's explanation is prohibitively large (Wright, 1929). Instead, he offered the more mechanistic explanation that enzymatic functions of genes would evolve insensitivity to the loss of one copy. Following from Fisher's perspective, quantitative genetics takes statistical perspective on epistasis. It is summarized as nonlinear sources of variation. There are many possible genetic modifications and rearrangements that have no detectable effect on phenotype. This observation is necessarily made at a specific level of phenotypic resolution with buffering taking place somewhere lower down. A nucleotide substitution in a coding region of the genome will clearly change the phenotype that is the DNA sequence. It may or may not also alter the structure of its mRNA transcript; the amino acid charged by its codon; the secondary or tertiary folding of the resulting protein; etc.

There are potentially numerous mechanisms that have evolved purposefully or fortuitously to prevent phenotypic effects from trickling upward from the level of DNA sequence to macroscopic phenotype. Researchers have speculated that redundancy in the genetic code has evolved for the purposes of reducing the potential for error (Haig and Hurst, 1999; Ardell, 1998). Strong G-C binding within RNA molecules stabilize structures both with respect to thermodynamic noise and mutation (Ancel and Fontana, 2000). Similarly, cooperativity in genetic networks is thought to have a double impact, simultaneously conferring robustness toward environmental fluctuations and genetic changes (von Dassow *et al.*, 2000). Knock-out experiments suggest that chaperone proteins may also serve as an important mutational buffer (Rutherford and Lindquist, 1998). There are countless examples of human behavioral interventions that dampen the phenotypic effects of mutation, for example, vision correction, dietary changes, and medication.

B. The Evolution of Epistatic Constraints

Mutational robustness clearly exists, but its origins are hard to explain. In this section, we briefly describe two theories for the evolution of mutational robustness. For a more comprehensive treatment, see De Visser *et al.* (2003).

A simple explanation for mutational robustness is that it evolves in direct response to natural selection against deleterious mutations (van Nimwegen *et al.*, 1999). Lineages that are prone to producing deleterious offspring will be at a long-run disadvantage to those that are not. If such mutations are rare, however, this second-order selection will be insufficient to bring about mutational robustness.

An alternative theory, first proposed by Wagner, is that mutational robustness may be correlated with environmental robustness and thus arise as a by-product of selection for stability toward the environment (Wagner *et al.*, 1997). As briefly mentioned above, our RNA work supports this theory. Recall that Figure 6-3 illustrates *typical* evolutionary trajectories for three different fitness functions. Unlike the population experiencing selection for a novel shape, the population undergoing selection for both shape and thermostability gets trapped in an evolutionary dead-end before reaching the desired phenotype. This fate is brought about by the correlation between thermodynamic stability (for which we select) and mutational stability (for which we do not select). The population becomes so mutationally buffered that beneficial variation cannot arise.

V. VIABILITY CONSTRAINTS

Variation is only useful to a population if it can be maintained, at least temporarily. Mutations that are fatal or have large deleterious impacts will not persist, even though they might be beneficial upon a future environmental change or additional mutations. Population genetics teaches us further that neutral mutations and even beneficial mutations run the risk of being lost by genetic drift, particularly in small populations. Whether a variant can arise and persist in a population for long enough to have an evolutionary impact thus depends on the magnitude and direction of its immediate fitness impact and population demographics.

The variational potential of a population also depends on its evolutionary history. Populations that have evolved mutational buffering will be less likely to produce beneficial mutations of sufficient magnitude to withstand genetic drift. In general, the variational potential of a population will be a function of the current location of a population in genotype space (Figure 6-5). Populations that have equilibrated under strong directional selection will be much less likely to produce viable variants than, for example, populations that have recently experienced a change of environment (of fitness function) and are in the early stages of directional selection. From RNA, we have learned that a change in environment is not the only event that can increase the likelihood of viable mutations. RNA molecules undergo episodic evolution (Figure 6-3) where long periods of stasis are punctuated by rapid phenotypic innovation (Fontana and Schuster, 1998; Ancel and Fontana, 2000). After a long dry spell in which most mutations are deleterious or neutral (Figure 6-5), a single mutation can bring the population into a tunnel through genotype space in which beneficial variants are unusually abundant.

The fate of a mutation may depend on not only its primary phenotypic target, but also its pleiotropic effects on other phenotypes. For example, a mutation in a tRNA gene is likely to impact numerous phenotypes as it ricochets through positive feedback loops consisting of DNA polymerases, RNA polymerases, ribosomal proteins, aminoacyl-tRNA synthetases, tRNA processing enzymes, and other molecules. There are countless other examples of genes that impact multiple phenotypes. The map from phenotype to fitness is complicated by pleiotropy (Figure 6-1). A mutation that "improves" a single phenotype may never persist if it has an overall deleterious effect on the fitness of the organism.

VI. MODULARITY: A WAY OUT OF THE CONSTRAINTS

It may seem as if variation is insurmountably constrained by the inability of mutation to sample the enormity of genotype space, by mutational robustness, and by pleiotropy. Yet the diversity of life suggests that this is decidedly not the case. Although constrained, variation is sufficiently abundant to fuel evolutionary change. Whether or not it is sufficient to bring populations to "optimal" phenotypes is unknown.

Some biological systems may have evolved a way around the variational challenges of mutational robustness and pleiotropy. In RNA, we found that extremely robust molecules were also extremely modular. They can be easily partitioned into subunits that maintain their structural integrity in the face of thermodynamic perturbations or genetic changes in the remaining portions of the molecule. Modularity creates robust subunits that can be mixed and matched by recombination to produce novel phenotypes. Thus, in the face of extreme mutational buffering and extensive intramodular pleiotropy (that is, a change in one base can completely disrupt the structural contributions of all other bases participating in the module), novel units of variation emerge.

In general, modularity is the organization of phenotypes into functionally and genetically independent subunits. Modularity is thought to be a prerequisite for the evolution of complex phenotypes and perhaps for adaptive evolution in general (Lewontin, 1970). The existence and evolutionary importance of modularity has been postulated for proteins (Gilbert, 1978) and developmental pathways (Raff, 1996). The diversity of protein structures (or lack thereof) indicates an evolutionary history of recombining structural motifs from a limited repertoire (Fuchs and Buta, 1997; Gilbert *et al.*, 1997). A commonly cited example of developmental modularity arises from the Hox gene clusters (Raff, 1996). These genes, which have been studied for almost 100 years, are conserved across diverse taxa, including vertebrates, arthropods, and nematodes, and are involved in determining the anterior–posterior body axes. Mutations in genes in the Hox cluster lead to transformation of one segment to another, for example, from an antennae to a leg in *Drosophila* (Kaufman, 1990). This suggests that interactions within the Hox complex are such that the genes involved in leg formation and in antennae formation are unaffected by changes in their genetic context. Halder *et al.* more recently demonstrated the modularity of eye formation in *Drosophila* by successfully inducing eyes on the antennae, wings, and legs of *Drosophila* (1995).

Complex molecules and complex organismal body plans may arise from a mixing and matching, duplicating, and mutating of independent modules. Modularity, as it shifts the syntax of genetic variation, opens new avenues for phenotypic innovation (Wagner and Altenberg, 1996; Bonner, 1998; Hartwell *et al.*, 1999). Though this advantage is compelling, it does not explain the origins of modularity in the first place. Until a sufficiently diverse set of modules and appropriate recombinational mechanisms are in place, it is not clear that natural selection would favor such organization.

Our RNA study illustrated that modularity may not have to evolve for the purpose of rescuing variation, but can arise instead as a side effect of natural selection for phenotypic stability (Ancel and Fontana, 2001). The evolution of thermodynamic stability is achieved through increasingly modular structural components. In this way, modularity appears as a by-product of selection for environmental robustness, arising independently of any future evolutionary advantages it might provide.

The lessons from the RNA model system may have broader implications. We conclude by drawing a rough analogy between RNA folding and organismal development. Interactions between nucleotides influence the kinetic pathway of the molecule and its robustness to both the environment and mutations. Similarly, interactions between genes determine the outcome and stability of developmental pathways. The relationship between thermostability and mutational stability found in RNA also holds in proteins (Bussemaker *et al.*, 1997; Vendruscolo *et al.*, 1997; Bornberg-Bauer and Chan, 1999). Although we have yet to establish a similar connection between environmental and genetic sensitivity for more complex phenotypes, there is anecdotal evidence in the form of phenocopies, which are traits that can be produced either by genetic mutation or environmental perturbation (Goldschmidt, 1938).

The RNA folding model is a transparent evolutionary system that has afforded new insight into the facilitative and inhibitive roles of mutational robustness. Variation in RNA is limited at three levels—by mutational mechanisms, epistatic interactions that buffer the impact of mutations, and mutant (in)viability—but can be rescued, to some degree, by modularity. Testing and generalizing these ideas may bring us closer to a single conceptual framework for understanding variation.

ACKNOWLEDGMENTS

The author extends her deepest gratitude to Matthew Cowperthwaite, a Ph.D. student at the University of Texas at Austin, for running and analyzing some of the RNA simulations presented in this chapter and for preparing Figures 6-2–6-7. This work was supported in part by a National Science Foundation grant (DEB-0303636).

REFERENCES

Aligue R., Akhavan-Niak, H., and Russell, P. (1994). A role of Hsp90 in cell cycle control: Wee1 tyrosine kinase activity requires interaction with Hsp90. *EMBO Journal* **13**, 6099–6106.

Ancel, L. W., and Fontana, W. (2000). Plasticity, evolvability, and modularity in RNA. *Journal of Experimental Zoology* **288**, 242–283.

Ancel, L. W., and Fontana, W. (2001). Evolutionary lock-in and the origin of modularity in RNA structure. In *Modularity: Understanding the Development and Evolution of Complex Natural Systems* (W. Callebaut and D. Rasskin-Gutman, eds.) Cambridge, MA: MIT Press.

Ardell, D. H. (1998). On error minimization in a sequential origin of the genetic code. *Journal of Molecular Evolution* **47**, 1–13.

Bois, P., Collick, A., Brown, J., and Jeffreys, A. J. (1997). Human minisatellite MS32 (D1S8) displays somatic but not germline instability in transgenic mice. *Human Molecular Genetics* **6**, 1565–1571.

Bornberg-Bauer, E., and Chan, H. S. (1999). Modeling evolutionary landscapes: Mutational stability, topology, and superfunnels in sequence space. *Proceedings of the National Academy of Sciences, USA* **96**, 10689–10694.

Bucci, C., Lavitola, A., Salvatore, P., Del Giudice, L., Massardo, D. R., Bruni, C. B., and Alifano, P. (1999). Hypermutation in pathogenic bacteria: Frequent phase variation in meningococci is a phenotypic trait of a specialized mutator biotype. *Molecular Cell* **3**, 435–445.

Bussemaker, H. J., Thirumalai, D., and Bhattacharjee, J. K. (1997). Thermodynamic stability of folded proteins against mutations. *Physics Review Letters* **79**, 3530–3533.

Cannone, J. J., Subramanian, S., Schnare, M. N., Collett, J. R., D'Souza, L. M., Du, Y., Feng, B., Lin, N., Madabusi, L. V., Müller, K. M., Pande, N., Shang, Z., Yu, N., and Gutell R. R. (2002).

The Comparative RNA Web (CRW) Site: An online database of comparative sequence and structure information for ribosomal, intron, and other RNAs: Correction. *BMC Bioinformatics* **3**, 15.

Chao, L., and Cox, E. C. (1983). Competition between high and low mutating strains of *Escherichia coli*. *Evolution* **37**, 125–134.

Cox, L., and Cox, E. C. (1994). *Multidimensional Scaling*. London: Chapman & Hall.

Crotty, S., Cameron, C. E., and Andino, R. (2001). RNA virus error catastrophe: Direct molecular test by using ribavirin. *Proceedings of the National Academy of Sciences USA* **98**, 6895–6900.

de Visser, J., Arjan, G. M., Hermisson, J., Wagner, G. P., Meyers, L. A., Bagheri-Chaichian, H., Blanchard, J. L., Chao, L., Cheverud, J. M., Elena, S. F., Fontana, W., Gibson, G., Hansen, T. F., Krakauer, D., Lewontin, R. C., Ofria, C., Rice, S. H., von Dassow, G., Wagner, A., and Whitlock, M. C. (2003). Perspective: Evolution and detection of genetic robustness. *Evolution* **57**, 1959–1972.

Denver, D. R., Swenson, S. L., and Lynch, M. (2003). An evolutionary analysis of the helix-hairpin-helix superfamily of DNA repair glycosylases. *Molecular Biology and Evolution* **20**, 1603–1011.

Domingo, E. (2003). Quasispecies and the development of new antiviral strategies. *Progress in Drug Research* **60**, 133–158.

Drake, J. W. (1999). The distribution of rates of spontaneous mutation over viruses, prokaryotes, and eukaryotes. *Annals of the New York Academy of Sciences* **870**, 100–107.

Drake, J. W., and Holland, J. J. (1999). Mutation rates among RNA viruses. *Proceedings of the National Academy of Sciences USA* **96**, 13910–13913.

Drake, J. W., Charlesworth, B., Charlesworth, D., and Crow, J. F. (1998). Rates of spontaneous mutation. *Genetics* **148**, 1667–1686.

Edgell, D. R., Malik, S. B., and Doolittle, W. F. (1998). Evidence of independent gene duplications during the evolution of archaeal and eukaryotic family B DNA polymerases. *Molecular Biology and Evolution* **15**, 1207–1217.

Eigen, M., McCaskill, J., and Schuster, P. (1989). The molecular quasispecies. *Advances in Chemical Physics* **75**, 149–263.

Eisen, J. A. (1998). A phylogenomic study of the MutS family of proteins. *Nucleic Acids Research* **26**, 4291–4300.

Ellington, A. D. (1994). RNA selection. Aptamers achieve the desired recognition. *Current Biology* **4**, 427–429.

Eshel, I., and Feldman, M. W. (1970). On the evolutionary effect of recombination. *Theoretical Population Biology* **1**, 88–100.

Feldman, M. W., Christiansen, F. B., and Brooks, L. D. (1980). Evolution of recombination in a constant environment. *Proceedings of the National Academy of Sciences USA* **77**, 4838–4841.

Filée, J., Forterre, P., and Sen-Lin, T. (2002). Evolution of DNA polymerase families: Evidences for multiple gene exchange between cellular and viral proteins. *Journal of Molecular Evolution* **54**, 763–773.

Fisher, R. A. (1930). *The Genetical Theory of Natural Selection*. Oxford, England: Clarendon Press.

Fontana, W., and Schuster, P. (1998). Continuity in evolution: On the nature of transitions. *Science* **280**, 1451–1455.

Fontana, W., Konings, D. A. M., Stadler, P. F., and Schuster, P. (1993). Statistics of RNA secondary structures. *Biopolymers* **33**, 1389–1404.

Gilbert, W. (1978). Why genes in pieces? *Nature* **501271**, 501.

Goldschmidt, R. B. (1938). *Physiological Genetics*. New York: McGraw-Hill.

Goodman, M. F. (2002). Error-prone repair DNA polymerases in prokaryotes and eukaryotes. *Annual Review of Biochemistry* **71**, 17–50.

Gupta, R. S. (1995). Phylogenetic analysis of the 90 kD heat shock family of protein sequences and an examination of the relationship among animals, plants, and fungi species. *Molecular Biology and Evolution* **12**, 1063–1073.

Haig, D., and Hurst, L. D. (1999). A quantitative measure of error minimization in the genetic code. *Journal of Molecular Evolution* **33**, 412–417.

Hartl, D. L., and Taubes, C. H. (1998). Towards a theory of evolutionary adaptation. *Genetical Research Cambridge* **102/103**, 525–533.

Haubold, B., Travisano, M., Rainey, P. B., and Hudson, R. R. (1998). Detecting linkage disequilibrium in bacterial populations. *Genetics* **150**, 1341–1348.

Ishii, K., Matsuda, H., Iwasa, Y., and Sasaki, A. (1989). "Evolutionary stable" mutation rate in a periodically changing environment. *Genetics* **121**, 163–174.

Jones, M., Wagner, R., and Radman, M. (1987). Mismatch repair of deaminated 5-methyl-cytosine. *Journal of Molecular Biology* **193**, 155–159.

Kimura, M. (1967). On the evolutionary adjustment of spontaneous mutation rates. *Genetics Research* **9**, 23–34.

Kondrashov, A. S. (1982). Selection against harmful mutations in large sexual and asexual populations. *Genetical Research Cambridge* **40**, 325–332.

Kondrashov, A. S. (1995). *Genetics Research* **66**, 53–70.

Kruskal, J. B., and Wish, M. (1978). *Multidimensional Scaling*. Beverly Hills, CA: Sage University Series.

Lane, D. P. (1992). Cancer: p53, guardian of the genome. *Nature* **358**, 15–16.

Leigh, E. G. (1970). Natural selection and mutability. *American Naturalist* **104**, 301–305.

Leigh, E. G. (1973). The evolution of mutation rates. *Genetics* **73**, 1–18.

Lewontin, R. C. (1970). The units of selection. *Annual Review of Ecology and Systematics* **1**, 1–17.

Liberman, U., and Feldman, M. W. (1986). Modifiers of mutation-rate: A general reduction principle. *Theoretical Population Biology* **30**, 125–142.

Liu, B., Nicolaides, N. C., Markowitz, S., Willson, J. K. V., Parsons, R. E., Jen, J., Papadopolous, N., Peltomaki, P., Delachapelle, A., Hamilton, S. R., Kinzler, K. W., and Vogelstein, B. (1995). Mismatch repair gene defects in sporadic colorectal cancers with microsatellite instability. *Nature Genetics* **9**, 48–55.

Loeb, L. A. (1991). Mutator phenotype may be required for multistage carcinogenesis. *Cancer Research* **51**, 3075–3079.

Mayer, M. P., and Bukau, B. (1999). Molecular chaperones: The busy life of Hsp90. *Current Biology* **9**, R322–R325.

Maynard Smith, J. (1968). Evolution in sexual and asexual populations. *American Naturalist* **102**, 469–473.

McCaskill, J. S. (1990). The equilibrium partition function and base pair binding probabilities for RNA secondary structure. *Biopolymers* **29**, 1105–1119.

Meyers, L. A., and Bull, J. J. (2002). Fighting change with change: Adaptive variation in an uncertain world. *Trends in Ecology and Evolution* **17**, 551–557.

Meyers, L. A., Levin, B., Richardson, A. R., and Stojiljkovic, I. (2003). Epidemiology, hypermutation, within-host evolution, and the virulence of *Neisseria meningitidis*. *Proceedings of the Royal Society of London, Series B, Biological Sciences* **270**, 1667–1677.

Miller, J. H. (1998). Mutators in *Escherichia coli*. *Mutation Research* **409**, 99–106.

Miller, J. H., Suthar, A., Tai, J., Yeung, A., Truong, C., and Stewart, J. L. (1999). Direct selection for mutators in *Escherichia coli*. *Journal of Bacteriology* **181**, 1576–1584.

Mornet, E., Chateau, C., Hirst, M. C., Thepot, F., Taillandier, A., Cibois, O., and Serre, J. L. (1996). Analysis of germline variation at the FMR1 CGG repeat shows variation in the normal-premutated borderline range. *Human Molecular Genetics* **5**, 821–825.

Moxon, E. R., Rainey, P. B., Nowak, M. A., and Lenski, R. E. (1994). Adaptive evolution of highly mutable loci in pathogenic bacteria. *Current Biology* **4**, 24–33.

Muller, H. J. (1932). Some genetic aspects of sex. *American Naturalist* **66**, 118–138.

Nelson, J. R., Lawrence, C. W., and Hinkle, D. C. (1996a). Thymine-thymine dimer bypass by yeast polymerase z. *Science* **272**, 1646–1649.

Nelson, J. R., Lawrence, C. W., and Hinkle, D. C. (1996b). Yeast REV1 encodes a deoxycytidyl transferase. *Nature* **382**, 729–731.

Nussinov. R., and Jacobson, A. B. (1980). Fast algorithm for predicting the secondary structure of single-stranded RNA. *Proceedings of the National Academy of Science USA* **77**, 6903–6913.

Oliver, A., Canton, R., Campo, P., Baquero, F., and Blazquez, J. (2000). High frequency of hypermutable *Pseudomonas aeruginosa* in cystic fibrosis lung infection. *Science* **288**, 1251–1254.

Otto, S. P., and Barton, N. H. (1997). The evolution of recombination: Removing the limits to natural selection. *Genetics* **147**, 879–906.

Palmisano, A. N., Winton, J. R., and Dickhoff, W. W. (1999). Sequence features and phylogenetic analysis of the stress protein Hsp90alpha in chinook salmon (*Oncorhynchus tshawytscha*), a poikilothermic vertebrate. *Biochemical and Biophysical Research Communications* **258**, 784–791.

Passarino, G., Cavalleri, G., Stecconi, R., Franceschi, C., Altomare, K., Dato, S., Greco, V., Cavalli-sforza, L., Underhill, P., and De Benedictis, G. (2003). Molecular variation of human Hsp90alpha and Hsp90beta genes in Caucasians. *Human Mutation* **21**, 554–555.

Ponticelli, A. S., Schultz, D. W., Taylor, A. F., and Smith, G. R. (1985). Chi-dependent DNA strand cleavage by RecBC enzyme. *Cell* **41**, 145–151.

Punnett, R. C. (1923). *Heredity in Poultry.* London: MacMillan.

Raff, R. A. (1996). *The Shape of Life: Genes, Development, and the Evolution of the Animal Form.* Chicago: University of Chicago Press.

Reidys, C. M., Stadler, P. F., and Schuster, P. (1997). Generic properties of combinatory maps: Neutral networks of RNA secondary structures. *Bulletin of Mathematical Biology* **59**, 339–397.

Reidys, C., Forst, C., and Schuster, P. (2001). Replication and mutation on neutral networks. *Bulletin of Mathematical Biology* **63**, 57–94.

Reuven, N. B., Arad, G., Maor-Shoshani, A., and Livneh, Z. (1999). The mutagenesis protein UmuC is a DNA polymerase activated by UmuD', RecA, and SSB and is specialized for translesion replication. *Journal of Biological Chemistry* **274**, 31763–31766.

Richardson, A. R., and Stojiljkovic, I. (2001). Mismatch repair and the regulation of phase variation in *Neisseria meningitidis. Molecular Microbiology* **40**, 645–655.

Rutherford, S., and Lindquist, S. (1998). Hsp90 as a capacitor for morphological evolution. *Nature* **396**, 336–342.

Stadler, B. M. R., Stadler, P. F., Wagner, G., and Fontana, W. (2001). The topology of the possible: Formal spaces underlying patterns of evolutionary change. *Journal of Theoretical Biology* **213**, 241–274.

Taddei, F., Radman, M., Maynard-Smith, J., Toupance, B., Gouyon, P. H., and Godelle, B. (1997). Role of mutator alleles in adaptive evolution. *Nature* **387**, 700–702.

Tanaka, M. M., and Feldman, M. W. (1999). Theoretical considerations of cross-immunity, recombination, and the evolution of new parasitic strains. *Journal of Theoretical Biology* **198**, 145–63.

Tang, M., Shen, X., Frank, E. G., O'Donnell, M., Woodgate, R., and Goodman, M. F. (1999). UmuD'2C is an error-prone DNA polymerase, *Escherichia coli* pol V. *Proceedings of the National Academy of Sciences USA* **96**, 8919–8924.

Tenaillon, O., Toupance, B., Le Nagard, H., Taddei, F., and Godelle, B. (1999). Mutators, population size, adaptive landscape, and the adaptation of asexual populations of bacteria. *Genetics* **152**, 485–493.

Tinoco, I., Jr., Uhlenbeck, O. C., and Levine, M. D. (1971). Estimation of secondary structure in ribonucleic acids. *Nature* **230**, 362–367.

Tinoco, I., Jr., Borer, P. N., Dengler, B., and Levine, M. D. (1973). Improved estimation of secondary structure in ribonucleic acids. *Nature, New Biology* **146**, 40–41.

Trobner, W., and Piechocki, R. (1985). Selective advantage of *polA1* mutator over *polA*+ strains of *Escherichia coli* in a chemostat. *Naturwissenschaften* **72**, 377–378.

Turner, D. H., Sugimoto, N., and Freier, S. M. (1989). RNA structure prediction. *Annual Review of Biophysics and Biophysical Chemistry* **17**, 167–192.

Van Nimwegen, E., Crutchfield, J. P., and Huynen, M. (1999). Neutral evolution of mutational robustness. *Proceedings of the National Academy of Sciences USA* **96**, 9716–9720.

Van Valen, L. (1973). A new evolutionary law. *Evolutionary Theory* **1**, 1–30.

Vendruscolo, M., Maritan, A., and Banavar, J. R. (1997). Stability threshold as a selection principle for protein design. *Physics Review Letters* 78, 3967–3970.

Vogler, D. W., and Kalisz, S. (2001). Sex among the flowers: The distribution of plant mating systems. *Evolution* 55, 202–204.

Von Dassow, G., Meir, E., Munro, E. M., and Odell, G. M. (2000). The segment polarity network is a robust developmental module. *Nature* 406, 188–192.

Wagner, G. P. (1988). The influence of variation and of developmental constraints on the rate of multivariate phenotypic evolution. *Journal of Evolutionary Biology* 1, 45–66.

Wagner, G. P., Booth, G., and Bagherichaichian, H. (1997). A population genetic theory of canalization. *Evolution* 51, 329–347.

Wagner, J., Gruz, P., Kim, S. R., Yamada, M., Matsui, K., Fuchs, R. P., and Nohmi, T. (1999). The dinB gene encodes a novel *E. coli* DNA polymerase, DNA pol IV, involved in mutagenesis. *Molecular Cell* 4, 281–286.

Waterman, M. (1978). *Secondary Structure of Single-Stranded Nucleic Acids*. New York: Academic Press.

Wright, S. (1929). Fisher's theory of dominance. *American Naturalist* 63, 274–279.

Wright, S. (1932). The roles of mutation, inbreeding, crossbreeding and selection in evolution. *Proceedings of the 6th International Congress of Genetics* 1, 356–366.

Wuchty, S., Fontana, W., Hofacker, I. L., and Schuster, P. (1999). Complete suboptimal folding of RNA and the stability of secondary structures. *Biopolymers* 49, 145–165.

Zuker, M. (1989). Computer prediction of RNA structure. *Methods in Enzymology* 180, 262–288.

Zuker, M. (1994). Prediction of RNA secondary structure by energy minimization. *Methods in Molecular Biology* 25, 267–294.

Zuker, M., and Stiegler, P. (1981). Optimal computer folding of larger RNA sequences using thermodynamics and auxiliary information. *Nucleic Acids Research* 9, 133–148.

CHAPTER 7

Developmental Origins of Variation

ELLEN LARSEN
Department of Zoology, University of Toronto, Toronto, Ontario, Canada

INTRODUCTION

Writing about the origin of developmental variation is an intriguing challenge for someone who has long been interested in the flip side of developmental variation, namely canalization. Both variation and canalization are of paramount interest because of their relationship to the patterns and processes of evolution. With respect to developmental variation, did Darwin not get it right? "Descent with modification" means that similarities between taxa are the result of common ancestry, and differences are the result of modifications in development wrought by hereditary changes. Although the problems inherent in identifying the paths of descent and recognizing modifications are many (see Hall, 2003, for a recent discussion of these issues), the emphasis on the importance of hereditary change leads to the conclusion that evolution results from the "selectable developmental variation" (West-Eberhard, 2003). Mutational processes propose variations, and selection disposes of them.

Although, historically, developmental variation and selection may have been twin pillars of evolution, the focus in the twentieth century was on selection. The reason for this has been discussed in a compelling way by West-Eberhard (2003) who points out that developmental variation took a back seat to selection when it was deemed to threaten another sacred tenet of neo-Darwinism, gradualism in evolution. Whatever the historical reasons for its temporary fall from view, the resurgence of interest in evolutionary developmental biology places developmental variation and its origins at the center of any understanding of the nature and dynamics of evolution. Conventional wisdom suggests that the origin of developmental variation results from mutational or environmental perturbations. Few can doubt that genetic variation exists in abundance and plays an important role in evolution. Intuitively, we appreciate genetic variation every time we observe denizens of city parks; both the variety of plumage color patterns in pigeons and the variety of humans we encounter suggests that genetic variation is common. Indeed, the chief message of twentieth-century population genetics was that there is so much genetic variation in natural populations, much of it hidden from gross phenotypic expression, we were hard put to explain how it was all maintained (Franklin and Lewontin, 1970). We have only to look at the developmental genetic literature to convince ourselves that there are literally thousands of genes with allelic variants providing a vast potential source of developmental variation. Indeed, recent authors have correlated changes in genes in natural populations with phenotypic changes either within a population (Gibson and Hogness, 1996; Dworkin *et al.*, 2003) or between related taxa (Sucena *et al.*, 2003). As Müller and Newman (2003) note, the gene-centered neo-Darwinian paradigm has much to recommend it, but it may be insufficient for understanding the origins of the phenotype; the results of genome projects showing a largely conserved repertoire of genes in organisms of vastly different form should inspire us to seek additional factors for the origins of developmental variation. It is telling that T. H. Morgan's 1907 text *Experimental Zoology* had chapters entitled "The Influence of External Condition in Causing Changes in the Structure of Animals" and "The Inherited Effects of Changes Induced by External Forms." Within a few decades, such topics would not appear prominently in a reputable text. After nearly a century of neglect, there appears to be a growing willingness to explore alternatives to a mutational origin for developmental and evolutionary variation (see West-Eberhard, 2003, Chapter 26).

A potentially important source of developmental variation is variation resulting from the dynamics of developmental processes themselves. If this variation exists but is not attributable to genetic change, then it might be a phenomenon of interest to developmental biologists, but can it bear on evolutionary change? The purpose of this chapter is to pose the question: Can developmental processes themselves generate variation that plays a role in evolutionary change? I call this potential source of variation "intrinsic developmental variation." I explore the question of whether or not intrinsic developmental variation exists and the conditions under which it might be revealed. For such variations to be a robust feature of evolutionary change, they must be "captured" genetically. In other words, the variation would have to precede the genetic changes, rather than being the result of them, as is usually assumed in current conceptions of the causal links in evolutionary change. If intrinsic developmental variation is an important factor in the evolution of development, then developmental biologists must devote more effort to understanding the origin of the "spontaneous" developmental variations we now ignore as impediments to doing controlled experiments.

I. DOES INTRINSIC DEVELOPMENTAL VARIATION EXIST?

Anecdotal evidence suggests that intrinsic variation in development is alive and well. The frustration expressed in the Harvard Law of Biology "under the most carefully controlled conditions, biological material does whatever it damn well pleases" is familiar to most of us working at the bench; we struggle to eliminate genetic and environmental factors, and yet development is variable.

To illustrate the different levels of organization in which developmental variation may arise, I describe a small sample of fruit fly variants that appear in uniform environments among individuals with the same genotype. No doubt, other developmental biologists will bring to mind additional examples.

At the molecular level, the distribution of gene products may vary from organism to organism. An example to which we will return is *bicoid* in flies. *bicoid* is a maternal effect gene with a DNA-binding motif whose message is deposited in the anterior portion of the egg. It binds to important anterior determining genes such as *hunchback* (*hb*). When *bicoid* is removed, the head region is posteriorized. The protein localization boundary can vary in different embryos and extend to as far as 30% of the embryo (Houschmandeh *et al.*, 2001).

Despite this inherent variation, adult morphology seems unimpaired by the "sloppiness" of the localization. At the level of cell organelle, the oocyte nucleus migrates through the cell to its final position in the future anterior of the embryo. The paths of the nuclei during migration vary in different embryos, but they all end up in the same region of the developing egg (Roth *et al.*, 1999). At the cellular level, division patterns are rarely predictable (development in the nematode *Caenorhabditis elegans* is an exception). Variable patterns of cell division can be seen by observing mitotic patterns in bilateral structures. In several kinds of organisms, somatic recombination may be used to produce genetically marked mitotic clones during development that can be distinguished from other groups of cells by a convenient marker such as color. If one compares clone size and shape at a particular landmark—the anterior/posterior boundary in the wings of flies—they may vary in the wings of the same organism as well as between organisms. Nevertheless, the adult wings are usually bilaterally symmetric and have virtually identical shapes both within and between organisms.

Deviations from bilateral symmetry often provide evidence that intrinsic developmental variation is producing a phenotypic effect. For example, the work of Smith and Sondhi (1960) indicated that, in the *ocelliless* mutation in *Drosophila subobscura*, it was possible to select for loss of either left or right posterior ocellus, but it was not possible to specify which one. The authors suggested that genetic variation for asymmetry exists, but the direction of the asymmetry is not genetically specified. If so, we can conclude that intrinsic developmental variation may determine that only one ocellus will be present, but this will have little evolutionary impact because there is no genetic variation to fix the direction of the asymmetry. The explanations for these observations are as enigmatic as they are important. Lobster claws also show random left–right asymmetry, but this has its explanation in differential use (Govind and Pearce, 1989). If one claw is bound with an elastic band, the other claw will be the one to grow large.

Clearly, not all left–right asymmetries are indeterminate. Land slug contractile lung openings are found uniformly on one side of the head as a result of counterclockwise torsion (Brusca and Brusca, 1990), and narwhal males have a developed left tusk (but see Bateson [1894] for bilateral tusks and other anomalies of asymmetric species). Presumably there is genetic variation associated with the reliability of these asymmetries.

Curiously, one of the phenomena pointing most clearly in the direction of intrinsic developmental variation occurs when mutant strains arise. It is underappreciated that many mutations are not expressed to the same extent in all individuals, and in fact, a reproducible proportion of individuals in a stock homozygous for a mutant allele may appear to be wild type (variable penetrance); if wild-type-appearing parents are bred together, the characteristic proportion of mutant and wild phenotypes will be found in their progeny. Sometimes the frequency of mutant progeny can be raised or lowered in the stock by selection of mutant or wild-type phenotypes respectively, yet it is often difficult to achieve 100% penetrance. Indeed, when choosing mutant stocks for classroom purposes, we often select for higher penetrance and expressivity shortly before students receive them. One explanation for this variation is that the mutation produces a perturbation that the usual buffering mechanisms are unable to deal with, thus revealing intrinsic variation.

Often, mutations are buffered during selection for the wild-type phenotype, illustrating that genetic variation may exist for this buffering (here synonymous with canalization). Variation in response to teratogens or phenocopy agents can be interpreted similarly as a failure of canalization in the face of perturbation. The same genotypes under the same environmental conditions produce variable phenotypes. The almost ubiquitous existence of canalizing genes, which can modulate perturbations of genetic or environmental origin, points to the possibility that variation from intrinsic causes can often be fixed (either in the mutant or wild-type form) by genetic "capture."

Those of us who deal frequently with variable penetrance and expressivity are driven to wonder what is special about those unusual genetic variations that can be relied on to have 100% expressivity and penetrance. One explanation is that they act in a part of a network in which regulation is difficult. According to this kind of thinking, the place in a network rather than the nature of the gene product is crucial. An anatomic analogy might give us some intuition here.

We are concerned about heart attacks but not hand attacks. A reason for this is that, in most regions of our bodies, arteries feeding tissues are anastomosed, e.g., arranged in a ladderlike pattern (Figure 7-1A). In contrast, coronary arteries lack such connections (Figure 7-1B); in consequence, blockage of a coronary artery leads to the death of cells fed by the artery below the blockage. Blockages elsewhere affect little tissue because blood can flow in from connecting vessels. The structure of the arterial network, not the nature of arteries or the blockage, renders the heart and brain vulnerable to vascular mishaps. Likewise, genetic mutations may have greater or lesser effects depending on the nature of the network they are part of, not on the quality of the mutational change.

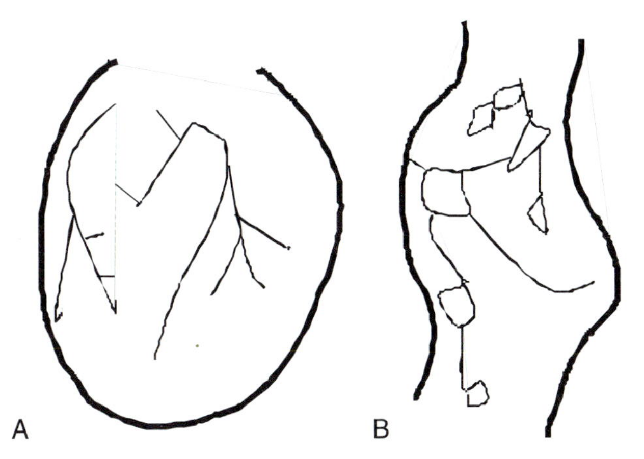

FIGURE 7-1. (A) Cartoon of nonanastomosing coronary arteries of the right heart ventricle and (B) of anastomosing arteries in the knee (after J. M. B. Grant, *The Atlas of Anatomy*, 1956).

II. INTRINSIC VARIATION IN DIFFERENT ENVIRONMENTS

Intrinsic developmental variation described so far has been phenotypic variation occurring under the same genotypic and environmental conditions. It was suggested that variation fitting this description is released when developing systems are perturbed and the usual buffering systems do not cope. This is a form of developmental plasticity and raises the question of the relationship of developmental plasticity to the evolution of development.

Developmental plasticity and its consequences have been reviewed at length in West-Eberhard (2003). I merely wish to point out here that a phenotype arising from a failure to buffer may be fixed genetically through selection, either as the sole phenotype in a variety of environments or as an alternative conditional phenotype, dependent on the environment (polyphenism) and thus contribute to evolution. This is the phenomenon of genetic assimilation (Waddington, 1953; Dworkin and Gibson, this book).

One may now ask, is there a difference between genetically modulated variation and intrinsic variation? Because organisms have an evolutionary history, intrinsic and genetic variations are intimately connected, and as stressed by Newman and Müller (2001), may be difficult to disentangle. It seems to me that there may be a difference between the two types of variation in that the appearance of intrinsic variation is biased by existing structures both material and abstract (regulatory networks) as well as by the physics of development. In contrast, mutationally derived variation is usually considered to be random (although Yampolsky and Stoltzfus [2001] discuss the possibility that mutationally derived variation is also biased). In any event, both the view in this chapter and the position of Yampolsky and Stoltzfus is that more effort should be devoted to "internalist" explanations for variation in evolution.

III. POTENTIAL ORIGINS OF INTRINSIC DEVELOPMENTAL VARIATION

A. Noise

The term *noise* is used in common parlance as well as in a technical sense. A feeling for noise in the technical sense can be gleaned from the following example.

Suppose we tracked the proliferation rate of individual blastomeres in a developing embryo. Each cell would have a particular rate that could be averaged to produce a population mean. The deviation of each cell rate from the mean of the population produces a measure of population variance. Suppose there was little noise in the system and all the cells divided virtually synchronously. In such a case, the population mean would mirror the individual means. Suppose, however, that there are two groups of cells, one that divides at one frequency and the other that does not divide at all. In this scenario, the population mean does not reflect this bimodal population, and the variance is high. In most cases, phenotypic variability appears to be stochastic (e.g., random), and this property is formally called "noise." A measure of this is:

population variance in phenotype/population mean.

Noise can be measured during development at every level of biologic organization from molecule to adult phenotype. Can we determine the processes that give rise to noise? Does noise at one level or time of development give rise to noise at higher levels or later times? Is noise a necessary aspect of development? Can it contribute to developmental variation with evolutionary consequences?

IV. AN EXAMPLE OF NOISE IN EUKARYOTIC TRANSCRIPTION

Baker's yeast, *Saccharomyces cerevisiae*, is an excellent system in which to search for noise emerging from basic eukaryotic gene expression systems that might provide a source of developmental variation in multicellular development. This yeast is well characterized genetically, and experiments can be done on clonal derivatives, reducing the probability that observed variations are due to mutations. For example, Blake *et al.* (2003) constructed an ingenious system of transcriptional regulation in which the reporter gene, green fluorescent protein (GFP), was positively driven by the nutrient galactose as well as by an inhibitor (anhydrotetracycline) of a transcriptional inhibitor (tetracycline). GFP expression was monitored by measuring fluorescence on a per-cell basis in a flow cytometer. By varying galactose and anhydrotetracycline, noise was varied from low to high. Under low-noise conditions, cells showed a unimodal distribution of GFP production, whereas distinctly bimodal distributions were found under high-noise conditions. Thus cell populations with low or high GFP synthesis can emerge from a uniform population by increasing noise in the transcriptional system. Increasing translational efficiency in this system amplified the cell-to-cell variation stemming from noisy transcription. These results suggest that nongenetic sources of cell variation can lead to cell differentiation. In this case, a threshold response to GFP quantity could divert the quantitative differences between cells into qualitatively different cell differentiation.

Is noise, as measured by stochastic gene activation, a feature of normal development? In their review of enhancer action, Fiering *et al.* (2000) suggest that enhancers may increase the probability of transcription of a gene rather than its rate of transcription. They bring together evidence suggesting that enhancers are regulators of inherently stochastic transcription. Not only are there stochastic aspects of transcription at the level of individual genes, it has long been held that inactivation of one X chromosome in each cell of female mammals is random (Lyon, 1961).

Is noise sufficiently reliable to generate predictable developmental trajectories? The work reviewed by Fiering *et al.*, described above, suggests that it is. Does noise always yield phenotypic consequences? The next example suggests that it may not.

V. NOISY *BICOID* GENE EXPRESSION IN FRUIT FLIES

It came as a surprise to many of us that one of the premier gradients for organizing early embryogenesis in flies, the bicoid (Bcd) protein gradient, was variable in terms of its posterior boundary (Houchmandzadeh *et al.*, 2002). The variation in the Bcd boundary location was on the order of five cell nuclei, whereas, surprisingly, one of the genes downstream of *bcd*, namely *hunchback (hb)*, had a precise protein boundary location with respect to embryo length with an error rate of less than one nucleus. Apparently, sloppiness of some gene expression, such as that of *bcd*, is tolerated in development and evolution if its consequences can be filtered out, in this case by *hb*. Although it is not understood how *hb* can be regulated by *bcd* as well as other genes to produce its precise expression pattern, this example shows that developmental variation at the level of gene expression may be more common than we might have thought and that such variation may have little developmental consequence.

VI. NOISE IN ASYMMETRY PRODUCTION

I have mentioned asymmetry as a form of developmental variation. The source of developmental and genetic variation resulting in asymmetry has been explored by Palmer *et al.* (1993). They consider subtle asymmetries, for example, fluctuating left–right asymmetry, presumably the result of randomness introduced by molecular movements, rates of physiologic processes, and cell division and growth. They also discuss macroscopic asymmetries such as the antisymmetry of lobster claws and directional asymmetry as found in the preponderance of human right-handedness. According to their analysis, the one factor common to the origin of all these asymmetries is "noise." For fluctuating asymmetry, this is the only developmental feature they associate with its origin.

VII. NOISY IMPLICATION FOR EVOLUTION

The lack of phenotypic consequence of some intrinsic developmental variation does not mean that such variation has no potential evolutionary consequence;

unless we are willing to argue that the enormous amount of hidden genetic variation revealed by twentieth-century population geneticists has no potential evolutionary significance either. Just as the effects of genetic variation may be suppressed by other genes, genes can also be selected to suppress or enhance intrinsic developmental variation as discussed above for enhancers. Put another way, genes can "capture" developmental variation or suppress it. Indeed, managing intrinsic developmental variation is an important function of canalization (see Chapters 9 and 21, this book). The possibility that genes "capture" forms produced by nongenetic processes and make them part of the genetically controlled morphologic repertoire was proposed by Newman (Newman, 1992, 2003; Newman and Müller, 2001) to have been important in the origin of biologic form. There is reason to suppose that variation arising from intrinsic developmental processes can be "captured" genetically. Experiments suggesting this have usually involved stressing organisms under conditions in which a small number of individuals show morphologic abnormalities. Continued selection of the abnormalities under stressful conditions may lead to the "assimilation" of the abnormal morphology even without stress (Waddington, 1952), and it demonstrates that genetic variation is often available to push developmental trajectories in new directions. The gene capture idea could well be tested in the yeast experimental system described above. Using flow cytometry, high and low GFP-producing lines could be established and continuously selected under conditions of high noise. A prediction of the gene capture scenario is that eventually, even in noisy conditions, the strains will no longer produce a bistable population of cells but rather either a high or low GFP-producing line depending on the selection regimen. Because the cells were originally clonal and the yeast is not undergoing meiosis, most genetic variation involved would be due to chromosomal rearrangements or mutation. In principle, the genetic changes involved could be determined because the yeast genome has been sequenced. So long as we are performing thought experiments, we should probably have several replicates of each selection line and ask if the genetic changes in each line are similar. Based on my experience with flies, I would guess that different genes for GFP production regulation would arise in the different lines.

VIII. NETWORKS

Our descriptions of development often involve the construction of "wiring diagrams" in which molecular pathways or even cellular and tissue inductive relationships are described. Such networks are an important abstraction in thinking about development because, fundamentally, development is about coordination, the coordination of cell behavior to make forms and the coordination of morphogenesis and both cell differentiation and patterns of differentiated cells. Networks provide a global way of describing unidirectional inductions and mutual interactions important for coordination. These are useful in understanding how qualitative differences between cells or regions may arise, but of particular interest here are emergent properties of these networks that depend primarily on the pattern of connectivity and not on the quality of the types of molecules, cells, or tissues involved.

As discussed earlier, in connection with blood vessel continuity, network structure affects the viability of the system after perturbation (in this case, blockage). We will review suggestions that different patterns of connectivity may be associated with different levels of noise (stochastic behavior) and that hence patterns of connection may be a source of intrinsic variation, which may themselves evolve in ways that modify the developmental variation produced. There are a variety of vantage points from which to consider networks, and we will discuss a few of these. As with the general subject of noise, we are at the beginning of a period of exploration of network consequences for development and evolution that involves both theory and experiment. The knowledge base we have erected in terms of genomics as well as of developmental pathways coupled with our increasing toolbox of transgenic techniques to alter network morphology provides a basis for experimental tests of network behavior predicted from models and simulations.

The assertion that pattern of network connectivity (topology) itself may be an important aspect of development, and its variation has been inferred from theoretical work, simulations of theoretical systems, and simulations of abstracted "real life" networks.

My first exposure to the possibility that pathway structure itself could be important was the work of Kacser (1957), who suggested from thermodynamic principles that, in a linear metabolic chain of substrates, the output of product would be insensitive to most fluctuations in enzyme efficiencies but that the situation would change if there were a branch point (Figure 7-2). In this case, changing enzyme efficiency near the branch point would modify the rate of product production. This work was followed up by Kacser and Burns (1981) with a study of metabolic mutants in *Neurospora* in which the theoretical predictions were largely borne out.

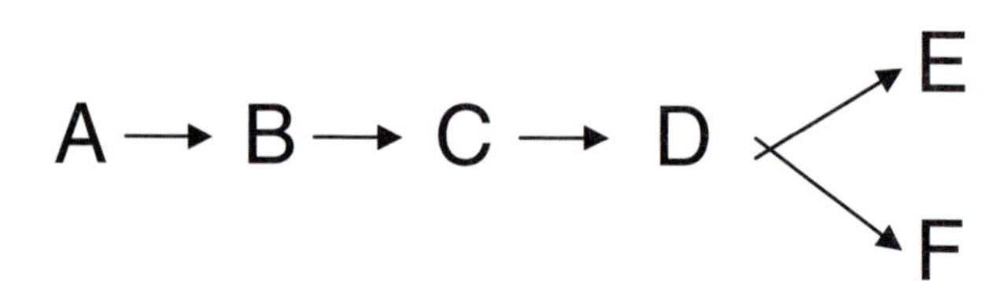

FIGURE 7-2. Pathway structure plays a role in pathway dynamics. If A–F are substrates and products in a pathway, mutations which reduce the rate at which Band C are made, will not affect the qualities of products E and F. If a mutation reduces the amount of D that is available, quantities of E and F produced will be affected.

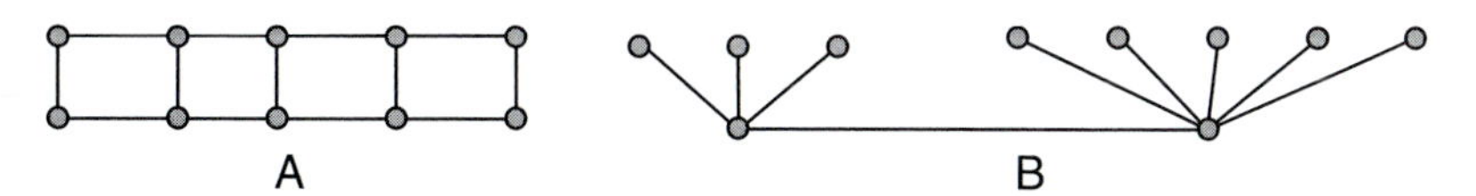

FIGURE 7-3. Two network configurations: (A) each node has two or three connections, (B) nodes have either two connections or three or more. In each case there are 10 nodes.

Another approach to biological networks was undertaken by Jeong *et al.* (2000) who explored the networks of core metabolic pathways in 43 organisms ranging from Archaea to eukaryotes and asked whether the connectivity observed consisted of substrate nodes that all had the same number of links or whether the topology comprised a few nodes (hubs) with a large number of links connecting nodes with fewer links to other nodes (Figure 7-3). The overwhelming conclusion was that the latter topology best describes metabolic networks. From what we know about signal transduction and transcription cascades in which a few pathways are used in a variety of different contexts (Hultmark, 1994; Wilkins, 2002), I would predict that the same will be found for these pathways in developmental contexts. Jeong *et al.* (2000) speculated that this pattern of connectedness is indicative of selection for "robust and error-tolerant networks."

Approaches to examining more developmental networks have been pursued with emphasis on well-studied pathways such as the segment polarity gene network in fruit flies. Starting with known relationships between segment polarity genes, Von Dassow *et al.* (2000) constructed a model of the network using differential equations, which included parameters such as protein and messenger RNA half-lives as well as binding rates and cooperativity coefficients. The authors then performed simulations using different parameter values and initial conditions to see what values mimicked the behavior of the system being modeled. To their surprise, a wide range of initial conditions and parameter values were capable of doing this, and they surmised that this robust behavior results from the topology of the network. These two examples suggest to me that contemporary networks have been selected for robustness and that probing their potential for producing developmental variation requires finding special cases in which noise is "allowed" to exert a phenotypic effect. The stochastic nature of transcriptional initiation might be such an opportunity as would a new environment in which the networks have not been selected.

The question of how the structure of developmental networks might evolve has been explored by Salazar-Ciudad *et al.* (2001a,b). They constructed a model involving a string of nuclei each with three interrelated genes, paracrine factors that can act on genes with activating or inhibiting consequences, and dynamic equations that they obey. To simulate evolution, they started with a 100 arrays of nuclei strings, ran the simulation, and looked for patterns in which nuclei differed in the production of a given gene product. They selected for the next generation of simulations the 50 arrays with a pattern closest to a preselected stripe pattern. After 100 simulations, they examined the relationships between the genes, the paracrine factors, and the number of stripes they produced. Two types of networks could be discerned, those in which components had mutual interactions (called emergent networks) and hierarchic networks in which the influence of the network elements flowed only in one direction. Further comparisons suggested that emergent networks produced more complex but also more variable patterns than did the hierarchic networks and for the same degree of pattern complexity (in their case, the number of stripes). They were simpler at a molecular level, requiring fewer genes. Some tantalizing speculations arise from these findings. We might expect that emergent networks could generate morphologic innovations and then be replaced by more robust hierarchic networks in evolution. On the other hand, traits showing rapid change in response to varying environments might be based on more plastic, emergent networks.

This selective sojourn into biological networks and their potential for generating developmental variation suggests that network structure has the potential to influence the extent and nature of developmental variation. Exploring network topology in developing systems can be facilitated by our developing technologic capabilities both with computational strategies that can infer network relationships from genome databases (Jeong *et al.*, 2001) or from RNA expression changes (Gardner *et al.*, 2003). A difficult challenge is to determine whether intrinsic developmental variations arising from networks/noise have contributed in the past and are contributing presently to the evolution of development. If these intrinsic sources of variation are, or have been, important, we must alter our worldview of evolution as being driven by mutation. In this scenario, mutation is only one force in evolution; intrinsic variation that produces advantageous phenotypes could be an additional factor if the phenotypes are "captured" by portions of standing genetic variation that reinforces the adaptive phenotype. This is not a new idea. As mentioned, it is in keeping with Newman's explanation of production of multicellular forms (2003) and with West-Eberhard's suggestion (2003) that developmental plasticity may first produce phenotypes that may then be honed and fixed by genetic variation. This question remains: Is the origin of developmental variation always "genes first," or is there frequently a source of intrinsic developmental variation producing a phenotype that is at first only indirectly dependent on genotype and then "captured" genetically, as a result of selection? The idea that phenotypes may arise before the alleles that stabilize them has never been easy to accept and accounts for the reluctance to accept the work of Baldwin (1896), who proposed that learned capabilities might have evolutionary consequences. Because "development first" is a substantial modification to our received wisdom about evolution, it is of sufficient importance to explore the possibility in greater depth.

IX. MORPHOGENETIC FIELDS: A POTENTIAL SOURCE OF VARIATION

The concept of the morphogenetic field as a unit (module) of development has enthralled and appalled generations of developmental biologists. Morphogenetic fields are defined operationally as those cells in a developing organism that, when transplanted elsewhere, will form a particular structure, say, a wing. The potential wing-forming region of a chick is greater than the region that usually produces the wing. When transplanted, this latter region produces a wing more frequently. The propensity to form a wing decreases in a graded fashion, hence the term *field* as analogous to physical fields. Biologists were enthralled because so many structures arise from primordia that express field properties (gradient of strength, ability to duplicate or regenerate) with self-organizing emergent features and appalled because they do not understand how these general properties arise, although sometimes particular gene expression can be correlated with the ability of cells to participate in a field (Technau *et al.*, 2000; Gilbert *et al.*, 1996).

My interest here is that morphogenetic fields are a potential locus for stimulating phenotypic change (an excitable medium in the terminology of some). Dramatic examples of malformations attributable to field responses are cyclopia in vertebrates caused by the failure of the normal "emancipation" of two eye fields from an initial eye primordium and extra frog legs, probably induced by parasites stimulating mirror image duplications in leg fields. We assume that morphogenetic field properties depend on cell–cell communication because surgical bifurcation can stimulate the production of two fields from one. Morphogenetic fields may be a reasonable place to look for intrinsic developmental disturbances because changes in cell behaviors, however triggered, may lead to modifications in the phenotype produced by the field. For example, enlargement of a field by excess cell proliferation or cell size might impair communication and lead to field duplications just as field bifurcation does (Buratovich and Bryant, 1995, 1997). Thus modifying the noise or the network organization of fields might provide a useful means to explore effects of intrinsic developmental variation. Because fields are inherently self-organizing and regulating for example, a partial field is able to develop a whole structure, there will be a tendency for field changes to be coordinated. The ability to induce eyes on antennae, legs, and wings by ectopically expressing the *eyeless* gene in different imaginal discs of fruit flies is a testament to the potential of perturbed fields to produce large phenotypic changes. Fields are potentially influenced by internal processes and genes or by external physical, chemical, or biologic (for example infection) perturbations. Genetic and environmental influences on *obake*, a mutation causing duplications of antenna and other head fields, were studied by Atallah *et al.* (2004). They show that specific mutations, natural genetic variation, and larval density affect the mean number of antenna duplications. Furthermore, the duplicating ability of fields offers the possibility of new morphologies, much as gene duplication has long been considered a source of gene family evolution. The formation of a biramous insect appendage in the uniramous insect taxon (Dworkin *et al.*, 2001) depended on the duplication of the antenna field. No doubt the Greeks had the power of morphogenetic fields in mind when they described Pegasus, a four-legged animal with additional wings.

X. IMPLICATIONS

One of the implications of a "development first" view of the origin of developmental variation affects the way we think of "mechanisms of development." In our current thinking, mechanisms revolve around the genes that, if altered, modify the developmental outcome. It is then natural to think of the genes as primary not only in evolution but in mechanisms of development as well. If asked how a limb develops, we first tend to recall signal transduction and transcription cascades rather than the cells and self-organizing fields in which these cascades play a role. If we find evidence that genes may also capture morphologies (analogous to turning an oral tradition into a written record), we will turn our attention from the words to the underlying story; words only have meaning in the context of a particular language at a particular time in history, and different words can tell the same story. Genes and genetic cascades are used for different purposes within an organism and in different taxa. It is as if we observe development like aliens coming to earth who try to understand the marvels of electric lights, stereos, televisions, computers, and all the other devices they see, by studying wall switches and fuse boxes. If we focus on intrinsic variation and its genetic capture, we will see development less as a canonical series of events and more as a dynamic where there is no single way of making a limb, but a dynamically changing group of ways with some similarities and some differences depending on context (rest of the genotype and environment). Sometimes these differences will produce the same phenotypes as other ways and sometimes different phenotypes.

From an evolutionary perspective, restrictions would be reduced if intrinsic developmental variation were an important source of evolutionary variation. For intrinsic developmental variation to have a plausible role in evolution, it must undergo "genetic capture." As suggested in this chapter, this might happen if an adaptive intrinsic developmental variant fortuitously contained alleles or allele combinations promoting that phenotype even at a level below a phenotypic expression threshold. In this way, the fortuitous genotype would increase in the population. The recurring nature of the developmental variant (because the structure of the developing system gives rise to a biased set of variants) would allow time for accumulation of additional alleles that in sum produce the phenotype "on their own." I am grateful to M. J. West-Eberhard for pointing out that Sewall Wright's ideas on random genetic drift in subpopulations could provide a genetic basis for this scenario. According to Wright's analysis (Wright, 1948), subpopulations can become genetically differentiated from one another because of random genetic drift, even in the absence of selection. As envisioned here, drift coupled with adaptive developmental variants could produce genetically based (adaptive) phenotypes more rapidly than if drift were operating on its own.

Even if plausible, is this a common mechanism for evolutionary change? It is usually impossible to recover the history of a change based on its present state, and I do not see a way of detecting a "signature" of developmental variants that have become genetically assimilated. Instead, I believe we must turn to the kind of laboratory experiments I suggested previously and hope that insights gained will allow us to explore the natural world more perceptively.

In his provocative book *Internal Factors in Evolution*, Whyte (1965) suggested that, in the history of physics, a nineteenth-century preoccupation with randomness in the universe gave way to an interest in order and structure in the twentieth. I suggest that, in twenty-first-century developmental biology, we combine the issues of order and randomness by investigating the effect of internal noise on the behavior of ordered structures to gain a more realistic view of the origin of developmental variation.

XI. SUMMARY

1. In considering the origin of developmental variation, we must consider the contribution of both genetic (mutational) sources and sources of variation arising from developmental processes themselves (intrinsic sources of developmental variation).
2. Phenotypic effects of intrinsic developmental variation are seen in phenomena ranging from fluctuating asymmetry to variable penetrance and expressivity of alleles.
3. Two sources of developmental variation are noise and perturbation of network structure.
4. From evolutionary perspectives, it is possible to conceive that variation arose first as intrinsic variation and then was captured genetically by abundant genetic variation.
5. Such a scenario provides an alternative to the "genes first" ideas of neo-Darwinism and provides additional avenues for rapid and saltational changes in the evolution of development as well as for gradual change.

ACKNOWLEDGMENTS

Many thanks to the following people whose suggestions and ideas have been incorporated into this manuscript: Joel Atallah, Ian Dworkin, Dorothea Godt, Brian K. Hall, Stuart Newman, Sue Varmuza, M. J. West-Eberhard, and Rudolf Winklbauer.

REFERENCES

Atallah, J., Cheng, U., Dworkin, I., Ing, B., Greene, A., and Larsen, E. (2004). The environmental and genetic regulation of *obake* expressivity: Morphogenetic fields as evolvable systems. *Evolution and Development*, **6**, 114–122.

Baldwin, M. J. (1896). A new factor in evolution. *The American Naturalist* **30**, 441–451.

Bateson, W. (1894). *Materials for the Study of Variation*. London: Macmillan.

Blake, W. J., Kaern, M., Cantor, R., and Collins, J. J. (2003). Noise in eukaryotic gene expression. *Nature* **422**, 633–637.

Brusca, R. C., and Brusca, G. J. (1990). *Invertebrates*. Sunderland, MA: Sinauer Associates.

Buratovich, M. A., and Bryant, P. J. (1995). Duplication of l(2)gd imaginal discs in *Drosophila* is mediated by ectopic expression of wg and dpp. *Developmental Biology* **168**, 452–463.

Buratovich, M. A., and Bryant, P. J. (1997). Enhancement of overgrowth by gene interactions in lethal (2) giant discs imaginal discs from *Drosophila melanogaster*. *Genetics* **147**, 657–670.

Dworkin, I., Tanda, S., and Larsen, E. (2001). Are entrenched characters developmentally constrained? Creating biramous limbs in an insect. *Evolution and Development* **3**, 424–431.

Dworkin, I., Palsson, A., Birdsall, K., and Gibson, G. (2003). Evidence that Egfr contributes genetic variation for photoreceptor determination in natural populations of *Drosophila melanogaster*. *Current Biology* **13**, 1888–1893.

Fiering, S., Whitelaw, E., and Martin, D. I. K. (2000). To be or not to be: The stochastic nature of enhancer action, *Bioessays* **22**, 381–387.

Franklin, I., and Lewontin, R. C. (1970). Is the gene the unit of selection? *Genetics*, **65**, 707–734.

Gardner, T. S., di Bernardo, D., Lorenz, D., and Collins, J. H. (2003). Inferring genetic networks and identifying compound mode of action via expression profiling. *Science* **301**, 102–105.

Gibson, G., and Hogness, D. (1996). Effects of polymorphism in the drosophila regulatory gene *Ultrabithorax* on homeotic stability. *Science* **271**, 200–203.

Gilbert, S. F., Opitz, J. M., and Raff, R. A. (1996). Resynthesizing evolutionary and developmental biology. *Developmental Biology* **173**, 357–372.

Govind, C., and Pearce, J. (1989). Critical period for determining claw asymmetry in developing lobsters. *Journal of Experimental Zoology* **249**, 31–35.

Grant, J. C. B. (1956). Critical period for determining claw asymmetry in developing lobsters. In *Atlas of Anatomy, by Regions*. 4th ed. Baltimore: Williams and Wilkins.

Hall, B. K. (2003). Descent with modification: The unity underlying homology and homoplasy as seen through an analysis of development and evolution. *Biological Reviews* **78**, 409–433.

Houchmandzadeh, B., Wieschaus, E., and Leibler, S. (2002). Establishment of developmental precision and proportions in the early *Drosophila* embryo. *Nature* **415**, 798–802.

Hultmark, D. (1994). Ancient relationships. *Nature* **367**, 116–117.

Jeong, H., Tombor, B., Albert, A., Ottvai, Z. N., and Barabasi, A. L. (2000). The large scale organization of metabolic networks. *Nature* **407**, 651–654.

Jeong, H., Masson, S. P., Barabasi, A.-L., and Oltvai, Z. N. (2001). Lethality and centrality in protein networks. *Nature* **411**, 41–42.

Kacser, H. (1957). Some physico-chemical aspects of biological organisation (Appendix). In *The Strategy of the Genes.* (C. H. Waddington, ed.), pp. 191–249. London: George Allen and Unwin.

Kacser, H., and Burns, J. A. (1981). The molecular basis of dominance. *Genetics* **97**, 639–666.

Lyon, M. (1961). Gene action in the X-chromosome of the mouse (*Mus musculus* L.). *Nature* **190**, 372–373.

Morgan, T. H. (1907). *Experimental Zoology.* London: Macmillan.

Muller, G. B., and Newman, S. A. (2003). Origination of organismal form: The forgotten cause in evolutionary theory. In *Origination of Organismal Form* (G. B. Muller and S. A. Newman, eds.), pp. 3–10. Cambridge, MA: The MIT Press.

Newman, S. A. (1992). Generic physical mechanisms of morphogenesis and pattern formation as determinants in the evolution of multicellular organization. *Journal of Biosciences* **17**, 193–215.

Newman, S. A. (2003). From physics to development: The evolution of morphogenetic mechanisms. In *Origination of Organismal Form* (G. B. Muller and S. A. Newman, eds.), pp. 221–239. Cambridge, MA: The MIT Press.

Newman, S. A., and Muller, G. B. (2001). Epigenetic mechanisms of character origination. In *The Character Concept in Evolutionary Biology* (G. P. Wagner, ed.), pp. 559–579. San Diego, CA: Academic Press.

Palmer, A. R., Strobeck, C., and Chippindale, A. K. (1993). Bilateral variation and the evolutionary origin of macroscopic asymmetries. *Genetica* **89**, 201–218.

Roth, S., Jordan, P., and Karess, R. (1999). Binuclear *Drosophila* oocytes: Consequences and implications for dorsal-ventral patterning in oogenesis and embryogenesis. *Development* **126**, 927–934.

Salazar-Ciudad, I., Newman, S. A., and Solé, R. (2001a). Phenotypic and dynamical transitions in model genetic networks:1. Emergence of patterns and genotype–phenotype relationships. *Evolution and Development* **3**, 84–94.

Salazar-Ciudad, I., Solé, R., and Newman, S. A. (2001b). Phenotypic and dynamical transitions in model genetic networks:2. Application to the mechanism of segmentation evolution. *Evolution and Development* **3**, 95–103.

Smith, J. M., and Sondhi, K. C. (1960). The genetics of a pattern. *Genetics* **45**, 1039–1050.

Sucena, E., Delon, I., Jones, I., Payre, F., and Stern, D. L. (2003). Regulatory evolution of shaven baby/ovo underlies multiple cases of morphological parallelism. *Nature* **424**, 935–938.

Technau, U., Cramer von Laue, C., Rentzch, F., Haobmayer, B., and Bode, H. R. (2000). Parameters of self-organization in Hydra aggregates. *Proceedings of the National Academy of Science USA* **97**, 12127–12131

Von Dassow, G., Meir, E., Munro, E. M., and Odell, G. M. (2000). The segment polarity network is a robust developmental module. *Nature* **406**, 188–192.

Waddington, C. H. (1952). Selection of the genetic basis for an acquired character. *Nature* **169**, 278.

Waddington, C. H. (1953). Genetic assimilation of an acquired character. *Evolution* **7**, 118–126.

West-Eberhard, M. J. (2003). *Developmental Plasticity and Evolution.* New York: Oxford University Press.

Whyte, L. L. (1965). *Internal Factors in Evolution.* London: Tavistock.

Wilkins, A. S. (2002). *The Evolution of Developmental Pathways.* Sunderland, MA: Sinauer Associates.

Wright, S. (1948). On the roles of directed and random changes in gene frequency in the genetics of populations. *Evolution* **2**, 279–295.

Yampolsky, L. Y., and Stoltzfus, A. (2001). Bias in the introduction of variation as an orienting factor in evolution. *Evolution and Development* **3**, 73–83.

CHAPTER 8

Canalization, Cryptic Variation, and Developmental Buffering: A Critical Examination and Analytical Perspective

IAN DWORKIN

Department of Genetics, North Carolina State University, Raleigh, North Carolina, USA

Variation

INTRODUCTION

In the folklore of evolutionary biology, one of the great wedges that occurred between the advocates of the Darwinian modern synthesis and other evolutionists concerned the fundamental issue of (heritable) variation. According to Darwin (1859), the variation that natural selection acted upon was, in general, quantitative. Bateson (1894) distinguished between continuous, meristic, and discontinuous variation and on the whole thought that the latter categories of variation were the targets of evolutionary forces. These opposing views diverged further during the development of population and quantitative genetics, where a fundamental assumption for most theoretical work (and statistical models) was that a very large number of loci, each with small (additive) effects, was responsible for trait expression. From this work, several models for the maintenance of genetic variation developed, such as mutation–selection balance, balancing selection, and overdominance, among others (see Hartl and Clark, 1997; Roff, 1997; for reviews). However, Waddington (1952, 1953) suggested an alternative mechanism to explain the maintenance of some genetic variation and with it an alternative model for the evolutionary process known as "genetic assimilation." The model of genetic assimilation predicts that in the face of unusual environment conditions, phenotypes can be genetically "captured" by the process of natural selection, if strong selection occurs. Implicit to this evolutionary model was a trove of hidden (cryptic) genetic variation for the trait, which was not generally observed (without the appropriate environmental stimulus), and a buffering mechanism, referred to as canalization, which helped to "store" the genetic variation. When the buffering mechanism failed (de-canalization), the cryptic genetic variation was released for selection to act upon. In the initial formulation of the model, if selection on this novel phenotype was strong (and consistent) enough, the new trait could itself then become canalized and be produced without the environmental stimulus. However, in later derivations of the model, it has been suggested that the assimilation process may not in fact occur by the mechanism as suggested by Waddington and that selection may act to change the threshold of trait expression or rare alleles that affect phenotypic penetrance are coselected and are in fact responsible for the genetic assimilation (Stern, 1958; Bateman, 1959). This last category is known as the "Baldwin" effect. This chapter will not deal any further with the mechanisms behind genetic assimilation and instead will focus on assessing canalization and cryptic genetic variation (see Scharloo, 1991, for review of the genetic assimilation controversy).

Regardless of the concerns with the mechanistic explanation of genetic assimilation, the plausibility of the phenomena of genetic assimilation as well as the existence of cryptic genetic variation were established via some empirical experiments. Waddington (1952, 1953) demonstrated that traits that were invariant under most (normal) environmental circumstances could be sensitized so as to express phenotypic variation for these traits. The classic example of Waddington's was the use of a high-temperature "heat-shock" in *Drosophila*, which resulted in some flies having lost their wing cross-veins. Waddington demonstrated that the cross-veinless phenotype could be selected upon, suggesting considerable hidden (cryptic) genetic variation for this trait (Waddington, 1952, 1953). Later work demonstrated that these observations could be extended to other environmental perturbations and traits (Waddington, 1956; Bateman, 1959) as well as to genetic perturbations (Rendel, 1959). However, all of these studies (and later ones) sufficiently demonstrated that the buffering mechanism (canalization) and the cryptic genetic variation being suppressed are intertwined (although this does not imply that the cryptic genetic variants are themselves responsible for the buffering).

I. A REVIEW OF THE REVIEWS

Given that a number of excellent reviews on canalization have appeared recently, I will provide a short overview of these papers, before proceeding with this chapter. Scharloo (1991) reviews the classic canalization literature spanning the 1950s through the 1980s. As I will discuss at length, there is little or no consensus as to the definitional issues regarding canalization and the apparently related biological processes of phenotypic plasticity and developmental stability. Two recent reviews (Debat and David, 2002; Nijhout and Davidowitz, 2003) discuss many of the definitional concerns, and historical constraints with the literature regarding canalization. Within the larger context of buffering, the mechanisms governing canalization and its evolution, De Visser *et al.* (2003) is quite comprehensive. For an alternative introduction to the principles of buffering, see Hartman *et al.* (2001).

In terms of exploring plausible mechanisms for genotype to phenotype mapping and canalization, Rutherford (2000) and Nijhout and Davidowitz (2003) both offer useful ideas to consider. Gibson and Wagner (2000) explore some of the methods for the detection of canalization and provide a general overview of the canalization literature. One of the few papers to focus almost exclusively on the evolutionary forces that may be responsible for canalization is the one by Meiklejohn and Hartl (2002). Several reviews focused more on the cryptic genetic variation part of the relationship (Wagner *et al.*, 1999; McLaren, 1999; Gottleib *et al.*, 2002).

For the remainder of this chapter, I will focus on some of the experimental and inferential issues involved with the empirical study of canalization and cryptic variation. I will not discuss at any length some of the theoretical work that has been done with respect to canalization, although this does not imply that I feel it should be in any way neglected. I simply do not have sufficient room to deal with this issue.

II. EMPIRICAL CONCERNS FOR THE STUDY OF CANALIZATION

Before delving into the statistical concerns involved with making inferences with regard to canalization, I feel it is important to address some issues about experimental design. In the next section, I will discuss the alternative definitions of canalization and the consequential effects on inference. However, regardless of the definition or metric used for canalization, there are certain experimental concerns to be addressed. As with so many subjects of inquiry, there is no single optimal design for experiments of canalization. However, there are a few specific areas of concern, which can have an impact on the interpretation of results garnered from the experiment. The three major concerns are as follows:

- The amount of genetic variation must be controlled between lines/populations.
- The need for multiple, independent samples (across genotypes, not individuals)
- Genetic background must be controlled for comparisons between treatments.

A. The Amount of Genetic Variation must be Controlled between Lines/Populations

It must always be kept at the forefront of the mind of researchers that canalization is not a property of a species or population, but of a genotype. For most species, each individual has a unique genotype, which means testing questions of canalization can be extremely difficult. If a population is studied under several environments, the results may have little to do with canalization, but the differential response of different genotypes (i.e., different genotypes in the population have differential fitness across environments). Thus it is important to employ some method to control for within-line genetic variation.

In most genetically tractable systems and in those species that reproduce clonally, it is possible to get closely related individuals via inbreeding, or controlled genetic crosses (i.e., via chromosome extraction procedures in *Drosophila*). For other species in which some genetic manipulation is possible, basic crossing schemes can be used to control for at least some of the genetic effects. This is of fundamental importance, so that not only can the same genotype be examined under multiple environments, but reasonable sampling of a given genotype can be performed.

B. The Need for Multiple, Independent Samples (Across Genotypes, Not Individuals)

This is one of the more common oversights in studies of canalization. Given that each line represents a single genetic sample, measuring multiple individuals within a line essentially increases the sampling (providing a better estimate) of a single measure. Thus, if two lines are used, the effective sample size can be considered two (not $2 * n$, where n = number of individuals sampled within line). Debat *et al.* (2000) used this approach (only two lines) when examining the patterns of within-individual variation and canalization. For some situations, this may be sufficient; however, it must generally be regarded with caution. For a feasible experiment, a balance must be found between the number of individuals sampled within a line (to get precise estimates) and number of independent lines to use (for statistical power). These issues are too complex to consider here, but needless to say depend highly on what central question is being addressed.

On a further note of caution, when choosing the multiple lines to use, it is important to consider the "independence" of the lines. In this context, independent means genetically unrelated. Otherwise, there can be "pseudoreplication" of the data, where the same genotype is being resampled, even though they are being treated as statistically independent. There are several studies that examine patterns of variation that may suffer from artifactual effects because of this type of problem (Woods *et al.*, 1999; Bourget, 2000). However, whether this approach leads to a "fatal" flaw is unclear and must be investigated further. It is also unclear how "unrelated" the individuals must be. Independent sampling of wild-caught individuals (even from the same population) will likely be sufficient for outcrossing species. However, this too must be empirically addressed.

C. Genetic Background must be Controlled for Comparisons between Treatments

This issue is pertinent to studies of genetic canalization where different mutants (or chromosomes) are being compared for their effect on canalization. For genetically tractable systems, it may be possible to have controlled genotypes where a given chromosome is crossed into many different genetic backgrounds. However, an approach that, although requiring some initial time and energy, is generally acceptable is to introgress the mutation via repeated backcrossing to an inbred line into the genetic backgrounds of interest. Several studies have demonstrated the efficacy of this approach in a number of model systems (Gibson and van Helden, 1997; Alonso-Blanco *et al.*, 1998; Polaczyk *et al.*, 1998; Gibson *et al.*, 1999; Bolivar *et al.*, 2001, Sershen *et al.*, 2002; Dworkin, 2005). In a species such as *D. melanogaster*, 10 generations of backcrossing the mutant (or other marker) leads to ~90% of the genome being replaced. The exact amount will depend largely on the rate of recombination in that region of the chromosome. This will allow ideal comparison between mutant and wild-type conspecifics (see Gibson and van Helden, 1997, for details on the procedure), who are essentially identical except for the allele of study. If genetic background is not controlled, then it will be unclear if the observed effect results from the mutation in question or some other loci in the genetic background of the mutant.

III. DEFINITIONS OF CANALIZATION

If one is to study a subject, it would be expected that there be a clear definition of the object of study. However, although there have been numerous studies on canalization, there is no agreement as to what defines it. Several studies have outlined the historical and logical reasons for the different definitions (Debat and David, 2001; Nijhout and Davidowitz, 2003). After providing a brief overview of these definitions, I will outline some empirical metrics for canalization given the different definitions.

Broadly speaking, there are two main definitions, which I call the reaction norm of the mean (RxNM) and the variation approach. Why should we care which definition of canalization we are using? As we will see, among other disparities, the different definitions lead to different metrics for the study of canalization. Before we continue with this discussion, several issues must be made clear about some definitions that we are using. First, when I speak of "line" effects, I am assuming that each individual within the line is genetically identical (or very close relatives), while different lines are sufficiently diverged from one another that they are "independent" in the statistical sense. Furthermore, the use of the term *environment* throughout this chapter refers not only to typical external macroenvironmental variables (such as temperature, salinity, or density), but also can refer to an internal environmental variable such as a mutation (i.e., the two environments could be wild type and mutant). The use of this common term *environment* to refer to both genetic and external causes of effects is not meant to imply a common mechanism (*sensu* Wagner *et al.*, 1997; Ancel and Fontana, 2000; Meiklejohn and Hartl, 2002), but is simply used as a matter of convenience.

Although I hope the reader will clearly recognize the distinction between the following definitions, at this point it is unclear which definition is in fact the correct one. Thus I suggest that, for future studies of canalization, both definitions (and derived metrics) be employed to address questions of canalization (and the relationship between the metrics).

IV. REACTION NORM OF THE MEAN (RxNM) DEFINITION OF CANALIZATION

Figure 8-1 (A, B) helps illustrate the first definition of canalization and its relationship to phenotypic plasticity. Figure 8-1A illustrates a "typical" reaction norm for a trait under study. Each line represents a genetic "line" (genotype) for which we have sampled multiple individuals in each of two environments (E1 and E2). Although we do not need to assume that there is no microenvironmental variance within E1 or E2, we do need to assume that it is equal. In Figure 8-1A, lines A, C, and E all have been observed with different means in each environment. Furthermore, these particular lines display what is commonly known as crossing of line means. However, lines B and D (Figure 8-1A) show virtually no difference in their trait means across environments (although the line means differ from one another). According to the RxNM definition, canalization is the opposite of phenotypic plasticity (Nijhout and Davidowitz, 2003). Lines B and D (Figure 8-1A) would be considered canalized with respect to environments E1 and E2, i.e., there is no environmental effect on trait expression.

How do we infer canalization for the RxNM approach? The majority of studies infer canalization by the decanalization of a system via an environmental (genetic or exogenous) stress that causes perturbation to normal trait expression. If we

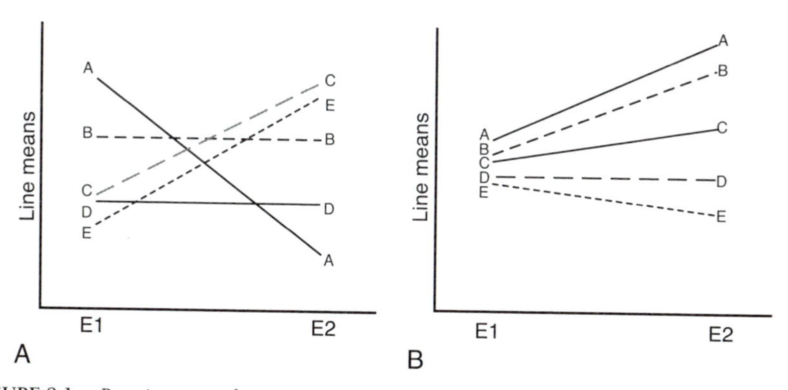

FIGURE 8-1. Reaction norm demonstrating phenotypic plasticity and canalization as opposite characteristics of the same phenomenon. (A) Lines A, C, and E all display classic phenotypic plasticity (change in trait mean across environments), while lines B and D show a form of canalization (no change in trait values). (B) The test of canalization for this metric is a change in the between-line (genetic) variation from one environment to another. E1 and E2 represent the two environments.

observe a change in phenotypic variation, we can ask whether we infer the release cryptic genetic variation for the trait. In the case of inferring canalization by decanalization, Figure 8-1B illustrates an idealized example of what we are seeking. In this example, lines A–E show relatively little between-line variation in E1. However, when individuals from these same lines are "exposed" to E2 (the stressful environment), we see a significant increase in the between-line variation (because of a release of cryptic genetic variation for trait expression). Fundamentally, when we are using the RxNM definition of canalization, this is what we are looking for (details on the process of statistical inference will be discussed in a following section). By this definition, the more canalized a line is, the less its mean should change across environments, and this in fact can become our definition of canalization for each line (denoted *C*). Specifically, we are interested in the unsigned deviation of the line means in environment E1 and E2 or

$$Cj = |\mu_{j,\,E1} - \mu_{j,\,E2}|$$

where *Cj* is the measure of canalization for line *j*, $\mu_{j,\,E1}$ is the line mean of *j* in E1.

V. THE VARIATION APPROACH TO CANALIZATION

The other major definition of canalization, which I simply define as the "variation" metric, is fundamentally concerned not with how the line mean changes across environments, but how the measure of variation *within* line changes (Stearns and Kawecki, 1994). This idea is illustrated in Figure 8-2(A, B). In Figure 8-2A, we observe the measures for two genetic lines (solid and dashed), where we have "error" bars, which represent some measure of within-line variation. The dashed line shows no difference in within-line variation in E1 or E2 (again the stressful environment). However, the solid line shows a significant increase in within-line variation in E2. We can illustrate this as a "reaction norm" (Figure 8-2B). However, it is important to note that the Y axis no longer represents line means, but within-line variation. The line means have been illustrated as invariant in Figure 8-2A for purposes of simplicity and do not necessarily reflect observed biological pattern.

Unlike the RxNM approach, for the within-line variation definition of canalization, the appropriate metric of canalization is less clear. For instance, if a trait is measured in a number of genetically distinct lines in a single environment, we may observe that some lines show more within-line variation than others (Figure 8-2A, environment E2). It could be argued that the lines that show lower levels of within-line variation are better canalized than other lines, even though all traits were measured in a single environment (even though this is far from the traditional definition of canalization and has been given other names; see Debat and David, 2001,

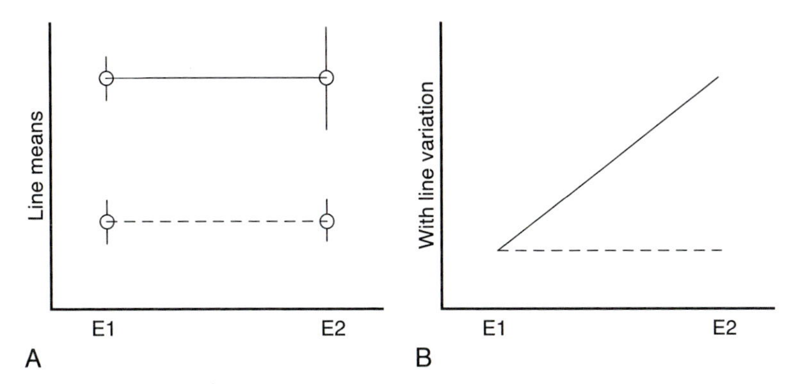

FIGURE 8-2. Variation approach to canalization. (A) Traditional reaction norm graph for two lines, demonstrating that, although the trait means do not change across environments, their within-line variance does, with the dashed line apparently being canalized (no change in within-line variation), while the solid line shows a significant increase in variation in E2 compared with E1. (B) Another view of the same phenomenon, but using a measure of within-line variation on the Y axis.

for a review of these definitions). Alternatively we can use an analogous approach to that for the RxNM, where we are not so much interested in the levels of within-line variation in a given environment, but how it changes across environments. The metric for this approach would be:

$$Vj = |CV_{j,\,E1} - CV_{j,\,E2}|$$

Where Vj is the measure of canalization for line j, $CV_{j,\,E1}$ is the coefficient of variation (or some other measure of within-line variation; see section on statistical inference) of j in E1.

VI. PARTITIONING SOURCES OF VARIATION

In an effort to make the distinctions between these definitions of canalization clearer, it may help to consider briefly the different sources of variation. Assume that we have a number of genetically distinct isogenic lines, where individuals within a line (who are all genetically identical) are all measured in a common set of environments. I will not demonstrate the statistical procedures of how to partition the variation (see Falconer and Mackay, 1996; Lynch and Walsh, 1998; and Palmer and Strobeck, 2003, for issues relating to V_{WI}). However, it is important to distinguish between the variation that is actually being measured and the sources of variation that are inferred (environmental or genetic).

A. Variation within Individual (V_{WI})

This is most commonly measured either on a serially repeating trait or more often using the two sides of a bilaterally symmetrical individual (i.e., measuring wing length on both the left and the right sides of a fly). The deviation of left versus right sides is a measure of asymmetry (often fluctuating asymmetry), which has a long and distinguished literature dealt with in Chapter 10 of this volume. There are a number of important biological issues that should be considered with respect to V_{WI}. First, there is no genetic variation (unless the somatic mutation rate is high), and second, there is no environmental variation (but see Nijhout and Davidowitz, 2003). It is generally assumed that V_{WI} is a proxy for the developmental noise of the organism (i.e., random developmental differences between left and right sides of the individual that cannot be controlled for).

B. Variation between Individuals, within Genotype (V_{BI})

The *CVj* discussed in the previous section is one particular metric of this. Some authors have suggested that this is an equivalent metric to V_{WI} if all individuals are raised in a common environment, with little or no uncontrolled variation, and there is some evidence to support this (Clarke, 1998).

C. Between-Line (Genetic) Variation (V_G)

Assuming all individuals (of all lines) have been raised in a common environments, this component represents the between-line variation. This is essentially what we are interested in for the RxNM metric of canalization (and how it changes across environments) when we infer canalization by its breakdown.

VII. INFERRING CANALIZATION: WHEN IS A TRAIT CANALIZED?

Given that the vast majority of studies of canalization have utilized (implicitly) the RxNM approach, we will spend some time dealing with inferences of canalization for the metric derived from it. When we perturb a system to cause decanalization (and infer canalization from this), we are actually measuring two related processes. First is the buffering component (canalization) of the system/trait. However, if there were no available "noise" in the system (i.e., genetic or environmental variation), then decanalization could not be observed (Nijhout and Davidowitz, 2003). Thus we are often measuring the release of the cryptic genetic variation (in the perturbed state) and inferring canalization (in the unperturbed state).

Let us begin with a hypothetical trait that shows "ideal" canalization (Figure 8-3). As can be seen in Figure 8-3, we have measured a number (four) of lines, in four environments (of which we can measure the magnitude of environmental effects). The between-line variation is greater in E1 and E4 than in either E2 or E3. This is an "ideal" trait for studies of canalization because we have environments in which we know *a priori* the environmental region for which the trait is in the "zone of canalization" where it varies little. The environmental effects of E1 and E4 are sufficient to perturb trait expression (decanalization), and we observe an increase in between-line variation in E1 or E4 (i.e., a release of cryptic genetic variation) relative to E2 or E3. If we observed this pattern of effects, then it would be relatively easy to make an argument about the canalization of the trait.

However, in the majority of studies dealing with canalization, only two environments are used. Usually one environment is considered "normal," while the second environment is "stressful." If *a priori* there were no information as to how a given environment affected the trait, then it would be possible that the experiment could run into some difficulties. If by chance the two environments compared were E1 and E2, E1 and E3, E2 and E4, or E3 and E4, then we would observe the increase in between-line variance and could make an inference about canalization from there. However, in comparisons of either E2 and E3 or E1 and E4, we would not observe any significant difference across environments for the between-line variance. From this, it would be possible to make two conclusions: The trait is canalized or it is not. This is not very useful information. With the comparison between E2 and E3, we are in fact observing a trait in its zone of canalization, but this is obviously not the case for E1 and E4. The reverse argument also holds.

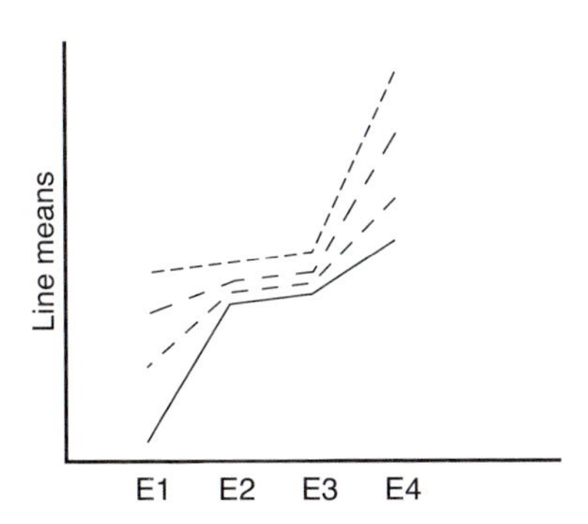

FIGURE 8-3. An idealized example of a canalized trait. For this trait, there is apparently a "zone of canalization" between environments E2 and E3, while the two extreme environments seem to cause a perturbation in trait expression resulting in an increase of between-line variation (a release of cryptic genetic variation). We assume that the environments E1 and E4 form some type of gradient of effect, even though we are measuring the traits in "discrete" environments.

Therefore, if only two environments are going to be used, it is important that there is prior information to inform which environments are chosen.

Although two environments are sufficient to make an inference about canalization, as wishfully concluded from the preceding argument, it is valuable to include more than just two environments. For external environmental effects (temperature, nutrients, density), this is usually a possibility. However, when the different environments being examined are in fact genetic in nature (i.e., different mutations), this can require additional introgressions (see the preceding text) that can be considerable work. As well, it is not always clear what sort of "gradient" of effects the mutations have. For future studies of canalization, it will be important to examine the shapes of the reaction norms within and outside of the "zone of canalization," and we recommend this type of approach for those interested.

VIII. WHAT ARE THE APPROPRIATE TESTS FOR MAKING STATISTICAL INFERENCES ABOUT CANALIZATION?

Clearly, given the multiple definitions of canalization, we are confronted by at least as many possible metrics. However, even within a given definition of canalization, it is not clear what is the best possible method for statistical inference. Although another chapter in this volume (Chapter 2) deals specifically with the statistics of variation, I will highlight a few approaches that have been used to address questions of canalization. We should point out at the outset that we are not in fact advocating any particular method, and a thorough analysis of the comparative properties of the various tests of variation is still required.

If we assume that canalization is best measured with respect to within-line (genotypic) variation, in a given environment, then perhaps an initial approach would be to use the coefficient of variation (*CV*) as a measure. The coefficient of variation is a standardized, dimensionless quantity measured by dividing the sample standard deviation by the sample mean (σ/μ, often multiplied by 100%). This approach has been used by Stearns and Kawecki (1994). However, there are certain issues with respect to the use of *CV*. Unlike an estimate of the mean for a trait, the sample size required for an accurate estimate can be quite large, and for small sample sizes, it is biased (Lande, 1977; Sokal and Braumann, 1980; Sokal and Rohlf, 1998), although there are corrections for this metric (Sokal and Braumann, 1980). If we simply compare individuals of a single genotype in two environments, then there are a number of options for how to make a statistical inference (and to date it is not clear which option is most appropriate for a given situation). Lewontin (1966) advocated computing an *F*-test statistic on the sample standard deviations computed from the log-transformed data, or on the CV^2 when

$CV < 30\%$. Sokal and Braumann (1980) suggested the use of a *t*-test using the standard errors of the corrected *CV*s. Zar (1999) advocates the use of an asymptotic test statistic (Miller, 1991). Finally, Schultz (1985) advocated the median form of the Levene's test. Thus we are left with a rather large set of possible tests to do. Clearly, it is not appropriate to perform all of these tests.

Schultz (1985) compared a number of these tests via simulation and advocated the median form of the Levene's statistic. However, an examination of the results from these simulations clearly shows that the median form of Levene's statistic can be overly conservative for two groups, given normally distributed data. In this example (i.e., normally distributed *CV*s), the *F*-test statistic on the CV^2 appears to be the most appropriate. Van Valen (1978) has argued that even when the data appears normal, one has to be very cautious about the *F* test because of its sensitivities to nonnormal distributions. However, until further work has been done to determine which of these tests are most appropriate, I would suggest referring to the results in Schultz (1985) to determine the best recourse depending upon the data at hand. In my experience, for the majority of cases, the different tests do not make wildly different inferences, unless the data is far from normal.

If trying to decide on the appropriate test for a two-group situation seems baroque, then for $k > 2$ group case, it can seem downright daunting. The majority of these tests are labeled as tests for the homogeneity of variances (one of the assumptions of analysis of variance [ANOVA] models). Sokal and Braumann (1980) suggest using either the Bartlett's or Levene's test on the log-transformed variates. An alternative is an extension of the asymptotic test for the equality of *CV*s discussed earlier (Feltz and Miller, 1996). Schultz (1985) demonstrates that the median form of Levene's test is again the most appropriate test for the $k = 3$ example, as compared with Bartlett's test, which is quite sensitive to nonnormality of the data (Zar, 1999). Unfortunately, to date there has been no comparison of each of these methods. One advantage of using the Levene's statistic (either the median or the mean form) is that as a metric it is easily cast in the format of an ANOVA, which allows complex models with interaction terms to be explored. For studies of canalization, where there is likely a line component and an environmental component, being able to examine this interaction term can be quite useful, if not imperative (depending upon the definition of canalization). Given that we infer canalization through a release in cryptic genetic variation, we are fundamentally interested in changes in the genetic variation. Within the context of an ANOVA, there has been some work that has specifically explored this issue (Aitkin, 1987; Foulley *et al.*, 1994; Sancristobal-Gaudy *et al.*, 1998; Sorensen and Waagepetersen, 2003). It is important to the field that all of the methods described in the preceding text are explored and compared to determine the most appropriate methods for future empirical studies, although it is likely that different methods will suit different designs.

IX. IN THE INTERIM ...

However, until the properties of the various tests have been more fully explored, I provisionally suggest a course of action. Given a study with k lines, and j environments (where k and $j > 1$), I suggest that the Levene's statistic on the log-transformed data be used, for both its relative simplicity and how readily it can be used for complex models. Even though the distribution of Levene's statistic will be a truncated normal, given that it is the unsigned deviation, it generally has the appropriate type I error when comparing groups with no differences (Schultz, 1985; Palmer and Strobeck, 1992; Palmer, 1994). Furthermore, permutation tests can be performed on the data if there are specific concerns about the distributions of Levene's statistic.

$$LS_{ijk} = |\mathrm{Log}(x_{ijk}) - E[\mathrm{Log}(x_{jk})]|$$

Where LS_{ijk} is the Levene's statistic for individual i, for line j in environment k. $E[\mathrm{Log}(x_{jk})]$ is the mean of the log-transformed data for all individuals for line j in environment k. Schultz (1985) suggests that the mean approach also tends to be anticonservative when the data is not normally distributed and suggests the use of the median $\{Md[\mathrm{Log}(x_{jk})]\}$ as an alternative measure of central tendency. It is worth computing and comparing, although they tend to give similar answers under many empirical circumstances. If Levene's statistic is used, then the measure of canalization across environments becomes (by substituting LS for CV):

$$Vj = |LS_{j,\,\mathrm{E1}} - LS_{j,\,\mathrm{E2}}|$$

As discussed earlier, it is not clear for this definition what the correct measure of canalization is. Therefore, I suggest following the ANOVA framework as per the RxNM approach described in the following text (using LS as the metric as opposed to the trait value for each individual).

X. ANALYSIS FOR THE RxNM APPROACH

When canalization is viewed as the "opposite" to phenotypic plasticity, then the framework for analysis is somewhat clearer. Given k lines (L) and j environments (E), we can start out within the framework of an ANOVA. With some variant of the model

$$Y_{ijk} = \mu + E + L + E \times L + \varepsilon$$

As with the analysis of phenotypic plasticity, we can begin by examining the significance of these model terms. Evidence for genetic variation for "plasticity" can be inferred if there is a significant $E \times L$ term for the model. If this term is not significant, but both E and L are, then there is evidence for plasticity of the trait and genetic variation for the trait itself, but not for genetic variation for plasticity of the trait (Figure 8-4).

If there is a significant $E \times L$ term, then we need to determine whether there is in fact evidence for canalization. It is important to recognize that a significant $E \times L$ term can arise from different processes (Robertson, 1959; Gibson and van Helden, 1997; Lynch and Walsh, 1998; Gibson and Wagner, 2000), but not all are evidence for canalization. Specifically, we want to separate out cases where the $E \times L$ term arose because of significant line crossing (change in relative ranks) across environment (Figure 8-1A), as opposed to a change in the scaling of the line means across environments (Figure 8-1B). For this latter case (i.e., Figure 8-1B), this will result in a perfect correlation across environments for the line means. For studies on phenotypic plasticity, the $E \times L$ term is usually partitioned into the two components (Robertson, 1959; see Lynch and Walsh, 1998, for review). However, knowing what proportion of the variation results from line crossing versus scaling effects is not itself of interest for canalization. We are specifically interested in whether this scaling effect (i.e., the increase in between-line variance) is significant.

One line of evidence that is consistent with canalization of a trait is a release of cryptic genetic variation in the "stressful" environment. This is inferred by the increase in between-line variation in the stressful environment (Gibson and van Helden, 1997; Gibson and Wagner, 2000). Thus, *a priori,* we recognize the need for some supplementary tests to determine whether there is a release of cryptic genetic variation. Here we return to many of the same statistical issues (and variety of tests) that we used to examine patterns of variation in the preceding section.

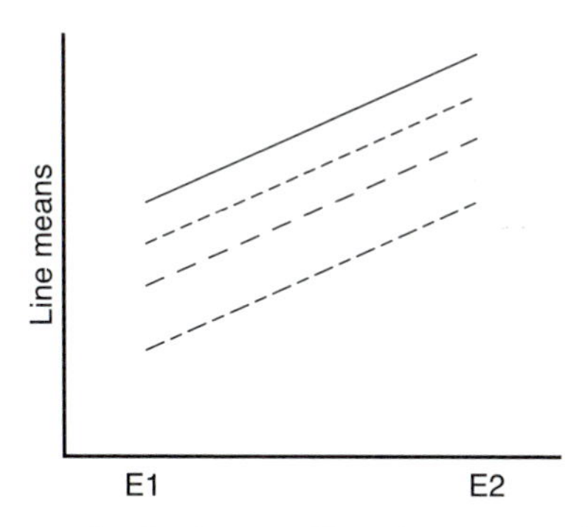

FIGURE 8-4. A reaction norm plot showing no evidence for a genotype by environment (genotype–environment interactions [GEI] or $E \times L$ effect in the model in the text) contribution. Although there is evidence for line (genetic) effects because the line means differ, and for environmental effects (because the mean obviously differs across environments), all of the slopes are identical.

Gibson and van Helden (1997) used the *F*-test approach, by comparing phenotypic variance across environments, as well as comparing the variances from between-line means (an estimate of genetic variation, but see Lynch and Walsh, 1998, for concerns with this approach). If using this approach, then we suggest the use of the CV^2 as opposed to the variances to cast the *F*-tests (Lewontin, 1966; Schultz, 1985). If this test is significant, then there is evidence for an increase in phenotypic variation in one environment over the other. The *F*-test used in this context suffers from all of the same inadequacies as discussed earlier. Demonstrating an increase in phenotypic variation is not sufficient to conclude canalization for the trait. It is the test of the variation of between-line means (again using CV^2) that will (if significant) infer a release in cryptic genetic variation (and thus provide evidence for canalization). It should be pointed out that, unless a large number of lines are used, this criterion may be hard to meet as formal significance will be unlikely (even if there is indeed cryptic genetic variation). Therefore, one of the other tests described in the preceding section may in fact be preferred (such as Levene's test). Of course all the tests can be employed, but formal significance must be adjusted (by Bonferroni or other such approaches) to control for multiple tests. In the interim, we again suggest the use of Levene's test to test for an increase in the between-line variation. In this case, the test is perhaps best framed as follows:

$$LS_{jk} = |\mathrm{Log}(\mu_{jk}) - E[\mathrm{Log}(\mu_k)]|$$

where LS_{jk} is the Levene's statistic for line j in environment k. $\mathrm{Log}(\mu_{jk})$ is the log-transformed line mean for j in environment k, and $E[\mathrm{Log}(\mu_k)]$ is the mean of the log-transformed line means in environment k. If this approach is used, and there are two environments (k = E1, E2), then a paired t test may in fact be the most logical test using the pairs ($LS_{j,\,\mathrm{E1}}$ and $LS_{j,\,\mathrm{E2}}$).

XI. THE ANALYSIS OF CRYPTIC GENETIC VARIATION

Implicit from the first studies of canalization was the realization that there was a significant amount of standing hidden (cryptic) genetic variation for traits. Waddington (1953) demonstrated that the wing venation phenocopies produced via a temperature heat shock could be selected upon to increase or decrease their frequencies (and ultimately the phenotypes could be expressed without the environmental stimulus). Of some consequence, Waddington (1953) demonstrated that some of the cryptic genetic variation for this trait was allelic to known developmental mutants (whose phenotype was being phenocopied). Subsequent work demonstrated that this was also the case for other phenotypes and genes

(Waddington, 1956; Gibson and Hogness, 1996). Given that there is evident abundance of cryptic genetic variation, we are left with an interest in the genetic architecture of cryptic genetic variation for traits. Several questions immediately arise from this.

Some of the standard questions pertaining to the genetic architecture of a trait include (see Mackay, 2001, for a complete list):

- How many genes are involved with trait expression?
- What is the distribution of the (magnitude of) effects of each of the genes?
- How do the genes interact with each other and the environment with respect to trait expression?
- What is the mutation rate of the genes involved with trait expression?

In some sense, these questions (and ones that derive from them) are addressed because they inform us of the evolutionary potential of the trait (evolvability).

However, with regards to cryptic genetic variation for a trait, there are a few more questions that we must ask. First and foremost, we are interested in whether the genetic architecture is in any way different than that for trait expression itself (i.e., is the genetic architecture of bristle number different than the genetic architecture for the canalization of bristle number?). I will provide some evidence here that in fact there is no evidence for difference in genetic architecture (or molecular polymorphisms within genes) to suggest this. Second, we are interested in whether all cryptic genetic variation (not just that caused by genetic perturbations) is caused by epistasis. If we can in fact examine the actual polymorphisms within genes that are responsible for the cryptic genetic variation, we can try to understand the evolutionary forces that have been maintaining them (i.e., do those regions of the gene seem to be under specific evolutionary forces, or are they evolving in a more or less neutral fashion, are they common or rare alleles?).

As evolutionary biologists, we are interested in cryptic genetic variation for a number of reasons. Does the existence of cryptic genetic variation allow for an increased rate of evolutionary response (i.e., can a trait evolve faster with the presence of cryptic genetic variation than without it)? Second, why do we not observe the effects of cryptic genetic variation under most environmentally relevant circumstances?

XII. MAPPING CRYPTIC GENETIC VARIANTS

Although the early studies demonstrated some of the genetic basis of cryptic genetic variation, if we are interested in the evolutionary dynamics of the alleles responsible for this phenotypic variation, we must extend our analysis. Gibson and Hogness (1996) did just this by following single-stranded conformational polymorphism (SSCP) marker frequencies in the *ultrabithorax* (*Ubx*) region in a

D. melanogaster population that was selected for increased sensitivity to ether-induced haltere to wing homeotic transformations. They observed that the ether exposure was correlated with a decrease in the amount of *Ubx* transcript found in the haltere imaginal disc. In addition, selection for the homeotic phenotype was itself correlated with an increase in certain markers, which were likely in linkage disequilibrium with polymorphisms under strong selection. This was the first such demonstration of intermediate frequency polymorphisms being subject to selection for such a "cryptic" trait.

However, the design used in this study did not allow a test of whether those polymorphisms scored were in fact responsible for the phenotypic variation observed. A recent study has overcome this difficulty using another trait that demonstrates cryptic genetic variation, namely photoreceptor determination in *D. melanogaster*. Normally the number of photoreceptors per ommatidia is invariant at eight. However, when the system is perturbed genetically, the number of photoreceptors can be altered. Via the introgression of the dominant *Egfr*^{E1} allele into a panel of isofemale lines, the effect of genetic background on this allele was studied (Polaczyk *et al.*, 1998). It was demonstrated that (1) there was a considerable amount of phenotypic variation for this trait and (2) a large portion of this variation was genetic in nature (74% of the variation was estimated to be genetic). This result suggests that a surprising amount of hidden cryptic genetic variation is available (and that there is a sufficient mechanism of canalization to buffer against the effects of this variation).

Based on these results, we asked whether we could in fact map (to the nucleotide level) the polymorphisms responsible for this cryptic genetic variation. Given that previous studies have suggested that some of the cryptic genetic variation for traits was allelic to known genes whose phenotypes were being phenocopied, we decided to investigate how natural genetic variation in *Egfr* itself modifies the effect of the *Egfr*^{E1} allele (Dworkin *et al.*, 2003). In this study, we observed significant association between several single nucleotide polymorphisms and phenotypic variation for eye roughness (a proxy for photoreceptor number) and replicated the most significant associations with an independent sample and statistical approach. Interestingly, we also found some evidence to suggest that there was mutation–selection balance acting on this trait–gene association. Thus this provided evidence that at least one source of cryptic genetic variation for a trait was in fact in the locus whose function was perturbed. Interestingly, we have found no evidence for common (pleiotropic) polymorphisms involved with variation for eye roughness and two other candidate traits known to be affected by *Egfr* function, namely wing shape (Palsson and Gibson, 2004; Dworkin *et al.*, 2005) and the spacing between the dorsal appendages on the egg shell (Goering and Gibson, 2005). Unlike photoreceptor determination, both of these traits display natural genetic variation without sensitization. Does this suggest that the genetic basis for cryptic genetic variants is somewhat different than for "normal" varying traits?

XIII. IS THE GENETIC ARCHITECTURE OF CRYPTIC GENETIC VARIATION DIFFERENT FROM THAT OF OTHER GENETIC VARIATION INVOLVED WITH TRAIT EXPRESSION?

To address this question, we have employed a reanalysis of two studies that have examined the association between molecular polymorphisms in a number of candidate genes and sternopleural bristle number in *D. melanogaster*. These studies are ideal for a number of reasons. First, environmental variables such as "stressful" temperatures (greater than 29° C) and mutations have been shown to increase the phenotypic variance of sternopleural bristle number (Beardmore, 1960; Moreno, 1994; Lyman and Mackay, 1998; Bubliy *et al.*, 2000; Indrasamy *et al.*, 2000; Dworkin, 2005b). There is substantial evidence that the increase in the variation has a genetic component (Lyman and Mackay, 1998; Robin *et al.*, 2002; Dworkin, 2003b), at least for cases where the perturbation is genetic in nature. The second useful quality for our purpose is that two of the preceding studies also examined patterns of association between molecular polymorphisms in candidate genes and bristle number (Long *et al.*, 1998; Robin *et al.*, 2002). The particular design used in these studies allows us to address specific questions about the genetic architecture with regards to cryptic genetic variation. For both of these studies, chromosomes derived from wild populations of *D. melanogaster* were extracted and placed into a common laboratory background. More importantly, the "wild" alleles were then introgressed into the common laboratory background, so that except for the gene region of interest (and flanking sequence), there was no uncontrolled genetic variation. Each of these "wild" alleles were genotyped for a number of molecular markers for the genes in question (*Delta* for Long *et al.* [1998] and *hairy* for Robin *et al.* [2002]). What makes these experiments particularly suitable to address the question of the genetic architecture of cryptic genetic variation (and whether it is different from trait architecture in any particular way) is that well-characterized mutations in the candidate genes (*Delta* and *hairy*) were also independently introgressed into the common laboratory background (*Samarkand*). The "wild" alleles (introgressed chromosomal segments) were tested by crossing to both the laboratory stock allele (*Sam*) as well as being crossed to the mutant of interest (*Delta* or *hairy*), which was also congenic with *Sam*. Thus the ideal comparison of "congenic" chromosomes differing only in the mutant or wild-type (*Sam*) allele of interest were both used to test for association of bristle number with the molecular polymorphisms. For these results, a simple linear model was employed; both sexes were analyzed separately (no effect with respect to sternopleural bristles) with the molecular polymorphism as the independent variable and line means as dependent (conducted for each polymorphism).

Figure 8-5 illustrates the results for the association test for sternopleural bristles with polymorphisms in *Delta*, for both the *Sam* chromosome and the *Delta* (*Dl*) mutation. The strength of the association is monitored by the –log(*p*) value (thus a more significant value shows a higher peak). There are two important features to point out for this figure. First, the overall shape of the profiles for both the normal (*Sam*) and sensitized (*Dl*) backgrounds are similar, with both showing the same significant peak around the marker ha_8_6. Second, the associations from the sensitization crosses are generally stronger. Although the results are not presented here, the same qualitative results are found for the *hairy* gene region. Thus this evidence suggests that the same polymorphisms are responsible for both the natural and "cryptic" genetic variation for sternopleural bristle number. It generally appears that sensitization only amplifies the effects of the variation. Therefore for discontinuous traits (such as photoreceptor number) it simply lowers the threshold for the genetic variation in the trait that is otherwise suppressed, but does not reveal a qualitatively different type of variation. These results are consistent with the general quantitative genetic models for discontinuous traits, where trait expression is affected by an underlying normal distribution of effects, but the presence of a

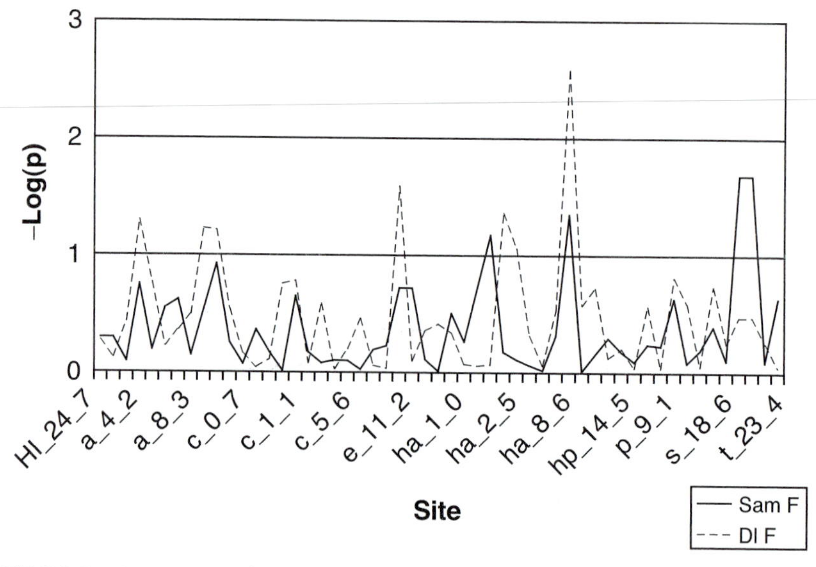

FIGURE 8-5. Association test for sternopleural bristle number with restriction site polymorphisms in the *Delta–Hairless* gene region. Y axis is the –log of the significance from the ANOVA. The strength of the association is stronger when the allele of interest is heterozygous over a mutation of *Delta* (*Dl F*) rather than over a "wild-type" *Samarkand* chromosome (that only differ genetically for the mutant allele). Although the magnitude of the effects are different, the two crosses show similar overall profiles.

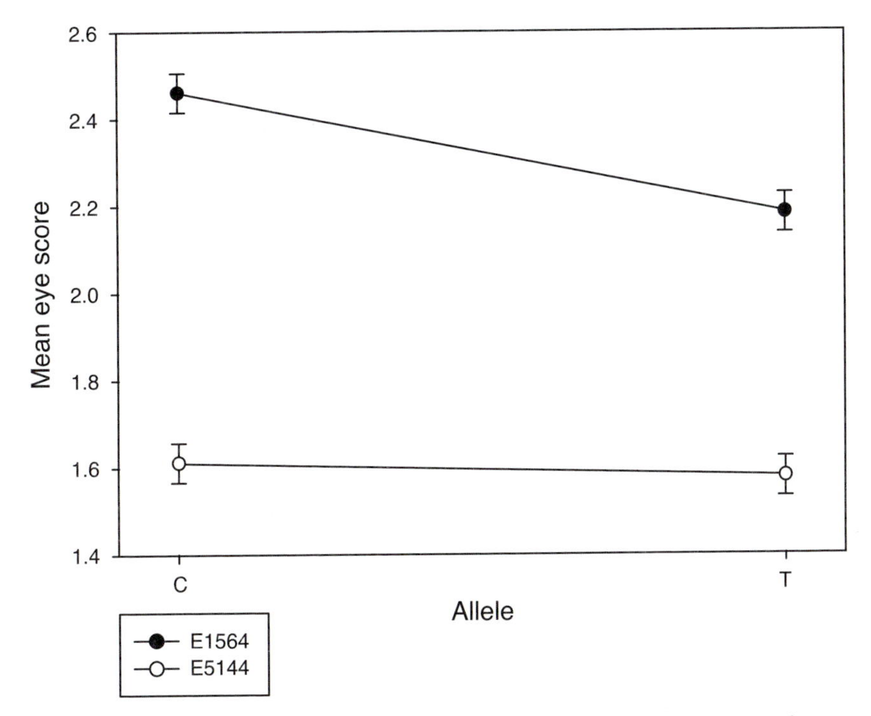

FIGURE 8-6. The effects of genetic background on single nucleotide polymorphism affects. The mean effect of substituting one allele for another depends heavily on genetic background. Not only is the mean effect on photoreceptor determination different for these two crosses, but there is evidence for a large scaling difference of the substitution of alleles.

threshold either hides the variation (for invariant traits) or sets up a situation where a quantitative gradient is interpreted digitally (dichotomous). If we return to the results for photoreceptor determination (Dworkin *et al.*, 2003), we can see some results consistent with this as well. Figure 8-6 displays a reaction norm profile across two different alleles for mean eye roughness for the two different crosses employed in this study. Not only does the cross utilizing line 1564 have a greater mean trait value than 5144, but as can be seen, there is evidence for a cross-X polymorphism effect, given that the slopes of the two lines are different from one another. This also points out another important feature. Even though crosses with line 5144 are sufficient to reveal some cryptic genetic variation for eye roughness, in general it produced much lower levels of between-line (genetic) variation. Thus the breakdown of the canalization is not an all or nothing proposition, and (at least for some traits, such as photoreceptor determination) there is a degree to its effects.

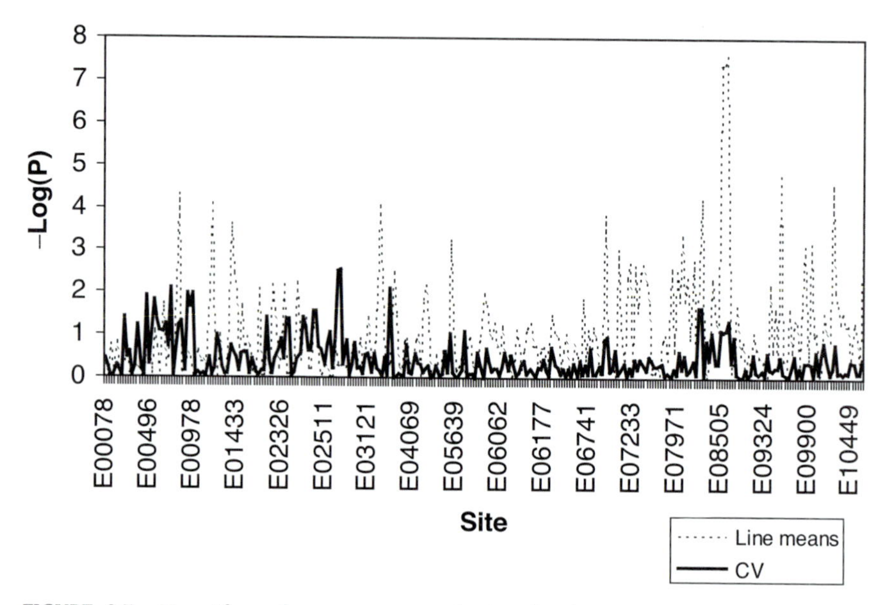

FIGURE 8-7. No evidence for a common mechanism for the two measures of canalization. Association between molecular polymorphisms in the *Egfr* gene in 210 inbred lines and line means or within-line coefficient of variation (*CV*). Not only is there no evidence for *Egfr* having an effect on the levels of within-line variation, but the profiles are quite different for each measure.

Finally, this sort of study can also be used to address some canalization-specific questions. For instance, what is the relationship between the two different measures of canalization (RxNM and variation)? If we assume that the degree of eye roughness in the crosses is inversely proportional to canalization (the greater the eye roughness score, the less canalization that line shows), then the association informs us about the RxNM approach to canalization. At the same time, we can use the coefficient of variation for each line to perform an association study for within-line variation. The results are displayed in Figure 8-7 (see Dworkin *et al.*, 2003, for more details about the primary analysis). As can be seen, the association using the RxNM metric of canalization shows several strong associations. However, the variation metric (*CV*) does not show any significant sites (after correction for multiple tests). Indeed, the profile of the associations is also quite distinct. This suggests that *Egfr* does not harbor natural genetic variation for the "variation" measure of canalization, unlike what we have observed for the RxNM approach. This is a relatively weak test of this idea. However, a recent study (Dworkin, 2005b) has explicitly addressed this question with sternopleural bristles and did not find evidence for a common genetic mechanism between these metrics of canalization.

XIV. NOW THAT I HAVE ALL OF THIS CRYPTIC GENETIC VARIATION, WHAT DO I DO WITH IT?

Considering all of the traits that have been observed to express cryptic genetic variation once sensitized, it is reasonable to ask whether or not this source of genetic variation has any evolutionary potential. To date this question has only been addressed in two empirical studies. Bubliy *et al.* (2000) investigated whether or not the increased phenotypic variation observed for sternopleural bristle number when flies are raised at high temperature (31° C) would increase the response to artificial selection as compared with 25° C controls. In this 10-generational experiment, they found no evidence for an increased response to selection. However, their lack of a "positive" result is in and of itself telling. Although the basic design of the experiment is sound, it was unlikely that the authors would have observed an effect (even if there is indeed a real effect). This is because of a conspiracy of factors that make observing a significant effect very difficult. First, as has been previously shown, the sampling variance on selection differentials is extremely large (Falconer and Mackay, 1996). For this study (Bubliy *et al.*, 2000), I estimated (from the available data) the sampling variance to be ~0.19 (at least as large as the difference in phenotypic variation between treatments). In relation to this, the average increase in phenotypic variation from temperature stress is relatively small (approximately a 12% increase in *CV* under the stressful temperature in Bubliy *et al.*, 2000). Thus to observe a significant effect if there is in fact a real difference could require a large increase in sample size to see a significant effect. Although the approach of Bubliy *et al.* (2000) was the correct one (in principle), other techniques over mass selection may be required. Alternatively, using a sensitization procedure (such as a mutation) that increases the genetic variation by a larger degree may also help to increase the probability of seeing an effect (if it exists).

However, there is an alternative approach that has addressed the same question of evolutionary potential of cryptic genetic variation. Lauder and Doebley (2002) have examined patterns of genetic variation between hybrids of teosinte (the maize progenitor species) and an inbred line of maize. They investigated a number of traits that are invariant in maize, teosinte, or both (but differ from maize to teosinte) and have shown that lines of teosinte harbor considerable cryptic genetic variation for traits (invariant in teosinte) that appear to those that have been selected upon during maize evolution. Interestingly, many of the quantitative trait locus (QTL) regions responsible for teosinte–maize differences were also found to also harbor cryptic genetic variation in teosinte. To my knowledge, this is the first study demonstrating a plausible evolutionary role for cryptic genetic variation. I hope that further work will be done to narrow down the QTL to candidate loci, for linkage disequilibrium (LD) mapping. If candidate polymorphisms are found that are responsible for some of the cryptic genetic variation, molecular population genetic analysis may help to reveal the evolutionary history of such alleles.

XV. THE FUTURE FOR STUDIES OF CANALIZATION

In this chapter, I have provided a quantitative framework for future studies of canalization. However, this does not mean that I am only advocating a traditional quantitative genetics framework for the study of canalization. In fact I suggest that a "quantitative developmental genetic" framework may be more revealing. What I suggest is that using the techniques of molecular and developmental genetics, in combination with the analytical framework of quantitative genetics, may provide clear and efficient routes to understanding canalization, especially from a mechanistic point of view. In the following text, I discuss a few studies that illustrate the potential for this approach.

Recent work has demonstrated that mutations in the *HSP90* gene reveal an extensive amount of phenotypic variation in a large number of seemingly developmentally unrelated (and otherwise invariant) traits in both *Drosophila* and *Arabidopsis* (Rutherford and Lindquist, 1998; Queltsch *et al.*, 2001). In addition, these studies demonstrated that this variation could be selected upon in the same manner as Waddington (1953, 1956). Although there have been numerous critiques about inferences made from this work (Wagner *et al.*, 1999, Meiklejohn and Hartl, 2002), it has provided stimuli for research into buffering mechanism in general. Perhaps one of the most important questions to come out of this work is whether or not there is evidence for "universal" mechanisms of canalization. Although this idea is contrary to previous evidence reporting little correlation between lines in their ability to buffer against the effects of different (but related) perturbations (Polaczyk *et al.*, 1998), it is an interesting idea. To date the work done with the *HSP90* gene has focused on qualitative and not quantitative traits. The latter would be a welcome addition both in terms of the types of traits studied and allowing for a more rigorous analytical procedure.

Along these lines, I would suggest that one type of experiment that could prove extremely valuable would be a mutagenesis, preferably using a tagged mutagen, where the traits studied would in fact be the canalization of several traits, i.e., examining patterns of within-individual (fluctuating asymmetry) and between-individual variation, as well as across environment variation (RxNM). If this sort of design is used for a number of independent traits, then it may help to distinguish whether the effects of genes such as *HSP90* are at the extreme of a distribution of genes with pleiotropic effects or whether they are in some way qualitatively different. Perhaps more importantly, this could help identify genes involved with buffering and determine if they are in some manner independent of trait expression.

There is at least one other approach that is worth considering. By quantitatively examining the spatial distribution of proteins for determinants of early embryonic polarity, Houchmandazdeh *et al.* (2002) observed a most interesting phenomenon. The maternally deposited *Bicoid* (*Bcd*) showed a considerable amount of phenotypic variation in its spatial distribution, specifically in the amount of protein observed at 50% of embryo length. A direct downstream target of *Bcd*, *Hunchback* (*Hb*) showed significantly less variation at this point along the embryo than *Bcd* (although a considerable amount of variation at other points along the embryo). This raises a number of questions. First, how is it that the noise in the *Bcd* signal is apparently filtered out with respect to *Hb* expression? Why is it that *Hb* only shows this effect at certain locations along the embryo? This type of study posits some suggestive ideas for the mechanism of buffering of this particular interaction, but we believe the implications for this study are much wider. First, it would be very interesting to have a handle on what sort of natural genetic variation (if any) is found for the canalization of *Hb* expression? Second, what are the phenotypic consequences of this buffering (or the lack there of)? How general a phenomenon is this? Neither this study nor the one of Rutherford and Lindquist (1998) used the type of controls as I outlined earlier in this chapter. So, for instance, it is unknown whether some of the variation observed in the *Bcd* gradient is genetic in nature or whether in fact it is all environmentally induced. Not only are these studies interesting in their own right, but within the framework of a rigorous approach for analysis as suggested in this chapter, it is likely that as a community we will be able to answer many of the outstanding questions in canalization research.

ACKNOWLEDGMENTS

I would like to thank the editors of this volume, B. Hallgrimsson and B. K. Hall, for allowing me to express my views on the study of canalization. Thanks to E. Larsen for initial discussions of many of these ideas and to A. Palsson, L. Goering, and J. Moser for comments on an earlier draft. Special thanks to Greg Gibson for extended discussion on canalization and how to measure it, which has lead to this chapter. Thanks also to R. Lyman and T. F. C. Mackay for providing data.

REFERENCES

Aitkin, M. (1987). Modelling variance heterogeneity in normal regression using GLIM. *Applied Statistics* **36**(3), 332–339.

Alonso-Blanco, C., El-Din el-Assal, S., Coupland, G., and Koornneef, M. (1998). Analysis of natural allelic variation at flowering time loci in the *Landsberg erecta* and Cape Verde Islands ecotypes of *Arabidopsis thaliana*. *Genetics* **149**, 749–764.

Ancel, L. W., and Fontana, W. (2000). Plasticity, evolvability and modularity in RNA. *Journal of Experimental Zoology (Molecular and Developmental Evolution)* **288**, 242–283.

Bateman, K. G. (1959). The genetic assimilation of four venation phenocopies. *Journal of Genetics* **56**, 443–474.

Bateson, W. (1894). *Materials for the Study of Variation Treated with Especial Regard to Discontinuity on the Origin of Species*. Baltimore: Johns Hopkins University Press (reprinted 1992).
Beardmore, J. A. (1960). Developmental stability in constant and fluctuating temperatures. *Heredity* **14**, 411–422.
Bolivar, V. J., Cook, M. N., and Flaherty, L. (2001). Mapping of quantitative trait loci with knockout/congenic strains. *Genome Research* **11**, 1549–1552.
Bourget, D. (2000). Fluctuating asymmetry and fitness in *Drosophila melanogaster. Journal of Evolutionary Biology* **13**, 515–521.
Bubliy, O. A., Loeschcke, V., and Imashevae, A. G. (2000). Effect of stressful and nonstressful growth temperatures on variation of sternopleural bristle number in *Drosophila melanogaster*. *Evolution* **54**(4), 1444–1449.
Clarke, G. M. (1998). The genetic basis of developmental stability. IV. Inter- and intra-individual character variation. *Heredity* **80**, 562–567.
Darwin, C. (1859). *The Origin of Species by Means of Natural Selection*. London: Murray.
Debat, V., and David, P. (2001). Mapping phenotypes: Canalization, plasticity and developmental stability. *Trends in Ecology and Evolution* **16**(10), 555–561.
Debat, V., Alibert, P., David, P., Paradis, E., and Auffray, J. C. (2000). Independence between developmental stability and canalization in the skull of the house mouse. *Proceedings of the Royal Society London Series B* **267**, 423–430.
de Visser, J. A. G. M., Hermisson, J., Wagner, G. P., Meyers, L. A., Bagheri-Chaichian, H.,. Blanchard, J. L., Chao, L., Cheverud, J. M., Elena, S. F., Fontana, W., Gibson, G., Hansen, T. F., Krakauer, D., Lewontin, R. C., Ofria, C., Rice, S. H., von Dassow, G., Wagner, A., and Whitlock, M. C. (2003). Perspective: Evolution and detection of genetic robustness. *Evolution* **57**(9), 1959–1972.
Dworkin, I. (2005). A study of canalization and developmental stability in the sternopleural bristle system of *Drosophila melanogaster*. *Evolution*. Acceptance pending revision.
Dworkin, I. (2005). Evidence for canalization of *distal-less* function in the leg of *Drosphilia melanogaster*. *Evolution and Development* **7**(2), 89–100.
Dworkin, I., Palsson, A., Birdsall, K., and Gibson, G. (2003). Evidence that *Egfr* contributes to cryptic genetic variation for photoreceptor determination in natural populations of *Drosophila melanogaster*. *Current Biology* **13**, 1888–1893.
Dworkin, I., Palsson, A., and Gibson, G. (2005). Replication of an *Egfr*-wing shape association in a wild-caught cohort of *Drosophilia melanogaster*. *Genetics*, in press.
Falconer, D. S., and Mackay, T. F. C. (1996). *Introduction to Quantitative Genetics*. 4th ed. New York: Longman.
Feltz, C., and Miller, G. E. (1996). An asymptotic test for the equality of coefficients of variation from *k* populations. *Statistics in Medicine* **15**, 647–658.
Foulley, J. L., Hebert, D., and Quass, R. L. (1994). Inferences on homogeneity of between-family components of variance and covariance among environments in balanced cross-classified designs. *Genetics Selection Evolution* **26**, 117–136.
Gibson, G., and Hogness, D. S. (1996). Effect of polymorphism in the *Drosophila* regulatory gene *Ultrabithorax* on homeotic stability. *Science* **271**, 200–203.
Gibson, G., and van Helden, S. (1997). Is function of the *Drosophila* homeotic gene *Ultrabithorax* canalized. *Genetics* **147**, 1155–1168.
Gibson, G., and Wagner, G. (2000). Canalization in evolutionary genetics. A stabilizing theory? *Bioessays* **22**, 372–380.
Gibson, G., Wemple, M., and van Helden, S. (1999). Potential variance affecting homeotic *Ultrabithorax* and *Antennapedia* phenotypes in *Drosophila melanogaster*. *Genetics* **151**, 1081–1091.
Gottlieb, T. M., Wade, M. J., and Rutherford, S. L. (2002). Potential genetic variance and domestication of maize. *Bioessays* **24**, 685–689.
Hartl, D. L., and Clark, A. G. (1997). *Principles of Population Genetics*. Sunderland, MA: Sinauer.

Hartman, J. L., Garvik, B., and Hartwell, L. (2001). Principles for the buffering of genetic variation. *Science* **291**, 1001–1004.

Houchmandzadeh, B., Weischaus, E., and Leibler, S. (2002). Establishment of developmental precision and proportions in the early *Drosophila* embryo. *Nature* **415**, 798–802.

Indrasamy, H., Woods, R. E., McKenzie, J. A., and Batterham, P. (2000). Fluctuating asymmetry for specific bristle characters in *Notch* mutants of *Drosophila melanogaster*. *Genetica* **109**, 151–159.

Lande, R. (1977). On comparing coefficients of variation. *Systematic Zoology* **26**, 214–216.

Lauter, N., and Doeley, J. (2002). Genetic variation for phenotypically invariant traits detected in teosinte: Implications for the evolution of novel forms. *Genetics* **160**, 333–342.

Lewontin, R. C. (1966). On the measurement of relative variability. *Systematic Zoology* **15**, 141–142.

Long, A. D., Lyman, R. F., Langley, C. H., and Mackay, T. F. C. (1998). Two sites in the *Delta* gene region contribute to naturally occurring variation in bristle number in *Drosophila melanogaster*. *Genetics* **149**, 999–1017.

Lyman, R. F., and Mackay, T. F. C. (1998). Candidate quantitative trait loci and naturally occurring phenotypic variation for bristle number in *Drosophila melanogaster*: The *Delta-hairless* gene region. *Genetics* **149**, 983–998.

Lynch, M., and Walsh, B. (1998). *Genetics and the Analysis of Quantitative Traits*. Sunderland, MA: Sinauer.

Mackay, T. F. C. (2001). The genetic architecture of quantitative traits. *Annual Review of Genetics* **35**, 303–339.

McLaren, A. (1999). Too late for the midwife toad: Stress, variability and Hsp90. *Trends in Genetics* **15**(5), 169–171.

Meiklejohn, C. D., and Hartl, D. H. (2002). A single mode of canalization. *Trends in Ecology and Evolution* **17**(10), 468–473.

Miller, G. E. (1991). Asymptotic test statistics for coefficients of variation. *Communication in Statistics: Theory and Methods* **20**(10), 3351–3363.

Moreno, G. (1994). Genetic architecture: Genetic behaviour and character evolution. *Annual Review of Ecological Systematics* **25**, 31–44.

Nijhout, H. F., and Davidowitz, G. (2003). Developmental perspectives on phenotypic instability, canalization, and fluctuating asymmetry. In *Developmental Instability: Causes and Consequences* (M. Polak, ed.). New York: Oxford.

Palmer, A. R. (1994). Fluctuating asymmetry analysis: A primer. In *Developmental Instability: Its Origins and Evolutionary Implications* (T. A. Markow, ed.). Dordrecht, Netherlands: Kluwer.

Palmer, A. R., and Strobeck, C. (1986). Fluctuating asymmetry: Measurement, analysis, patterns. *Annual Review of Ecological Systematics* **17**, 391–421.

Palmer, A. R., and Strobeck, C. (1992). Fluctuating asymmetry as a measure of developmental stability: Implications of non-normal distributions and power of statistical tests. *Acta Zoologica Fennica* **191**, 55–70.

Palmer, A. R., and Strobeck, C. (2003). Fluctuating asymmetry analyses revisited. In *Developmental Instability: Causes and Consequences* (M. Polak, ed.). New York: Oxford.

Palsson, A., and Gibson, G. (2004). Association between nucleotide variation in *Egfr* and wing shape in *Drosophila melanogaster*. Genetics **167**(3), 1187–1198.

Polaczyk, P. J., Gasperini, R., and Gibson, G. (1998). Naturally occurring genetic variation affects *Drosophila* photoreceptor determination. *Development Genes and Evolution* **207**, 462–470.

Queltsch, C., Sangster, T. A., and Lindquist, S. (2002). Hsp90 as a capacitor of phenotypic variation. *Nature* **417**, 618–624.

Rendel, J. M. (1959). Canalization of the scute phenotype in *Drosophila*. *Evolution* **13**, 425–439.

Robertson, A. (1959). The sampling variance of the genetic correlation coefficient. *Biometrics* **15**, 469–485.

Robin, C., Lyman, R. F., Long, A. D., Langley, C. H., and Mackay, T. F. C. (2002). Hairy: A quantitative trait locus for *Drosophila* sensory bristle number. *Genetics* **162**, 155–164.

Roff, D. A. (1997). *Evolutionary Quantitative Genetics*. New York: Chapman & Hall.

Rutherford, S. L. (2000). From genotype to phenotype: Buffering mechanisms and storage of genetic information. *Bioessays* **22**(12), 1095–1105.

Rutherford, S. L., and Lindquist, S. (1998). Hsp90 as a capacitor for morphological evolution. *Nature* **396**, 336–342.

SanCristobal-Gaudy, M., Elsen, J. M., Bodin, L., and Chevale, C. (1998). Prediction of the response to a selection for canalization of a continuous trait in animal breeding. *Genetics Selection Evolution* **30**, 423–451.

Scharloo, W. (1991). Canalization: Genetic and developmental aspects. *Annual Review of Ecology and Systematics* **22**, 65–93.

Schultz, B. B. (1985). Levene's test for relative variation. *Systematic Zoology* **34**(4), 449–456.

Sershen, H., Hashim, A., and Vadasz, C. (2002). Strain and sex differences in repeated ethanol treatment-induced motor activity in quasi-congenic mice. *Genes, Brain, and Behavior* **1**, 156–165.

Sokal, R. R., and Braumann, C. A. (1980). Significance tests for coefficients of variation and variability profiles. *Systematic Zoology* **29**, 50–66.

Sokal, R. R., and Rohlf, F. J. (1998). *Biometry.* 3rd ed. New York: Freeman.

Sorensen, D., and Waagepetersen, R. (2003). *Normal Linear Models with Genetically Structured Residual Variance Heterogeneity: A Case Study of Litter Size in Pigs.* Aalborg, Denmark: Aalborg University.

Stearns, S. C., and Kawecki, T. J. (1994). Fitness sensitivity and the canalization of life-history traits. *Evolution* **48**(5), 1438–1450.

Stern, C. (1958). Selection for sub-threshold differences and the origin of pseudoexogenous adaptations. *American Naturalist* **92**, 313–316.

Van Valen, L. (1978). The statistics of variation. *Evolutionary Theory* **4**, 33–43.

Waddington, C. H. (1952). Selection for the basis of an acquired character. *Nature* **169**, 278.

Waddington, C. H. (1953). Genetic assimilation of an acquired character. *Evolution* **7**, 118–126.

Waddington, C. H. (1956). Genetic assimilation of the bithorax phenotype. *Evolution* **10**, 1–13.

Wagner, G. P., Booth, G., and Bagheri-Chaichian, H. (1997). A population genetic theory of canalization. *Evolution* **51**(2), 329–347.

Wagner, G. P., Chiu, C., and Hansen, T. F. (1999). Is HSP90 a regulator of evolvability. *Journal of Experimental Zoology (Molecular and Developmental Evolution)* **285**, 116–118.

Woods, R. E., Sgro, C. M., Hercus, M. J., and Hoffman, A. A. (1999). The association between fluctuating asymmetry, trait variability, trait heritability, and stress: A multiply replicated experiment on combined stresses in *Drosophila melanogaster*. *Evolution* **53**(2), 493–505.

Zar, J. H. (1999). *Biostatistical Analysis.* 4th ed. Upper Saddle River, New Jersey: Prentice Hall.

CHAPTER 9

Mutation and Phenotypic Variation: Where is the Connection? Capacitators, Stressors, Phenotypic Variability, and Evolutionary Change

ARY A. HOFFMANN AND JOHN A. MCKENZIE
Centre for Environmental Stress and Adaptation Research, Department of Genetics, University of Melbourne, Parkville, Victoria, Australia

ABSTRACT

The idea that bursts of evolutionary change can occur when specific conditions generate heritable and selectable variation is not new, but its importance in producing evolutionary diversity remains debatable. There is abundant evidence in bacteria, but less in eukaryotes, for genes that increase mutation rates ("mutators") and for associations between mutation rates and extreme environmental factors ("stressors"), such as starvation, triggering the SOS system. Furthermore, environmentally enhanced transposition is known to increase mutation rates under some conditions, while genes influencing recombination rates ("recombinators") and stressor effects on recombination rates are known in microorganisms and eukaryotes. The impact of new mutations and recombinants on phenotypic variation depends on the extent to which traits are variable or canalized. In canalized traits, new mutations and recombinants may have little impact on phenotypic variance. However, stressors and some other genes ("capacitators") can alter the degree of canalization of traits. Any gene that influences the development of a particular canalized trait might act as a candidate capacitator. Candidate capacitators include the heat shock protein genes, in particular *Hsp90*, which influence the development of a broad range of traits. As in the case of mutators and recombinators, it is unclear if selection directly promotes the long-term persistence of capacitators. Instead, the evolution of capacitators could largely be driven by selection for developmental stability and selection for environmental canalization. One way of separating these effects is to investigate the impact of capacitator systems on different sources of phenotypic variability. This has recently been undertaken for *Hsp83* and for *Notch* mutants. Because there is normally ample genetic variation in traits maintained by mutation–selection balance, resulting in rapid and continuous selection responses for many traits that are not canalized, the role of stressors and other processes in the evolution of these traits is often considered to be minor. However, because of limited genetic/physiologic options for changing some traits, generators may be important in overcoming selection limits that apply to variable traits because of robust tradeoffs and low levels of genetic variability segregating in populations. A challenge for future work is to identify natural variation in eukaryotic capacitators, to link these and stressors to adaptive shifts in both canalized traits and variable traits at evolutionary limits, and to identify signatures that reflect the importance of variability generators in long-term evolution.

INTRODUCTION: VARIABILITY AND LIMITS

The idea that extreme environmental conditions and the action of some genes can elevate evolutionary rates by generating genetic and phenotypic variability has been around for many years. Both Schmalhausen (1949) and Waddington (1953) emphasized the role of stressful conditions in producing variability upon which selection could act. The evolutionary role of these processes was ignored for many years, probably because of the many studies that showed high levels of genetic variability in traits in unstressed populations and the high level of variability at the gene level. Lewontin (1974) commented on the general view that in *Drosophila* anything could be selected, reflecting high levels of heritable variation in populations. If any trait can be readily selected, there is little point in worrying about processes that generated variability. Instead the emphasis switched to processes that maintained genetic variation within populations. Nevertheless, even then researchers were aware of traits that exhibited little variability, the so-called "canalized" traits (for instance, Rendel, 1967). Such traits, including the number of scutellar and thoracic bristles in *Drosophila*, vibrissa number in mice, and corolla lobe number in some plants, were generally invariant within populations and species but often varied among species. There was also an awareness that a lack of selectable variation limited evolutionary potential: For example, although the evolution of heavy-metal resistance in several plant species illustrated rapid phenotypic shifts over relatively short distances, most species of plants failed to colonize contaminated tailings, and in unselected populations of such species, there was often no heritable variation for resistance (Gartside and McNeilly, 1974).

Evolutionary studies have tended to focus on descriptions of successful evolutionary changes rather than limits to variability that constrain organisms. Despite this, the fact that organisms have restricted distributions and occupy limited niches within their distributions attests to the universality of evolutionary limits. Although there are a number of reasons why organisms fail to adapt to changing conditions (Hoffmann and Blows, 1994), a lack of variability may be a common cause. Understanding these limits is important because organisms fail to adapt to the many anthropogenic changes and biodiversity continues to decline.

In this chapter, we briefly overview the processes that can generate phenotypic and genetic variability, starting with mutation, recombination, and stressful environments. We then consider the processes whereby hidden genetic variability can be expressed to alter rates of evolution in canalized traits and the connection between genetic variability, environmental variability, and developmental instability. This is followed by a discussion of whether variability generators are really required for evolutionary change, particularly for the evolution of traits that are normally not canalized. Finally, a research agenda is outlined for investigating some of the unresolved issues in this area.

I. MUTATORS, RECOMBINATORS, STRESSORS, AND GENETIC VARIABILITY

A. Mutation

Stressful conditions can elevate mutation rates. This phenomenon is well known in bacteria where error-prone pathways can be triggered by ultraviolet (UV) irradiation, chemicals, pH stress, and nutritional stresses (Walker, 1984; MacPhee, 1985;

Velkov, 1999; Foster, 2000). A number of error-prone pathways can be induced by stress in bacteria, including the SOS response, which is triggered by DNA damage and involves derepression of a number of genes, which include error-prone DNA polymerases and various genes involved in recombination and repair. When bacteria are starved or stressed in other ways, they are thought to enter a hypermutatable state that leads to elevated mutation rates. There is now evidence that bacteria under natural starvation conditions show elevated mutation rates. Bjedov *et al.* (2003) obtained natural isolates of *Escherichia coli* from a variety of environments and exposed these to starvation stress in their stationary growth phase, which followed an exponential growth phase. The isolates showed an increased rate of mutation leading to resistance to rifampicin and other drugs when under the starvation stress. The increase in mutation rate varied widely among isolates. The researchers concluded that mechanisms underlying the increase in mutagenesis were likely to be heterogeneous and involve a number of proteins.

Mutation rates in eukaryotes are also increased by stress. For instance, Goho and Bell (2000) exposed *Chlamydomonas* strains with a low level of genetic variation to temperature extremes, osmotic stress, low pH, starvation, and toxins. They then measured heritable changes to, and the variance in, fitness and showed that although mean fitness declined, the variance in fitness increased following stress. These findings were interpreted in terms of stresses elevating mutation rates, which led to an initial reduction in mean fitness but an increase in the variance in fitness. Stressors, including temperature extremes, chemical treatments, and low humidity, are also known to increase mutation rates in other eukaryotes (reviewed in Hoffmann and Parsons, 1991).

Elevated mutation rates under stress may be adaptive or may reflect the breakdown of processes that maintain and repair the genome. It is difficult to distinguish between these alternatives, although there is good evidence that high rates of mutation can be adaptive under some circumstances. For instance, Cox and Gibson (1974) showed some time ago in chemostat experiments that strains of *E. coli* with high mutation rates could be favored in novel and stressful environments. Other more recent examples are listed in Velkov (1999) and Sniegowski *et al.* (2000).

The adaptive advantage to organisms is likely to be greatest when organisms show elevated mutation rates specifically under stressful conditions. Consequently new variants that are, by chance, adapted to the new conditions are more likely. There is no need to interpret such results as "directed" mutagenesis. For example, directed mutagenesis in *E. coli* lac$^-$ mutants under starvation conditions reflects a normal Darwinian process; it only appears directed because amplification of the mutant gene occurs before a revertant mutation arising within the amplified cell subpopulation. The starvation stress has no direct effect on mutation rates at specific loci (Hendrickson *et al.*, 2002). This process appears to be controlled by the SOS response at random loci normally triggered as a general genome-wide response to stress (McKenzie *et al.*, 2000).

Some of the increase in mutation rates under stress may be associated with transposons. McClintock (1984) outlined the potential importance of transposons in triggering evolutionary change under a range of genomic and environmental stresses. In bacteria, excision of transpositions can be induced by UV irradiation and chemicals via the SOS stress response (reviewed in Velkov, 1999). In eukaryotes, stressors can trigger mobility of some transposons (Capy *et al.*, 2000). Excision and insertion of transposons could generate genetic variability particularly as transposons constitute a substantial fraction of the genome of organisms (Capy *et al.*, 1997). However, the mechanisms that trigger the mobility response and the links between this response and adaptive genetic change currently remain unclear.

Under what conditions will alleles that increase overall mutation rates be favored? Given that most mutations are deleterious, selection will act to reduce mutation rates, unless environments change continuously and new variants are continuously being selected. Even when new variants are selected under novel conditions, alleles that increase mutation rates will only increase in frequency if they are in linkage disequilibrium with new mutations having a high fitness in the new environment. Selection for increasing the mutation rate because of the generation of beneficial mutations may be weak compared with selection for decreasing it because of the generation of deleterious genes (Leigh, 1973; Johnson, 1999; Sniegowski *et al.*, 2000). The exception is in asexual lineages, because mutation in one part of the genome will influence all loci, compared with sexual organisms, where the association between an allele elevating rates and a favored allele will be lost unless there is tight linkage. In asexual organisms, an optimal mutation rate is expected to evolve. The rate depends on which beneficial mutations arise and are fixed (Orr, 2000). However, this long-term optimum may be rarely realized. Intervening periods of environmental constancy, when no new beneficial mutations are favored, and costs of deleterious mutations reduce the mutation rate. Therefore, when interspersed by periods when mutation rates increase as mutators hitchhike with beneficial genes, mutation rates in asexual populations may be continually changing.

In their review on mutators in populations, Sniegowski *et al.* (2000) emphasized that increases in the frequency of mutators may not necessarily be directly adaptive. Instead they may represent the consequences of adaptation. By hitchhiking along with a beneficial mutation, the mutator can reach fixation as a consequence of the beneficial mutant increasing in a population. Mutators could increase rates of subsequent evolution in populations if another environmental change occurred, but they will also generate additional deleterious mutations. There is a balance between these benefits and costs. The only way to reduce the costs is by restricting mutations to loci under selection and by restricting times of elevated mutation to stressful periods. In *E. coli*, at least, constitutive mutators are rarer than stress-inducible mutators (Bjedov *et al.*, 2003). Nevertheless, there are conditions when constitutive mutators will increase rates of adaptive change in populations, even when they are present at a low frequency (Taddei *et al.*, 1997; Travis and Travis, 2002).

The presence of specific pathways such as the SOS pathway that are associated with increased mutation rates under stress would seem to point to an adaptive mechanism whereby variability is generated via mutation under stress. However, conditions when such mechanisms are favored may be extremely limited. Roth *et al.* (2003) specifically addressed whether the hypermutable state of *E. coli* could have evolved as a mechanism to facilitate evolutionary change under stressed conditions. They found that mutagenesis rates would have to be 150 times higher than the normal rate for this model to operate. In addition, the deleterious mutations generated via this random process would make long-term persistence of this mechanism unlikely. The generation of deleterious mutations by random mutagenesis also appears to be a general problem for any mechanism that involves an increase of mutagenesis under stress. Although the increase in mutagenesis may provide a beneficial mutation that can overcome stressful conditions, this beneficial mutation will be accompanied by numerous deleterious mutations. Under changing environmental conditions, it is therefore difficult to see how a sudden increase in mutagenesis under one set of conditions can be favored by selection.

Nevertheless, there is evidence that different environments select for different stress-induced mutation rates. In the natural isolates investigated by Bjedov *et al.* (2003), those obtained from omnivores had a weaker inducible response than those from carnivores, suggesting that selective factors associated with diet were influencing mutation rates. Moreover, Bjedov *et al.* (2003) showed with computer simulations that inducible mutators can persist in populations if stress phases are relatively common. Rates of adaptive change can be increased when these mutators are present. There is also evidence that fluctuating natural conditions can select for increased rates of mutations. For instance, in the fungus *Sordaria fimicola*, Lamb *et al.* (1998) showed that isolates from a more climatically variable southern side of a canyon had mutation rates that were three times higher than isolates from the more constant northern side.

II. RECOMBINATION

As in the case of mutations, the effects of stress on recombination have been long known. The pioneering work of Plough (1917) in *Drosophila* showed that both high-temperature and low-temperature extremes led to an increase in recombination rate. Increases in recombination with stress are known for other organisms and other types of stresses (Hoffmann and Parsons, 1991). As in the case of mutation rate, effects generated by recombination are likely to be maladaptive. Some new combinations of alleles may have a high fitness. When there are initially negative associations among alleles at different loci, increased recombination will be selected because it enables favored alleles to come together in the same individual. Barton and Charlesworth (1998) reviewed the ways negative associations can be generated in populations and the alternative hypotheses for the evolution of recombination. They concluded that modifiers that increase recombination can be favored under a variety of conditions, particularly when there is strong directional selection in a population. Spatial environmental variability may enhance this process (Lenormand and Otto, 2000).

Changing, as opposed to constant, conditions can select for increased rates of recombination. In a well-known example (Zhuchenko *et al*., 1985), recombination rates were monitored at three chromosome 2 loci in *Drosophila* melanogaster populations exposed to a constant 25° C or to fluctuating temperatures between 15° C and 32° C for 26 generations. Recombination frequencies were consistently higher in populations exposed to the fluctuating rather than the constant conditions. There is evidence that recombination rates in natural populations from stressed or variable conditions can be higher than those from more constant conditions, as in the mole rat *Spalax ehrenbergi* (Nevo *et al*., 1995). In other systems, a higher incidence of stress and environmental variability is associated with a higher rate of sex and recombination (Grishkan *et al*., 2002). In the fungus *S. fimicola,* for instance, isolates from the stressful side of a canyon that experience variable conditions showed a higher rate of recombination than those from the less variable side with more favorable conditions (Saleem *et al*., 2001). As discussed in the preceding text, a similar observation was made for mutation rate (Lamb *et al*., 1998). However these observations could be caused by a variety of factors unrelated to the environment, such as indirect effects of habitat conditions on population size.

An increase in the rate of recombination can enhance evolutionary rates. Regions of chromosomes with high rates of recombination can show high rates of sequence divergence, rearrangement, and the development of multigene families (Perry and Ashworth, 1999; Akhunov *et al*., 2003). There are specific cases where directional selection for a particular trait has increased rates of recombination, presumably because favorable genes can then be brought together more quickly (for an example, see Korol and Iliadi, 1994). However, when recombination rates are manipulated, the response to selection is not necessarily increased (see, for example, Bourguet *et al*., 2003). Although recombination may therefore increase rates of evolution in the long-term, it may be difficult in the short-term for rates of recombination to evolve in direct response to selection on a quantitative trait.

Do recombinators exist? Mechanisms that influence mutation rates may also be associated with altered levels of recombination because these often share common elements. In fact, recombination between similar repeat sequences of DNA can lead to DNA changes via conversion or reciprocal strand exchange and to more drastic changes involving deletions and duplications. The SOS response in bacteria involves induction of several genes involved in recombination (McKenzie *et al*., 2000). Therefore changes in recombination rates may occur because of selection for other functions related to DNA repair, and changes in evolutionary rates could be a by-product of this selection (Michod and Levin, 1988). Even the increase in recombination under stress might be the result of selection for DNA repair because more DNA damage is likely under stressful conditions.

Over the long-term, rates of recombination along a chromosome could be adjusted by selection. For instance, essential genes appear to be located on parts of chromosomes that show low levels of recombination (Pal and Hurst, 2003), whereas multigene families with high rates of evolution occur in distal parts of chromosomes (Akhunov *et al.*, 2003). *Drosophila* and other Diptera exhibit inversions that limit recombination in some parts of the genome, and the frequency of inversion polymorphism consistently varies from central to marginal populations in Dipteran species (Lewontin, 1974; Brussard, 1984). Although the reason for this pattern is not entirely clear, it suggests that recombination rates are under selection.

In summary, mutators and recombinators can increase in frequency under limited conditions and can have an impact on the evolution of some traits under directional selection. Some environmental conditions favor relatively higher levels of mutation and recombination. However, it is difficult to know what the impact of mutators and recombinators has been on evolution in the absence of some signature of these processes in an organism's DNA or physiology. An intriguing case where a signature may be present relates to variation in stress response genes in *E. coli.* Rocha *et al.* (2002) found that the incidence of repeat DNA units in stress response genes of *E. coli* was higher compared with genes in the rest of the genome. Because repeat units can generate phenotypic variability by inducing misrepair of DNA, or by influencing RNA and protein synthesis, the authors suggest that the high incidence of short repeats may reflect a history of selection on these genes. The phenotypic variability resulting from the high number of short repeats may favor variants that are adaptive under stress conditions.

III. THE IMPACT OF NEW MUTANTS AND RECOMBINANTS: CANALIZATION AND CAPACITATORS

Two processes govern variation in the development of organisms. One of these is developmental stability, the process that buffers the development of organisms along a trajectory within the same environment. Following Debat and David (2001), we define developmental stability (DS) as the mechanisms that keep the phenotype constant despite developmental noise that produces slight variation among homologous parts within individuals and DI as developmental instability, the process that produces developmental noise because of minor perturbations in development. Any random deviations in the development of bilaterally symmetric traits can be expected to result from developmental accidents measured by asymmetry (Van Valen, 1962). DI and DS are therefore normally measured by fluctuating asymmetry (FA), the degree to which the left and right sides of an organism covary. In addition, other repeat units within an organism, such as the variance in internode length and leaf length in plants, can be used to measure DI (Freeman *et al.*, 2003).

The second process is canalization, the process by which a structure develops to a target phenotype under various genetic and environmental conditions. Canalization is normally divided into two components: genetic canalization, the developmental process that reduces the sensitivity of a structure to allelic variation, and environmental canalization, the process that reduces the sensitivity of a structure to environmental variation (Debat and David, 2001). Canalization is measured by examining variation among individuals rather than within them, although some authors have used canalization and DS interchangeably. Traits that are highly canalized show low levels of variation. Strong stabilizing or directional selection should lead to a high level of canalization because the same phenotype is always favored by natural selection. Because trait canalization is expected to depend on the intensity of selection, traits that are closely related to fitness are expected to be more canalized than those more distantly related. Stearns *et al.* (1995) found evidence for such an association in lines of *D. melanogaster* reared under varying environments; variation among lines and among environments was highest for late fecundity and life span and lowest for age at eclosion, while early fecundity and dry weight fell between these extremes. These patterns were similar for variation measured within a line (environmental canalization) and between lines (genetic canalization) and reflected how closely related the traits were to overall fitness.

An enormous body of empirical and theoretical literature has developed around these concepts. The central issue is whether the processes of canalization and developmental stability are controlled by the same genes and the same regulatory mechanisms. Despite this effort, the issue remains largely unresolved. As emphasized in several chapters of a recent comprehensive volume on this area (Polak, 2003), the issue will only be resolved by elucidating the genetic and regulatory mechanisms underlying both processes and maintaining a consistent terminology. The entire area remains confused because of inconsistent definitions and interchanging of the terms *canalization* and *developmental stability*.

One issue is whether DS and canalization are viewed as separate processes or part of the same process, whether mechanisms generating variability within individuals are related to those generating differences among individuals. At the level of individual traits, this certainly holds. Clarke (1998) and Windig and Nylin (2000) showed that characters with high levels of variability within individuals also exhibited high levels of variability among individuals. However, at other levels, this association can break down. For instance, environmental conditions can influence both DI and canalization, but conditions that affect the DI of traits as measured by FA are often quite separate from those that influence variance among individuals (Hoffmann and Woods, 2001). Part of this may relate to the sources of variation among individuals; when environmental variation rather than phenotypic variation is considered, there may be instances of variation within individuals and between individuals being linked. This was demonstrated for wing traits but not a bristle trait in a clonal strain of *D. mercatorum* where the environmental variance could be separated from other sources of variation (Kristensen *et al.*, 2003). A problem is that different sources of phenotypic variance can vary in an unpredictable manner.

The link between canalization and DI, as well as genetic and environmental canalization, will determine the way selection influences these processes. Mieklejohn and Hartl (2002) have argued that selection will be most intense for environmental canalization, because organisms will most often be selected to produce the same phenotype under a range of environmental conditions. They suggest that selection for genetic canalization will be much weaker than for environmental canalization because only rarely will genetic variants be expressed. This argument depends on the two processes being tightly linked.

IV. IN SEARCH OF CAPACITATORS: GENES THAT INFLUENCE DEVELOPMENTAL STABILITY AND CANALIZATION

In a widely cited paper, Rutherford and Lindquist (1998) outlined evidence that *Hsp90* may act as a capacitator of evolutionary change. They investigated mutants in the *Hsp83* gene in *D. melanogaster*, which codes for *Hsp90*, for morphological abnormalities. The *Hsp90* mutant strains, as well as a specific inhibitor of *Hsp90* activity, the drug geldanamycin, led to the expression of morphological abnormalities in several traits (Rutherford and Lindquist, 1998). Once new variants of these genes were selected for several generations, the expression of the abnormal variants no longer depended on the mutant background. Rutherford and Lindquist argued that a similar process could occur under stress when levels of *Hsp90* could be naturally suppressed in a transient manner.

These changes are reminiscent of the evolution of threshold traits. Under a simple model of a threshold trait (Rendel, 1967), an allele underlying the phenotypic variation underlying a trait is only expressed when a particular threshold is exceeded (Figure 9-1A). Once this happens, the allele can be selected. The stress serves to move the underlying distribution of a trait to pass the threshold, either by increasing the underlying variance of the trait or by moving the distribution in one or other direction. Thus, stressful conditions, or suppression of *Hsp90* levels in mutants, or a specific inhibitor of *Hsp90* such as geldanamycin, could influence the position of the threshold in relation to the underlying distribution of the trait.

The work of Rutherford and Lindquist (1998) was followed by a paper demonstrating similar *Hsp90* capacitator functions in the plant *Arabidopsis thaliana* (Queitsch *et al.*, 2002). Because *A. thaliana* contains multiple copies of *Hsp90*, geldanamycin and another drug, radicicol, were used to inhibit *Hsp90* function. These drugs produced several abnormalities related to the morphology of leaves, cotyledons, roots, and whole seedlings. It was also found that the type of abnormality produced depended on the genetic background of the line tested and that the specific abnormality produced in a line was already evident at a very low frequency in untreated seedlings. Queitsch *et al.* (2002) therefore concluded that the drugs uncovered preexisting predispositions rather than drug specific effects, and they showed that similar effects could be uncovered by a mild temperature treatment. They also found that at least some of the variants uncovered by the inhibitor treatments were genetically based and could be selected. Finally, these authors argued that *Hsp90* affected developmental stability. This was based on testing the effects of *Hsp90* inhibition on F1 progeny of crosses between different strains. These were expected to be more developmentally unstable than genomes of the parental strains. They predicted that the incidence of abnormalities in the F1s should increase if *Hsp90* was suppressed and *Hsp90* influenced DS. Abnormalities were relatively more common in the F1s in support of this conjecture, although this does not represent a shift in DS, as defined in the preceding text, because variation between, rather than within, plants was scored. Although the authors discuss phenotypic variance, it is not clear if there are changes in abnormality frequencies or more continuously variable traits. It would be interesting to examine *Hsp90* effects on DI in *A. thaliana* by looking at variation within plants, particularly as such measures have recently been described in this plant and found to increase under heavy-metal stress (Tan-Kristanto *et al.*, 2003).

Rutherford (2003) has made a more general case that protein chaperones could act as capacitators, arguing that they can help "release" variation in invariant traits that are normally neutral. Rutherford (2003) suggested that chaperone-buffering systems have evolved both as a consequence of group and also individual selection. Because chaperones act to mediate the activity of proteins by allowing mutant unstable proteins to form active configurations, they may allow novel functions to evolve. *Hsp90* normally targets cell cycle regulators and developmental regulators. It also appears to have the capacity to interact with mutant alleles, allowing their products to be expressed and function as normal proteins. In this sense, *Hsp90* can help to maintain trait thresholds. However, under stressful conditions when chaperones become limiting, the mutant proteins may become expressed, leading to the expression of trait variability. *Hsp90* is therefore expected to alter the expression of variability in traits for which there is a store of unexpressed variability—invariant traits where there is strong buffering of phenotypic expression. It does this by targeting the signaling network of cells, which in turn influences a range of developmental processes.

Under stressful conditions, the expression of variability in traits may be increased and chaperone functioning decreased, leading to an increased expression of phenotypic variability and allowing for the selection of new morphological variants. Rutherford (2003) likened this to an increase in mutation under stress, decreasing population fitness on average but increasing the likelihood of rare beneficial mutants. In the absence of stress, the variation is not expressed, but in its presence, it is expressed, because thresholds for expression are shifted. In support of this model, there is evidence in bacteria that high levels of proteins related to *Hsp90* buffer the expression of deleterious mutations.

Rutherford (2003) also linked effects of *Hsp70* to the expression of morphological variation. Roberts and Feder (1999) found that heat stress could generate morphological abnormalities in *D. melanogaster*, and that some protection from this abnormality formation was provided by extra copies of the *Hsp70* gene. However, in this case, inheritance of the morphological variants was not demonstrated. Thus, although *Hsp70* can influence environmental canalization, it is not clear that it affects genetic canalization.

Although *Hsp83* is a good candidate for a capacitator, there are a number of hurdles to overcome before its evolutionary significance can be established. These relate to the extent to which shifts in trait means, rather than canalization, might explain some of the results, natural variation in *Hsp90* segregating in populations, the presence of stressors, and the persistence of natural variants in populations. The stressor issue relates to the fact that, although stressors may reduce levels of *Hsp90* and increase levels of variability, this connection has not yet been demonstrated with respect to stressors that occur naturally. The selection experiments demonstrating *Hsp90*-mediated assimilation all involved impairment by mutant *Hsp83* alleles or the presence of a specific *Hsp90* inhibitor, not stressors that might be experienced in nature such as temperature extremes and starvation. If a connection could be made to natural stressors, this could provide evidence of an important mechanism whereby stressors influence evolutionary rates.

The persistence issue relates to the question of whether variants of *Hsp83* that suppress *Hsp90* levels are maintained in populations. These variants might only persist when their effects are evident under stressful conditions, when an organism is facing a markedly reduced reproductive output and perhaps death. *Hsp90* variants with broader effects may well disappear rapidly from populations because canalized phenotypes are favored under most conditions. Although Rutherford (2003) invoked group selection to account for the persistence of *Hsp90* variants, they still have to persist in populations during periods without stress. This makes them prone to the difficulties discussed with respect to mutators (in the preceding text); their presence is directly selected over short periods, they need to be in linkage disequilibrium with favored alleles, and their beneficial effects in generating new variants is countered by other new variants that are not favored by selection.

The issue of trait means relates to the fact that *Hsp83* mutants (and geldanamycin) have effects on the mean of many traits including relatively canalized traits. Given these effects, are proteins such as *Hsp90* really capacitators or simply proteins that directionally alter the expression of traits by changing underlying thresholds?

This problem can be illustrated with respect to a different gene in *Drosophila*, *Notch*. The *Notch* signaling pathway plays a crucial role in achieving the pattern of thoracic bristles on the notum of fly (Simpson *et al.*, 1999). In *D. melanogaster*, there are 13 macrochaete (large) bristles (11 thoracic bristles and 2 scutellar bristles) on each heminotum, and these are arranged in stereotypic positions that rarely vary between individuals (Simpson *et al.*, 1999). This pattern is highly conserved. It is seen throughout the 2000 or so species of the Drosophilidae (Grimaldi, 1987). Therefore the stereotypic patterns of bristles are likely to be important (Simpson *et al.*, 1999).

The *Notch* gene influences bristle number, which is normally invariant. *Notch* mutants can increase or decrease bristle number, particularly scutellar bristle number, depending on the mutant. To see whether they influenced phenotypic variation within individuals, Indrasamy *et al.* (2000) scored them for DS. They found that strains with mutants at the *Notch* locus often differed from a wild-type strain for FA, as measured by the absolute value of differences between the left (L) and right (R) measurements of a fly ($S|L - R|/N$). Differences remained when asymmetry was adjusted for mean character size by division with $S([(L + R)/2]/N)$ and after backcrossing, which was used to isolate *Notch*-specific effects. Differences in scutellar bristle number in the strains tend to involve either an increase or a decrease in number depending on the mutant, and these shifts usually involved the addition or loss of a single bristle. As these individuals are all asymmetric, there is an increase in FA as soon as bristle numbers are shifted outside the invariant norm. The *Notch* variants therefore altered the expression of variants away from the canalized phenotype. In this sense, they acted as capacitators as selection of deviant phenotypes can now occur, perhaps involving background alleles that lead to the expression of the bristle variant and eventually providing deviant phenotypes in the absence of the *Notch* alleles. An appropriate chemical inhibitor could presumably have the same effect as the mutants. However, does this make *Notch* a trait-specific capacitator or just one of the genes underlying the threshold trait? If we go to a finer level and look at specific scutellar bristles in the *Notch* mutants, there is evidence that there are changes at only a few sites (H. Indrasamy, 2003). Thus *Notch* mutants are having a specific effect on bristle number at one or two sites on the scutellum.

The degree of canalization of bristles may depend on their position on the notum, particularly as bristles at different positions are determined by developmental processes occurring at different times and affected by different genes (Scharloo, 1991). Because the development of macrochaetes is well understood and bristle development at different positions can be partly independent, it may be more meaningful to focus on variability at individual bristle positions rather than on summed numbers (Scharloo, 1991). The traditional model of canalization may be appropriate for some threshold traits, but it has been criticized in the context of the development of scutellar bristles (Whittle, 1998). Any deviations from the canalized phenotype of the scutellum and thorax would involve loss, gain, or duplication of bristles at specific locations.

These arguments do not belittle the potential role of *Notch* or other genes in generating variability in many invariant traits. In this sense, any gene that influences the canalized traits may be viewed as a capacitator of that trait, because it exposes the trait to selection. What these arguments establish is a degree of caution in the interpretation and rigor in the definition of capacitators. The effects of *Notch* on invariant bristle traits may simply reflect a direct role of *Notch* in the development of the trait.

Like *Notch*, *Hsp83* probably does not generate developmental noise in traits. To investigate effects on DS, Milton *et al.* (2003) scored the FA of *Hsp83* mutants for traits that were not canalized, including several bristle traits and wing size. By computing asymmetry in several traits, Milton *et al.* (2003) could compute overall composite FA across traits as well as single trait asymmetry. This is important because composite asymmetry is likely to be a better measure of developmental stability than single-trait asymmetry (Leung *et al.*, 2000) and can differ between treatments even when single-trait measures mostly do not differ (Woods *et al.*, 1999; Hewa-Kapuge and Hoffmann, 2001). Milton *et al.* (2003) showed that there were no effects of the mutants on the DS of traits. Moreover, they also showed that geldanamycin, the specific inhibitor of *Hsp90*, did not significantly influence the FA of traits (Table 9-1). Assuming that FA is a good monitor of developmental disturbance, this would indicate that *Hsp90* does not cause an overall increase in disturbance despite its broad developmental role.

There is considerable debate as to whether the process of canalization is different from the processes maintaining developmental stability as measured by asymmetry (Debat and David, 2001). Milton *et al.* (2003) therefore also examined the canalization of the traits by scoring trait variances and coefficients of variation. They found that these measures of variability did not generally increase for flies treated with the geldanamycin inhibitor or in the mutant. However, *Hsp90* did

TABLE 9-1. **Effect of Geldanamycin Treatment on Means, Variances, and Fluctuating Asymmetry (FA) of Quantitative Traits in Outbred *D. Melanogaster*. Geldanamycin (GA) is an Inhibitor of Hsp90 and is Applied at the Larval Stage. Modified from Milton *et al.* (2003).**

	Control			Inhibitor (GA)		
	Mean (L+R)	Variance	FA	Mean (L+R)	Variance	FA
Sternopleural bristles	19.32	2.46	0.92	18.20	1.62	0.88
Orbital bristles	18.20	2.02	0.94	18.18	1.40	0.66
Ocellar bristles	7.46	0.80	0.54	7.02	0.85	0.62
Vibrissae and carnia	29.28	3.78	0.92	25.96	2.64	1.02
Wing centroid	3130.0	6008.4	3.46	3073.2	4067.6	3.97

have direct effects on the means of traits, including changes in bristle number, wing size, and wing shape. Therefore, while low levels of *Hsp90* can release variability in some traits, it does not generally increase the variability of many traits that are variable. In contrast, *Hsp83* mutants do show an increase in variability of a variety of traits that are canalized. As in the case of the *Notch* mutants, it would be interesting to see if *Hsp90* altered canalized bristle traits such as the scutellars in highly specific ways on particular backgrounds. Do changes in bristle numbers reflect changes at specific positions or a general increase/decrease at specific positions?

Therefore, although *Hsp83* mutants can alter means of an impressive array of traits, the gene products may not usually alter variances or *CVs*. *Hsp90* acts early in development and has a range of phenotypic effects that can facilitate selection on invariant traits. It is a decanalizer in that it has the capacity to generate variants at traits that are generally invariant but does not generally increase variability. A number of genes have this capacity, but *Hsp83* seems unusual in its effects on numerous traits. *Hsp90* suppression may facilitate the production of variation following exposure to environmental stress, but this remains a model. Natural variation in *Hsp90* as well as stressors relevant to natural environments need to be assessed.

Finally, there are models not involving *Hsp90* for the impact of stressors on rates of evolutionary change. Several years ago, it was shown that the expression of variation at many genes was buffered by enzyme dynamics; metabolic flux through biochemical pathways is related in a nonlinear way to enzyme activity such that large changes at relatively high activity levels have little impact on flux. Small changes at low activity levels have a large impact (Kacser and Burns, 1981). It was suggested by Hartl *et al.* (1985) that this relationship could be altered in unfavorable conditions such that differences in activity even at high levels would have an effect on flux, leading to the expression of genetic variation (Figure 9-1, ii). There are a number of other reasons that the expression of genetic variation might be increased under stressful conditions (Hoffmann and Parsons, 1991).

The best example of a DS gene relevant to natural conditions remains the case of insecticide resistance and asymmetry in the sheep blowfly, *Lucilia cuprina*. In this system, there is a modifier gene that affects the expression of asymmetry in bristle traits of individuals that are resistant to the pesticides diazinon and malathion (reviewed in McKenzie, 2003). Wild-type individuals have a lower level of bristle asymmetry than organophosphorus-resistant individuals unless the modifier gene is present. Individuals resistant to dieldrin, a cyclodiene, also have elevated asymmetry scores relative to wild type. These scores are not influenced by the modifier. Bristle asymmetry can also be modified in this system by exposure to stressful environments but only when resistance is absent or the modifier and resistance are present. Temperature interruption experiments suggest that the asymmetry develops between the prepupal and pupal stages of development. The effects of the resistance genes on asymmetry are specific to bristles because the FA

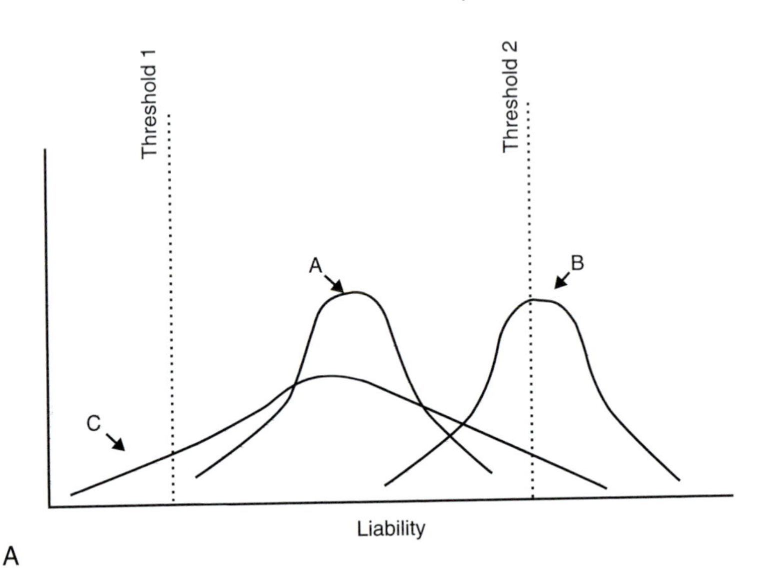

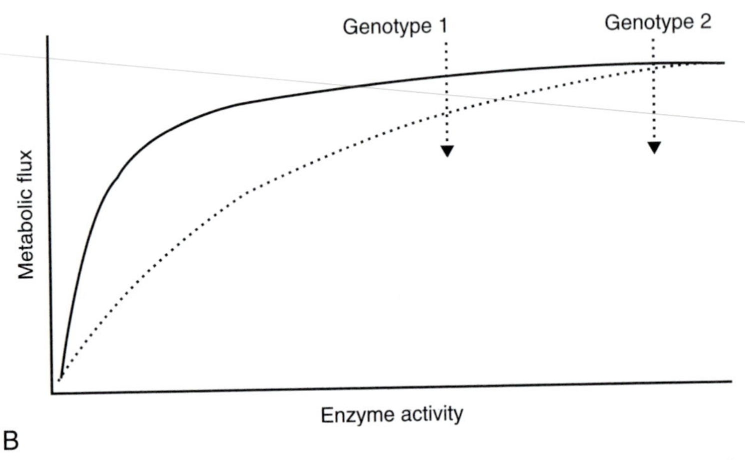

FIGURE 9-1. Models for the evolution of canalized traits. (i) Threshold model for the effects of changes in liability means and variances on the expression of a canalized trait when there are two thresholds. (A) Almost all liability values fall between the two thresholds and within the canalized zone, so almost no trait variation is expressed. (B, C) Liability mean or variance has shifted to encompass the thresholds, resulting in the expression of variability. Note also that no developmental instability (DI) is expressed in traits unless trait values fall around the thresholds. (ii) Association between changes in flux through a biochemical pathway and enzyme activity in a normal (solid line) and stressed (hatched line) situation. Differences between two genotypes in activity only influence flux in the stressed situation. (From Hartl *et al.*, 1985).

of wing traits is not altered (Clarke *et al.*, 2000), suggesting that the phenotypic effects of the genes are confined to cellular bases of the development of bristles rather than some overall process generating phenotypic variation within individuals. Clarke *et al.* (2000) also showed that wing character asymmetry was influenced by temperature extremes, indicating that stressors were capable of altering expression of microenvironmental variation in this trait as well as in bristle FA.

Progress has been made in identifying the candidate gene modifying asymmetry in this system. The modifier was mapped to a chromosome region (McKenzie and Game, 1987). The *Scalloped wing* (*Scl*) gene of *L. cuprina*, which is homologous to the *Notch* gene of *D. melanogaster*, was identified as a possible candidate (Davies *et al.*, 1996). However, subsequent analysis (K. Freebairn, 2000) has revealed that the modifier locus is closely linked to *Scl*.

Finally, the genotypes with a high level of asymmetry in *L. cuprina* are not associated with an increase in trait variability. The data presented in Figure 9-2 indicate the association between variation among individuals (as measured by standard deviations) and variation within individuals (as measured by fluctuating asymmetry) for different genotypes of *L. cuprina*. Some of these genotypes carry the resistance alleles, and these have an increased level of FA of the bristle traits (especially for two of the traits) when the modifier allele is absent. However, there is no tendency for the genotypes with higher FA also to exhibit relatively larger standard deviations. There is also no association between FA and standard deviation for any of the wing traits. Therefore differences between genotypes for bristle FA are not associated with differences in variation among individuals. Note however that for wings there is a positive association at the trait level; traits with a high FA also tend to have a high standard deviation (Figure 9-2) in agreement with Clarke (1998). In this case, the candidate DI gene appears to generate effects on DI that are independent of the canalization process as well as effects on the means of traits.

V. CAPACITATORS, STRESSORS, AND QUANTITATIVE VARIATION

To relate data to natural populations where defined mutant strains do not exist and specific inhibitors are not identified, a quantitative approach for assessing natural variation is needed. Is there evidence for quantitative genetic variation in levels of within individual variability and for variability between individuals? Genetic variation in developmental stability is a contentious issue. Overall, DI as scored by FA tends to have a low heritability. This conclusion was based on a Bayesian study of literature that does not suffer from confounding effects (Van Dongen, 2000) and counters the findings of an earlier review (Moller and Thornhill, 1997). However, whether the heritability of FA reflects the heritability of DI is another matter. Because individuals have only two sides, FA measures can represent a poor estimate

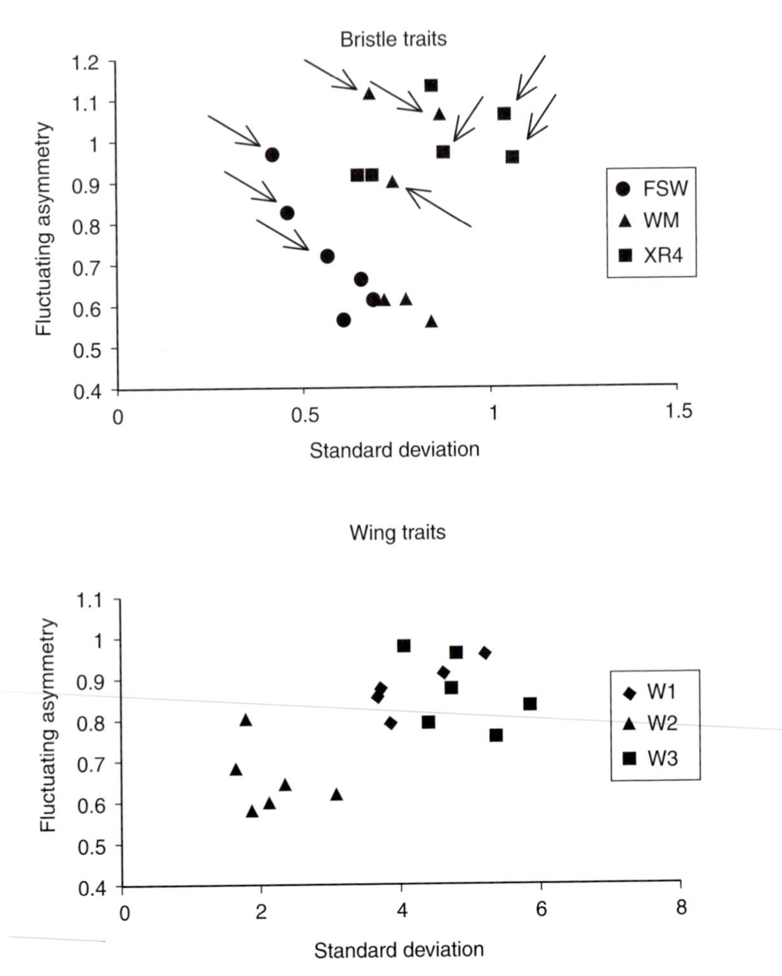

FIGURE 9-2. Association between variation within individuals (measured as fluctuating asymmetry) and variation among individuals (measured as the standard deviation) for three meristic bristle traits and three wing traits measured on individuals belonging to different genotypes of *Lucilia cuprina* with or without the *Rop1* and *Rdl* resistance genes and associated modifier. The bristle traits (*FSB, WMB,* and *XR4*) are described in Clarke and McKenzie (1987), while the three wing traits (*W1, W2,* and *W3*) correspond to intersection points between veins R1 and R2+3, the distance between the r-m junction of R4+5 and the distal cross-vein, and the length of R4+5 (see Clarke *et al.*, 2000). Arrows indicate genotypes expected to have high asymmetry values because of the absence of modifier genes and presence of resistance genes.

of DI; a low heritability of FA might still indicate a high heritability of DI (Whitlock, 1996; Van Dongen, 1998; Gangestad and Thornhill, 2003). FA might still represent a useful measure of the heritability of DI if selection experiments are used to increase levels of FA (Fuller and Houle, 2002) or if some adjustment is made for the fact that FA is only based on measurements of two sides. For instance, Van Dongen and Lens (2000) suggest that the heritability of developmental stability is around 14%. Unfortunately, all these estimates are subject to large errors from several sources.

Perhaps the best evidence on the heritability of DI comes from plants. Repeated structures such as floral parts and leaves are common, and variation among leaves or floral parts within a plant, reflecting DI, can therefore be estimated more accurately. It is also often easier to measure canalization in plants because controlled crosses can be undertaken among genotypes to generate individuals that only differ in the microenvironment that they have experienced. A good example of this approach is provided by Paxman (1956), who undertook diallel crosses between five varieties of tobacco (*Nicotiana rustica*) and measured floral as well as leaf characters. Paxman characterized floral asymmetry as the variance in pistil and stamen lengths of three flowers measured for each plant and found that there were differences among varieties for the variances, reflecting differences in DI. The DI of leaf shape was measured by first plotting shape against node number to reflect the change in leaf shape with node position and then fitting a regression line and obtaining an estimate of the variance around this line as an estimate of DI. This also showed differences among the varieties for DI, suggesting genetic variation for DI. However, there was no association between DI in the flowers and DI in the leaves, reflecting the absence of a universal DI measure affecting all plant parts.

The heritability of canalization can potentially be explored at two levels. Estimation may be made of canalization per se or of the heritability of plasticity where the plastic responses of organisms to two or more environments are monitored. This is beyond the scope of the present discussion. Plastic responses can show heritable variation (Roff and Bradford, 2000), but this is often low, particularly when compared with the heritabilities of trait means (Simons and Johnston, 2000; Laurila *et al.*, 2002). The heritability of canalization is generally not assessed in quantitative genetic studies; therefore we do not know if parents that produce offspring that are variable for a particular quantitative trait also have a tendency to produce offspring with variable offspring. Some traits are likely to be selected because of variability. Thus, in populations exposed to variable environments, it may often be useful to produce offspring that are bet hedgers, expressing variable phenotypes (Philippi and Seger, 1989). Studies of defined strains provide some evidence that genotypes can differ in their levels of variability; for instance, differences among inbred and outbred strains for expression of phenotypic variability are well known (Falconer, 1989).

Although stressors may influence the expression of variation in mostly invariant traits, much less is known about their impact on genetic variance in quantitative traits. As reviewed in Hoffmann and Merila (1999), there are good cases where genetic variance appears to be increased under stressful conditions, as in the case of temperature extremes and sternopleural bristle number in *Drosophila*. However, in other cases, the opposite trend is apparent. For instance, in many bird species, there is a decrease in genetic variance and an increase in environmental variance of size-related traits with nutritional stress. These inconsistent trends are unsurprising because they are likely to reflect the many ways that unfavorable conditions can interact with genetic variation. Hoffmann and Merila (1999) list a number of hypotheses about how environmental conditions can influence the heritability and other measures of the ability of traits to evolve. These include changes in levels of genetic variance resulting from effects on recombination and mutation, changes in levels of environmental variance, different histories of selection in environments, and effects of environmental conditions on phenotypic variance as conditions become limiting.

Stressful conditions also point to highly specific phenotypic changes in the variance and asymmetry of traits (Woods *et al*., 1999; Hoffmann and Woods, 2001). In many studies, environmental conditions only influence variability in one or a few traits. For instance, Kristensen *et al.* (2003) examined the impact of high-temperature stress in a parthenogenetic strain of *D. mercatorum* so differences between individuals are only the result of environmental sources of variation. They found changes with stress in wing trait asymmetry but not bristle trait asymmetry, indicating that environmental factors were trait specific. Moreover, conditions that alter the variance of traits among individuals do not alter levels of fluctuating asymmetry in that trait and vice versa (Hoffmann and Woods, 2001). For instance, Woods *et al.* (1999) found that combined stress conditions consistently increased FA in one trait (orbital bristle number) but not in a different bristle trait or in three wing traits, whereas the same stress increased phenotypic variability in the wing traits as measured by coefficients of variation.

These results suggest that stressors as generators act in a trait-specific manner; any change in variability within and between individuals will often be character specific, so rates of adaptation will only be increased for a specific set of traits. There is no set of conditions that increases the expression of variability in all quantitative traits. Moreover, changes in phenotypic variation are not always tied to changes in variation within organisms.

In summary, there is only limited evidence for the expression of quantitative variation for DI as measured by FA, but this is a poor measure of DI. The expression of trait variability among individuals under constant conditions may have a heritable component, but this has rarely been investigated. Stressful conditions can affect the expression of the phenotypic variance by influencing the genetic variance and environmental variance, but patterns are likely to be stress and trait specific.

VI. DO WE NEED VARIABILITY GENERATORS?

Selection can sometimes produce highly predictable evolutionary changes, utilizing variation produced by mutation to generate very specific genetic changes from a limited set of options. In the pesticide resistance literature, there are several examples where laboratory mutagenesis has produced resistance with the same genetic basis as evolved in field populations (McKenzie and Batterham, 1998). In the sheep blowfly, *L. cuprina*, resistance to dieldrin involves an allelic substitution at the *Rdl* locus, while resistance to diazinon involves a substitution at the *Rop-1* locus that codes for a carboxylesterase. In the laboratory, chemical mutagenesis was used to generate strains resistant to dieldrin and diazinon. Resistance in these strains mapped to the same locus as in the laboratory and was even associated with the same amino acid change. The evolution of resistance to these chemicals is therefore highly predictable from laboratory studies.

In cases where strong selection pressures produce highly specific genetic changes, alleles that are favored by selection are probably segregating in populations at a low frequency before selection being applied. These alleles are presumably being maintained by a balance between mutation generating the alleles and selection removing them because of low fitness of the alleles. For instance, Garcia-Villada *et al.* (2002) examined evolutionary changes in a microalga exposed to the contaminant 2, 4, 6 trinitrotoluene (TNT). They found TNT-resistant variants segregating at a low frequency in the population before selection. The resistant variants had a low fitness in the absence of TNT, including a low photosynthetic rate and low competitive ability, suggesting that resistant mutations were being maintained in the base population at a very low frequency by mutation–selection balance. In this situation, the response to selection depends only on the initial frequency of the allele, whether the allele shows dominance, and the chromosomal location of the allele. Orr and Betancourt (2001) have shown that when previously deleterious mutations segregate in a population, and these are selected, their probability of fixation in a population is independent of dominance and occurs relatively more rapidly if the genes are on autosomes compared with the X chromosome. These patterns differ from alleles that are introduced as new mutations into a population, where dominant mutations are more likely to go to fixation than recessive mutations, because they are expressed phenotypically in the heterozygous form.

In systems such as these, there is no requirement for evolutionary capacitators or recombinators to generate rapid evolutionary change. Mutators might increase the frequency of resistant alleles in an unselected population, but strong selection ensures that resistance alleles will increase in frequency even if they are rare initially. Variability generators are more likely to play a role in cases where evolutionary responses are not readily attained. In the case of chemical resistance, these are the situations where resistance does not evolve. For instance, when laboratory mutagenesis was applied to *L. cuprina* to generate strains resistant to the insect growth regulator cyromazine, strains with only a low level of resistance were generated (Yen *et al.*, 1996). These strains had a low viability, and the combination of low resistance levels plus these fitness costs probably explained the absence of cyromazine resistance in field populations of *L. cuprina*. Populations appear to be at a selection limit that cannot be easily broken in the absence of variability generators.

Selection limits that result from an absence of genetic variability are likely to be common for chemical resistance because there are numerous pests and diseases that have not evolved resistance to agrichemicals despite strong selection pressures. In the literature on heavy-metal resistance in plants, there are also instances where selection responses have failed to occur because of the absence of genetic and phenotypic variability. In plant species that have successfully colonized mine soil, tolerant individuals can normally be found at a low frequency in populations from uncontaminated areas, whereas these individuals are absent in species that have not colonized contaminated areas (Gartside and McNeilly, 1974; Macnair, 1997). This suggests that genetic variation is lacking in populations that are unable to evolve increased tolerance (Bradshaw, 1991). In an extension of this work, Al-Hiyaly *et al.* (1988) considered zinc contamination from coated electricity pylons and showed that *Agrostis* capillaries populations around the pylons differed in zinc tolerance; some had evolved tolerance, whereas others had not. Al-Hiyaly *et al.* (1993) then showed that, in an area where tolerance had failed to evolve, there was no evidence for tolerance among plants from the surrounding uncontaminated area. In contrast, tolerance was present in four other populations from uncontaminated sites surrounding pylons where resistance had evolved. These findings suggest that appropriate genetic variation in base populations is needed for tolerance evolution. The absence of variability in a population predicts the lack of evolutionary response to the stress.

In the case of heavy-metal resistance and pesticide resistance, responses to selection usually involve one or a few genes. A lack of genetic variability may reflect the limited genetic options at these restricted sets of loci, but do quantitative traits that are controlled by a number of loci also show a lack of variability? Can selection limits resulting from a lack of genetic variability be important even in these cases? Much of the emphasis in the genetic study of quantitative genetics is on mutation–selection balance, determining the level of heritable variation that can be maintained in quantitative traits when there is stabilizing selection acting on a trait and removing variation, while new variation is continually being introduced by mutation. For fitness-related traits, the mutational variance introduced into populations is around 10^{-3} times the environmental variance (Houle *et al.*, 1996). This is based on estimates of fitness traits measured in a limited number of organisms (mostly *D. melanogaster* and *Caenorhabditis elegans*) under fairly optimal conditions. With these levels of variation and despite weak stabilizing selection, intermediate and even high heritabilities typically seen in populations can be maintained

(Zhang and Hill 2002; Zhang *et al.*, 2002). With new variation that is being continually introduced into populations via mutation selection balance, there is ample phenotypic and genetic variation available for evolutionary shifts. Impressive and continuous responses to selection can be obtained for large numbers of traits—in *D. melanogaster* examples include responses to selection for desiccation resistance (Gibbs *et al.*, 1997), wing shape (Weber, 1990), and ethanol resistance (Weber and Diggins, 1990).

However, in some cases, selection responses can be difficult to achieve. A recent example concerns the response to selection for desiccation resistance in a rain forest species (*Drosophila birchii*). This species is restricted to patches of rain forest along the east coast of Australia, which is increasingly being threatened by fragmentation from development. *D. birchii* is particularly sensitive to desiccation, and clinal variation in resistance suggests a pattern of past adaptation in this trait (Hoffmann *et al.*, 2003). When the most resistant geographic population was intensely selected for increased resistance, there was no response for this trait even after 30 generations; the trait seemed to be at a selection limit. Parent–offspring comparisons indicated no detectable heritable variation for this trait despite high levels of heritable variation in two morphological traits and high levels of molecular variation at microsatellite loci. These results contrast markedly with heritable variation for this trait and rapid responses to selection in other *Drosophila* species such as *D. melanogaster.* In fact, desiccation resistance is among traits with the highest heritability in *Drosophila* (Hoffmann, 2000).

There are also other studies in which selection experiments suggest limits to evolutionary change. In *Drosophila* these include the lack of selection response for altered embryonic development rate (Neyfakh and Hartl, 1993) and selection plateaus for increased heat resistance (Hoffmann *et al.*, 1997; Gilchrist and Huey, 1999). In addition there is some evidence in the literature suggesting a lack of heritable variation for traits relevant to the ecology of species. For instance, in the frog *Rana temporaria*, there was no detectable additive genetic variance for mass at metamorphosis (Sommer and Pearman, 2003), while Baer and Travis (2000) were unable to select for increased thermal tolerance in a live-bearing fish, *Heterandria formosa.* Perhaps conclusions about high levels of variability based on a few traits and a limited number of readily culturable species do not generalize to other situations. Where attempts have been made to select a range of related species for the same trait, the responses to selection in the different species (or conspecific populations) can be quite disparate, as in the case of *Drosophila* species and populations selected for increased knock-down resistance to ethanol fumes (Cohan and Hoffmann, 1989). As in the case of heavy metals where there is a clear-cut case of species being unable to colonize contaminated soil, more attention should be given to cases where quantitative traits are important ecologically and where selection limits could have ramifications for the distribution and abundance of species.

In the *D. birchii* example discussed in the preceding text, an increase in desiccation resistance may allow this species to extend its range outside of rain forest areas.

When selection responses do not occur, factors apart from a lack of genetic variability may be involved, such as selection being limited by trade-offs among traits. For instance, laboratory selection experiments have shown that increased life span is accompanied by a cost to reproduction (Harshman and Haberer, 2000). This is an example of a robust trade-off—one that is repeatedly detected in selection experiments undertaken by different investigators with different stocks and therefore likely to reflect a genetic limit. Whether trade-offs constrain long-term evolution is still not clear; long-term selection experiments suggest that even robust trade-offs can often be broken down over time (Phelan *et al.*, 2003), although there are a number of reasons why conclusions from laboratory selection experiments may not apply to natural populations (Harshman and Hoffmann, 2000). Traits may also fail to change under selection for other reasons. Selection responses can be prevented because of maternal effects that impact negatively on a trait. In the case of trade-offs and maternal effects, genetic variation for a trait may segregate in a base population, but selection responses are curtailed unless there is genetic variation that is decoupled from the trade-off or maternal effect. Variability generators, including capacitators, might play a role in producing variation of the right type if there are canalized processes.

Finally, recent data on size suggest that evolutionary responses can fail to occur in quantitative traits even when there is phenotypic variation and when selection acts in one direction on phenotypic variation. This can occur when selection acts on the environmental component of the phenotypic variance rather than on the genetic component (Merila *et al.*, 2001). As a consequence, means of some traits in vertebrate populations are static or even changing in a direction opposite to the one predicted by selection. For example, Kruuk *et al.* (2002) found heritable variation for antler size in red deer as well as directional selection for increased size. However, there was no evolutionary shift in antler size over almost 30 years despite the heritable variation; it appears that only the environmental component of variation in antler size was under selection and that this component was probably related to the nutritional state of the organisms. Genetic variation in these studies may be present, but the phenotypic variance relevant to the fitness of these organisms presumably has a heritability of zero. This reflects a lack of genetic variability in underlying traits; if genetic variation in populations altered metabolic efficiency, this would probably result in a genetic component for dealing with nutritional stress. Selection on variation in body size resulting from exposure to different nutrition conditions would then lead to a genetically based change in the population.

In summary, although genetic variation in populations is commonly assumed to be present for quantitative and threshold traits controlled by a number of loci, this might not be the case for organisms with restricted distributions. Moreover, where genetic variation has been detected, selection responses might still be limiting because of effects such as trade-offs or the genetic component of phenotypic variation having no impact on fitness. In all of these cases, variability generators could help to facilitate evolutionary change, but their impact has not yet been evaluated.

VII. CONCLUDING REMARKS: EXPERIMENTAL PROGRAMS FOR DEFINING THE ROLE OF VARIABILITY GENERATORS

There are two ways of assessing the role of mutators, recombinators, stressors, and capacitators in evolution. The first is an empirical approach and involves following adaptive changes in populations exposed to different conditions. The second approach involves looking for signatures indicating that variability generators have had a role in past adaptive shifts. This is difficult because it is not clear what the signatures look like.

The empirical approach is typified by experiments such as those carried out by Flexon and Rodell (1982) and by Bourguet *et al.* (2003). In the former, recombination rates were scored after directional selection had produced an adaptive shift in resistance to a pesticide. In the latter, recombination rates were varied before selection was started to determine the role of altered recombination in generating an adaptive shift.

The empirical approach has been widely used to assess the role of mutators and recombinators in evolutionary change. Where it has not been widely applied is in the study of selection limits. If we take a population that is at a selection limit (either because of the absence of genetic variation or because of a strong trade off), can this limit be broken by treatments that influence the expression of capacitators? For instance, if we take *D. birchii* populations that are at a selection limit for desiccation resistance because they have not responded to selection for many generations, can the geldanamycin inhibitor or another inhibitor be used to break this selection limit? Could the limit be broken by exposure to a stressor that is known to increase the expression of genetic variation for desiccation resistance? Can irradiation or mutagens be used to trigger evolutionary changes in traits at limits by inducing high rates of mutations, or can limits be broken by introducing mutators directly? Once limits are broken, a number of follow-up studies could be undertaken to assess the nature of changes in capacitators and other processes affecting variability. Was a limit broken because of effects of capacitators or stressors on trait means, DI, or trait variances? Are shifts in means accompanied by shifts in DI and in canalization?

These types of experiments should be carried out in species that are ecologically restricted in terms of habitat or distribution, but with more widespread relatives that have evolved to occupy a greater range of habitats (for example, insect species restricted to particular host plants or rain forest habitats but with relatives that utilize a range of host plants and that are found outside rain forests). This may eventually allow the genetic changes in breaking limits to be compared with genetic differences between the related species.

Whether signatures can eventually be identified that reflect a role of variability generators remains an open question. If a stressor induces mutations by mobilizing a particular transposon, it may be possible to link the distribution of the transposon to the genetic basis of adaptive differences between species. If genes that contribute to trait divergence among taxa can be identified, these might be located in genomic regions with high recombination rates in groups of related taxa that show a high level of divergence, but in regions with low recombination in groups that show low divergence. If *Hsp90* or other capacitators control evolutionary shifts in canalized traits, relatively lower levels of expression of *Hsp90* or other capacitators might be expected in groups of taxa that have diverged further for canalized traits over a particular evolutionary period. Perhaps phenotypic rates of divergence might be higher in groups of taxa that are more frequently exposed to the types of stressors that suppress levels of capacitators.

When identifying signatures, it will often be difficult to separate the direct and indirect effects of variability generators. For instance, stressful conditions not only have the potential to increase mutation and recombination rates and alter patterns of variability in traits, but they will also influence population size and disrupt gene flow among local populations. In intermittently stressful environments where rates of evolutionary divergence are often high, it will be difficult to separate the direct effects of stressors on genomic variation from the indirect effects on population processes (Hoffmann and Parsons, 1997). The approaches suggested in the preceding text may be a way of doing this.

REFERENCES

Akhunov, E. D., Goodyear, A. W., Geng, S., Qi, L. L., Echalier, B., Gill, B. S., Miftahudin, Gustafson, J. P., Lazo, G., Chao, S., Anderson, O. D., Linkiewicz, A. M., Dubcovsky, J., La Rota, M., Sorrells, M. E., Zhang, D., Nguyen, H. T., Kalavacharla, V., Hossain, K., Kianian, S. F., Peng, J., Lapitan, N. L., Gonzalez-Hernandez, J. L., Anderson, J. A., Choi, D. W., Close, T. J., Dilbirligi, M., Gill, K. S., Walker-Simmons, M. K., Steber, C., McGuire, P. E., Qualset, C. O., and Dvorak, J. (2003). The organization and rate of evolution of wheat genomes are correlated with recombination rates along chromosome arms. *Genome Research* **13**, 753–763.

Al-Hiyaly, S. A., McNeilly, T., and Bradshaw, A. D. (1988). The effects of zinc contamination from electricity pylons: Evolution in a replicated situation. *New Phytologist* **110**, 571–580.

Al-Hiyaly, S. A., McNeilly, T., Bradshaw, A. D., and Mortimer, A. M. (1993). The effect of zinc contamination from electricity pylons: Genetic constraints on selection for zinc tolerance. *Heredity* **70**, 22–32.

Baer, C. F., and Travis, J. (2000). Direct and correlated responses to artificial selection on acute thermal stress tolerance in a livebearing fish. *Evolution* **54**, 238–244.

Barton, N. H., and Charlesworth, B. (1998). Why sex and recombination? *Science* **281**, 1986–1990.

Bjedov, I., Tenaillon, O., Gerard, B., Souza, V., Denamur, E., Radman, M., Taddei, F., and Matic, I. (2003). Stress-induced mutagenesis in bacteria. *Science* **300**, 1404–1409.

Bourguet, D., Gair, J., Mattice, M., and Whitlock, M. C. (2003). Genetic recombination and adaptation to fluctuating environments: Selection for geotaxis in *Drosophila melanogaster. Heredity* **91**, 78–84.

Bradshaw, A. D. (1991). Genostasis and the limits to evolution. *Philosophical Transactions of the Royal Society of London Series B: Biological Sciences* **333**, 289–305.

Brussard, P. F. (1984). Geographic patterns and environmental gradients: The central-marginal model in *Drosophila* revisited. *Annual Review of Ecology and Systematics* **15**, 25–64.

Capy, P., Bazin, C., Higuet, D., and Langin, T. (1997). *Dynamic and Evolution of Transposable Elements.* Austin, TX: R. G. Landes.

Capy, P., Gasperi, G., Biemont, C., and Bazin, C. (2000). Stress and transposable elements: Co-evolution or useful parasites? *Heredity* **85**, 101–106.

Clarke, G. M. (1998). The genetic basis of developmental stability. V. Inter- and intra-individual character variation. *Heredity* **80**, 562–567.

Clarke, G. M., and McKenzie, J. A. (1987). Developmental stability of insecticide resistant phenotypes in blowfly: A result of canalizing natural selection. *Nature* **325**, 345–346.

Clarke, G. M., Yen, J. L., and McKenzie, J. A. (2000). Wings and bristles: Character specificity of the asymmetry phenotype in insecticide-resistant strains of *Lucilia cuprina. Proceedings of the Royal Society of London Series B: Biological Sciences* **267**, 1815–1818.

Cohan, F. M., and Hoffmann, A. A. (1989). Uniform selection as a diversifying force in evolution: Evidence from *Drosophila. American Naturalist* **134**, 613–637.

Cox, E. C., and Gibson, T. C. (1974). Selection for high mutation rates in chemostats. *Genetics* **77**, 169–184.

Davies, A. G., Game, A. Y., Chen, Z. Z., Williams, T. J., Goodall, S., Yen, J. L., McKenzie, J. A., and Batterham, P. (1996). Scalloped wings is the *Lucilia cuprina Notch* homologue and a candidate for the modifier of fitness and asymmetry of diazinon resistance. *Genetics* **143**, 1321–1337.

Debat, V., and David, P. (2001). Mapping phenotypes: Canalization, plasticity, and developmental stability. *Trends in Ecology and Evolution* **16**, 555–561.

Falconer, D. S. (1989). *Introduction to Quantitative Genetics.* Burnt Mill, England: Longman.

Flexon, P. B., and Rodell, C. F. (1982). Genetic recombination and directional selection for DDT resistance in *Drosophila melanogaster. Nature* **298**, 672–674.

Foster, P. L. (2000). Adaptive mutation: Implications for evolution. *Bioessays* **22**, 1067–1074.

Freeman, D. C., Graham, J. H., Emlen, J. M., Tracy, M., Hough, R. A., and Escos, J. (2003). Plant developmental instability: New measures, applications, and regulation. In *Developmental Instability: Causes and Consequences* (M. Polak, ed.), pp. 367–386. New York: Oxford University Press.

Fuller, R. C., and Houle, D. (2002). Detecting genetic variation in developmental instability by artificial selection on fluctuating asymmetry. *Journal of Evolutionary Biology* **15**, 954–960.

Gangestad, S. W., and Thornhill, R. (2003). Fluctuating asymmetry, developmental instability, and fitness: Toward model-based interpretation. In *Developmental Instability: Causes and Consequences* (M. Polak, ed.), pp. 62–80. New York: Oxford University Press.

Garcia-Villada, L., Lopez-Rodas, V., Banares-Espana, E., Flores-Moya, A., Agrelo, M., Martin-Otero, L., and Costas, E. (2002). Evolution of microalgae in highly stressing environments: An experimental model analyzing the rapid adaptation of *Dictyosphaerium chlorelloides* (Chlorophyceae) from sensitivity to resistance against 2, 4, 6-trinitrotoluene by rare preselective mutations. *Journal of Phycology* **38**, 1074–1081.

Gartside, D. W., and McNeilly, T. (1974). The potential for evolution of heavy metal tolerance in plants. II. Copper tolerance in normal populations of different plant species. *Heredity* **32**, 335–348.

Gibbs, A. G., Chippindale, A. K., and Rose, M. R. (1997). Physiological mechanisms of evolved desiccation resistance in *Drosophila melanogaster. Journal of Experimental Biology* **200**, 1821–1832.

Gilchrist, G. W., and Huey, R. B. (1999). The direct response of *Drosophila melanogaster* to selection on knockdown temperature. *Heredity* **83**, 15–29.

Goho, S., and Bell, G. (2000). Mild environmental stress elicits mutations affecting fitness in *Chlamydomonas. Proceedings of the Royal Society of London Series B: Biological Sciences* **267**, 123–129.

Grimaldi, D. A. (1987). Amber fossil Drosophilidae (Diptera), with particular reference to the Hispaniolan taxa. *American Museum Novitates* **2880**, 1–23.

Grishkan, I., Korol, A. B., Nevo, E., and Wasser, S. P. (2002). Ecological stress and sex evolution in soil microfungi. *Proceedings of the Royal Society of London Series B: Biological Sciences* **270**, 13–18.

Harshman, L. G., and Haberer, B. A. (2000). Oxidative stress resistance: A robust correlated response to selection in extended longevity lines of *Drosophila melanogaster? Journals of Gerontology Series A: Biological Sciences and Medical Sciences* **55**, B415–B417.

Harshman, L. G., and Hoffmann, A. A. (2000). Laboratory selection experiments using *Drosophila:* What do they really tell us? *Trends in Ecology and Evolution* **15**, 32–36.

Hartl, D. L., Dijkhuizen, D. E., and Dean, A. M. (1985). Limits of adaptation: The evolution of selective neutrality. *Genetics* **111**, 655–674.

Hendrickson, A., Slechta, E. S., Bergthorsson, U., Andersson, D. I., and Roth, J. R. (2002). Amplification-mutagenesis: Evidence that directed adaptive mutation and general hypermutability result from growth with a selected gene amplification. *Proceedings of the National Academy of Sciences USA* **99**, 2164–2169.

Hewa-Kapuge, S., and Hoffmann, A. A. (2001). Composite asymmetry as an indicator of quality in the beneficial wasp *Trichogramma* nr. *brassicae* (Hymenoptera: Trichogrammatidae). *Journal of Economic Entomology* **94**, 826–830.

Hoffmann, A. A. (2000). Laboratory and field heritabilities: Some lessons from *Drosophila.* In *Adaptive Genetic Variation in the Wild.* (T. A. Mousseau, B. Sinervo, and J. A. Endler, eds.). New York: Oxford University Press.

Hoffmann, A. A., and Blows, M. W. (1994). Species borders: Ecological and evolutionary perspectives. *Trends in Ecology and Evolution* **9**, 223–227.

Hoffmann, A. A., and Merila, J. (1999) Heritable variation and evolution under favorable and unfavorable conditions. *Trends in Ecology and Evolution* **14**, 96–101.

Hoffmann, A. A., and Parsons, P. A. (1991). *Evolutionary Genetics and Environmental Stress.* Oxford, England: Oxford University Press.

Hoffmann, A. A., and Parsons, P. A. (1997). *Extreme Environmental Change and Evolution.* Cambridge, England: Cambridge University Press.

Hoffmann, A. A., and Woods, R. E. (2001). Trait variability and stress: Canalization, developmental stability, and the need for a broad approach. *Ecology Letters* **4**, 97–101.

Hoffmann, A. A., Dagher, H., Hercus, M., and Berrigan, D. (1997). Comparing different measures of heat resistance in selected lines of *Drosophila melanogaster. Journal of Insect Physiology* **43**, 393–405.

Hoffmann, A. A., Hallas, R., Dean, J., and Schiffer, M. (2003). Low potential for climatic stress adaptation in a rainforest *Drosophila* species. *Science* **301**, 100–102.

Houle, D., Morikawa, B., and Lynch, M. (1996). Comparing mutational variabilities. *Genetics* **143**, 1467–1483.

Indrasamy, H., Woods, R. E., McKenzie, J. A., and Batterham, P. (2000). Fluctuating asymmetry for specific bristle characters in *Notch* mutants of *Drosophila melanogaster. Genetica* **109**, 151–159.

Johnson, T. (1999). Beneficial mutations, hitchhiking, and the evolution of mutation rates in sexual populations. *Genetics* **151**, 1621–1631.

Kacser, H., and Burns, J. A. (1981). The molecular basis of dominance. *Genetics* **97**, 639–666.

Korol, A. B., and Iliadi, K. G. (1994). Increased recombination frequencies resulting from directional selection for geotaxis in *Drosophila. Heredity* **72**, 64–68.

Kristensen, T. N., Pertoldi, C., Andersen, D. H., and Loeschcke, V. (2003). The use of fluctuating asymmetry and phenotypic variability as indicators of developmental instability: A test of a new method employing clonal organisms and high temperature stress. *Evolutionary Ecology Research* **5**, 53–68.

Kruuk, L. E. B., Slate, J., Pemberton, J. M., Brotherstone, S., Guinness, F., and Clutton-Brock, T. (2002). Antler size in red deer: Heritability and selection but no evolution. *Evolution* **56**, 1683–1695.

Lamb, B. C., Saleem, M., Scott, W., Thapa, N., and Nevo, E. (1998). Inherited and environmentally induced differences in mutation frequencies between wild strains of *Sordaria fimicola* from "Evolution Canyon." *Genetics* **149**, 87–99.

Laurila, A., Karttunen, S., and Merila, J. (2002). Adaptive phenotypic plasticity and genetics of larval life histories in two *Rana temporaria* populations. *Evolution* **56**, 617–627.

Leigh, E. G. (1973). The evolution of mutation rates. *Genetics* **73**(Suppl.), 1–18.

Lenormand, T., and Otto, S. (2000). The evolution of recombination in a heterogeneous environment. *Genetics* **156**, 423–438.

Leung, B., Forbes, M. R., and Houle, D. (2000). Fluctuating asymmetry as a bioindicator of stress: Comparing efficacy of analyses involving multiple traits. *American Naturalist* **155**, 101–115.

Lewontin, R. C. (1974). *The Genetic Basis of Evolutionary Change*. New York: Columbia University Press.

Macnair, M. (1997). The evolution of plants in metal-contaminated environments. In *Environmental Stress, Adaptation, and Evolution* (R. Bijlsma and V. Loeschcke, eds.), pp. 3–24. Basel, Switzerland: Birkhauser.

MacPhee, D. G. (1985). Indications that mutagenesis in *Salmonella* may be subject to catabolite repression. *Mutation Research* **151**, 35–41.

McClintock, B. (1984). The significance of responses of the genome to challenge. *Science* **226**, 792–801.

McKenzie, G. J., Harris, R. S., Lee, P. L., and Rosenberg, S. M. (2000). The SOS response regulates adaptive mutation. *Proceedings of the National Academy of Sciences USA* **97**, 6646–6651.

McKenzie, J. A. (2003). The analysis of the asymmetry phenotype: Single genes and the environment. In *Developmental Instability: Causes and Consequences* (M. Polak, ed.), pp. 135–141. New York: Oxford University Press.

McKenzie, J. A., and Batterham, P. (1998). Predicting insecticide resistance: Mutagenesis, selection, and response. *Philosophical Transactions of the Royal Society of London Series B: Biological Sciences* **353**, 1729–1734.

McKenzie, J. A., and Game, A. Y. (1987). Diazinon resistance in *Lucilia cuprina:* Mapping of a fitness modifier. *Heredity* **59**, 381–391.

Meiklejohn, C. D., and Hartl, D. L. (2002). A single mode of canalization. *Trends in Ecology and Evolution* **17**, 468–473.

Merila, J., Sheldon, B. C., and Kruuk, L. E. B. (2001). Explaining stasis: Microevolutionary studies in natural populations. *Genetica* **112**, 199–222.

Michod, R. E., and Levin, B. R. (1988). *The Evolution of Sex*. Sunderland, MA: Sinauer.

Milton, C. C., Huynh, B., Batterham, P., Rutherford, S. L., and Hoffmann, A. A. (2003). Quantitative trait symmetry independent of Hsp90 buffering: Distinct modes of genetic canalization and developmental stability. *Proceedings of the National Academy of Sciences, USA,* **100**, 13396–13401.

Moller, A. P., and Thornhill, R. (1997). A meta-analysis of the heritability of developmental stability. *Journal of Evolutionary Biology* **10**, 1–16.

Nevo, E., Filippucci, M. G., Redi, C., Simson, S., Heth, G., and Beiles, A. (1995). Karyotype and genetic evolution in speciation of subterranean mole-rats of the genus *Spalax* in turkey. *Biological Journal of the Linnean Society* **54**, 203–229.

Neyfakh, A. A., and Hartl, D. L. (1993). Genetic control of the rate of embryonic development: Selection for faster development at elevated temperatures. *Evolution* **47**, 1625–1631.

Orr, H. A. (2000). The rate of adaptation in asexuals. *Genetics* **155**, 961–968.

Orr, H. A., and Betancourt, A. J. (2001). Haldane's sieve and adaptation from the standing genetic variation. *Genetics* **157**, 875–884.

Pal, C., and Hurst, L. D. (2003). Evidence for co-evolution of gene order and recombination rate. *Nature Genetics* **33**, 392–395.

Paxman, G. J. (1956). Differentiation and stability in the development of *Nicotiana rustica. Annals of Botany.* **20**, 331–347.

Perry, J., and Ashworth, A. (1999). Evolutionary rate of a gene affected by chromosomal position. *Current Biology* **9**, 987–989.

Phelan, J. P., Archer, M. A., Beckman, K. A., Chippindale, A. K., Nusbaum, T. J., and Rose, M. R. (2003). Breakdown in correlations during laboratory evolution. I. Comparative analyses of *Drosophila populations*. *Evolution* **57**, 527–535.

Philippi, T., and Seger, J. (1989). Hedging one's evolutionary bets, revisited. *Trends in Ecology and Evolution* **4**, 41–44.

Plough, H. H. (1917). The effect of temperature on crossing over in *Drosophila*. *Journal of Experimental Zoology* **24**, 148–209.

Polak, M. (ed.) (2003). *Developmental Stability: Causes and Consequences*. New York: Oxford University Press.

Queitsch, C., Sangster, T. A., and Lindquist, S. (2002). Hsp90 as a capacitor of phenotypic variation. *Nature* **417**, 618–624.

Rendel, J. M. (1967). *Canalisation and Gene Control*. London: Logos Press.

Roberts, S. P., and Feder, M. E. (1999). Natural hyperthermia and expression of the heat shock protein Hsp70 affect developmental abnormalities in *Drosophila melanogaster*. *Oecologia* **121**, 323–329.

Rocha, E. P. C., Matic, I., and Taddei, F. (2002). Over-representation of repeats in stress response genes: A strategy to increase versatility under stressful conditions? *Nucleic Acids Research* **30**, 1886–1894.

Roff, D. A., and Bradford, M. J. (2000). A quantitative genetic analysis of phenotypic plasticity of diapause induction in the cricket *Allonemobius socius*. *Heredity* **84**, 193–200.

Roth, J. R., Kofoid, E., Roth, F. P., Berg, O. G., Seger, J., and Andersson, D. I. (2003). Regulating general mutation rates: Examination of the hypermutable state model for Cairnsian adaptive mutation. *Genetics* **163**, 1483–1496.

Rutherford, S. L. (2003). Between genotype and phenotype: Protein chaperones and evolvability. *Nature Reviews Genetics* **4**, 263–274.

Rutherford, S. L., and Lindquist, S. (1998). Hsp90 as a capacitor for morphological evolution. *Nature* **396**, 336–342.

Saleem, M., Lamb, B. C., and Nevo, E. (2001). Inherited differences in crossing over and gene conversion frequencies between wild strains of *Sordaria fimicola* from "Evolution Canyon." *Genetics* **159**, 1573–1593.

Scharloo, W. (1991). Canalization: Genetic and developmental aspects. *Annual Review of Ecology and Systematics* **22**, 65–93.

Schmalhausen, I. I. (1949). *Factors of Evolution*. Philadelphia: Blakiston.

Simons, A. M., and Johnston, M. O. (2000). Plasticity and the genetics of reproductive behaviour in the monocarpic perennial, *Lobelia inflata* (Indian tobacco). *Heredity* **85**, 356–365.

Simpson, P., Woehl, R., and Usui, K. (1999). The development and evolution of bristle patterns in *Diptera*. *Development* **126**, 1349–1364.

Sniegowski, P. D., Gerrish, P. J., Johnson, T., and Shaver, A. (2000). The evolution of mutation rates: Separating causes from consequences. *Bioessays* **22**, 1057–1066.

Sommer, S., and Pearman, P. B. (2003). Quantitative genetic analysis of larval life history traits in two alpine populations of *Rana temporaria*. *Genetica* **118**, 1–10.

Stearns, S. C., Kaiser, M., and Kawecki, T. J. (1995). The differential genetic and environmental canalization of fitness components in *Drosophila melanogaster*. *Journal of Evolutionary Biology* **8**, 539–557.

Taddei, F., Radman, M., Maynard-Smith, J., Toupance, B., Gouyon, P. H., and Godelle, B. (1997). Role of mutator alleles in adaptive evolution. *Nature* **387**, 700–702.

Tan-Kristanto, A., Hoffmann, A. A., Woods, R., Batterham, P., Cobbett, C., and Sinclair, C. (2003). Translational asymmetry as a sensitive indicator of cadmium stress in plants: A laboratory test with wild-type and mutant *Arabidopsis thaliana*. *New Phytologist* **159**, 471–477.

Travis, J. M. J., and Travis, E. J. (2002). Mutator dynamics in fluctuating environments. *Proceedings of the Royal Society of London Series B: Biological Sciences* **269**, 591–597.

Van Dongen, S. (1998). How repeatable is the estimation of developmental stability by fluctuating asymmetry? *Proceedings of the Royal Society of London, Series B* **265**, 1423–1427.

Van Dongen, S. (2000). The heritability of fluctuating asymmetry: A Bayesian hierarchical model. *Annales Zoologici Fennici* **37**, 15–23.

Van Dongen, S., and Lens, L. (2000). The evolutionary potential of developmental instability. *Journal of Evolutionary Biology* **13**, 326–335.

Van Valen, L. (1962). A study of fluctuating asymmetry. *Evolution* **16**, 125–142.

Velkov, V. V. (1999). How environmental factors regulate mutagenesis and gene transfer in microorganisms. *Journal of Bioscience* **24**, 529–559.

Waddington, C. H. (1953). The genetic assimilation of an acquired character. *Evolution* **7**, 118–126.

Walker, G. C. (1984). Mutagenesis and inducible responses to deoxyribonucleic acid damage in *Escherichia coli*. *Microbiological Reviews* **48**, 60–93.

Weber, K. E. (1990). Selection on wing allometry in *Drosophila melanogaster*. *Genetics* **126**, 975–989.

Weber, K. E., and Diggins, L. T. (1990). Increased selection responses in larger populations. II. Selection for ethanol vapor resistance in *Drosophila melanogaster* at two population sizes. *Genetics* **125**, 585–597.

Whitlock, M. (1996). The heritability of fluctuating asymmetry and the genetic control of developmental stability. *Proceedings of the Royal Society of London, Series B* **263**, 849–853.

Whittle, J. R. S. (1998). How is developmental stability sustained in the face of genetic variation. *International Journal of Developmental Biology* **42**, 495–499.

Windig, J. J., and Nylin, S. (2000). How to compare fluctuating asymmetry of different traits. *Journal of Evolutionary Biology* **13**, 29–37.

Woods, R. E., Sgrò, C. M., Hercus, M. J., and Hoffmann, A. A. (1999). The association between fluctuating asymmetry, trait variability, trait heritability, and stress: A multiply-replicated experiment on combined stresses in *Drosophila melanogaster*. *Evolution* **53**, 493–505.

Yen, J. L., Batterham, P., Gelder, B., and McKenzie, J. A. (1996). Predicting resistance and managing susceptibility to cyromazine in the Australian sheep blowfly *Lucilia cuprina*. *Australian Journal of Experimental Agriculture* **36**, 413–420.

Zhang, X. S., and Hill, W. G. (2002). Joint effects of pleiotropic selection and stabilizing selection on the maintenance of quantitative genetic variation at mutation–selection balance. *Genetics* **162**, 459–471.

Zhang, X. S., Wang, J. L., and Hill, W. G. (2002). Pleiotropic model of maintenance of quantitative genetic variation at mutation–selection balance. *Genetics* **161**, 419–433.

Zhuchenko, A. A., Korol, A. B., and Kovtyukh, L. P. (1985). Change of the crossing-over frequency in *Drosophila* during selection for resistance to temperature fluctuations. *Genetics* **67**, 73–78.

CHAPTER 10

Within Individual Variation: Developmental Noise versus Developmental Stability

KATHERINE E. WILLMORE* AND BENEDIKT HALLGRÍMSSON[†]

*Department of Medical Sciences, University of Calgary, Health Sciences Centre, Calgary, Alberta, Canada

[†]Department of Cell Biology and Anatomy, Joint Injury and Arthritis Research Group, University of Calgary, Calgary, Alberta, Canada

INTRODUCTION

Phenotypic variation is ubiquitous. Within organisms this variation is modulated by two opposing forces: developmental stability and developmental noise. Developmental stability is the tendency for development to follow the same trajectory under identical genetic and environmental conditions; that is, developmental stability is the absence or minimization of phenotypic variation that may arise because of perturbations within an individual's developmental trajectory. Canalization is a closely related phenomenon. Canalization also refers to the buffering of development against perturbations. However, canalization reduces the effects of genetic and environmental insults and thereby reduces variation between individuals (Hallgrímsson *et al.*, 2002). It is currently an open question as to whether or not developmental stability and canalization are modulated by the same mechanisms; however, the prevailing view is that the mechanisms that confer developmental stability are separate from those involved in canalization. Developmental noise can be defined as perturbations from within an individual and therefore arise under the same genetic and environmental conditions. Therefore, developmental noise acts to disrupt the developmental trajectory, and developmental stability works to maintain development along this pathway.

Quantifying the degree of developmental stability is problematic, because it requires the measurement of lack of variation caused by perturbations from within an individual. Therefore, the phenotypic effects of its inverse, developmental instability, are measured. To accomplish this, we must partition the phenotypic variation caused by extrinsic factors such as the genotype and environment from variation arising from intrinsic perturbations. The most common measure used is fluctuating asymmetry (FA). FA refers to minor deviations of symmetry that, when measured for a population, are random in direction and are normally distributed with a mean of 0 (Van Valen, 1962). The logic is that the sides of symmetric structures when compared within individuals have the same genotype and presumably experience the same environmental conditions. For bilaterally symmetric traits, it is assumed that symmetry is the ideal or target state and that any deviations from symmetry are viewed as deviations from the normal developmental trajectory and thus are a measure of developmental instability. Traits with higher levels of FA have a higher degree of developmental instability and therefore also a lower degree of developmental stability.

Although convenient, the use of FA to measure developmental instability is problematic. First, FA is based on the assumption that perfect symmetry is the ideal and that any deviations from perfect symmetry represent a breakdown of the normal developmental trajectory. This assumption is intuitive, but lacks proof. Second, the deviations measured by FA are quite subtle, on the order of 1% of trait size, often rendering detection of this asymmetry difficult to impossible (Lens *et al.*, 2002). Finally, FA is the resultant phenotypic variation that represents the opposition between developmental noise arising within a structure and the developmental stability of that structure (Lens *et al.*, 2002). Therefore, a trait could have a low degree of FA because it experienced very little noise or because its level of developmental stability is high. Conversely, a structure could exhibit a large degree of asymmetry because it was subjected to high levels of developmental noise or because its developmental stability is low. This difficulty in distinguishing between the *causes* of FA and the *mechanisms* that buffer against the perturbations mirrors the confusion that is often found in the literature on developmental stability. Often the terms *fluctuating asymmetry, developmental noise,* and *developmental instability* are used interchangeably, contrary to their separate meanings. In this chapter, we attempt to emphasize the importance of differentiating between the causes of developmental noise and the mechanisms that create developmental stability. To break open the black box that still envelops the mechanisms that confer fidelity of development, it is necessary to understand the factors that can disturb the system and the mechanisms that maintain normal development. Therefore, this chapter is divided into two main sections, the first describing the causes of developmental noise and the second outlining the mechanisms used to buffer developmental noise to confer developmental stability.

In addition, our aim is to highlight the different levels that developmental noise and developmental stability work within the organism. Traditionally, developmental stability has been studied at the level of the organism, and developmental noise has been vaguely described as unknown thermodynamic noise (Reeve and Robertson, 1953). However, with the recent advances made in molecular biology and renewed interest in genotype to phenotype mapping, it has been possible to study developmental stability at the molecular level and to describe more accurately the factors that cause developmental noise. We review both the causes of noise and the mechanisms of stability at three phenotypic levels: the molecular level, the developmental systems level, and the level of the trait or organism. The purpose of this chapter is not to be an exhaustive review of the many causes of developmental noise and the mechanisms used to buffer against such random perturbations. Rather, we hope to underscore the importance of differentiating between developmental stability and developmental noise as well as the importance of understanding both phenomena from an entire systems approach.

I. CAUSES OF DEVELOPMENTAL NOISE

Developmental noise is usually defined as perturbations that arise from random fluctuations at the molecular and cellular level (Palmer, 1996). However, the complex interactions involved in the development and functioning of multicellular organisms can amplify the noise originating at the molecular scale. Therefore, in this chapter, any mechanism that amplifies the effects of random perturbations is considered a cause of developmental noise.

A. Causes of Developmental Noise at the Molecular Level

Considerable morphologic variation is often observed among cells in a uniform population, including clonal cell populations (McAdams and Arkin, 1999). As these cells experience essentially the same genetic and environmental influences, the resulting variation must arise from developmental noise intrinsic to the cellular processes. Random perturbations arising from the stochastic nature of molecular and cellular processes tend to act locally, but the developmental noise incurred at this level can have cascading effects at levels of higher phenotypic complexity (Lens *et al.*, 2002). Our understanding of the exact causes of developmental noise is still incomplete. However, much of the noise arises from imperfections in the molecular and cellular machinery itself. In the following text, we attempt to summarize the many known causes of stochastic molecular and cellular processes.

1. Somatic Mutation

DNA can be damaged in a variety of ways, and this damage is often caused by intrinsic processes. The stochastic nature of such damage can lead to random perturbations that can be perpetuated at levels of higher complexity potentially affecting the phenotype if left unrepaired. Therefore, random errors at the level of DNA are potential causes of developmental noise for the organism.

Incorrect DNA replication can lead to base-pair mismatches and insertion/deletion loops (Mohrenweiser *et al.*, 2003). If left uncorrected, these incorrect copies of DNA can have cascading effects because subsequent replication events will perpetuate the errors (Alberts *et al.*, 1998). Double-strand breaks are a particularly harmful form of DNA damage that can occur by a failure in DNA replication (Mohrenweiser *et al.*, 2003). Double-strand breaks can lead to extreme consequences for the system, including cell death and the formation of cancer cells (Mills *et al.*, 2003). Heterologous DNA exchanges can occur if there are errors during recombination (Mohrenweiser *et al.*, 2003).

Reactive oxygen species formed from normal cellular metabolic processes can also cause several forms of DNA damage (Alberts *et al.*, 1998; Mills *et al.*, 2003; Mohrenweiser *et al.*, 2003). These potentially harmful metabolic by-products include oxygen free radicals from mitochondria, protons from oxidation–reduction cycles, and ethoxy compounds from oxidized lipids (Kaufmann and Paules, 1996). Depurination and deamination cause lesions in the DNA structure that arise spontaneously presumably from reacting with cellular metabolites (Alberts *et al.*, 1998). Similarly, base excision and double-strand breaks can occur as a consequence of DNA's vulnerability to hydrolytic decomposition, often caused by normal by-products of metabolism (Mohrenweiser *et al.*, 2003).

It is difficult to quantify an average rate for somatic mutations because it varies among different tissue types and under different circumstances. However, Frank and Nowak (2004) conservatively estimate the number of human colon stem cell mutations that would occur by the age of 60. Based on a conservative estimate of skin cell division in the colon, they suggest that by the age of 60 there would be at least 3000 cell divisions. They suppose that for each cell division in the colon the mutation rate is 10^{-7}; with 10^{5} genes, there will be 10^{-2} mutations in the coding region of the genes. After 3000 cell divisions estimated to occur over 60 years in the colon, they approximate that an individual would have experienced 30 mutational events. The colon is divided into several compartments, and with several stem lineages, the actual number of mutational events would be in the hundreds (Frank and Nowak, 2004). This estimation is quite conservative, but still underscores the high incidence of mutations in somatic cells and the potential damaging effects of these mutations if left uncorrected.

Damage to DNA can disrupt the fidelity of RNA transcription as well as its translation into proteins. In the following text, we outline some of the many other ways that noise can arise at the RNA level.

2. Causes at the Level of RNA and Protein Production

RNA transcription and translation have an error rate that is three orders of magnitude greater than that of DNA replication, and is therefore, a significant source of developmental noise intrinsic to the cell (Alberts *et al.*, 1998).

In addition to the errors that can occur during the process, transcription is an inherently stochastic process, leading to variable amounts of products at variable time intervals (McAdams and Arkin, 1999). Recent research indicates that transcription in eukaryotes is regulated by a binary "on" versus "off" switching mechanism, based on probability (McAdams and Arkin, 1999; Fiering *et al.*, 2000; Blake *et al.*, 2003; Klingenberg 2004a). Fiering and colleagues (2000) found evidence in support of this binary regulation in a study of the effect of enhancers on transcription. They found that gene expression in myofibers was stochastic, that within a given nucleus, a gene was either on or off, and that gene enhancers increase the probability that a gene will be activated. A result of this binary regulation is that transcription occurs in bursts of activity with intervals of inactivity of random duration (Klingenberg, 2004a). The products of transcription as well as the proteins from their subsequent translation will likewise be produced in sporadic bursts, enhancing the effects of the noise created during transcription (McAdams and Arkin, 1999; Swain *et al.*, 2002; Blake *et al.*, 2003; Klingenberg, 2004a). The production of proteins in this sporadic fashion at random time intervals, leads to variable amounts of protein product in the cell at any given time (McAdams and Arkin, 1997, 1999). Stochastic protein production can lead to large differences in subsequent regulatory cascades across a population of cells and is another potential source of developmental noise (McAdams and Arkin, 1997).

Models based on stochastic transcription and translation predict that developmental noise increases as the amount of transcript decreases (Elowitz *et al.*, 2002). Therefore, slower transcription rates (Alberts *et al.*, 1998; Blake *et al.*, 2003), longer durations between transcriptional activity, and shorter half-lives of products (Klingenberg, 2004a) will potentially lead to increased noise. In addition, reduced numbers of products extrinsic to transcription that are involved in its regulation may also increase developmental noise. These extrinsic factors include the number of RNA polymerases, ribosomes, and proteins (Swain *et al.*, 2002).

The endoplasmic reticulum (ER) is the gateway for proteins to the secretory pathway (Fewell *et al.*, 2001; Sitia and Braakman, 2003). The ER provides a particularly favorable environment for the folding and maturation of proteins, often acting as a final check for the fidelity of proteins before their entry to the secretory pathway (Fewell *et al.*, 2001; Ellgaard and Helenius, 2003; Sitia and Braakman, 2003). Quality control in the ER is particularly important as the final locations of secreted proteins are often lacking chaperones or other folding factors that could fix their configuration if they were to become damaged (Ellgaard and Helenius, 2003). The proteins secreted from the ER are involved in a variety of important processes such as nutrient storage, gene regulation, and some extracellular carrier functions for ligands (Ellgaard and Helenius, 2003). The functional importance of proteins produced in the ER, coupled with their potentially far-reaching effects as a result of entering the secretory pathway, would enable a cascade of downstream perturbations if the stability of these proteins were disrupted. Several mechanisms exist to ensure the fidelity of proteins produced in the ER (see the following text under mechanisms). However, proteins that have somehow evaded the quality control mechanisms of the ER have a great potential to cause several and potentially severe downstream disruptions. One possible cause for the release of damaged or nonfunctional proteins from the ER arises from the primary method of quality control. The ER does not account for the functionality of proteins. Rather, quality control relies on the correct conformation of proteins, assuming that proteins that are properly folded are functional, potentially leading to the secretion of nonfunctional, but correctly folded proteins (Ellgaard and Helenius, 2003).

The presence of disulfide bonds is characteristic of secretory proteins, and the ER provides a unique environment that supports disulfide bond formation (Fewell *et al.*, 2001). Under normal circumstances, disulfide bonds are beneficial, and there are several factors within the ER to ensure their formation. However, the processes involved in disulfide bond formation are quite slow in comparison with other folding processes, and they may therefore slow the production and the subsequent secretion of proteins (Fewell *et al.*, 2001). If there were a disruption to disulfide bond formation within the ER, time-sensitive downstream processes could be halted or slowed because of a reduced number of secretory proteins.

Proteins act to regulate the further production of other proteins as well as to modulate the expression of genes. Therefore, any stochasticity at the level of

RNA and protein production can have cascading consequences throughout the developmental system.

3. Causes at the Level of Genes and Genetic Networks

Stochastic gene expression may lead to random variation within a system. Because fluctuations in gene expression can produce heterogeneity within a clonal cellular population (McAdams and Arkin, 1999; Elowitz *et al.*, 2002; Blake *et al.*, 2003), any process that alters gene expression is a potential cause of developmental noise. Gene expression is influenced by numerous factors, including molecules within the cell, mutations causing dominant negative effects and haploinsufficiency, signaling molecules from surrounding cells and the environment, and epistasis.

Various molecules within the cell modulate gene expression. RNA polymerases, proteins, ribosomes, and mRNA degrading machinery are all examples of molecular species that can influence gene expression within the cell (Elowitz *et al.*, 2002; Swain *et al.*, 2002). The concentration, activity level, and state of these molecules can create fluctuations in gene expression. Therefore, random production of these molecules, unequal partitioning of cellular components during cell division (McAdams and Arkin, 1999), and varying degrees of proximity of these molecules to different genes can cause fluctuations in gene expression.

Decreases in gene dosage are known to increase the degree of variability of gene expression and subsequent processes in the regulatory and developmental cascade, leading to potential phenotypic variation (Klingenberg, 2004a). Gene dosage can be reduced by mutations causing haploinsufficiency and dominant negative effects. Haploinsufficiency can effectively reduce gene activity, because a single dose of the gene can be insufficient for normal development and therefore unable to produce the wild type (Klingenberg, 2004a). Dominant negative effects reduce gene dosage by interfering with the normal form of a gene, thereby inhibiting its function (Klingenberg, 2004a). The stochastic effects of haploinsufficiency on the phenotype can be observed in neurofibromatosis type 1 (NF1) patients. NF1 patients exhibit café-au-lait macules as a result of NF1 haploinsufficiency of melanocytes (Kemkemer *et al.*, 2002). Therefore, this irregular distribution of pigmentation appears to result from the stochastic expression of melanocyte genes.

Gene interactions within and between regulatory networks are another source of variable gene expression. Epistasis is simply the phenotypic effect of these interactions among different loci and can partially explain how modifications of one gene can affect the stability of subsequent reactions and potentially the phenotype (Routman and Cheverud, 1997; Klingenberg, 2004a). A recent study by Leamy *et al.* (2002) found significant epistasis in FA for centroid size in mouse mandibles. Because fluctuating asymmetry is one measure of the phenotypic effects of developmental noise, this study illustrates how epistasis is a potential source of developmental noise. Galis and colleagues (2002) likewise found a positive association between gene interactions and increased noise in segment polarity gene networks. They found that the effects of mutations were amplified via pleiotropy during embryogenesis.

Variable gene expression is a particularly important mechanism for developmental noise because many developmental processes are dependent on a binary threshold. Fluctuations in gene expression will cause fluctuations in gene products that are necessary for subsequent processes in development and the regulation of development. If concentrations of products produced by gene expression are not sufficient to reach the threshold required for cascading processes, development will be altered, leading to random phenotypic effects. Klingenberg (2003) uses an example of a mutation in *Drosophila melanogaster* to illustrate how variable gene expression can affect the stability of the phenotype. As described in Klingenberg (2003), flies homozygous for the *abrupt*^{1} mutation (*ab*^{1}) exhibit shortened L5 wing veins. Homozygous *ab*^{1} flies are known to have variable gene expression, whereby some mutant flies appear to be wild type with the normal development of L5 veins. Presumably, the mutation reduces gene expression so that the gene activity is below the threshold for normal L5 wing vein development. However, some individuals will surpass this threshold simply by chance because of the variable gene expression found in *ab*^{1} flies, and the L5 vein will develop normally. This example demonstrates how factors that influence gene expression are potential sources of random phenotypic effects and may therefore cause developmental noise.

Noise at the molecular level is common and can arise in a multitude of ways. Molecular and cellular machinery are highly interactive, and therefore minute disruptions of development within the cell can have significant consequences even at the gross phenotypic level.

B. Causes of Developmental Noise at the Developmental Systems Level

Eukaryotic development is a nonlinear and a highly interactive process. Perturbations at one point in the developmental process can lead to a cascade of noise, and because of the nonlinear nature of development, the consequences of this noise are often unpredictable. In this section, we attempt to outline some of the causes of developmental noise that arise at the developmental systems level.

Components of development interact with each other in a variety of ways. Signaling is a common mechanism and can occur between different cells within a structure, or it can occur between different structures (Salazar-Ciudad *et al.*, 2003). Cells can have inductive effects on other cells by secreting diffusible molecules or through direct interactions with adjacent cells through gap junctions (Salazar-Ciudad *et al.*, 2003). These inductive interactions can occur in hierarchic systems. For instance, a signal from cell A is received and interpreted by cell B, inducing some

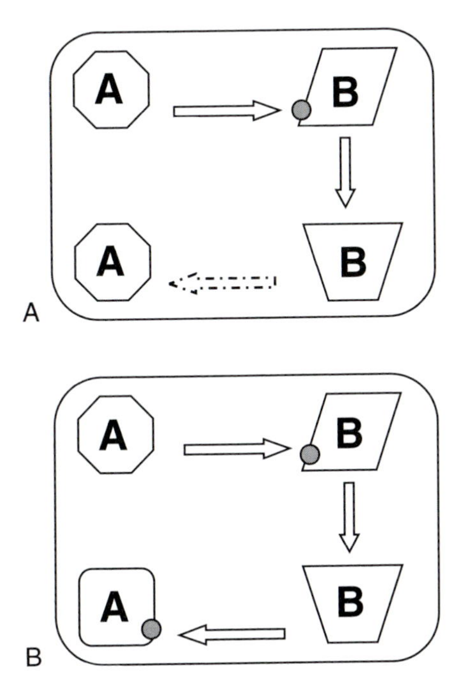

FIGURE 10-1. Simplified examples of two types of inductive systems. (A) Represents a hierarchic inductive system where cell A signals (arrow) cell B. Cell B is able to receive the signal because of a receptor (grey circle), and this signal instigates a conformational change in cell B. Cell B responds by returning a signal, which is not received by cell A, and there is no subsequent shape change in cell A. (B) Represents an emergent inductive signal, where the sequence of events are as described in (A) except that cell A does have a receptor for the returned signal from cell B and undergoes a conformational change as well.

change in cell B (Figure 10-1A). However, any signal emitted from cell B in response will have no such inductive effect on cell A (Salazar-Ciudad *et al.*, 2003). In contrast, emergent inductive systems allow cell B to reply to cell A via some signal, and this signal will induce some effect in cell A (Figure 10-1B). Similarly, there is evidence of left–right signaling between bilateral body structures (Klingenberg and Nijhout, 1998). This communication between cells and between structures is part of the normal developmental process, but can also lead to an accumulation and dissemination of random errors if the regulation of these interactions goes awry. Incorrect signaling could have severe effects for cell positional information. Cells are able to recognize their position within a group of cells by concentration gradients of regulator molecules (Bodnar, 1997; Furusawa and Kaneko, 2003). Inaccurate cell signaling could lead to changes in concentration gradients used by the system to recognize cell position and may create a cascade of incorrect interactions. Communication via feedback can also lead to increased noise if the conditions are not ideal. For instance, it has been suggested that positive feedback is utilized in synchronizing growth between bilateral structures and is dependent on the time lag between communications, the strength of the signal, and the presence of other feedback systems (Emlen *et al.*, 1993). Assuming that this sort of cross-talk between bilateral structures indeed exists, overcompensation on the side that is lagging in growth can occur if the time lapse between communications is large and the feedback signal is quite strong. This overcompensation could lead to increased asymmetry rather than the desired return to symmetry (Emlen *et al.*, 1993).

Noise can occur during development because of competition between cells or structures for some resource that is growth limiting (Klingenberg and Nijhout, 1998). This resource is usually a growth factor or a nutrient that is necessary in a certain quantity for normal growth and development. Klingenberg and Nijhout (1998) demonstrated that competition between body structures for a limited resource is a potential cause of morphologic asymmetry. They tested the hypothesis that structures compete for resources by removing the hindwing imaginal discs of developing buckeye butterflies. When both left and right hindwing imaginal discs were removed, structures that developed from surrounding imaginal discs were heavier on both sides. When only the left or right hindwing disc was removed, the structures developing from the surrounding imaginal discs on the treated side were affected only, leading to increased asymmetry between left and right traits. Further evidence that these changes in morphology resulted from competition was that the structures that showed the strongest response to the removed imaginal disc were those traits that develop in closest proximity to the hindwing disc.

Another potential cause of noise during development is differential morphogen concentration and diffusion. Klingenberg and Nijhout (1999) showed that random fluctuations in morphogen release or migration through the intercellular matrix can lead to increased noise. Likewise, Graham and colleagues (1993) demonstrated that oscillations in morphogen concentration could lead to abnormal asymmetry.

The developmental mechanisms described in the preceding text have their main effects on noise by altering the state of cells. Salazar-Ciudad and colleagues (2003) describe another variety of processes that alter the relative arrangement of cells but do not affect their states. These mechanisms include apoptosis, cellular contraction, differential cellular adhesion, differential migration, and random matrix swelling. All of these mechanisms alter the patterning of cells and are part of normal developmental processes. However, if the regulation of these mechanisms is somehow disrupted, aberrant cell patterns could result.

All of the preceding mechanisms can have cascading consequences as a result of the common use of thresholds in development. Graham and colleagues (1993) explain the effects of thresholds in development in terms of chaos theory. Chaos, in very simplistic terms, describes how a minute disruption of one component or process can lead to very large disruptions of other components or processes and that these disruptions are unpredictable (Emlen *et al.*, 1993). Likewise, a threshold in development can transform a minor change in the concentration of some developmental component into a much larger change in concentration of another component, and the consequences of these changes are often unpredictable (Graham *et al.*, 1993).

The nonlinear nature of development amplifies all of the preceding imperfections arising at the developmental systems level, as well as the stochasticity produced at the cellular and molecular level. If a developmental pathway is depicted as a nonlinear curve as in Figure 10-2, the effect of any perturbation is dependent on the location along the curve. This nonlinearity can arise from temporal differences because different structures are more sensitive to perturbations at different developmental stages. Therefore these perturbations will have varying effects depending on the stage of development. Another source of nonlinearity of development is the genetic background. Different genetic backgrounds will cause individuals to fall at different locations along a developmental pathway and therefore will affect their sensitivity to fluctuations in initial conditions. Finally, interactions between different developmental pathways can create nonlinearity.

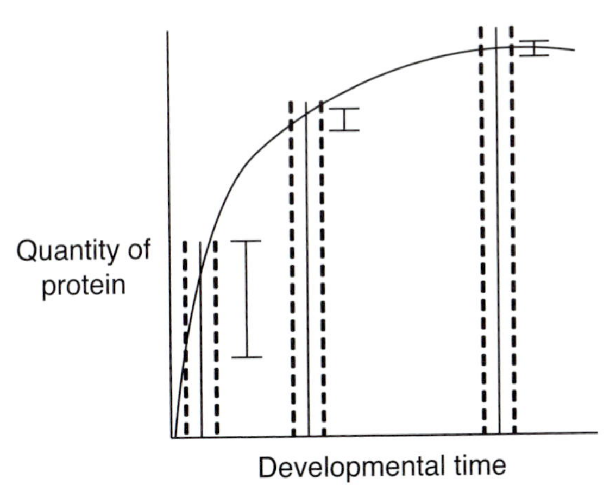

FIGURE 10-2. Representation of a nonlinear developmental pathway for the production of a protein that is dependent on developmental time. The curved line represents the developmental pathway. The vertical solid lines represent different points in development, and the flanking dashed vertical lines represent minor fluctuations in this timing. The three crossbars illustrate the differences in the amount of protein produced at different points during development when subjected to the same fluctuations in timing of events.

Nonlinearity can have cascading effects on other downstream developmental pathways and processes. As discussed in the preceding text, many developmental processes rely on a threshold phenomena, and varying sensitivity to perturbations could affect the quantity of some product or the rate at which it is produced. Depending on the severity of these differences, the probability that a dependent threshold will be reached will likewise vary.

C. Causes of Developmental Noise at the Organismal Level

As mentioned earlier, it may seem unusual to talk about the causes of developmental noise at the level of the organism. Although it is less intuitive at the organismal level, we suggest that interactions between different tissues and between different traits that were developed under a regimen that was subjected to various types of developmental noise could potentially create new random perturbations or perhaps simply amplify the consequences of stochastic variation introduced at the molecular or developmental level. Either way, the piecing together of the developmental processes that construct the functional organism, via epigenetic interactions, adds yet another potential layer of developmental noise.

The term *epigenetics* has a variety of definitions. Here, we use the definition of Atchley and Hall (1991), which describes epigenetics as heritable factors that can arise from within the individual's genome or by interactions between the genome of the individual and the maternal genome. An example is the relationship between muscle and bone. The different factors that define muscular structure and function are heritable and can modify the structure and function of bone (Atchley and Hall, 1991). A large body of research has focused on this interaction, particularly on the effects of the muscles of mastication on cranial morphology in mammals. In a series of papers, Herring and colleagues have investigated the different strains in the skull produced by the muscles of mastication and how these strains modulate cranial growth and structure. They have found that strains measured in the cranium are directly related to the patterns of muscle contraction during chewing (Rafferty and Herring, 1999; Herring and Teng, 2000). The polarity of this mechanical loading has also been related to cell proliferation in bone; that is, tension tends to promote cell replication at sutures, and compression inhibits cell proliferation (Herring, 1993; Herring *et al.*, 2002). These interactions can create or amplify developmental noise in the strains created by the right and left muscular counterparts. As with handedness, most mammals have a preferred side of chewing. Increased usage of one side will most likely lead to increased muscle mass and power on the favored side and will result in differential strain patterns in the skull. As strain patterns have been related to cranial morphology, asymmetric usage or strength of the muscles of mastication could lead to asymmetric skull form.

Machida *et al.* (2003) found such a relationship in human patients that showed mandibular asymmetry. They observed that differential reflex activity of the masseter muscle was capable of creating major deviations in mandibular symmetry.

Cranial morphology is affected by other epigenetic factors as well. Vinter and colleagues (1997) found a positive relationship between loss of teeth and increased variation in the mandible. This change in shape of the mandible has cascading effects on cranial morphology as the direction of pull from the muscles of mastication is also changed. The displacement of the temporomandibular joint (TMJ) in rabbits has likewise shown a positive correlation with increased cranial variability (Legrell and Isberg, 1998, 1999). Legrell and Isberg (1998, 1999) found that the ramus of the mandible was shortened on the side of the displaced TMJ and that there was an increase in bone growth along the inferior border on the affected side. Again, these changes in mandibular structure will in turn affect cranial morphology.

Random perturbations incurred at any one of the preceding phenotypic levels can lead to unstable development. Fortunately, an emergent property of complex systems is the development of a multitude of buffering mechanisms that can correct or compensate for the effects of this noise.

II. MECHANISMS OF DEVELOPMENTAL STABILITY

Fidelity of development is the norm despite the numerous perturbations to which any developing trait is subjected. Therefore, mechanisms that buffer development against this noise must exist. Recently, questions have been raised as to the nature of these mechanisms. Are these mechanisms an adaptive response to ensure developmental stability? Alternatively, are these mechanisms emergent properties of the system that just fortuitously infer developmental fidelity? Currently, these questions cannot be answered, and therefore the following is a discussion of several of the mechanisms used by the developing system at all phenotypic levels against stochastic perturbations. There is no inference as to whether these mechanisms exist for the sole purpose of creating developmental stability or whether they exist as emergent properties of a complex system. The goal of this section is simply to point out the varied ways that development is buffered at different phenotypic levels.

A. Mechanisms for Developmental Stability at the Molecular Level

Stochastic fluctuations at the molecular level can have profound effects on the phenotype through cascading processes of regulatory and developmental networks. Therefore, it is important that much of the noise induced at this level is buffered, and the earlier this buffering occurs within the developmental cascades, the less effect the noise will have. At the molecular level there are many buffering mechanisms, rendering the random errors incurred during DNA replication and recombination, transcription and translation of RNA, and the stochastic expression of genes undetectable at the gross phenotypic level. The following is a description of many of these buffering processes. It is interesting to note that many of the processes of developmental stability can also be sources of developmental noise.

1. Mechanisms at the Level of DNA

DNA is susceptible to damage by intrinsic perturbations, and this damage can have severe consequences including cell death, neoplastic transformation, and chromosomal abnormalities (Mills *et al.*, 2003). Fortunately, there are several mechanisms inherent at this level that repair the damage before such consequences can occur.

Mismatch repair is a process whereby random errors such as base-pair mismatches, insertion/deletion loops, and heterologies that can occur during DNA replication and recombination are fixed. Mismatch repair can repair either the mismatches or the insertion/deletion loops, or it can prevent heterologous DNA exchanges (Mohrenweiser *et al.*, 2003).

Double-strand breaks can occur during DNA replication, through interactions with intrinsic metabolites, and by exposure to external mutagens. Double-strand breaks do not usually have the chance to affect subsequent reactions adversely because they are repaired by two mechanisms: homologous recombination and nonhomologous end joining (Mills *et al.*, 2003; Mohrenweiser *et al.*, 2003). Homologous recombination repair uses the intact sister chromatid as a template and copies the undamaged sequence information to join to the broken segment of DNA (Mohrenweiser *et al.*, 2003). Nonhomologous end joining is a homology-independent method that either adds or deletes sequence information to restore the continuity and stability of the broken chromosome (Mohrenweiser *et al.*, 2003).

DNA is prone to the random excision of bases and nucleotides, often as a consequence of interacting with reactive metabolites. Two separate mechanisms exist that repair these defects. Base excision is corrected by the aptly named mechanism, base excision repair, whereby chemical modifications to DNA bases are recognized and repaired (Mohrenweiser *et al.*, 2003). Nucleotide excision repair is quite similar but usually functions to fix damage that creates large structural defects in the normal Watson–Crick base pairing (Mohrenweiser *et al.*, 2003).

It is important that these repair mechanisms are given a chance to fix the damaged DNA as the defects are recognized. To ensure that damaged DNA is not transmitted by DNA replication of the defective DNA, cells have developed a mechanism called damage recognition cell cycle delay response (Mohrenweiser *et al.*, 2003). This mechanism coordinates repair mechanisms with DNA replication and cell division through a series of checkpoints (Kaufmann and Paules, 1996;

Mohrenweiser *et al.*, 2003). Checkpoints function throughout all phases of the cell cycle to ensure that the necessary events in the cell cycle are complete before moving into the next cycle and to provide more time for the repair (Kaufmann and Paules, 1996).

Occasionally, DNA damage will manage to escape the preceding repair mechanisms, and replication of this defective DNA will occur. As a final effort to allow replication and cell division to continue, bypass polymerases ameliorate the effects of the defective sequence functioning as a mechanism for tolerance (Mohrenweiser *et al.*, 2003). This tolerance can be beneficial for the cell, but it can also act to amplify errors and lead to reduced stability.

Reducing errors at the DNA level is extremely important, because all genetic information is encoded within the DNA. However, several mechanisms exist at the level of RNA and protein production that can potentially compensate for errors in the DNA structure.

2. Mechanisms at the Level of RNA and Protein Production

Common causes of noise at the level of protein production include fewer transcripts, irregular bursts of protein synthesis, and low concentration of regulatory proteins. Therefore, any process that can create more transcripts, reduce the stochasticity of protein production, and increase the number of regulatory proteins and that can generally increase the fidelity of transcription, translation, and protein production will be a process that infers stability of development.

The number of transcripts found within the cell at any given time can be increased in two ways. The first way is to produce more transcripts in a certain amount of time, and the second method is to lengthen the life span of the transcripts. Fiering and colleagues (2000) discuss how enhancers can increase the concentration of transcript within the cell. They suggest that enhancers increase the probability of gene transcription rather than increasing the rate of transcription as was previously believed to be their role. Therefore, enhancers aid the production of a higher number of transcripts by increasing the probability of transcriptional activity. Other factors that determine if, and how often, transcription will occur are the strength of the promoter and binding strengths of the ribosome and RNAse to the mRNA transcript (McAdams and Arkin, 1997). If the binding strengths of these molecules are ideal, the fidelity of transcription will increase, as will the rate of transcription. Increasing the half-life of RNA transcripts also raises the concentration of transcript within the cell at a given point in time. Lengthening the life span of transcripts can be achieved by reduced activity of RNA degradation machinery.

Increasing the rate and the probability of transcription will also decrease the effect of “bursty” protein production. McAdams and Arkin (1997, 1999) discuss the inherent stochasticity of protein production and how these random bursts of activity produce variable numbers of proteins at variable time intervals, thereby creating noise at the molecular level. Any process that enhances the fidelity of transcription and translation will promote more consistent protein production and therefore ensure a more constant concentration of protein within the cell at any given time.

Structural properties of the cell also increase the fidelity of protein production. The nuclear envelope functions to separate the DNA and transcriptional machinery from the translational apparatus, thereby preventing the accidental translation of incorrect transcripts (Bird, 1995).

It was previously noted that the error rate in transcription and translation is quite high. Therefore, it is necessary to have many mechanisms that can correct for these errors. However, as molecular processes are highly interactive, there is still the opportunity for uncorrected errors to be buffered at higher levels within the genetic network of a developmental process.

The correct conformation of proteins is essential to their proper activity and functioning. Proteins are considered to be properly folded if they are in their native conformation, which is the pattern of folding that is most energetically favorable (Ellgaard and Helenius, 2003). Extreme consequences of incorrect protein folding are diseases such as cystic fibrosis and Creutzfeldt-Jacob disease (Brown *et al.*, 1997; Sitia and Braakman, 2003). Several molecular and chemical chaperones exist that help to ensure the proper folding of proteins.

Heat shock proteins are molecular chaperones and have received a great deal of attention in recent literature on developmental stability and canalization. Heat shock proteins are able to recognize and bind to damaged proteins and either aid in their refolding or in their degradation (Klingenberg, 2003). The heat shock protein, Hsp90, has been particularly well studied and has been found to be ubiquitous throughout all eukaryotic organisms (Queitsch *et al.*, 2002). Hsp90 interacts with proteins that are involved in signal transduction pathways under nonstressed physiologic conditions (Klingenberg, 2003). These proteins include a wide variety but very specific group of unstable proteins such as transcription factors and kinases, and Hsp90 helps to maintain their activity until they are stabilized by shape changes (Queitsch *et al.*, 2002). However, when the system is exposed to some stress, such as extreme temperature, Hsp90 functions in the general stress response, abandoning its usual specialized function (Klingenberg, 2003). Experiments by Rutherford and Lindquist (1998) demonstrated the potential buffering role of Hsp90 for developmental stability in *Drosophila*. They found that mutants for Hsp90 produced a variety of phenotypic effects and that these effects varied between different genetic backgrounds. These results are explained by the role that Hsp90 generally plays in ensuring the fidelity of proteins in signal transduction pathways. Variation in genes within these pathways and their inherent stability and reliance on Hsp90 activity will lead to the variable response of phenotypic structures to the mutation within individuals as well as between different genetic backgrounds. Similar phenotypic results were found when the flies were subjected to a stressful temperature environment (Rutherford and Lindquist, 1998). Increased temperature during development of *Drosophila* resulted in an increased incidence of aberrant phenotypes. The increased temperature induced the general stress response of Hsp90 leaving the buffering of the proteins involved in signal transduction pathways more vulnerable to disruption. Both of these experiments show that by altering the usual function of Hsp90 either by mutation or by exposure to an external stress, several phenotypic abnormalities result, and that these defects show differential penetrance depending on the trait observed and the genetic background. Therefore, it has been suggested that Hsp90 is a potential mechanism of developmental stability under normal physiologic conditions.

Chemical chaperones have also been found to be important in the fidelity of protein folding. As is the case with molecular chaperones, chemical chaperones are not incorporated into the resulting properly folded protein, nor do they give any direct information for the conformational changes. Chaperones, both chemical and molecular, increase the fidelity of protein folding by increasing the probability that newly synthesized proteins enter a correct pathway for folding (Brown *et al.*, 1997). In a study involving three mutant cell lines, Brown and colleagues (1997) found that exposure to the chemical chaperones glyceral, trimethylamine N-oxide, and deuterated water, resulted in the wild-type phenotype of these mutants under nonpermissive temperatures. Therefore, chemical chaperones are another potential mechanism used by the system to ensure developmental fidelity.

Quality control of proteins within the ER has been shown to be a highly effective mechanism of protein fidelity and the subsequent stability of downstream processes. As previously mentioned, the ER is the point of entry for proteins into the secretory pathway. The potential for disruptive polygenic effects from the release of damaged or misfolded proteins from the ER into these pathways provides strong selection for the proper folding of proteins synthesized in the ER. The ER contains many chaperones and enzymes that aid folding (Sitia and Braakman, 2003). These folding factors are highly redundant so that if a misfolded protein is missed by one chaperone or enzyme, or if one of the folding factors is damaged or missing, another factor will be able to recognize and bind to the aberrant protein (Ellgaard and Molinari, 1999). There are two main types of quality control within the ER, a primary response and a secondary response. The primary response is based on biophysical properties that are shared by all types of proteins (Ellgaard and Molinari, 1999; Ellgaard and Helenius, 2003; Sitia and Braakman, 2003). Secondary quality control processes differ from primary mechanisms in that they are cell-type specific (Ellgaard and Molinari, 1999; Ellgaard and Helenius, 2003; Sitia and Braakman, 2003). The general strategy of the primary response is to retain misfolded proteins in the ER until they can be properly folded or degraded (Ellgaard and Molinari, 1999). Misfolded proteins are recognized by folding chaperones by the presence of hydrophobic patches on the surface of the proteins, lack of compactness, and mobile loops (Ellgaard and Molinari, 1999). The folding chaperones and enzymes bind to the unassembled proteins, ensuring their retention within the ER as they attempt to aid in the refolding of the proteins. However, if after several attempts the proteins remain misfolded, the ER-associated degradation response (ERAD) is instigated (Ellgaard and Molinari, 1999; Ellgaard and Helenius, 2003; Sitia and Braakman, 2003). ERAD involves several processes. First, terminally misfolded proteins are recognized by folding factors within the ER, and then they are retro-translocated (removed from the ER) through the Sec61 channel into the cytosol (Ellgaard and Molinari, 1999). Within the cytosol, the aberrant proteins are deglycosylated and polyubiquitinated to allow the desired proteasomal degradation (Ellgaard and Molinari, 1999). Another primary quality control process is the unfolded protein response (UPR). If a large number of terminally misfolded proteins accrues within the ER, increased synthesis of folding chaperones is induced to deal with the backlog of proteins (Sitia and Braakman, 2003).

Within and outside of the ER, there are many folding chaperones and enzymes that help to ensure that proteins are in their correct conformation and are important factors of developmental stability.

3. Mechanisms at the Level of Genes and Genetic Networks

Genes are not stand-alone entities, but are embedded within networks, allowing for a vast number of interactions within and between cells. Genetic networks are an emergent property of a complex system and tend to dampen the effects of noise encountered by a particular gene. As discussed in a previous section, gene interactions can also magnify the noise encountered by a single gene if the perturbation is quite strong or goes undetected. However, in this section, we discuss the many ways in which genetic networks and gene interactions enhance the fidelity of the developmental system.

As mentioned previously, epistasis is the interaction between different loci, whereby an allele at one locus can influence the phenotypic effects of an allele at a separate locus (Wagner *et al.*, 1998; Klingenberg, 2004a). Because of complex genetic networks, interactions between genes can occur not only between genes that affect the same phenotype, but also between genes with unrelated functions (Wagner, 2000). Multiple interactions can effectively dampen the phenotypic effects of any abnormalities of one or a few genes within the network (Galis *et al.*, 2002). Common mechanisms used within genetic networks to infer stability are feedback loops, inhibition, and checkpoints. Becskei and Serrano (2000) demonstrated that autoregulatory negative feedback systems increase stability. They found that feedback within genetic circuits of *Escherichia coli* resulted in a two-fold increase in stability, by constraining the range over which network components can fluctuate. Regulation within genetic networks of eukaryotes is strongly influenced by inhibitions (Kirschner and Gerhart, 1998). In fact, these inhibitions are often required to relieve the effects of another inhibition. Inhibitions are often used as a component of another mechanism of stability, checkpoints (Hartwell and

Weinert, 1989). As was discussed as a mechanism used to ensure fidelity of DNA replication and recombination, checkpoints within genetic networks ensure coordination of cascading events within the network (McAdams and Arkin, 1999). If conditions for further progress within a regulatory cascade are insufficient, inhibitions can essentially halt the cascade until the proper conditions are met, increasing the stability of the resulting phenotype (McAdams and Arkin, 1999).

Redundancy is a special case of epistasis, whereby two or more genes have overlapping functions and can either fully or partially substitute for each other (Thomas, 1993; Wagner, 1999). The effects of a mutation of one gene can therefore be masked by the function of another gene. Genes that are redundant for the same function are called paralogous genes (Wilkins, 1997). Genes with a variety of functions have demonstrated redundancy, including genes that encode for extracellular proteins, homeotic genes, protein kinases and enzymes, transcriptional regulators, signal transduction proteins, and enzymes of metabolic pathways (Wagner, 1999; Krakauer and Plotkin, 2002). Redundancy often arises from duplication events. Duplication is common and generally results from unequal cross-over during meiosis (Shimeld, 1999). A less common mechanism of duplication is polyploidy, in which the entire gene complement of an organism is duplicated (Shimeld, 1999). Redundancy can also occur via correlated gene functions, heat shock proteins, alternative metabolic pathways, and tRNA suppressors (Krakauer and Plotkin, 2002). Redundancy of genes and genetic processes enhance the fidelity of development and can potentially widen the range of conditions under which normal development can occur (Thomas, 1993). Redundancy also increases gene expression for a certain function and can have a cumulative effect (Thomas, 1993); that is, redundant genes can create a cumulative contribution to the production of a certain gene product and therefore increase the probability that any concentration threshold for that gene product will be met. This cumulative function of redundant genes is found in ribosomal RNA genes. Multiple copies of these genes allow for the high rate of ribosomal RNA production that is necessary for optimal growth where a single copy of the gene would be insufficient (Thomas, 1993).

Another potential, but as yet unconfirmed, mechanism of producing developmental stability is through the action of "buffering genes." Specific genes have been found that reduce variation found between individuals, but the role of these genes in the dampening of variation within individuals is debated. The molecular chaperone Hsp90 is found throughout the animal and plant kingdoms and functions to increase the fidelity of protein folding under stressful situations, ensuring proper protein conformation (Queitsch *et al.*, 2002). Although the main function of Hsp90, to promote accurate protein folding under stressful temperature conditions, is conserved, its role in developmental stability appears to differ among animals and plants. Queitsch and colleagues (2002) found that Hsp90 was able to modulate the effects of stochastic perturbations and produce developmental stability in the plant *Arabidopsis thaliana*. However, there has been no such association between developmental stability and Hsp90 found in *Drosophila* (Milton *et al.*, 2003). A common conclusion in the recent literature is that there are no specific developmental stability genes based on the negative results found for Hsp90 in fruit flies (Arjan *et al.*, 2003; Milton *et al.*, 2003). However, because Hsp90 was found to affect developmental stability in *Arabidopsis*, it seems hasty to discount the potential for other developmental stability genes in animals. Furthermore, recent work on bacteria hints at the possibility of individual genes responsible for increased stability. Fares and colleagues (2002) have found that GroEL, a heat shock protein found in abundance in endosymbiotic bacteria, reduces the effects of mutations and decreases variation. This study is focused on the effects of GroEL on variation between individuals, and therefore it does not refute or corroborate the suggestion of developmental stability genes. It does, however, point to another potential candidate gene that could be examined in the future. In *E. coli,* the deletion of a gene called *RecA* was found to decrease developmental stability (Elowitz *et al.*, 2002). RecA functions to rescue stalled replication forks, helping to ensure a consistent copy number between different parts of the chromosome. The reduced stability found in strains of *E. coli* lacking the *RecA* gene could result from transient copy number among different parts of the chromosome (Elowitz *et al.*, 2002). It is still an open question as to whether or not specific genes that confer developmental stability exist, but continued search for other candidate genes should help to settle the controversy.

All of these mechanisms are effective on their own, but generally they work together to produce stable development, that is, more genetic network cascades buffer noise by a combination of feedback loops, inhibition, checkpoints, redundancy, and potentially specific genes.

B. Mechanisms for Developmental Stability at the Developmental Systems Level

Normal developmental processes are highly interactive, and these interactions are nonlinear. The varied mechanisms used for these interactions can occasionally create noise as discussed in an earlier section. Generally, however, communication between different cells and developmental pathways offer robustness against random perturbations, allowing the affected component to compensate for any defect. The following is a discussion of how normal interactions during development can lead to increased developmental stability.

Cell signaling is an important mechanism that helps regulate cell positioning and differentiation (Heitzler and Simpson, 1991; Furusawa and Kaneko, 2003). Concentration gradients of diffusible molecules allow cells to determine their relative position within a system. This positional information can be used by the cell to determine whether or not to differentiate, and if so, the type of cell it should become (Heitzler and Simpson, 1991; Furusawa and Kaneko, 2003). Cell signaling through lateral inhibition was shown to be necessary for the proper differentiation of neural and epidermal cells in *Drosophila* (Heitzler and Simpson, 1991). In *Drosophila*, a group of cells exist around the future site of macrochaete, and each of these cells is equipotential. However, in the wild type, only one of these cells becomes a neural precursor, and the rest differentiate into epidermal cells. All the cells in the cluster express the genes of the achaete-scute (ac-sc) complex, and it is the expression of these genes that results in the neural differentiation. Therefore, there must be some inhibitory signal that keeps all but one of the cells in the cluster from expressing ac-sc. Heitzler and Simpson (1991) found that the genes *Notch, Delta,* and *shaggy* are all responsible for the inhibition of the ac-sc complex in *Drosophila*. The clusters of cells expressing the ac-sc genes developed into neural cells if flies were mutant for any one of the *Notch, Delta,* or *shaggy* genes. The result is a tuft of macrochaete in mutant flies, whereas in the wild type only one macrochaete develops. Correct cellular differentiation is extremely important to maintain stable development; therefore, it is necessary that communication between cells is accurate.

Communication between developmental components via feedback mechanisms is another process that can maintain developmental stability. Feedback is a mechanism that a system can use to monitor itself (Freeman, 2000). Such feedback systems are used to synchronize growth between bilateral traits within an organism via positive feedback loops that promote catch-up growth (Emlen *et al.*, 1993). Yuge and Yamana (1989) demonstrated this type of communication between right and left sides in frog embryos. They tested for the presence of feedback between left and right sides of *Xenopus laevis* embryos by removing all of the cells that normally are used in the development of the right dorsal side of the embryo. They found that cells from the left side of the embryo migrated to the right, resulting in normal morphology, both internally and externally. This resultant symmetry after such an extreme perturbation points to the presence of feedback between bilateral sides of the embryo.

A particularly important source of integration of development comes from morphogenetic fields. Morphogenetic fields are regions in the embryo that form modules of developmental precursors that are independent of one another (Kirschner and Gerhart, 1998; Klingenberg, 2004b). Cells within these fields are more highly integrated than are cells between fields. This tight integration within embryonic fields allows for the fine-tuning of development of particular regions of the organism increasing the probability of stable development. Partitioning the embryo into compartments according to developmental precursors also allows for better regulation of development, because a disturbance in one region of the embryo is less likely to have an adverse effect within a separate region (Kirschner and Gerhart, 1998). In a series of papers, Klingenberg describes how interactions occur within and between these morphogenetic fields (Klingenberg and Zaklan, 2000; Klingenberg, 2002, 2004a; Klingenberg *et al.*, 2001, 2003). The two types of interactions he defines are direct and parallel. Direct interactions are intrinsic to the developmental system and are therefore relevant to a discussion on developmental stability. Parallel interactions, however, result from some stimulus extrinsic to the developmental system and will not be described here. In direct interaction pathways, a stimulus affects different traits in a similar manner because multiple traits develop from the same precursor or because one component within a system induces another component directly through cell signaling (Klingenberg *et al.*, 2003). Direct interactions created by development from the same precursor tend work within morphogenetic fields, promoting integration within these regions. Direct interactions through inductive signaling tends to create integration between modules. Klingenberg and Zaklan (2000) tested for these interactions by measuring the covariance of FA of wing veins within *D. melanogaster*. The wings of fruit flies are divided into anterior and posterior compartments that develop from separate cell lineages. Therefore, it is reasonable to hypothesize that there will be greater integration within each compartment than between the compartments. However, it was found that the entire wing acted as an integrated module. This integration of the entire wing is explained by the fact that the boundary between anterior and posterior compartments is the origin of a morphogen gradient that determines wing vein position (Klingenberg and Zaklan, 2000). Presumably, the direct interactions between compartments caused by inductive signaling of morphogens were able to integrate variation throughout the wing, because fly wings are sufficiently compact for the morphogen gradient to impact the entire structure.

Mechanisms of noise buffering at the developmental systems level are extremely important, because any errors that escape correction at this level are likely to have an effect at the gross phenotypic level. Organismal consequences of developmental instability can negatively affect the fitness of an individual and therefore have subsequent consequences for survival.

C. Mechanisms for Developmental Stability at the Organismal Level

To ensure the proper functioning of traits at the organismal level, it is necessary that these traits have the optimum phenotype. Mechanisms at the organismal level that correct or act to compensate for structural defects accrued during earlier developmental processes are the last steps in ensuring proper function at the gross phenotypic level.

An emergent property within organisms is morphologic integration. As defined by Hallgrímsson *et al.* (2002), morphologic integration is the study of covariation in organismal structure. This area of study originated with the work of Olson and Miller (1958), who used patterns of correlation coefficients to determine patterns of covariation between morphologic structures. Cheverud (1982, 1984, 1995) furthered this area of research to show that cranial structures in *Macaca mulatta* that shared common functional and/or developmental input covaried. Structures that are integrated in such a way are constrained in the degree to which they can vary. Therefore, structures that are integrated tend to vary less, and this reduced variability tends to lead to more consistent development of the structures involved. For instance, structures that share a common mechanical effect show reduced variability (Hallgrímsson *et al.*, 2003). Corruccini and Beecher (1984) demonstrated such an effect in a study of cranial integration in baboons. They found that when baboons were fed a soft diet, integration between facial structures was reduced, and variation in these structures increased. The soft diet caused a reduction in the influence of the mechanical stimulus of mastication on the different facial structures, resulting in reduced integration of the face and the subsequent increase in variation. Probably the most effective and powerful mechanism that ensures developmental stability is stabilizing selection, arguably the most common form of natural selection (Schmaulhausen, 1949; Waddington, 1957). Schmaulhausen describes how selection maintains fidelity of development in the following statement: "The process of slow and stabilizing selection is always and continuously causing the development of regulating mechanisms which protect the slowly changing norm against disturbances by external influences" (Schmaulhausen, 1949). Selection favors phenotypes that cluster near the optimum for a given environment. Therefore, phenotypes found at the ends of a distribution of possible forms are selected against and removed from the gene pool. Selection can therefore be thought to ensure consistent development by purging those genotypes from the gene pool that lead to deviant phenotypes.

Several studies have demonstrated this relationship between natural selection and decreased developmental instability. In the fruit fly *Drosophila buzzatii*, it was found that males with a high degree asymmetry for wing length had a corresponding lower mating success (Santos, 2001). Wing length in *Drosophila* is an important fitness factor and therefore led to less competitive flies. Bergstrom and Reimchen (2003) showed that, under circumstances of reduced selection pressure as a result of reduced predation, threespine sticklebacks exhibited a higher degree of variation in structural lateral plates. The structural lateral plates in sticklebacks confer increased protection to the fish after capture by prey. In habitats where predation was lower, there was a corresponding increase in variation. Similarly, asymmetry was found to affect escape performance in the lizard *Psammodromus algirus* (Martín and López, 2001). Asymmetry of the femoral segment of the hindlimb in these lizards significantly reduced overall escape speed. Lizards rely on fast escape to avoid predation, and therefore natural selection will favor those individuals that are more symmetric. Domestication of animals also allows us to view asymmetry under conditions of relaxed selection pressure. Increased asymmetry is commonly found in domesticated animals in comparison with their wild counterparts (Parsons, 1990). Domestic animals rely less heavily on competitively seeking out food, or escaping predation or injury, and therefore, selection for symmetry is less advantageous. Although modern medicine modulates the effects of natural selection, developmental stability is still favored in humans. Individuals with extreme congenital malformations have a corresponding increase in morbidity, and if the dysmorphology has severe consequences on individual fitness, natural selection will effectively remove the individual from the population (Livshits and Kobyliansky, 1991).

III. IMPLICATIONS

No developmental process is perfect, and as a result, we see phenotypic variation not just among individuals with different genotypes or environmental histories, but also within individuals. Developmental stability and its effects on phenotypic variation have relevant implications in evolutionary biology and medicine. To understand the consequences of this variation, we must have a full understanding of how this variation is modulated by developmental stability. Therefore, it is necessary to know where noise arises and how this noise can be amplified or reduced. We need to determine what types of imperfections in which processes lead to the most obvious phenotypic effects. Similarly, it is crucial to determine and to understand the mechanisms that confer robustness for this noise. Understanding developmental stability at only one phenotypic level reveals only part of the overall role of developmental fidelity in evolution and biomedicine. Therefore, it is necessary to appreciate the contributions to developmental stability made at all phenotypic levels.

Developmental stability modulates the degree of phenotypic variation and therefore affects the evolvability of traits (Gibson and Wagner, 2000). A greater understanding of the causes of stability and instability of development will provide insight into which traits are likely to vary and therefore evolve and the potential mechanisms responsible for the origins of novel phenotypes. Likewise, knowledge of the processes that confer stable development is relevant to the study of congenital anomalies. Developmental instability can theoretically acerbate or even cause malformations in development. By gaining a more thorough understanding of the processes that can create inconsistencies in development as well as the mechanisms that confer developmental fidelity at the population level, it may be possible to predict individuals within a population that are more susceptible to developing dysmorphologies. Furthermore, there is potential to determine what types of malformations are most apt to occur within a population. It has been suggested that the suite of malformations exhibited by individuals with syndromes such as Down syndrome are malformations that occur relatively frequently but individually in the rest of the population (Shapiro, 1975, 1983). This phenomenon is termed “amplified developmental instability,” whereby individuals who are already developmentally compromised, such as individuals with Down syndrome, will develop other malformations that are common in the population with a higher frequency, resulting in a suite of anomalies. There is a great deal of variation in the symptoms expressed by different individuals with the same syndrome that makes treatment for these individuals a major challenge. Again, by understanding more thoroughly the potential causes of developmental noise and the mechanisms used to buffer against such perturbations, we should gain insight into useful modes of treatment for individuals with syndromes.

If developmental stability is to be used as a tool for understanding how variation arises and is suppressed within populations, it will be necessary to continue to integrate the fields of molecular and developmental biology with morphology. Several studies have used this integrative approach (Routman and Cheverud, 1997; Klingenberg and Nijhout, 1998, 1999; Hallgrímsson *et al.*, 2002; Queitsch *et al.*, 2002), and as our understanding of molecular and developmental networks expands, such research should prove to be quite fruitful.

REFERENCES

Alberts, B., Bray, D., Johnson, A., Lewis, J., Raff, M., Roberts, K., and Walter, P. (1998). *Essential Cell Biology: An Introduction to the Molecular Biology of the Cell*. New York: Garland.

Arjan, J., De Visser, G. M., Hermisson, J., Wagner, G. P., Meyers, L. A., Bagheri-Chaichian, H., Blanchard, J. L., Chao, L., Cheverud, J. M., Elena, S. F., Fontana, W., Gibson, G., Hansen, T. F., Krakauer, D., Lewontin, R. C., Ofria, C., Rice, S. H., Von Dassow, G., Wagner, A., and Whitlock, M. C. (2003). Perspective: Evolution and detection of genetic robustness. *Evolution* **57**, 1959–1972.

Atchley, W. R., and Hall, B. K. (1991). A model for development and evolution of complex morphological structures. *Biological Review* **66**, 101–157.

Becskei, A., and Serrano, L. (2000). Engineering stability in gene networks by autoregulation. *Nature* **405**, 590–593.

Bergstrom, C. A., and Reimchen, T. E. (2003). Asymmetry in structural defenses: Insights into selective predation in the wild. *Evolution* **57**, 2128–2138.

Bird, A. P. (1995). Gene number, noise reduction, and biological complexity. *Trends in Genetics* **11**, 94–100.

Blake, W. J., Kaern, M., Cantor, C. R., and Collins, J. J. (2003). Noise in eukaryotic gene expression. *Nature* **422**, 633–637.

Bodnar, J. W. (1997). Programming the *Drosophila* embryo. *Journal of Theoretical Biology* **188**, 391–445.

Brown, C. R., Hong-Brown, L. Q., and Welch, W. J. (1997). Correcting temperature-sensitive protein folding defects. *Journal of Clinical Investigation* **99**, 1432–1444.

Cheverud, J. M. (1982). Phenotypic, genetic, and environmental integration in the cranium. *Evolution* **36**, 499–516.

Cheverud, J. M. (1984). Quantitative genetics and developmental constraints on evolution by selection. *Journal of Theoretical Biology* **110**, 155–171.

Cheverud, J. M. (1995). Morphological integration in the saddle-back tamarin (*Saguinus fuscicollis*) cranium. *The American Naturalist* **14**, 63–89.

Corruccini, R. S., and Beecher, R. M. (1984). Occlusofacial morphological integration lowered in baboons raised on soft diet. *Journal of Craniofacial Genetics and Developmental Biology* **4**, 135–142.

Ellgaard, L., and Helenius, A. (2003). Quality control in the endoplasmic reticulum. *Nature Reviews: Molecular Cell Biology* **4**, 181–191.

Ellgaard, L., and Molinari, M. (1999). Setting the standards: Quality control in the secretory pathway. *Science* **286**, 1882–1888.

Elowitz, M. B., Levine, A. J., Siggia, E. D., and Swain, P. S. (2002). Stochastic gene expression in a single cell. *Science* **297**, 1183–1186.

Emlen, J. M., Freeman, D. C., and Graham J. H. (1993). Nonlinear growth dynamics and the origin of fluctuating asymmetry. *Genetica* **89**, 77–96.

Fares, M. A., Ruiz-González, M. X., Moya, A., Elena, S. F., and Barrio, E. (2002). GroEL buffers against deleterious mutations. *Nature* **417**, 398.

Fewell, S. W., Travers, K. J., Weissman, J. S., and Brodsky, J. L. (2001). The action of molecular chaperones in the early secretory pathway. *Annual Review of Genetics* **35**, 149–191.

Fiering, S., Whitelaw, E., and Martin, D. I. K. (2000). To be or not to be active: The stochastic nature of enhancer action. *Bioessays* **22**, 81–387.

Frank, S. A, and Nowak, M. A. (2004). Problems of somatic mutation and cancer. *Bioessays* **26**, 291–299.

Freeman, M. (2000). Feedback control of intercellular signaling in development. *Nature* **408**, 313–319.

Furusawa, C., and Kaneko, K. (2003). Robust development as a consequence of generated positional information. *Journal of Theoretical Biology* **224**, 413–435.

Galis, F., Van Dooren, T. J. M., and Metz, J. A. J. (2002). Conservation of the segmented germband stage: Robustness or pleiotropy? *Trends in Genetics* **18**, 504–509.

Gibson, G., and Wagner, G. (2000). Canalization in evolutionary genetics: A stabilizing theory? *Bioessays* **22**, 372–380.

Graham, J. H., Freeman, D. C., and Emlen, J. M. (1993). Antisymmetry, directional asymmetry, and dynamic morphogenesis. *Genetica* **89**, 121–137.

Hallgrímsson, B., Willmore, K., and Hall, B. K. (2002). Canalization, developmental stability, and morphological integration in primate limbs. *Yearbook of Physical Anthropology* **45**, 131–158.

Hallgrímsson, B., Miyake, T., Wilmore, K., and Hall, B. K. (2003). Embryological origins of developmental stability: Size, shape, and fluctuating asymmetry in prenatal random bred mice. *Journal of Experimental Zoology (Molecular and Developmental Evolution)* **296B**, 40–57.

Hartwell, L. H., and Weinert, T. A. (1989). Checkpoints: Controls that ensure the order of cell cycle events. *Science* **246**, 629–634.

Heitzler, P., and Simpson, P. (1991). The choice of cell fate in the epidermis of *Drosophila*. *Cell* **64**, 1083–1092.

Herring, S. W. (1993). Epigenetic and functional influences on skull growth. In *The Skull,* Vol. 1 (J. Hanken and B. K. Hall, eds.). Chicago: University of Chicago Press.

Herring, S. W., and Teng, S. (2000). Strain in the braincase and its sutures during function. *American Journal of Physical Anthropology* **112**, 575–593.

Herring, S. W., Decker, J. D., Liu, Z. -J., and Ma, T. (2002). Temporomandibular joint in miniature pigs: Anatomy, cell replication, and relation to loading. *The Anatomical Record* **266**, 152–166.

Kaufmann, W. K., and Paules, R. S. (1996). DNA damage and cell cycle checkpoints. *FASEB Journal* **10**, 238–247.

Kemkemer, R., Schrank, S., Vogel, W., Gruler, H., and Kaufmann, D. (2002). Increased noise as an effect of haploinsufficiency of the tumor-suppressor gene neurofibromatosis type 1 in vitro. *Proceedings of the National Academy of Sciences USA* **99**, 13783–13788.

Kirschner, M., and Gerhart, J. (1998). Evolvability. *Proceedings of the National Academy of Sciences USA* **95**, 8420–8427.

Klingenberg, C. P. (2002). Morphometrics and the role of the phenotype in studies of the evolution of developmental mechanisms. *Gene* **287**, 3–10.

Klingenberg, C. P. (2003). A developmental perspective on developmental instability: Theory, models and mechanisms. In *Developmental Instability: Causes and Consequences* (M. Polak, ed.). New York: Oxford University Press.

Klingenberg, C. P. (2004a). Dominance, nonlinear developmental mapping, and developmental stability. In *The Biology of Genetic Dominance* (R. A. Veitia, ed.). Georgetown, TX: Landes.

Klingenberg, C. P. (2004b). Integration, modules, and development: Molecules to morphology to evolution. In *Phenotypic Integration: Studying the Ecology and Evolution of Complex Phenotypes* (M. Pigliucci and K. Preston, eds.). New York: Oxford University Press, 213–230.

Klingenberg, C. P., and Nijhout, H. F. (1998). Competition among growing organs and developmental control of morphological asymmetry. *Proceedings of the Royal Society of London Series B: Biological Sciences* **265**, 1135–1139.

Klingenberg, C. P., and Nijhout, H. F. (1999). Genetics of fluctuating asymmetry: A developmental model of developmental instability. *Evolution* **53**, 358–375.

Klingenberg, C. P., and Zaklan, S. D. (2000). Morphological integration between developmental compartments in the *Drosophila* wing. *Evolution* **54**, 1273–1285.

Klingenberg, C. P., Badyaev, A. V., Sowry, S. M., and Beckwith, N. J. (2001). Inferring developmental modularity from morphological integration: Analysis of individual variation and asymmetry in bumblebee wings. *The American Naturalist* **157**, 11–23.

Klingenberg, C. P., Mebus, K., and Auffray, J. -C. (2003). Developmental integration in a complex morphological structure: How distinct are the modules in the mouse mandible? *Evolution and Development* **5**, 522–531.

Krakauer, D. C., and Plotkin, J. B. (2002). Redundancy, antiredundancy, and the robustness of genomes. *Proceedings of the National Academy of Sciences USA* **99**, 1405–1409.

Leamy, L. J., Routman, E. J., and Cheverud, J. M. (2002). An epistatic genetic basis for fluctuating asymmetry of mandible size in mice. *Evolution* **56**, 642–653.

Legrell, P. E., and Isberg, A. (1998). Mandibular height asymmetry following experimentally induced temporomandibular joint disk displacement in rabbits. *Oral Surgery, Oral Medicine, Oral Pathology, Oral Radiology, and Endodontics* **86**, 280–285.

Legrell, P. E, and Isberg, A. (1999). Mandibular length and midline asymmetry after experimentally induced temporomandibular joint disk displacement in rabbits. *American Journal of Orthodontics and Dentofacial Orthopedics* **115**, 247–253.

Lens, L., Van Dongen, S., Kark, S., and Matthysen, E. (2002). Fluctuating asymmetry as an indicator of fitness: Can we bridge the gap between studies? *Biological Review* **77**, 27–38.

Livshits, G., and Kobyliansky, E. (1991). Fluctuating asymmetry as a possible measure of developmental homeostasis in humans: A review. *Human Biology* **63**, 441–466.

Machida, N., Yamada, K., Takata, Y., and Yamada, Y. (2003). Relationship between facial asymmetry and masseter reflex activity. *American Association of Oral and Maxillofacial Surgeons* **61**, 298–303.

Martín, J., and López, P. (2001). Hindlimb asymmetry reduces escape performance in the lizard *Psammodromus algirus*. *Physiological and Biochemical Zoology* **74**, 619–624.

McAdams, H. H., and Arkin, A. (1997). Stochastic mechanisms in gene expression. *Proceedings of the National Academy of Sciences USA* **94**, 814–819.

McAdams, H. H., and Arkin, A. (1999). It's a noisy business! Genetic regulation at the nanomolar scale. *Trends in Genetics* **15**, 65–69.

Mills, K. D., Ferguson, D. O., and Alt, F. W. (2003). The role of DNA breaks in genomic instability and tumorigenesis. *Immunological Reviews* **194**, 77–94.

Milton, C. C., Huynh, B., Batterham, P., Rutherford, S. L., and Hoffmann, A. A. (2003). Quantitative trait symmetry independent of Hsp90 buffering: Distinct modes of genetic canalization and developmental stability. *Proceedings of the National Academy of Sciences USA* **100**, 13396–13401.

Mohrenweiser, H. W., Wilson, D. M., and Jones, I. M. (2003). Challenges and complexities in estimating both the functional impact and the disease risk associated with the extensive genetic variation in human DNA repair genes. *Mutation Research* **526**, 93–125.

Olson, E. C., and Miller, R. L. (1958). *Morphological Integration*. Chicago: University of Chicago Press.

Palmer, A. R. (1996). Waltzing with asymmetry: Is fluctuating asymmetry a powerful new tool for biologists or just an alluring new dance step? *Bioscience* **46**, 518–532.

Parsons, P. A. (1990). Fluctuating asymmetry: An epigenetic measure of stress. *Biological Review* **65**, 131–145.

Queitsch, C., Sangster, T. A., and Lindquist, S. (2002). Hsp90 as a capacitor of phenotypic variation. *Nature* **417**, 618–624.

Rafferty, K. L., and Herring, S. W. (1999). Craniofacial sutures: Morphology, growth, and in vivo masticatory strains. *Journal of Morphology* **242**, 167–179.

Reeve, E. C. R., and Robertson, F. W. (1953). Analysis of environmental variability in quantitative inheritance. *Nature* **171**, 874–875.

Routman, E. J., and Cheverud, J. M. (1997). Gene effects on quantitative trait: Two-locus epistatic effects measured at microsatellite markers and at estimated QTL. *Evolution* **51**, 1654–1662.

Rutherford, S. L., and Lindquist, S. (1998). Hsp90 as a capacitor for morphological evolution. *Nature* **396**, 336–342.

Salazar-Ciudad, I., Jernvall, J., and Newman, S. A. (2003). Mechanisms of pattern formation in development and evolution. *Development* **130**, 2027–2037.

Santos, M. (2001). Fluctuating asymmetry is nongenetically related to mating success in *Drosophila buzzatii*. *Evolution* **55**, 2248–2256.

Schmaulhausen, I. I. (1949). *Factors of Evolution*. Chicago: University of Chicago Press.

Shapiro, B. L. (1975). Amplified developmental instability in Down's syndrome. *Annals of Human Genetics* **38**, 429–437.

Shapiro, B. L. (1983). Down syndrome: A disruption of homeostasis. *American Journal of Medical Genetics* **14**, 241–269.

Shimeld, S. M. (1999). Gene function, gene networks, and the fate of duplicated genes. *Cell and Developmental Biology* **10**, 549–553.

Sitia, R., and Braakman, I. (2003). Quality control in the endoplasmic reticulum protein factory. *Nature* **426**, 891–894.

Swain, P. S., Elowitz, M. B., and Siggia, E. D. (2002). Intrinsic and extrinsic contributions to stochasticity in gene expression. *Proceedings of the National Academy of Sciences USA* **99**, 12795–12800.

Thomas, J. H. (1993). Thinking about genetic redundancy. *Trends in Genetics* **9**, 305–309.

Van Valen, L. M. (1962). A study of fluctuating asymmetry. *Evolution* **16**, 125–142.

Vinter, I., Krmpotic-Nemanic, J., Ivankovic, D., and Jalsovec, D. (1997). The influence of the dentition on the shape of the mandible. *Collegium Antropologicum* **21**, 555–560.

Waddington, C. H. (1957). *The Strategy of the Genes*. New York: MacMillan.

Wagner, A. (1999). Redundant gene functions and natural selection. *Journal of Evolutionary Biology* **12**, 1–16.

Wagner, A. (2000). Robustness against mutations in genetic networks of yeast. *Nature Genetics* **24**, 355–361.

Wagner, G. P., Laubichler, M. D., and Bagheri-Chaichian, H. (1998). Genetic measurement theory of epistatic effects. *Genetica* **102–103**, 569–580.

Wilkins, A. S. (1997). Canalization: A molecular genetic perspective. *Bioessays* **19**, 257–262.

Yuge, M., and Yamana, K. (1989). Regulation of the dorsal axial structures in cell-deficient embryos of *Xenopus laevis*. *Development, Growth, and Differentiation* **31**, 315–324.

CHAPTER 11

Developmental Constraints, Modules, and Evolvability

CHRISTIAN PETER KLINGENBERG

Faculty of Life Sciences, The University of Manchester, Manchester, United Kingdom

ABSTRACT

This chapter discusses the developmental origins and evolutionary implications of covariation between traits. These are important factors influencing the evolutionary potential of morphological traits. Strong covariation can constitute an evolutionary constraint because some character combinations are more likely to evolve than others. Modularity is a widespread feature of organismal organization: Groups of traits covary with each other but are relatively independent of other groups of traits. This modularity results from a similar organization of developmental systems, where signaling interactions primarily take place within spatially distinct fields. The covariation among morphological traits can result from direct interactions of the developmental pathways that produce the traits, which take place within developmental modules, or from parallel variation of separate pathways in response to the simultaneous influence of an external factor. These two origins of covariation among traits have different implications for pleiotropy of genes, the evolution of pleiotropy, the total genetic covariance structure, and the resulting evolutionary constraints.

INTRODUCTION

The question about the relative importance of intrinsic and extrinsic factors as determinants of evolution has existed throughout the history of evolutionary thought and has led to lively debates under different headings (Gould, 2002). In recent years, this debate has mainly concentrated on the question of whether natural selection primarily drives evolution, which would therefore proceed by adaptive optimization of characters, or whether the direction and rate of evolution are affected substantially by factors intrinsic to the organism and its function (Gould and Lewontin, 1979; Alberch, 1989; Parker and Maynard Smith, 1990; Rose and Lauder, 1996; Gould, 2002). With the increased interest in the influence of developmental processes on evolutionary change (Kirschner and Gerhart, 1998; Arthur, 2001), a new perspective on this topic has opened up, and questions have been asked about evolvability, the intrinsic tendency of organisms to produce new variation (Wagner and Altenberg, 1996). Are all features of the phenotype equally variable, or are there features that are inherently more variable than others and therefore more likely to evolve rapidly? If there is such a tendency of organisms to produce particular kinds of variation, does it evolve itself?

A topic that has long attracted particular attention is the integration among the traits of an organism (Olson and Miller, 1958; Cheverud, 1982, 1996; Wagner, 1996). The central question is whether variation is coordinated among traits so that each organism is a fully integrated ensemble or whether the organism is a composite of several subunits that are more or less free to vary independently from one another. Strong integration of parts may lead to an organism that is functionally better coordinated. However, such global coordination may limit the potential for future evolution because different functional systems cannot evolve separately. Each adaptive improvement would come at the cost of a deterioration of some other aspect of performance (Kirschner and Gerhart, 1998).

In this chapter, I examine the topics of evolvability and constraint as well as integration and modularity from a perspective that unites development and evolutionary quantitative genetics. Much of the emphasis is on the question of how integrated morphological variation originates in the development of the respective traits. The explicit consideration of the developmental origin of morphological covariation (e.g., Riska, 1986; Klingenberg, 2004) offers a possible alternative to the hypotheses on the origin of integration and modularity by adaptive evolution (Cheverud, 1984, 1996; Wagner, 1996; Wagner and Altenberg, 1996). I argue that the organization of morphological structures into developmental modules is a crucial factor for the genetic architecture of these traits and the evolution of genetic covariance structure.

I. EVOLVABILITY AND CONSTRAINTS

Evolvability is the potential of a population to respond to selection or to undergo nonadaptive evolution by drift. By contrast, a constraint is something that limits or biases the potential for evolutionary change, and it is therefore a force opposing or channeling evolution, reducing evolvability in at least some directions of the phenotypic space (e.g., Maynard Smith *et al.*, 1985).

Most of the discussion on constraints has focused on their role in limiting the evolvability of traits in particular directions, making certain phenotypes inaccessible to evolution or at least more difficult to attain. In contrast to this predominant usage, Gould (1989; 2002, pp. 1025–1061) went to considerable length to emphasize that constraint is not necessarily a "negative" force limiting or reducing the evolvability in some directions of phenotypic space, but that it can be a "positive" force biasing or channeling evolutionary variation toward certain directions. As prime examples of constraints, Gould (2002, pp. 1037–1051) mentions allometry and heterochrony, for which ontogeny provides a strong directionality in the phenotypic space, because evolution by changes of growth control can easily achieve an extension or truncation of an ancestral growth trajectory (e.g., Klingenberg 1998). Evolution by this mechanism will therefore be constrained (in the positive sense) to changes in the direction of a conserved ontogenetic trajectory, but it is also constrained (in the negative sense) from achieving changes in directions perpendicular to this growth trajectory. Arthur (2001) suggested the term *developmental drive* for the positive sense of constraint. Note, however, that this drive is bidirectional, because variation is biased to be along a particular axis in space, but that variation along this axis is equally distributed in both directions and not just in one of them, as is the case for other kinds of drive that are unidirectional, for instance, meiotic drive in population genetics (Hartl and Clark, 1997, pp. 247–250) or the different types of drive in macroevolution (Gould, 2002, pp. 717–731).

Constraints or biases of variation can be absolute or relative (Figure 11-1). An absolute constraint is the situation where there is no variation at all in some directions of phenotypic space, and all variation is contained entirely in a subspace,

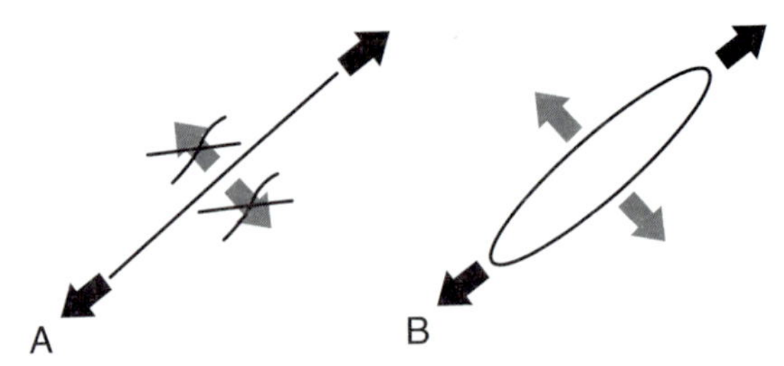

FIGURE 11-1. Constraints on phenotypic variation. (A) Absolute constraint. There is no variation in one direction of the phenotypic space, the plane of the graph. Evolution can therefore only occur along the single axis of variation (black arrows), but not in the direction perpendicular to it. (B) Relative constraint. There is more variation in the direction of the long axis of the ellipse, and the phenotype is therefore more evolvable (black arrows) than in the direction perpendicular to it (gray arrows).

for example, in a single line in a two-dimensional graph (Figure 11-1A) or in a plane of a three-dimensional space. In this case, phenotypes lying outside the subspace are inaccessible to evolution. An absolute constraint will always be associated with a gap in the distribution of traits or with a clear limit of that distribution (Alberch, 1989).

An interesting way to examine whether a "missing" morphology may result from an absolute constraint is to examine developmental anomalies such as teratologic and mutant phenotypes. Because these phenotypes are clearly not functional and often lethal, the regularities in the phenotypes are caused by the developmental system, and selection for functionality can be ruled out. These regularities and patterns can reveal the system's potential to generate novel forms, which is a precondition for evolvability, and may reveal a "logic of monsters" (Alberch, 1989). Similarly, mutants can be used to "engineer" novel phenotypes and to explore the limits of developmental systems. Dworkin *et al.* (2001; see also Larsen, 2003) combined *Drosophila* mutants to flies with antennae resembling the biramous condition seen in the limbs of crustaceans and ancestral arthropods. The problem with this approach is, of course, whether such an engineered condition really corresponds to a purportedly constrained morphology, that is, whether the boundaries that define absolute constraints have been identified correctly.

Relative constraints (Figure 11-1B) concern differences in the amounts of variation available in different directions of the phenotypic space. The question is not whether or not a particular phenotype can be reached at all, but how easily a population will evolve in different directions of the phenotypic space. If selection for a single optimal phenotype is maintained for a sufficient number of generations, a population will eventually overcome relative constraints to achieve the phenotypic optimum. The importance of relative constraints is therefore not primarily in precluding particular phenotypes as the endpoint of selection, but such constraints are a major factor determining the evolutionary trajectory by which the population will reach this endpoint. If several alternative optima exist as separate peaks of the fitness landscape, relative constraints may be decisive in determining which peak the population will reach.

Absolute constraints are fairly rare, because most phenotypic traits show considerable genetic variation (e.g., Roff, 1997). However, no systematic searches for absolute constraints have been done with rigorous multivariate methods. Such searches face considerable technical and statistical difficulties so that large experimental designs will be required. Weber (1990, 1992) applied artificial selection for different shape changes in *Drosophila* wings and obtained significant responses for all of them. Likewise, Beldade *et al.* (2002b) conducted selection experiments on the relative sizes of different eyespots on butterfly wings and found a substantial response to selection, even for the trait combinations that initially had been expected to be constrained.

The structure of genetic variances and covariances reflects these constraints (Lande, 1979; Cheverud, 1984; Arnold, 1992; Roff, 1997; Steppan *et al.*, 2002). A variety of experimental designs is available to estimate the genetic covariance matrix (**G** matrix), which contains the genetic variances and covariances among the traits of interest (e.g., Lynch and Walsh, 1998). Constraints can therefore be diagnosed and quantified by analyzing the **G** matrix with the methods of multivariate statistics.

Absolute constraints can be shown if the **G** matrix has one or more directions that are devoid of genetic variation (Figure 11-1A), or in algebraic terms, if the **G** matrix is singular. There are several statistical tools to assess this condition. A method that is available in most standard statistical software packages is principal component analysis (e.g., Jolliffe, 2002) of the **G** matrix: If there are one or more principal components that do not account for any variation (eigenvalues of zero), there will be absolute constraints. The reverse is not necessarily true, however, because absolute constraints may not be linear. The presence of nonlinear absolute constraints may not yield a singular **G** matrix. Such nonlinear absolute constraints may be difficult to demonstrate empirically. In general, given the considerable difficulties involved in estimating **G** matrices (Lynch and Walsh, 1998), inferring genetic constraints from them is a challenging task. It may be helpful to note, therefore, that absolute constraints also can be inferred if the phenotypic covariance matrix (**P** matrix) is singular, because this automatically implies that the **G** matrix must be singular. It usually will be easier to show that there are principal components of the **P** matrix that are not associated with any variation, although this method cannot reveal all cases where the **G** matrix is singular. All these inferences about the **G** matrix are technically demanding in their application.

Relative constraints result from a situation in which the amounts of variation in different directions of the phenotypic space are unequal (Figure 11-1B). Accordingly, the propensity for evolutionary change by selection will be greater for some trait combinations than for others, and evolutionary change by drift will tend to favor the same trait combinations. Relative constraints can be assessed by the differences of the eigenvalues of the **G** matrix. There are strong relative constraints if some principal components take up a disproportionate share of the total variation, whereas others account only for minor amounts of variation (e.g., Klingenberg and Leamy, 2001). Because the response to natural selection depends on the **P** and **G** matrices jointly as well as the direction in which selection is applied, the relative magnitudes of eigenvalues in either matrix alone are not always sufficient to assess the severity of constraints.

As can be seen from the diagrams in Figure 11-1, the constraints result from an association between the coordinates of the plots. For absolute constraints, one or more variables are completely determined by the remaining ones, and not every combination of traits is available. For instance, if all the variation is on a single straight line in a plane, then any phenotype that is at a distance to the line is inaccessible (Figure 11-1A). In the case of a relative constraint, however, this association is a stochastic one, where the value of one variable can only be predicted from the value of the other variable or variables with some uncertainty (Figure 11-1B). For both absolute and relative constraints, associations of traits are of critical importance, because it is covariation between traits that limits evolution in some directions and makes certain combinations of traits difficult to achieve. Therefore adaptive evolution of individual traits can be constrained by the integration of traits within the entire organism.

II. INTEGRATION AND MODULARITY

The idea that the parts of organisms are coordinated to form a functional whole goes back to the early nineteenth century when Cuvier stated it as the "principle of correlation" (e.g., Mayr, 1982, p. 460 f.) and more recently has been discussed under the heading of morphological integration (Olson and Miller, 1958; Cheverud, 1996; Hallgrímsson *et al.*, 2002). This coordination has a functional and possibly adaptive basis as well as a developmental and mechanistic basis, and the two contexts are often distinguished by terms such as *functional integration* or *developmental integration*. Here, I will discuss primarily the developmental basis of morphological integration, and functional aspects will play only a secondary role.

Organisms are not completely and equally integrated throughout, but they are organized into distinct parts, or modules (Figure 11-2). Modules are assemblages of parts that are tightly integrated internally by relatively many and strong interactions but relatively independent of one another because there are only relatively few or weak interactions between modules (Cheverud, 1996; Raff, 1996; Wagner, 1996; Wagner and Altenberg, 1996). The development and morphology of different modules can therefore evolve independently, at least to some extent, without disrupting function at the level of the whole organism. This modular type of organization recurs at different levels from the interactions in genetic regulatory networks to the morphological structure of whole bodies and their parts (e.g., von Dassow and Munro, 1999; Winther, 2001). In this chapter, the focus will be this organismal level of organization, where modular structure results from interactions among the developmental processes that build morphological parts (Cheverud, 1996; Klingenberg, 2003a; Klingenberg *et al.*, 2003).

Modularity is a hierarchical concept—there can be modules within modules, depending on the level of organization (e.g., in Figure 11-2, modules 1 and 2 together constitute the higher-level module 4). For instance, this structural hierarchy can reflect successive rounds of patterning that progressively subdivide the developmental field into finer regions corresponding more and more to the anatomic details of the prospective morphological structure (Davidson, 1993; Wilkins, 2002, Chapter 8). As a consequence, modules arising in this manner at a later time are therefore within modules that have originated earlier. The subdivision of embryonic fields is not the only process giving rise to modularity, and processes

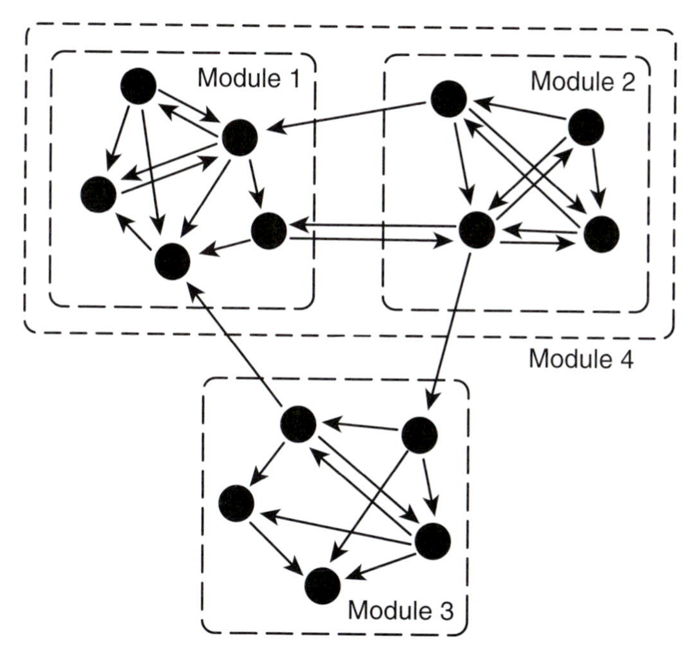

FIGURE 11-2. Modules and developmental interactions. A module is a set of traits (circles) that is rendered internally coherent by multiple interactions (arrows) among the constituent traits and relatively independent from other modules because there are fewer or weaker interactions between modules. Modularity is hierarchical in that modules at one level can be the traits that make up a module at a higher level of organization.

occurring later in ontogeny, for instance, bone remodeling (Herring, 1993; Enlow and Hans, 1996), can potentially have major effects on the patterns of modularity.

Integration manifests itself as the covariation among morphological traits. As such it can be analyzed statistically using morphometric data. Different methods have been used since the inception of this approach, including the analysis of correlations among distance measurements (e.g., Olson and Miller, 1958; Cheverud, 1982; Leamy and Atchley, 1984; Zelditch, 1987; Cheverud, 1995) or among the positions of morphological landmark points (Klingenberg and Zaklan, 2000; Klingenberg *et al.*, 2001a; Bookstein *et al.*, 2003; Klingenberg *et al.*, 2003). The patterns of covariation among traits can provide important information about modularity: Modules are expected to be sets of traits that are highly integrated internally and relatively independent among each other, whereas structures that are completely integrated will show high degrees of morphological covariation in all their parts.

III. DEVELOPMENTAL ORIGINS OF COVARIATION AMONG TRAITS

Different mechanisms can contribute to covariation among traits, including genetic and environmental factors acting on the traits through various mechanisms. To understand the influence of integration and modularity on evolvability and constraints, it is important to understand the developmental basis by which the covariation comes about and how this question can be addressed by empirical studies (Klingenberg, 2003a, 2004).

Two conditions are required for covariation of two or more traits: There must be variation, and there must be a mechanism creating an association between the traits so that the variation affects them jointly. Depending on the particular circumstances, the source of variation and the mechanism causing the association between traits may be separate processes, or the source of variation itself may be responsible for the association. The developmental basis of the mechanisms that create this association can have a substantial influence on the evolutionary effects of integration and modularity. It is therefore important to examine these mechanisms in some detail.

There are two main types of mechanisms that give rise to covariation of morphological traits: direct developmental interaction between the developmental pathways that produce the traits (Figure 11-3A, B) and parallel variation in pathways that are separate from each other (Figure 11-3C). In the first type, there is a direct interaction between two pathways that creates the association between them by passing on the variation to both pathways. The variation can be passed down from earlier ("upstream") steps in the pathway and therefore need not be associated with the mechanism of interaction that generates the association

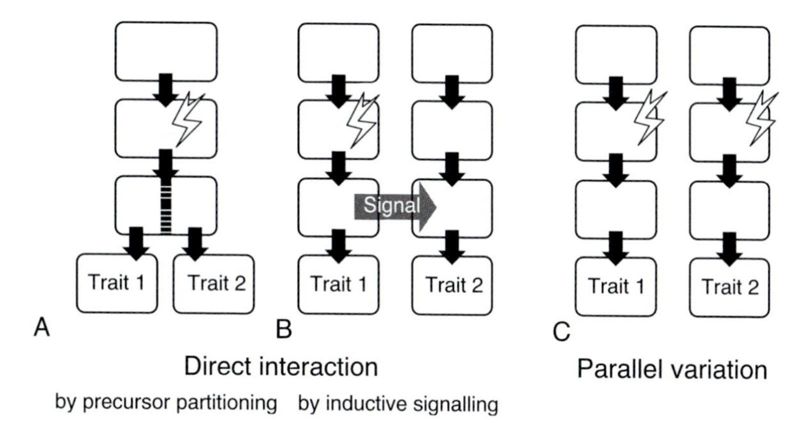

FIGURE 11-3. Mechanisms that generate covariation between traits. The diagrams depict developmental pathways that produce traits in several steps. Variation in a pathway (lightning bolts) can be transmitted through the pathways and thereby affects the respective traits. In (A) and (B), this variation is shared between traits because a direct interaction of the pathways transmits the variation directly. This interaction can be the bifurcation of a single pathway into two pathways by partitioning of a developmental precursor (A), a signal going from one pathway to the other (B), or other forms of developmental interaction. In (C), the two pathways are separate and there is no direct interaction between them. In this case, covariation is entirely the result of the simultaneous effect of the same source of variation on each pathway separately (separate lightning bolts in the two pathways).

between the traits. In contrast, for parallel variation of pathways, two different pathways are affected by a factor causing variation in both of them simultaneously, and the factor responsible for the association between traits is therefore the source of variation itself.

Direct interaction between developmental pathways can occur in a variety of ways. One example is the division of a precursor tissue into two distinct populations, which may correspond to distinct anatomic parts (Figure 11-3A). Variation that arises in the common upstream part of the pathway (lightning bolt in Figure 11-3A) is passed down through the partitioning step to both parts simultaneously, creating a positive covariance between them (Riska, 1986). In contrast, variability in the ratio of partitioning itself will produce a negative association between the two traits (Riska, 1986), as they will compete directly for the precursor tissue (e.g., Nijhout and Emlen, 1998). Another type of direct interaction is inductive signaling from one pathway or tissue to another, which can transmit variation from the pathway emitting the signal so that it simultaneously affects both pathways (Figure 11-3B). Signaling interactions are of fundamental importance for many patterning processes in development (e.g., Francis-West *et al.*, 1998; Gurdon and Bourillot, 2001), and accordingly, there is considerable opportunity for this mechanism of integration. A hallmark of covariation caused by direct interaction of developmental pathways is that variation arising within the pathways themselves (lightning bolts in Figure 11-3A, B) can be transmitted to other pathways. This intrinsic developmental variation is therefore a source of covariation between traits.

In contrast, the mechanism of parallel variation in separate pathways relies on the simultaneous action of an outside factor (lightning bolts in Figure 11-3C) on multiple pathways to generate covariation among traits. For this type of mechanism, therefore, the source of variation itself also provides the basis for the association between traits. The variation can come from environmental factors that have effects on several developmental pathways simultaneously, for instance, temperature differences or nutritional factors. Alternatively, allelic variation of genes that are involved in both pathways can generate parallel variation, even if they are expressed at different anatomic locations or different stages of development. An example would be the *Distal-less* (*Dll*) gene, which is involved both in the development of the distal parts of the legs and antennae and of the color patterns on the wings of butterflies (Panganiban *et al.*, 1994; Brakefield *et al.*, 1996; Beldade *et al.*, 2002a). If alleles of *Dll* differ in their developmental activities in these different contexts jointly, then variation at the *Dll* locus will have simultaneous effects on multiple traits and can generate covariation between them. For covariation by parallel variation of separate pathways, the origin of variation lies outside the individuals, for example, if individuals experience differences in environmental conditions or carry different alleles of a gene. The association between traits arises because this variation affects the development of multiple traits in each individual simultaneously. In neither of these cases is there a direct interaction between the pathways that produce the traits. Variation arising within any pathway only affects downstream steps within that pathway itself but has no effects on other pathways. Therefore, it does not produce covariation between traits.

A possibility to distinguish between the two main origins of covariation among traits is therefore to control rigorously for genetic and environmental variation. Eliminating these components of variation will also eliminate covariation by parallel variation of separate pathways. Any covariation between traits, under these conditions, will therefore be from direct connections between pathways.

A particularly convenient way to achieve control over external variation is to focus on fluctuating asymmetry, the small random differences between bilateral structures of the left and right body sides (Palmer and Strobeck, 1986; Møller and Swaddle, 1997). The two body sides share the same genome and nearly the same environment, at least for most mobile organisms, and focusing on asymmetries therefore provides a means to minimize external variation. The differences between left and right sides are caused by random fluctuations in developmental processes, for example, from relatively small numbers of certain molecules involved in transcriptional regulation or signaling (McAdams and Arkin, 1999; Klingenberg, 2003b). Because this variation is random, a systematic association that is manifest statistically as a correlation between the asymmetries of traits can only occur if the effects of the perturbations are transmitted from a source of variation to the traits, which in turn requires a direct interaction between the respective developmental pathways. Moreover, because this spontaneous variation is intrinsic to the developmental pathways, it can only be transmitted between pathways by direct interaction. Accordingly, covariation of fluctuating asymmetries among traits is therefore exclusively the result of direct interactions between pathways (Klingenberg, 2003a). A number of empirical studies have used this method to examine the developmental basis of morphological integration and modularity (Klingenberg and Zaklan, 2000; Klingenberg *et al.*, 2001a, 2003).

The two sources of covariation are not mutually exclusive but can operate side by side. Simultaneous external influences can cause parallel variation in pathways that are also linked by a direct developmental interaction. Still, the analysis of covariation in the asymmetries of different traits will indicate the contribution of direct interaction to the total trait covariation.

IV. DEVELOPMENTAL INTERACTIONS AND PLEIOTROPY

The different origins of morphological covariation apply to broad classes of variation, and in particular, they also apply to variation that results from allelic differences that are involved in the developmental pathways. As a result, the respective loci have pleiotropic effects on multiple traits.

The mechanisms that produce pleiotropy have been the subject of study throughout the history of genetics, and the explanations used to account for it have reflected the changing concepts of developmental genetics (e.g., Grüneberg, 1938; Hadorn, 1945; Pyeritz, 1989; Hodgkin, 1998; Nadeau *et al.*, 2003). Different types of pleiotropy have been distinguished, which partly match the current distinction of sources of morphological covariation. Grüneberg (1938) defined "genuine" pleiotropy as an ideal case in which a gene produces its effect on different traits by distinct mechanisms, but he expressed doubt about its existence. He opposed this mode to the more common "spurious" pleiotropy, in which the gene affects multiple characters by the same mechanism or where intermediate causes are involved. Similarly, Hadorn (1945, p. 91) distinguished "primary" and "secondary" pleiotropy. Primary pleiotropy is the direct result of the constitution of the cells that are the precursors of the traits, whereas secondary pleiotropy is generated by the transmission of effects from other cells to the separate populations of progenitor cells of different traits. In this distinction, primary pleiotropy closely corresponds to parallel variation of separate developmental pathways resulting from simultaneous allelic effects, and secondary pleiotropy results from direct interaction of the pathways that generate the traits. A much more elaborate classification of pleiotropy, subdividing primary pleiotropy into several types according to the biochemical modes by which parallel effects in separate pathways come about, has been advanced more recently (Hodgkin, 1998). Here, I will concentrate on the two main categories.

Pleiotropy by direct interactions of developmental pathways has implications for the phenotypic effects of the respective loci. Accordingly, the mapping of functional and causal relationships in the developmental networks, which has been pursued since the inception of studies of pleiotropy (Grüneberg, 1938; Pyeritz, 1989), remains an activity that is central to this subject (Davidson *et al.*, 2003; Nadeau *et al.*, 2003). The particular nature of the connection between pathways will impart a specific pattern on the effects of multiple genes that are upstream of the connection. Variation is transmitted along developmental pathways in a manner specific to each developmental step rather than the source from which it originated. For instance, increases of several different activating transcription factors or decreases of inhibiting factors might all lead to increased expression of a gene, and in both cases, the response will be a greater concentration of that gene's product. Accordingly, each pathway "lumps together" the variation from the upstream steps so that the individual inputs may not be identifiable. The patterns of covariation among traits are generated through the interactions of different pathways and will be shared by those loci that contribute to the variation at the step where the interaction takes place and upstream of it. Because variation may be attenuated or amplified as it is transmitted through a pathway, and because each pathway may interact with several others at different steps, different loci may contribute differently to the interactions at different steps. Therefore, if a pathway interacts with several other pathways at various steps, the patterns of phenotypic effects may differ among groups of loci depending on the steps in which they participate. Nevertheless, there will be similarities between the effects of the loci that are active in the same steps.

In contrast, pleiotropy by parallel variation of separate pathways stems from allelic differences that jointly affect two or more developmental pathways in which the respective loci are active. The activity of a locus in multiple developmental contexts can come about by various mechanisms (Pyeritz, 1989; Hodgkin, 1998), but it is likely that the modular organization of *cis*-regulatory elements and the history of cooption of genes to new functions facilitate this multiple deployment of genes (e.g., Davidson, 2001; Wilkins, 2002; Levine and Tjian, 2003). Many genes that have important roles in development are expressed in different locations and at different stages of development (Carroll *et al.*, 2001; Davidson, 2001; Wilkins, 2002). For example, the *Distal-less* gene in butterflies is involved in generating the distal parts of the legs and antennae and also the colored eyespots on the wings (Carroll *et al.*, 1994; Panganiban *et al.*, 1994; Beldade *et al.*, 2002a). If alleles at such a locus differ among each other in their activity in different contexts jointly, they can cause pleiotropy by parallel variation of pathways. Examples of such effects are abundant and include the numerous syndromes caused by single mutations known in human medical genetics (e.g., Pyeritz, 1989; Jabs, 2002).

For pleiotropy of this kind, each locus provides not only the source of variation by differences in the developmental activity between alleles, but through the joint activity in two or more pathways, the gene is also directly responsible for its contribution to the covariation between the traits derived from the pathways. Accordingly, the patterns of covariance among traits can be as diverse as the loci exhibiting pleiotropy of this kind, and no grouping of loci by their patterns of phenotypic covariation can be expected.

In summary, the difference between these two mechanisms of generating pleiotropy is that direct interactions between pathways establish groups of loci that have similar patterns of trait covariation, whereas pleiotropy by parallel variation of pathways will result in phenotypic patterns that are individually different from locus to locus.

Empirical studies on the role of the two processes in generating pleiotropy of single loci are technically difficult and have only just begun to be undertaken. One possibility is to use multivariate approaches to mapping of quantitative trait loci (QTLs; Leamy *et al.*, 1999; Klingenberg *et al.*, 2001b; Workman *et al.*, 2002) and to analyze the patterns of QTL effects. These can then be related to patterns of trait covariation for fluctuating asymmetry, which provide a "standard" for the patterns generated by direct interaction of pathways (Klingenberg, 2003a, 2004). This approach has been taken for the study of shape variation in the mouse mandible, where both the patterns of covariation for QTL effects (Klingenberg *et al.*, 2004) and those for fluctuating asymmetry (Klingenberg *et al.*, 2003) show similar patterns of a partial separation between anterior and posterior modules. However, these studies are fraught with a number of inherent difficulties, such as weak statistical power with a limited number of QTLs (e.g., Flint and Mott, 2001), and further studies addressing these issues are needed.

V. EVOLUTION OF PLEIOTROPY AND DEVELOPMENTAL INTERACTIONS

The patterns of pleiotropy can undergo evolutionary change by changes in the developmental mechanisms that produce pleiotropy. The two main groups of mechanisms that produce covariation between traits differ in their evolutionary flexibility and in the implications of evolutionary changes for the developmental system itself. I will address these questions by considering the consequences of changes, for instance by mutation, that would alter the patterns of pleiotropy.

Evolution of the patterns of pleiotropy through direct interactions of pathways must occur by changes in the interactions themselves. For instance, the ratio of partitioning of a developmental precursor may change (Figure 11-4A), or there may be a change of an inductive signaling process from one pathway to another (Figure 11-4B). These changes may come about by new mutations of large effect,

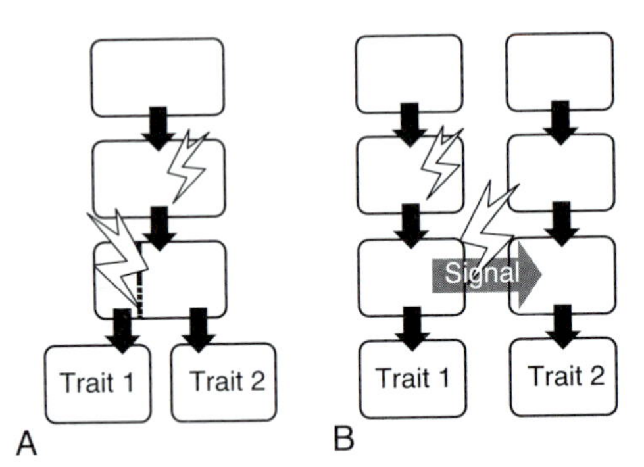

FIGURE 11-4. Genetic change in direct interactions of developmental pathways. (A) A change in the partitioning of a precursor (large lightning bolt) leads to a change in the patterns of covariation produced in response to variation transmitted from the "upstream part" of the pathway (small lightning bolt). For instance, this could be a change in the pleiotropic effects of a gene participating in the pathway. (B) Similarly, a change in the signaling interaction (large lightning bolt) may produce a change of the pattern of covariation in response to variation in the pathway (small lightning bolt).

or they can be based on the standing variation in a population, by changes in allele frequencies for loci that affect the interactions. The changes may be based on a variety of mechanisms. For instance, the partitioning of a cell population into different components may rely on the readout of morphogen gradients (Gurdon and Bourillot, 2001). Changes can affect the morphogen secretion, the mechanisms of morphogen transport, or the abundance of receptors on the responding cells and the intracellular signal transduction mechanisms that elicit their response. Similar cellular and molecular components may also be involved in changes in inductive signaling among pathways, strengthening or weakening the sensitivity of the target pathway to the signals from a source.

Any change of the interactions among developmental pathways will affect the patterns of pleiotropy of all loci that are upstream of the interaction as well as the integration of nongenetic components of variation. These changes thus affect the modular structure of the developmental system as a whole, and such a reorganization of the system is likely to have substantial effects on the phenotype. Examples of these effects can be found among major teratologies, which may be caused by the failure of specific developmental interactions (Alberch, 1989; Wilkie and Morriss-Kay, 2001; Cohen, 2002). Because of these serious consequences, changes in the connections among pathways will normally be selected against if the traits that depend on the respective developmental system are under stabilizing selection. There may even be selection for modifiers that stabilize the interactions among developmental pathways. Remarkable stability of the molecular mechanisms that set up or respond to morphogen gradients has been demonstrated empirically for embryonic patterning in *Drosophila* (e.g., Eldar *et al.*, 2002; Houchmandzadeh *et al.*, 2002) and may occur in different developmental contexts as well. As a result of such robustness, the patterns of pleiotropy for multiple upstream loci are stabilized simultaneously.

An example of evolutionary changes in a signaling interaction and their far-reaching consequences is the loss of eyes in cave populations of the fish *Astyanax mexicanus* that have originated from surface-living forms on multiple occasions (Jeffery, 2001; Jeffery *et al.*, 2003). In cave fish, the lens cells undergo apoptosis, the eye primordia stop developing, degenerate, and are eventually covered by skin. A cave fish lens transplanted to a surface fish embryo undergoes apoptosis, whereas a surface lens transplanted to a surface fish stimulates the development of a complete eye (Yamamoto and Jeffery, 2000). Lens transplants also alter the growth of the orbital bones surrounding the eyes and the size of the olfactory pit, whereas other craniofacial differences between cave and surface fish appear to be independent of the eye (Yamamoto *et al.*, 2003). A change in signaling from the lens to other parts of the eye and adjacent structures therefore is responsible for much of the difference between the cave and surface forms. Given the dramatic effects of the change, it is plausible that this signaling interaction is under stabilizing selection in the surface populations.

The difference between cave and surface fish is visible much earlier in embryonic development as a difference in *Pax6* expression patterns (Jeffery, 2001). The restricted *Pax6* expression in cave fish, in turn, appears to result from increased signaling by *sonic hedgehog* (*Shh*) from the midline, which suppresses *Pax6* (Yamamoto *et al.*, 2001; Jeffery *et al.*, 2003). Injection of *Shh* mRNA into early embryos of surface fish results in a reduction of *Pax6* expression and eye regeneration similar to those in cave fish (Yamamoto *et al.*, 2001; Jeffery *et al.*, 2003). Therefore, the early change in Shh signaling appears to cause the later degeneration of the lens and loss of its signaling activity. There are therefore two direct developmental interactions involved, with the Shh signal from the midline upstream of the lens signal. There is indirect evidence for stabilizing selection on Shh activity, because both increases and decreases can cause malformations. Knock-outs of the *Shh* gene in mice have been shown to cause cyclopia (a single eye positioned in the midline; Chiang *et al.*, 1996) and mutations with reduced Shh activity have been shown to cause human holoprosencephaly (various degrees of underdevelopment of the midline of the brain and face, with cyclopia as an extreme; Nanni *et al.*, 1999; Schell-Apacik *et al.*, 2003). In contrast, experimental increase of Shh activity in craniofacial primordial of the chick produces an expansion of midline structures and may be related to hypertelorism in humans (enlarged distance between the eyes and overdevelopment of midline features; Hu and Helms, 1999). Because of the serious developmental consequences of both decreases and increases of Shh activity, it is reasonable to think that the signaling level is under stabilizing selection. With an ecological change such as the transition between surface and cave living for the fish *Astyanax*, this selective regimen may change (Jeffery, 2001; Yamamoto *et al.*, 2003). The result is that a whole suite of important changes, affecting developmental processes and morphological traits, will take place as a response to a single genetic change in a crucial developmental interaction.

A very different set of conditions for the evolution of pleiotropic effects exists when pleiotropy originates by parallel variation in separate developmental pathways. Because the source of variation is itself the basis for the coupling of effects between traits, the pleiotropic effects of each locus are free to change independently of other loci. It is to be expected that *cis*-regulatory control of gene expression (e.g., Davidson, 2001; Levine and Tjian, 2003) is an important mechanism determining these effects and that many evolutionary changes of pleiotropic effects will result from the evolution of these control mechanisms (Wray *et al.*, 2003).

For simplicity, I will outline the changes of pleiotropic effects by parallel variation for a hypothetical gene with two separate *cis*-regulatory elements (promoters) that control expression of the gene in two different developmental pathways (Figure 11-5). If an allelic difference is located in the coding region of the gene and affects the developmental activity of the protein product, this difference will have an effect on both pathways simultaneously and will show pleiotropy of the resulting traits (Figure 11-5A). In contrast, an allelic difference that is located in one of the regulatory regions will only change the level of gene expression in the respective pathway but will leave the other pathway unaffected (Figure 11-5B, C). Accordingly, such allelic differences in single regulatory regions do not have pleiotropic effects on the resulting traits. This applies equally to differences resulting from new mutations and to differences between alternative "wild-type" alleles occurring in populations. New mutations or recombination events in the coding and regulatory regions of a gene can produce new patterns of joint effects on different traits. Because the regulation of each gene can evolve more or less independently, no particularly strong effects opposing evolutionary change are to be expected. Studies of real examples of genetic variation for gene regulatory sequences have found considerable variation and evolutionary potential (e.g., Rockman and Wray, 2002; Romano and Wray, 2003; Wray *et al.*, 2003). Therefore, it can be expected that pleiotropy by parallel variation will exhibit a considerable evolutionary flexibility.

In reality, the processes of gene expression contain additional complexities beyond the simplified description given here (Figure 11-5), which add more detail but do not change the principal conclusions drawn here (e.g., Davidson, 2001; Levine and Tjian, 2003). Not all allelic differences in the protein coding regions of a gene may produce pleiotropic effects, for instance, because alternative splicing of transcripts may restrict the effects of allelic differences to special developmental contexts and therefore to one pathway or another. These processes provide extra possibilities for the evolution of pleiotropy by parallel variation and therefore reinforce the conclusion drawn from the simplified model.

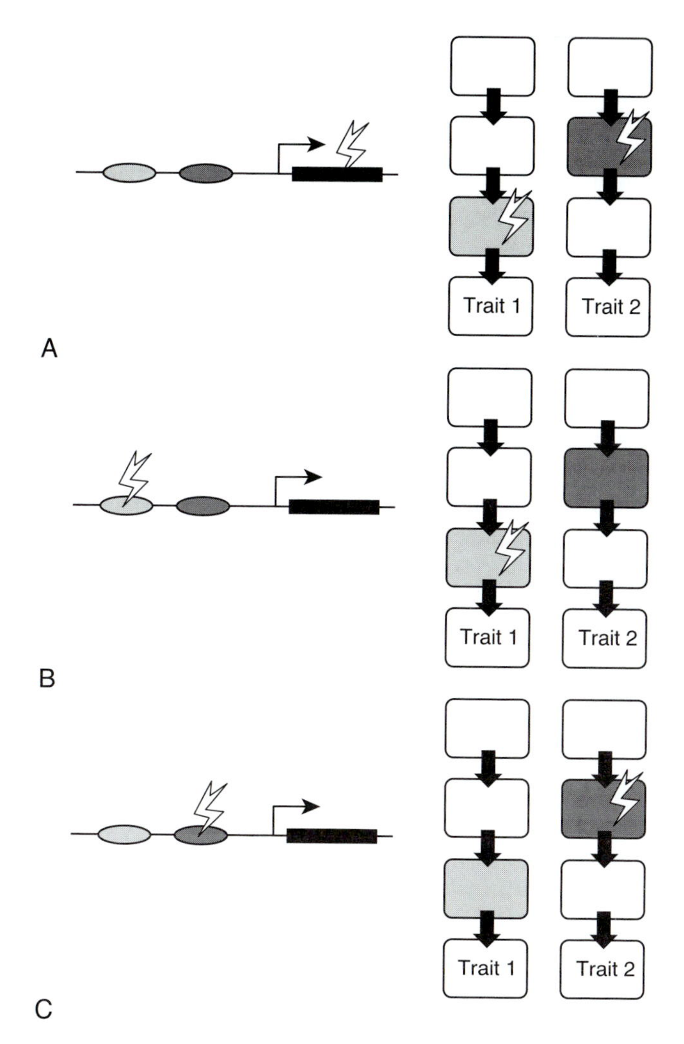

FIGURE 11-5. Mutation of regulatory and coding sequences of a gene and their consequences on pleiotropy of the gene by parallel variation of pathways. The gene has two regulatory modules (ellipses) that activate transcription in two different developmental contexts (corresponding shading of the regulatory module and the respective step in one of the pathways). (A) A mutation in the coding region of the gene. This mutation will have an effect on the gene product itself and will therefore affect both pathways simultaneously, thereby producing pleiotropy between the traits originating from the pathways (unless there are complicating factors such as alternative splicing, etc.). (B) Mutation is one of the regulatory modules (light gray shading). The mutation produces an effect only in one of the pathways. Therefore only one trait is affected, and the mutation does not have a pleiotropic effect. (C) The mutation is in the other regulatory module (dark gray), and the effect is just the reverse of the situation in (B). There is no pleiotropy either.

VI. MODULARITY OF PLEIOTROPIC EFFECTS: INHERENT IN DEVELOPMENTAL SYSTEMS OR EVOLVED PROPERTY?

The preceding discussion of pleiotropy has focused on its developmental origins, and likewise, modularity has been defined exclusively in terms of direct developmental interactions. This perspective is viewing modularity as an outcome of the developmental system, which is an intrinsic feature of the organism that can produce patterns of variation and constraints (Seilacher, 1974; Gould and Lewontin, 1979; Alberch, 1989; Gould, 2002).

This explanation for the origin of modularity of genetic variation is an alternative to the perspective advanced by Wagner and Altenberg (1996) and Wagner (1996), which is related to ideas published by Cheverud (1982, 1984), Riedl (1975), and Olson and Miller (1958). This explanation is based on the assumption that selection favors integration within complexes of traits that serve a particular function. For instance, integrated variation of upper and lower jaws will be necessary to ensure proper occlusion and therefore to support functions such as biting and chewing effectively. Accordingly, selection is expected to favor pleiotropy among those traits that belong to the same functional units. Because different functions can pose different adaptive demands, the same reasoning suggests that pleiotropic effects among traits that serve different functions would be disfavored. Selection will tend, on the one hand, to extend the pleiotropic effects of genes to the sets of traits serving particular functions, and on the other hand, to break up pleiotropic complexes of traits that are involved in different functions (integration and parcellation in the terminology of Wagner and Altenberg, 1996). As a result of this adaptive process, separate sets of loci will have effects on the sets of traits associated with different functions. The genetic modularity will match the subdivision of morphological structures into functional units. The genetic modules are distinct sets of loci, each internally connected by a network of pleiotropic effects, which will map directly to functional modules, sets of traits related by shared functions (Wagner and Altenberg, 1996). The crucial point of this view is that modularity is the outcome of selection for variation that can accommodate groups of traits serving different functions.

This theory assumes that the sets of traits affected by pleiotropy can evolve. This means that the loci accounting for the genetic variation of the traits possess alleles that differ in the distribution of pleiotropic effects. This assumption is the standard in theoretical quantitative genetics, which like many other areas of genetics, is primarily based on the differential gene concept. This concept defines genes as units of genomic change associated with phenotypic change without specifically considering the processes involved (Gilbert, 2000; Schwartz, 2000). Therefore, the primary aim is to construct a connection between allelic variation and the associated phenotypic differences. The resulting "genotype–phenotype map" (Wagner and

Altenberg, 1996) is an abstract mapping between genes and phenotypic characters that does not take into account the developmental mechanisms by which genes exert their effects on the phenotype (diagrams represent the mapping by straight arrows from genes to characters; e.g., Wagner, 1996; Wagner and Altenberg, 1996). This may be surprising given the prominent place development takes in the discussion of these models (Cheverud, 1982, 1984, 1996; Wagner, 1996; Wagner and Altenberg, 1996). This point is of far more than just symbolic importance, because the developmental system that mediates between genes and phenotypes will determine the extent to which pleiotropic effects are variable among alternative alleles and new mutations of the loci that affect the traits.

If pleiotropy originates by parallel variation of separate developmental pathways, there should be plentiful variation in the patterns of pleiotropy and ample opportunity for evolutionary change (Figure 11-5). In contrast, if pleiotropy mainly originates from direct interaction of developmental pathways, there may not be much variation in the pleiotropic patterns among alleles, because the interaction imparts similar patterns of pleiotropy to all upstream loci. Even after a change of the interaction itself, variation in pleiotropic patterns may not increase because the change of the interaction may simply yield a switch to a new pattern of pleiotropy that might still apply to all upstream loci simultaneously. Pleiotropy by direct interaction of developmental pathways therefore produces conditions that are less favorable for the adaptive evolution of pleiotropy at individual loci. The developmental origin of pleiotropy therefore clearly matters for the evolution of modularity. This consideration on how pleiotropy can evolve, however, does not address the question of whether modularity is an evolved property.

The empirical evidence is indecisive. A considerable amount of work has been done on integration and modularity in the mouse mandible (Atchley and Hall, 1991). Studies of the overall genetic variation have produced evidence for a degree of subdivision into anterior and posterior modules (Atchley *et al.*, 1985; Cheverud *et al.*, 1991; Klingenberg and Leamy, 2001), consistent with differences in the functions and embryonic precursors of the different parts (Atchley and Hall, 1991; Atchley, 1993; Tomo *et al.*, 1997). Several studies have examined the pleiotropic effects of individual QTLs and found patterns consistent with a subdivision into two modules (Cheverud *et al.*, 1997; Mezey *et al.*, 2000; Ehrich *et al.*, 2003; Klingenberg *et al.*, 2004). However, the patterns of overall genetic variation and of individual QTL effects were consistent with the patterns of correlated fluctuating asymmetry that are indicative of direct developmental interactions (Leamy, 1993; Klingenberg *et al.*, 2003). This evidence is consistent both with the hypotheses of modularity as an adaptively evolved property and with the alternative that modularity is an automatic outcome of the developmental system. Distinguishing between these two hypotheses poses substantial challenges for empirical studies.

In many ways, the question of whether modularity has originated by adaptive evolution or as an automatic outcome of developmental systems parallels the debate about the origin of dominance (Kacser and Burns, 1981; Orr, 1991; Porteous, 1996; Mayo and Bürger, 1997; Bourguet and Raymond, 1998; Bourguet, 1999; Omholt *et al.*, 2000). Just as is the case for dominance, it is clear that modularity is associated with most developmental systems and that it can evolve. The conditions are therefore met for both alternatives. To resolve the question of the relative importance of the two factors decisively, special experimental systems will be required. A possible approach is to use systems where one of the factors has been ruled out, as in the analysis of novel phenotypes such as mutants and teratologies, which are nonfunctional and therefore not the product of adaptation (Alberch, 1989; Dworkin *et al.*, 2001; Monteiro *et al.*, 2003). Another possibility is to study the evolution of systems where the functional and developmental units are clearly incongruent. Such studies have yet to be undertaken.

VII. FROM PLEIOTROPIC GENE EFFECTS TO G MATRICES

Regardless of the mechanisms that generate pleiotropic effects of single loci, a population's potential to respond to selection or to evolve by drift depends on the aggregate effect of all the loci affecting a set of traits. It is therefore important to evaluate the consequences of developmental changes for the overall quantitative genetic setup of phenotypic traits. To understand the consequences for evolvability and constraints of morphological traits, this section will examine how the effects of individual loci combine to the overall patterns of the genetic covariance matrix **G** and therefore the potential for evolutionary change (Lande, 1979; Roff, 1997; Steppan *et al.*, 2002). The modes by which pleiotropy is produced have different implications for the genetic covariance structure.

A direct interaction of pathways simultaneously imparts its pattern of pleiotropy to all the loci upstream in the pathway, from which it receives an input of variation. As a result, all these loci have more or less congruent patterns of pleiotropy, which they contribute to the overall pattern of genetic variation. After adding these effects from all loci, the interaction may therefore have a substantial influence on the structure of the total genetic variation, as it can be characterized by the **G** matrix. Because these patterns of genetic integration are also expected to coincide with the subdivision of morphological structures into developmental modules, the **G** matrix is expected to reflect this modular structure.

In contrast, the loci whose pleiotropy is based on parallel variation of separate pathways generate a diversity of different patterns, where each locus may have its own characteristic pleiotropic pattern. When the effects of all loci are combined, this diversity of different pleiotropic patterns will tend to "dilute" the effects of direct interaction. As a result, the patterns of overall genetic variation may only coincide to some degree with the modules defined by direct developmental interaction, but no complete separation is to be expected (e.g., Klingenberg and Leamy, 2001; Klingenberg *et al.*, 2003).

In addition to these effects of pleiotropy, another contribution to the **G** matrix comes from loci that may not have pleiotropic effects at all, but jointly affect multiple traits because of linkage disequilibrium between them (Falconer and Mackay, 1996; Lynch and Walsh, 1998). Because linkage disequilibrium depends on the population structure, this contribution to the genetic covariances is expected to be highly variable over time. Moreover, the contribution of linkage disequilibrium to genetic covariance structure has no relationship to the modular structure of phenotypic traits. The relative importance of pleiotropy and linkage disequilibrium for the **G** matrix has rarely been investigated, but a study of genetic correlations among floral traits in wild radish suggested that pleiotropy was the primary factor (Conner, 2002).

VIII. G MATRICES, CONSTRAINTS, AND EVOLUTIONARY DYNAMICS

The components of genetic covariance from the different developmental origins differ substantially in their effects on evolutionary dynamics. How pleiotropy of individual loci can evolve has been discussed in the preceding text, and this section will expand on those arguments to explore evolutionary change of entire **G** matrices and its implications for the dynamics of evolution of the mean phenotype.

Because pleiotropy by parallel variation is specific to each locus, its effects on the **G** matrices will conform to the situation implied by the models of the evolution of genetic covariance structure, where each locus can have alleles with different pleiotropic patterns, which are subject to selection (e.g., Lande, 1980). Therefore, this contribution to genetic covariance structure will be evolutionarily malleable, and modularity can easily evolve to reflect the functional subdivision of morphological structures (Cheverud, 1982, 1996; Wagner and Altenberg, 1996). Because of the flexible nature of this source of pleiotropy, absolute or lasting genetic constraints are not expected to originate in this manner. These evolutionary changes will occur in a more or less gradual manner, depending on the magnitudes of effects of individual alleles. The influence of linkage disequilibrium on the genetic covariance structure is likely to be similar, but will be even more transient.

Direct developmental interactions not only differ from the other components of genetic covariance because they tend to shape the **G** matrix in a specific way corresponding to the developmental modularity, but they also have particular implications for evolutionary dynamics. As mentioned in the preceding text (see "Evolution of Pleiotropy and Developmental Interactions"), it is likely that fundamental changes of the developmental architecture in a complex of traits are usually selected against and that direct interactions therefore provide a sort of buffering for the patterns of pleiotropy. There is little direct evidence on this subject, and empirical studies are clearly needed (even though they may be technically challenging). However, examples such as that of the *sonic hedgehog* gene, where mutations have serious adverse effects regardless of whether they reduce or increase signaling activity (Hu and Helms, 1999; Schell-Apacik *et al.*, 2003), provide indirect evidence for stabilizing selection. Moreover, there are indications for evolutionary conservation of the developmental function of gene regulatory systems even despite divergence in regulatory sequences, suggesting that the function of the whole systems is under stabilizing selection (e.g., Ludwig *et al.*, 2000; Romano and Wray, 2003). Overall, it is likely that direct developmental interactions are evolutionarily conservative.

This evolutionary conservatism of direct developmental interactions also tends to render the resulting patterns of pleiotropy resistant to change. Moreover, because all loci upstream of a direct interaction will obtain the same pleiotropic pattern from it, these stable patterns may make a substantial contribution to the total genetic variation. As a result, the **G** matrix will be fairly stable over evolutionary time, which may be manifest in comparative genetic studies as a similarity of **G** matrices of related populations or even species (Kohn and Atchley, 1988; Roff, 1997, 2000; Steppan *et al.*, 2002). Any genetic constraints may have a sustained influence on the evolutionary trajectories of populations.

Once changes in direct developmental interactions signaling mechanisms do occur, however, the consequences can be momentous. The change in *sonic hedgehog* that appears to account for the loss of eyes in cave fish and its manifold effects on craniofacial morphology is a clear example of this (Jeffery, 2001; Yamamoto *et al.*, 2001, 2003). Because the direct interactions influence the effects of all genes that act upstream of the interaction, a change in the interaction will trigger a change in the pleiotropic effects of all these loci. Such a concerted change of the pleiotropic patterns of multiple loci can have a substantial effect on the total genetic covariation among traits, which results from the combined effects of all loci in the whole system. Hence, a single change of developmental interactions can precipitate a fundamental change not only of the average morphology, but also of the standing stock and organization of genetic variation in the population and thus may substantially alter its potential to respond to selection.

As a result of a change in a direct developmental interaction between pathways, the **G** matrix may change considerably and, along with it, the potential for evolutionary change. Reorganization of the developmental system is therefore likely to be accompanied by a reorganization of the genetic constraints and evolvability of the structure. Because a single change is potentially sufficient to produce this reorganization, changes in the **G** matrix may be rapid. If reorganization is favored by selection and progresses through the population as a selective sweep, then the change of the **G** matrix may appear to be instantaneous on an evolutionary time scale.

If that is the case, then the evolution of patterns of pleiotropy may proceed in a punctuated manner in which extended periods of stabilizing selection and no change of the pleiotropic relationships are interspersed with short episodes of developmental reorganization of the modular structure of developmental systems that may coincide with dramatic changes in the genetic and phenotypic covariance structure. Phases of strong directional selection on the mean phenotype might override the effects of stabilizing selection and favor the reorganization of the developmental interactions among developmental pathways. Such episodes of change of the covariance structure would remove genetic constraints of the evolution of the population average phenotype and release new phenotypic variation. The molding of phenotypic variation by developmental interactions between pathways therefore can act in a manner analogous to the evolutionary "capacitance" by some mechanisms of phenotypic buffering (Rutherford and Lindquist, 1998; de Visser *et al.*, 2003; Rutherford, 2003). The release of variation in response to the developmental reorganization may be a factor contributing to classical punctuated evolution as it has been characterized from the fossil record (reviewed by Gould, 2002).

IX. PERSPECTIVE: DEVELOPMENTAL PROCESSES AND EVOLUTIONARY CONSTRAINTS

This discussion of the developmental origins of morphological covariation and modularity has substantial implications for the interpretation of constraints in evolution. A number of authors have discussed whether modularity enhances evolvability or is necessary for it (Cheverud, 1996; Wagner, 1996; Wagner and Altenberg, 1996; Hansen, 2003). My approach here is slightly different from these discussions, as I consider the ways in which modularity and pleiotropy come about in developmental systems. I have outlined in the preceding text that the mechanisms responsible for generating pleiotropy differ considerably in their evolutionary behavior, both in terms of the likelihood of change and, once a change occurs, in how profound its effects will be. The primary conclusion that emerges from this line of reasoning is that the developmental origin of the covariation among traits is a factor of prime importance for evolutionary quantitative genetics.

Patterns of pleiotropy that originate by the direct interaction of developmental pathways are likely to be evolutionarily conservative, but once a change occurs, it will simultaneously affect multiple upstream loci in a coordinated manner and will therefore have a profound effect on the **G** matrix. As a result of this change in the modular structure of the system, the patterns of genetic constraints will also change, and the potential for response to selection will be altered as well. A single change of the developmental system may thus release an avalanche of changes in the evolutionary potential of the traits, which may manifest itself as a punctuated change in the evolutionary behavior of the evolving lineage.

ACKNOWLEDGMENTS

I thank Casper Breuker, Nelly Gidaszewksi, and Yoshie Jintsu for helpful comments on an earlier version of the manuscript.

REFERENCES

Alberch, P. (1989). The logic of monsters: Evidence for internal constraint in development and evolution. *Geobios* **12**, 21–57.

Arnold, S. J. (1992). Constraints on phenotypic evolution. *American Naturalist* **140**, S85–S107.

Arthur, W. (2001). Developmental drive: An important determinant of the direction of phenotypic evolution. *Evolution and Development* **3**, 271–278.

Atchley, W. R. (1993). Genetic and developmental aspects of variability in the mammalian mandible. In *The Skull* (J. Hanken and B. K. Hall, eds.), pp. 207–247. Chicago: University of Chicago Press.

Atchley, W. R., and Hall, B. K. (1991). A model for development and evolution of complex morphological structures. *Biological Review* **66**, 101–157.

Atchley, W. R., Plummer, A. A., and Riska, B. (1985). Genetics of mandible form in the mouse. *Genetics* **111**, 555–577.

Beldade, P., Brakefield, P. M., and Long, A. D. (2002a). Contribution of *Distal-less* to quantitative variation in butterfly eyespots. *Nature* **415**, 315–318.

Beldade, P., Koops, K., and Brakefield, P. M. (2002b). Developmental constraints versus flexibility in morphological evolution. *Nature* **416**, 844–847.

Bookstein, F. L., Gunz, P., Mitteroecker, P., Prossinger, H., Schaefer, K., and Seidler, H. (2003). Cranial integration in *Homo*: Singular warps analysis of the midsagittal plane in ontogeny and evolution. *Journal of Human Evolution* **44**, 167–187.

Bourguet, D. (1999). The evolution of dominance. *Heredity* **83**, 1–4.

Bourguet, D., and Raymond, M. (1998). The molecular basis of dominance relationships: The case of some recent adaptive genes. *Journal of Evolutionary Biology* **11**, 103–122.

Brakefield, P. M., Gates, J., Keys, D., Kesbeke, F., Wijngaarden, P. J., Monteiro, A., French, V., and Carroll, S. B. (1996). Development, plasticity and evolution of butterfly eyespot patterns. *Nature* **384**, 236–242.

Carroll, S. B., Gates, J., Keys, D. N., Paddock, S. W., Panganiban, G. E. F., Selegue, J. E., and Williams, J. A. (1994). Pattern formation and eyespot determination in butterfly wings. *Science* **265**, 109–114.

Carroll, S. B., Grenier, J. K., and Weatherbee, S. D. (2001). *From DNA to Diversity: Molecular Genetics and the Evolution of Animal Design.* Malden, MA: Blackwell Science.

Cheverud, J. M. (1982). Phenotypic, genetic, and environmental morphological integration in the cranium. *Evolution* **36**, 499–516.

Cheverud, J. M. (1984). Quantitative genetics and developmental constraints on evolution by selection. *Journal of Theoretical Biology* **110**, 155–171.

Cheverud, J. M. (1995). Morphological integration in the saddle-back tamarin (*Saguinus fuscicollis*) cranium. *American Naturalist* **145**, 63–89.

Cheverud, J. M. (1996). Developmental integration and the evolution of pleiotropy. *American Zoology* **36**, 44–50.

Cheverud, J. M., Hartman, S. E., Richtsmeier, J. T., and Atchley, W. R. (1991). A quantitative genetic analysis of localized morphology in mandibles of inbred mice using finite element scaling. *Journal of Craniofacial Genetics and Developmental Biology* **11**, 122–137.

Cheverud, J. M., Routman, E. J., and Irschick, D. J. (1997). Pleiotropic effects of individual gene loci on mandibular morphology. *Evolution* **51**, 2006–2016.

Chiang, C., Litingtung, Y., Lee, E., Young, K. E., Corden, J. L., Westphal, H., and Beachy, P. A. (1996). Cyclopia and defective axial patterning in mice lacking *sonic hedgehog* gene function. *Nature* **383**, 407–413.

Cohen, M. M., Jr. (2002). Malformations of the craniofacial region: Evolutionary, embryonic, genetic, and clinical perspectives. *American Journal of Medical Genetics (Seminars in Medical Genetics)* **115**, 245–268.

Conner, J. K. (2002). Genetic mechanisms of floral trait correlations in a natural population. *Nature* **420**, 407–410.

Davidson, E. H. (1993). Later embryogenesis: Regulatory circuitry in morphogenetic fields. *Development* **118**, 665–690.

Davidson, E. H. (2001). *Genomic Regulatory Systems: Development and Evolution.* San Diego, CA: Academic Press.

Davidson, E. H., McClay, D. R., and Hood, L. (2003). Regulatory gene networks and the properties of the developmental process. *Proceedings of the National Academy of Science USA* **100**, 1475–1480.

de Visser, J. A. G. M., Hermisson, J., Wagner, G. P., Meyers, L. A., Bagheri-Chaichian, H., Blanchard, J. L., Chao, L., Cheverud, J. M., Elena, S. F., Fontana, W., Gibson, G., Hansen, T. F., Krakauer, D. C., Lewontin, R. C., Ofria, C., Rice, S. H., von Dassow, G., Wagner, A., and Whitlock, M. C. (2003). Evolution and detection of genetic robustness. *Evolution* **57**, 1959–1972.

Dworkin, I. M., Tanda, S., and Larsen, E. (2001). Are entrenched characters developmentally constrained? Creating biramous limbs in an insect. *Evolution and Development* **3**, 424–431.

Ehrich, T. H., Vaughn, T. T., Koreishi, S. F., Linsey, R. B., Pletscher, L. S., and Cheverud, J. M. (2003). Pleiotropic effects on mandibular morphology I. Developmental morphological integration and differential dominance. *Journal of Experimental Zoology (Molecular and Developmental Evolution)* **296B**, 58–79.

Eldar, A., Dorfman, R., Weiss, D., Ashe, H., Shilo, B. Z., and Barkai, N. (2002). Robustness of the BMP morphogen gradient in *Drosophila* embryonic patterning. *Nature* **419**, 304–308.

Enlow, D. H., and Hans, M. G. (1996). *Essentials of Facial Growth.* Philadelphia: Saunders.

Falconer, D. S., and Mackay, T. F. C. (1996). *Introduction to Quantitative Genetics.* Essex, England: Longman.

Flint, J., and Mott, R. (2001). Finding the molecular basis of quantitative traits: Successes and pitfalls. *Nature Reviews: Genetics* **2**, 437–445.

Francis-West, P., Ladher, R., Barlow, A., and Graveson, A. (1998). Signalling interactions during facial development. *Mechanisms of Development* **75**, 3–28.

Gilbert, S. F. (2000). Genes classical and genes developmental: The different use of genes in evolutionary syntheses. In *The Concept of the Gene in Development and Evolution: Historical and Epistemological Perspectives* (P. J. Beurton, R. Falk, and H. J. Rheinberger, eds.), pp. 178–192. Cambridge, England: Cambridge University Press.

Gould, S. J. (1989). A developmental constraint in *Cerion*, with comments on the definition and interpretation of constraint in evolution. *Evolution* **43**, 516–539.

Gould, S. J. (2002). *The Structure of Evolutionary Theory.* Cambridge, MA: Harvard.

Gould, S. J., and Lewontin, R. C. (1979). The spandrels of San Marco and the Panglossian paradigm: A critique of the adaptationist programme. *Proceedings of the Royal Society London Series B: Biological Sciences* **205**, 581–598.

Grüneberg, H. (1938). An analysis of the "pleiotropic" effects of a new lethal mutation in the rat (*Mus norvegicus*). *Proceedings of the Royal Society London Series B: Biological Sciences* **125**, 123–144.

Gurdon, J. B., and Bourillot, P. Y. (2001). Morphogen gradient interpretation. *Nature* **413**, 797–803.

Hadorn, E. (1945). Zur Pleiotropie der Genwirkung. *Archiv. der Julius Klaus-Stift. für Verebungsforsch.* Supplement **20**, 82–95.

Hallgrímsson, B., Willmore, K., and Hall, B. K. (2002). Canalization, developmental stability, and morphological integration in primate limbs. *Yearbook of Physical Anthropology* **45**, 131–158.

Hansen, T. F. (2003). Is modularity necessary for evolvability? Remarks on the relationship between pleiotropy and evolvability. *Biosystems* **69**, 83–94.

Hartl, D. L., and Clark, A. G. (1997). *Principles of Population Genetics*. Sunderland, MA: Sinauer Associates.

Herring, S. W. (1993). Epigenetic and functional influences on skull growth. In *The Skull* (J. Hanken and B. K. Hall, eds.), pp. 153–206. Chicago: University of Chicago Press.

Hodgkin, J. (1998). Seven types of pleiotropy. *International Journal of Developmental Biology* **42**, 501–505.

Houchmandzadeh, B., Wieschaus, E., and Leibler, S. (2002). Establishment of developmental precision and proportions in the early *Drosophila* embryo. *Nature* **415**, 798–802.

Hu, D., and Helms, J. A. (1999). The role of sonic hedgehog in normal and abnormal craniofacial morphogenesis. *Development* **126**, 4873–4884.

Jabs, E. W. (2002). Genetic etiologies of craniosynostosis. In *Understanding Craniofacial Anomalies: The Etiopathogenesis of Craniosynostoses and Facial Clefting* (M. P. Mooney and M. I. Siegel, eds.), pp. 125–146. New York: Wiley-Liss.

Jeffery, W. R. (2001). Cavefish as a model system in evolutionary developmental biology. *Developmental Biology* **231**, 1–12.

Jeffery, W. R., Strickler, A. G., and Yamamoto, Y. (2003). To see or not to see: Evolution of eye degeneration in Mexican blind cavefish. *Integrative and Comparative Biology* **43**, 531–541.

Jolliffe, I. T. (2002). *Principal Component Analysis*. New York: Springer-Verlag.

Kacser, H., and Burns, J. A. 1981. The molecular basis of dominance. *Genetics* **97**, 639–666.

Kirschner, M., and Gerhart, J. (1998). Evolvability. *Proceedings of the National Academy of Science USA* **95**, 8420–8427.

Klingenberg, C. P. (1998). Heterochrony and allometry: The analysis of evolutionary change in ontogeny. *Biological Reviews* **73**, 79–123.

Klingenberg, C. P. (2003a). Developmental instability as a research tool: Using patterns of fluctuating asymmetry to infer the developmental origins of morphological integration. In *Developmental Instability: Causes and Consequences* (M. Polak, ed), pp. 427–442. New York: Oxford University Press.

Klingenberg, C. P. (2003b). A developmental perspective on developmental instability: Theory, models and mechanisms. In *Developmental Instability: Causes and Consequences* (M. Polak, ed), pp. 14–34. New York: Oxford University Press.

Klingenberg, C. P. (2004). Integration, modules, and development: Molecules to morphology to evolution. In *Phenotypic Integration: Studying the Ecology and Evolution of Complex Phenotypes* (M. Pigliucci and K. Preston, eds.) pp. 213–230. New York: Oxford University Press.

Klingenberg, C. P., and Leamy, L. J. (2001). Quantitative genetics of geometric shape in the mouse mandible. *Evolution* **55**, 2342–2352.

Klingenberg, C. P., and Zaklan, S. D. (2000). Morphological integration between developmental compartments in the *Drosophila* wing. *Evolution* **54**, 1273–1285.

Klingenberg, C. P., Badyaev, A. V., Sowry, S. M., and Beckwith, N. J. (2001a). Inferring developmental modularity from morphological integration: Analysis of individual variation and asymmetry in bumblebee wings. *American Naturalist* **157**, 11–23.

Klingenberg, C. P., Leamy, L. J., Routman, E. J., and Cheverud, J. M. (2001b). Genetic architecture of mandible shape in mice: Effects of quantitative trait loci analyzed by geometric morphometrics. *Genetics* **157**, 785–802.

Klingenberg, C. P., Mebus, K., and Auffray, J.-C. (2003). Developmental integration in a complex morphological structure: How distinct are the modules in the mouse mandible? *Evolution and Development* **5**, 522–531.

Klingenberg, C. P., Leamy, L. J., and Cheverud, J. M. (2004). Integration and modularity of quantitative trait locus effects on geometric shape in the mouse mandible. *Genetics* **166**, 1909–1921.

Kohn, L. A. P., and Atchley, W. R. (1988). How similar are genetic correlation structures? Data from mice and rats. *Evolution* **42**, 467–481.

Lande, R. (1979). Quantitative genetic analysis of multivariate evolution, applied to brain:body size allometry. *Evolution* **33**, 402–416.

Lande, R. (1980). The genetic covariance between characters maintained by pleiotropic mutations. *Genetics* **94**, 203–215.

Larsen, E. (2003). Genes, cell behavior, and the evolution of form. In *Origination of Organismal Form: Beyond the Gene in Developmental and Evolutionary Biology* (G. B. Müller and S. A. Newman, eds.), pp. 119–131. Cambridge, MA: MIT Press.

Leamy, L. (1993). Morphological integration of fluctuating asymmetry in the mouse mandible. *Genetics* **89**, 139–153.

Leamy, L., and Atchley, W. R. (1984). Morphometric integration in the rat (*Rattus* sp). scapula. *Journal of Zoology* **202**, 43–56.

Leamy, L. J., Routman, E. J., and Cheverud, J. M. (1999). Quantitative trait loci for early- and late-developing skull characters in mice: A test of the genetic independence model of morphological integration. *Americal Naturalist* **153**, 201–214.

Levine, M., and Tjian, R. (2003). Transcription regulation and animal diversity. *Nature* **424**, 147–151.

Ludwig, M. Z., Bergman, C., Patel, N. H., and Kreitman, M. (2000). Evidence for stabilizing selection in a eukaryotic enhancer element. *Nature* **403**, 564–567.

Lynch, M., and Walsh, B. (1998). *Genetics and Analysis of Quantitative Traits*. Sunderland, MA: Sinauer.

Maynard Smith, J., Burian, R., Kauffman, S., Alberch, P., Campbell, J., Goodwin, B., Lande, R., Raup, D., and Wolpert, L. (1985). Developmental constraints and evolution. *Quarterly Review of Biology* **60**, 265–287.

Mayo, O., and Bürger, R. (1997). The evolution of dominance: A theory whose time has passed? *Biological Review* **72**, 97–110.

Mayr, E. (1982). *The Growth of Biological Thought: Diversity, Evolution, and Inheritance*. Cambridge, MA: Harvard University Press.

McAdams, H. H., and Arkin, A. (1999). It's a noisy business! Genetic regulation at the nanomolecular scale. *Trends in Genetics* **15**, 65–69.

Mezey, J. G., Cheverud, J. M., and Wagner, G. P. (2000). Is the genotype-phenotype map modular? A statistical approach using mouse quantitative trait loci data. *Genetics* **156**, 305–311.

Møller, A. P., and Swaddle, J. P. (1997). *Asymmetry, Developmental Stability, and Evolution*. Oxford, England: Oxford University Press.

Monteiro, A., Prijs, J., Bax, M., Hakkaart, T., and Brakefield, P. M. (2003). Mutants highlight the modular control of butterfly eyespot patterns. *Evolution and Development* **5**, 180–187.

Nadeau, J. H., Burrage, L. C., Restivo, J., Pao, Y. H., Churchill, G. A., and Hoit, B. D. (2003). Pleiotropy, homeostasis, and functional networks based on assays of cardiovascular traits in genetically randomized populations. *Genome Research* **13**, 2082–2091.

Nanni, L., Ming, J. E., Bocian, M., Steinhaus, K., Bianchi, D. W., de Die-Smulders, C., Giannotti, A., Imaizumi, K., Jones, K. L., Del Campo, M., Martin, R. A., Meinecke, P., Pierpont, M. E. M., Robin, N. H., Young, I. D., Roessler, E., and Muenke, M. (1999). The mutational spectrum of the *sonic hedgehog* gene in holoprosencephaly: *SHH* mutations cause a significant proportion of autosomal dominant holoprosencephaly. *Human Molecular Genetics* **8**, 2479–2488.

Nijhout, H. F., and Emlen, D. J. (1998). Competition among body parts in the development and evolution of insect morphology. *Proceedings of the National Academy of Science USA* **95**, 3685–3689.

Olson, E. C., and Miller, R. L. (1958). *Morphological Integration*. Chicago: University of Chicago Press.

Omholt, S. W., Plathe, E., Øyehaug, L., and Xiang, K. (2000). Gene regulatory networks generating the phenomena of additivity, dominance, and epistasis. *Genetics* **155**, 969–980.

Orr, H. A. (1991). A test of Fisher's theory of dominance. *Proceedings of the National Academy of Science USA* **88**, 11413–11415.

Palmer, A. R., and Strobeck, C. (1986). Fluctuating asymmetry: Measurement, analysis, patterns. *Annual Review of Ecology and Systematics* **17**, 391–421.

Panganiban, G., Nagy, L., and Carroll, S. B. (1994). The role of the *Distal-less* gene in the development and evolution of insect limbs. *Current Biology* **4**, 671–675.

Parker, G. A., and Maynard Smith, J. (1990). Optimality theory in evolutionary biology. *Nature* **348**, 27–33.

Porteous, J. W. (1996). Dominance: One hundred and fifteen years after Mendel's paper. *Journal of Theoretical Biology* **182**, 223–232.

Pyeritz, R. E. (1989). Pleiotropy revisited: Molecular explanations of a classic concept. *American Journal of Medical Genetics* **34**, 124–134.

Raff, R. A. (1996). *The Shape of Life: Genes, Development, and the Evolution of Animal Form*. Chicago: University of Chicago Press.

Riedl, R. (1975). *Die Ordnung des Lebendigen: Systembedingungen der Evolution*. Berlin: Parey.

Riska, B. (1986). Some models for development, growth, and morphometric correlation. *Evolution* **40**, 1303–1311.

Rockman, M. V., and Wray, G. A. (2002). Abundant raw material for *cis*-regulatory evolution in humans. *Molecular Biology and Evolution* **19**, 1991–2004.

Roff, D. A. (1997). *Evolutionary Quantitative Genetics*. New York: Chapman & Hall.

Roff, D. A. (2000). The evolution of the **G** matrix: Selection or drift? *Heredity* **84**, 135–142.

Romano, L. A., and Wray, G. A. (2003). Conservation of *Endo16* expression in sea urchins despite evolutionary divergence in both cis and trans-acting components of transcriptional regulation. *Development* **130**, 4187–4199.

Rose, M. R., and Lauder, G. V., eds. (1996). *Adaptation*. San Diego, CA: Academic Press.

Rutherford, S. L. (2003). Between genotype and phenotype: Protein chaperones and evolvability. *Nature Reviews: Genetics* **4**, 263–274.

Rutherford, S. L., and Lindquist, S. (1998). Hsp90 as a capacitor for morphological evolution. *Nature* **396**, 336–342.

Schell-Apacik, C., Rivero, M., Knepper, J. L., Roessler, E., Muenke, M., and Ming, J. E. (2003). *Sonic Hedgehog* mutations causing human holoprosencephaly impair neural patterning activity. *Human Genetics* **113**, 170–177.

Schwartz, S. (2000). The differential concept of the gene: past and present. In *The Concept of the Gene in Development and Evolution: Historical and Epistemological Perspectives* (P. J. Beurton, R. Falk, and H. J. Rheinberger, eds.), pp. 26–39. Cambridge, England: Cambridge University Press.

Seilacher, A. (1974). Fabricational noise in adaptive morphology. *Systematic Zoology* **22**, 451–465.

Steppan, S. J., Phillips, P. C., and Houle, D. (2002). Comparative quantitative genetics: Evolution of the **G** matrix. *Trends in Ecology and Evolution* **17**, 320–327.

Tomo, S., Ogita, M., and Tomo, I. (1997). Development of mandibular cartilages in the rat. *Anatomical Record* **249**, 233–239.

von Dassow, G., and Munro, E. (1999). Modularity in animal development and evolution: Elements of a conceptual framework for EvoDevo. *Journal of Experimental Zoology (Molecular Development and Evolution)* **285**, 307–325.

Wagner, G. P. (1996). Homologues, natural kinds and the evolution of modularity. *American Zoology* **36**, 36–43.

Wagner, G. P., and Altenberg, L. (1996). Complex adaptations and the evolution of evolvability. *Evolution* **50**, 967–976.

Weber, K. E. (1990). Selection on wing allometry in *Drosophila melanogaster*. *Genetics* **126**, 975–989.

Weber, K. E. (1992). How small are the smallest selectable domains of form? *Genetics* **130**, 345–353.

Wilkie, A. O. M., and Morriss-Kay, G. M. (2001). Genetics of craniofacial development and malformation. *National Review of Genetics* **2**, 458–468.

Wilkins, A. S. (2002). *The Evolution of Developmental Pathways*. Sunderland, MA: Sinauer Associates.

Winther, R. G. (2001). Varieties of modules: Kinds, levels, origins, and behaviors. *Journal of Experimental Zoology (Molecular Development and Zoology)* **291**, 116–129.

Workman, M. S., Leamy, L. J., Routman, E. J., and Cheverud, J. M. (2002). Analysis of quantitative trait locus effects on the size and shape of mandibular molars in mice. *Genetics* **160**, 1573–1586.

Wray, G. A., Hahn, M. W., Abouheif, E., Balhoff, J. W., Pizer, M., Rockman, M. V., and Romano, L. A. (2003). The evolution of transcriptional regulation in eukaryotes. *Molecular Biology and Evolution* **20**, 1377–1419.

Yamamoto, Y., and Jeffery, W. R. (2000). Central role for the lens in cave fish eye degeneration. *Science* **289**, 631–633.

Yamamoto, Y., Stock, D. W., and Jeffery, W. R. (2001). *Sonic hedgehog* controls the eyeless phenotype in cavefish. *Developmental Biology* **235**, 240.

Yamamoto, Y., Espinasa, L., Stock, D. W., and Jeffery, W. R. (2003). Development and evolution of craniofacial patterning is mediated by eye-dependent and -independent processes in the cavefish *Astyanax*. *Evolution and Development* **5**, 435–446.

Zelditch, M. L. (1987). Evaluating models of developmental integration in the laboratory rat using confirmatory factor analysis. *Systematic Zoology* **36**, 368–380.

CHAPTER 12

Developmental Regulation of Variability

MIRIAM LEAH ZELDITCH
Museum of Paleontology, University of Michigan, Ann Arbor, Michigan, USA

INTRODUCTION

Phenotypic variability is often taken for granted by evolutionary biologists even though variability is generally regarded as limited and nonrandom. Efforts to understand processes limiting and structuring variation have tended to focus on evolutionary mechanisms, especially stabilizing selection (e.g., Schmaulhausen, 1949; Waddington, 1957; Scharloo *et al.*, 1967; Cheverud, 1984; Kawecki, 2000). In theory, stabilizing selection leads to the evolution of developmental regulatory systems that buffer phenotypes against environmental or genetic perturbations. Recent theoretical investigations find that such developmental buffering (i.e., canalization) can arise as an emergent property of complex coregulatory systems (Wagner, 1996; Siegal and Bergman, 2002). Whether those emergent properties offer a general and reasonable alternative to stabilizing selection is presently difficult to say because we know very little about how most organisms regulate their variability.

Our ignorance about processes generating and regulating variability is not surprising in light of the sparse data on the subject. Although many studies estimate the variation of adult morphologies, few trace that variation to its causes or examine the ontogenetic dynamics of variation. In the absence of data, it is hard to say what a theory even *ought* to explain. In addition, it is hard to imagine what a general theory of variability might look like, given the diversity of developmental systems. Any theory that is broad enough to encompass just a fraction of that diversity might not make precise or even interesting predictions. However, there are some commonalities among otherwise disparate developmental systems that could enable formulating general and interesting theories about kinds of developmental regulation.

My focus herein is on the variability of mammalian craniofacial morphology, a model system for studies of the developmental basis of complex morphologies and their integration (e.g., Cheverud, 1982, 1995; Zelditch, 1988; Zelditch and Carmichael, 1989; Atchley and Hall, 1991; Klingenberg *et al.*, 2003). One important characteristic of skull morphology is that it is dynamic, changing continually over the lifetime of an individual. Dramatic changes in both size and shape occur long after organ primordia become visible, and those result from spatiotemporally organized patterns of growth. Even that spatiotemporal organization changes from age to age (Zelditch *et al.*, 1992, 2003), and so does the tissue itself. Throughout ontogeny, bone grows, reshapes, and maintains itself by balancing production with removal of tissue. We might therefore anticipate that the structure of skeletal variation is also dynamic, and empirical studies find it is (e.g., Zelditch, 1988; Zelditch and Carmichael, 1989; Cane, 1993).

Skeletal variability is not merely dynamic, it also seems to exhibit a predictable ontogenetic trend—decreasing substantially and rapidly early in postnatal growth. This reduction in variance suggests that skeletal morphology is intrinsically regulated. One aspect of skeletal morphology has long served as a paradigm for a developmentally canalized trait: body size (Waddington, 1952; Prader *et al.*, 1963; Tanner, 1963; Boersma and Wit, 1997). Skull shape also appears to be canalized because it too decreases in variability over ontogeny (Nonaka and Nakata, 1988; Zelditch *et al.*, 1993, 2004; Hingst-Zaher *et al.*, 2000). The ontogenetic reduction of variance has important and surprising implications because it suggests that canalization involves a form of error correction by compensatory growth. In the case of body size, it is easy to understand how compensatory growth might correct for deviations from the norm; early efforts to model the system proposed that organisms monitor body size, growth, and puberty via neuroendocrine mechanisms (e.g., Kennedy and Mitra, 1963; Tanner, 1963; Frisch and Revelle, 1971). However, it is difficult to imagine how developmental processes might monitor skull shape and recognize that errors have been made, much less how development would later correct them.

My first aim herein is to summarize and reconcile patterns found in the empirical literature, allaying the suspicion that the reduction in variance of skull shape might be just an artifact of measuring fragile, malleable infant skulls. After documenting the same pattern in another case, one unlikely to be prone to such artifacts, I consider what the magnitude and structure of variation indicates regarding candidate mechanisms. I then suggest four that could correct for developmental errors even when those errors cannot be recognized as such by developmental processes. All four hypothesize that something is regulated, but not skull shape, meaning that the canalization of skull shape is a developmental epiphenomenon.

I. EMPIRICAL PATTERNS

To my knowledge, only four studies examine the ontogenetic dynamics of variance of skeletal shape (Nonaka and Nokata, 1988; Zelditch *et al.*, 1993, 2004; Hingst-Zaher *et al.*, 2000), and one of them did not actually quantify variance (Zelditch *et al.*, 1993). All take a similar mathematical approach to shape, so all results should be at least proportional. E. Hingst-Zaher generously provided me with her data, so results for cotton rats (*Sigmodon fulviventer*), house mice (*Mus musculus domesticus*), and *Calomys expulsus* are based on exactly the same formulae, methods, and even algorithms. Nonaka and Nakata's analysis of Norway rats (*Rattus norvegicus*) used a similar approach. The analyses are all landmark-based geometric analyses, meaning that the data comprise configurations of coordinates of landmarks, superimposed by removing the variation unrelated to shape (i.e., that arising from position, scale, and orientation of specimens). The two most recent studies (Hingst-Zaher *et al.*, 2000; Zelditch *et al.*, 2004) use the Procrustes generalized least squares (GLS) method; Nonaka and Nokata's (1988) analysis predates the routine use of that method, so theirs relies on pairwise comparisons between specimens. After landmarks are optimally superimposed, the variance of skull shape can be computed by summarizing the variances of individual landmarks, which gives the same result as summing the variances of partial warp scores (including the uniform component) or summing the squared (Procrustes) distance between each individual and the mean divided by the appropriate degrees of freedom. Regardless of the variables used in the computation, the same results are obtained. To enable statistical testing of the changes in variance, standard errors of the variances were estimated by resampling (for more details on methods, see Zelditch *et al.*, 2004). Nonaka and Nokata did not estimate the total phenotypic variance, so I calculated it from their published variance components.

From the scatter around the landmarks of the cotton rat skull, it is visually obvious that variance decreases dramatically and quickly (Figure 12-1). I should

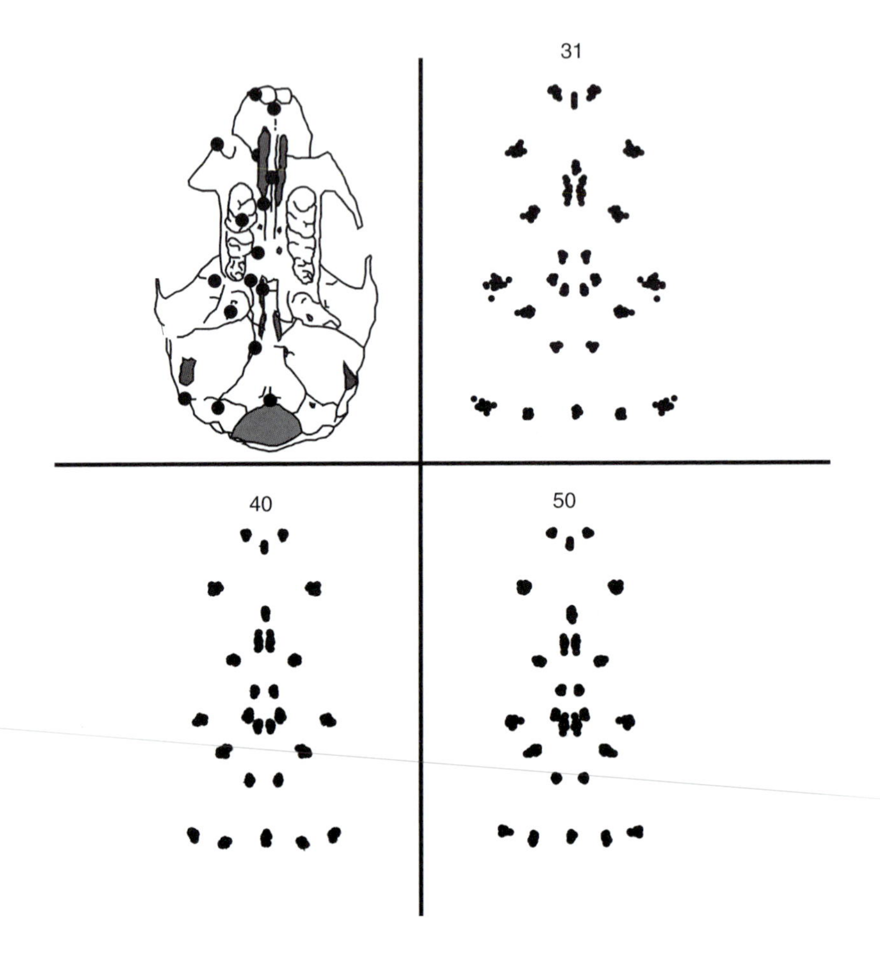

FIGURE 12-1. Variation of cotton rat craniofacial shape in palatal view. Landmark configurations of individuals within an age class are superimposed using the Procrustes generalized least squares (GLS) superimposition. When superimposing the landmarks, corresponding points on the right and left sides are averaged because they cannot be regarded as independent; variance of overall shape is estimated as the sum of the variances across all landmarks. To ease visual interpretation, the landmarks for the half skull are reflected across the midline. The (gestational) ages of each sample is given above the configuration.

note that the variances are not actually located *at* the landmarks. First, the superimposition procedure tends to smooth out the variance over all landmarks, and second, the variation is not *at* a point but rather in the location of that point relative to others. That visual impression of decreasing variance is confirmed by the quantitative analysis (Figure 12-2A). Figure 12-2 summarizes the results of all studies,

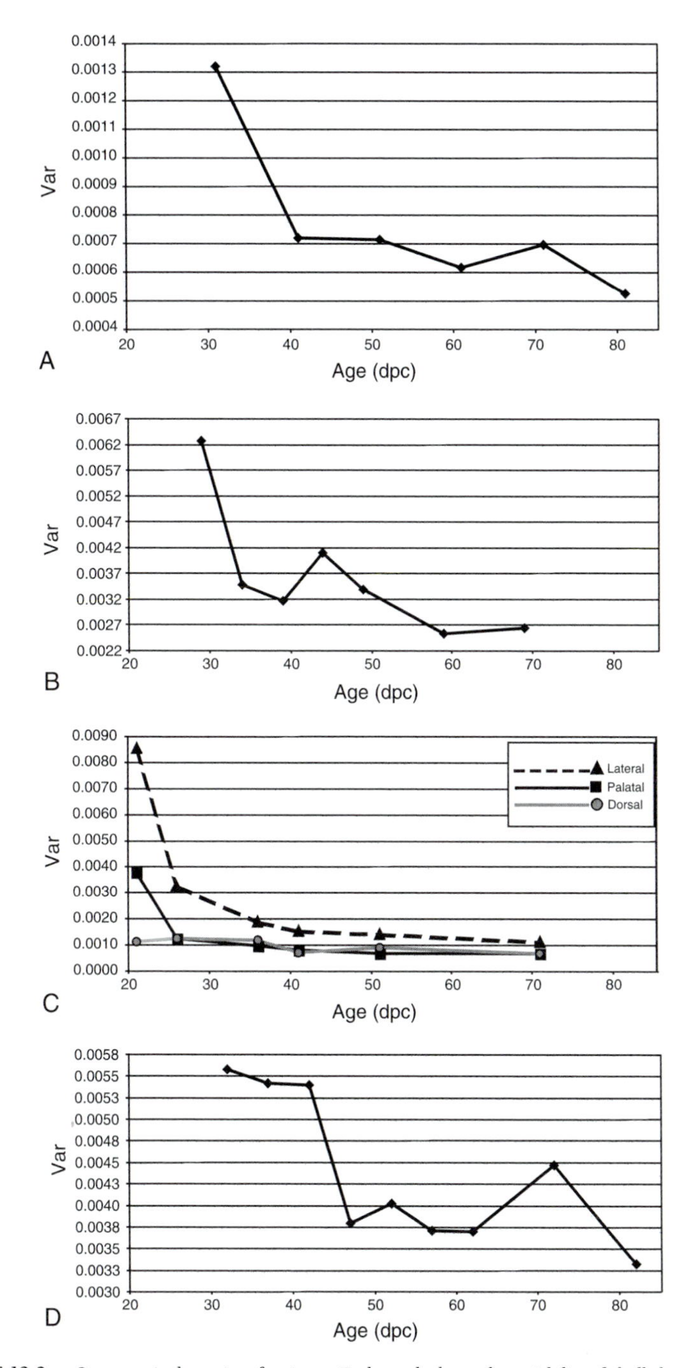

FIGURE 12-2. Ontogenetic dynamics of variance. Each graph shows the variability of skull shape relative to postnatal age. (A) Variance in skull shape of the cotton rat, *Sigmodon fulviventer*, palatal view, from Zelditch *et al.*, 2004; (B) variance in skull shape of the house mouse *Mus musculus domesticus*, palatal view, from Zelditch *et al.*, 2004. (C) variance in skull shape of *Calomys expulsus*, palatal, dorsal, and lateral views; (D) variance of skull shape of laboratory rats in lateral view from Nonaka and Nakata, 1988. (From Nonaka, K. and Nakata, M. [1988]. Genetic and environmental factors in the longitudinal growth of rats. 3. Craniofacial shape change. Journal of Craniofacial Genetics and Developmental Biology 8:337–344. Blackwell Publishing.)

including the separate analyses of three different views of skulls of *Calomys expulsus* (Figure 12-2C). Variance is plotted against gestational age because this scale seems more reasonable than postnatal age for comparing species that differ strikingly in gestation length and in degree of maturity at birth.

There are striking similarities among cotton rats, house mice, and *C. expulsus* in the ontogenetic dynamics of variance (Figure 12-2A, B, C respectively). Although not identical in all details, all three species undergo a dramatic drop in variance between the two youngest ages sampled; after 40 days postconception (40 dpc), variance stabilizes except for minor fluctuations (such as a secondary peak at 39 days in house mice, followed by a secondary decrease, Figure 12-2B). These three species differ somewhat in the timing of variance reduction, but that minor disparity pales in comparison with the contrast between them and the pattern found in Norway rats (Figure 12-2D). Even in this species, variance declines, but not until weaning (43 dpc).

A variety of biological explanations could account for the differences visible in Figure 12-2, but there is the troubling possibility that the explanation is not biological at all. It is possible that the very high variability of infant cotton rats, house mice, and *C. expulsus* results from preservational artifacts. Analyses of cotton rats and house mice are based on cleaned and dried skeletal material, and those of the two youngest cohorts of *C. expulsus* are based on cleared and stained skulls (the skulls of older *C. expulsus* are cleaned and dried). Cleaning and drying malleable infant skulls might deform them, even though we could not find any evidence of that in the spatial distribution of the variance (Zelditch *et al.*, 2004). Clearing and staining is likely to deform skulls because the process softens tissues and loosens the ligaments holding bones together. Should the variability of the youngest samples result primarily from preservational artifacts, there would be virtually no evidence of any ontogenetic decrease in variance. There would still be evidence of a dramatic postweaning decrease in variance (of Norway rat skulls), but no evidence of any earlier reduction in variance, and none of the other datasets indicates that weaning has a profound impact on variability of skull morphology. However, the analysis of Norway rats is the only one based on landmarks visualized on x-rays (Nonaka and Nakata, 1988). It is also the only study that repeatedly measured the same growing animals. One other statistical concern is the exceptionally small sample size of the youngest age-classes of *C. expulses* ($N \leq 5$).

Given that the analysis of Norway rats is least likely to be prone to preservational artifacts and to the statistical vagaries of cross-sectional sampling, it is possible that the postweaning reduction in variance is the sole biologically meaningful pattern in all these data. Variances of the oldest samples of Norway rats might be inflated by the impact of repeatedly anesthetizing and irradiating growing rats. However, that would not account for a delayed reduction of variance. To address this issue of potential artifacts, I examine a fifth case, another sample of male Norway rats, exploiting a published dataset of landmark coordinates. The primary question is whether variance drops dramatically early in postnatal growth or, instead, begins decreasing after weaning.

II. THE ONTOGENY OF VARIATION IN MALE NORWAY RAT CRANIAL SHAPE

This analysis is based on data from Vilmann and Moss, who conducted a series of studies on the ontogenetic changes in size, proportions, and angles of the rat skull (e.g., Moss and Vilmann, 1978; Vilmann and Moss, 1980, 1981). Coordinates of eight cranial landmarks (Figure 12-3) were digitized from tracings of the x-rays and provided to Bookstein, who used them to explore a variety of morphometric techniques, as have many other workers (see Bookstein, 1991, and references cited therein, and Monteiro, 1999). The data are published in Bookstein (1991) and can be obtained electronically from the Stony Brook morphometrics Web site (http://life.bio.sunysb.edu/morph/).

This particular sample consists of 21 males, x-rayed at 7, 14, 21, 30, 40, 60, 90, and 150 days (postnatal age). A few samples are incomplete; the 40-day-old and 90-day-old samples comprise 20 rats, and the 150-day-old comprises only 19. No information is available either on the particular strain used in the study, on the anesthesia used (if any), or on the protocols for standardizing the position of the rats. I assume that these are Wistar rats because Vilmann typically analyzed that strain. To make this analysis comparable with the others, I use a gestational age scale.

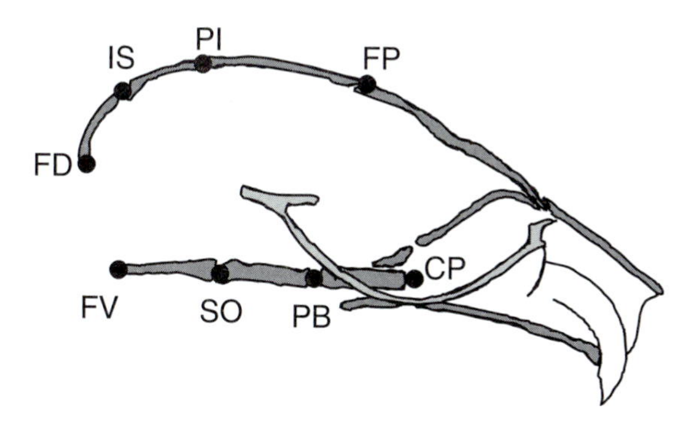

FIGURE 12-3. Eight cranial landmarks measured on 21 male laboratory rats in sagittal view. FV: Ventral extreme of the foramen magnum measured at the posterior end of the basioccipital (basion); SO: Sphenoid–occipital synchondrosis (measured at the midpoint between the two bones); PB: Basisphenoid–presphenoid suture (measured at the midpoint between the two bones); CP: Cribiform plate–presphenoid synchondrosis (measured at the midpoint between the two bones along the presphenoid axis); FP: Frontal–parietal (coronal) suture (bregma); PI: Parietal–interparietal suture (lambda); IS: Interparietal–supraoccipital suture; FD: Dorsal extreme of the foramen magnum measured on the supraoccipital bone along the midline (opisthion).

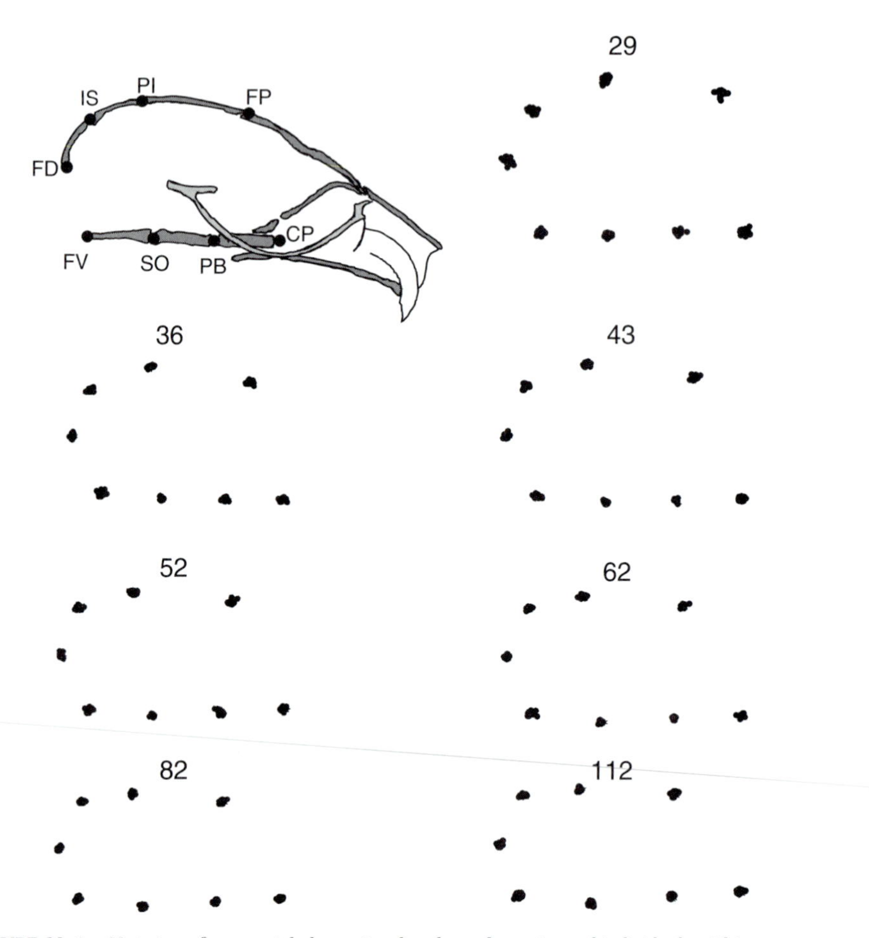

FIGURE 12-4. Variation of rat cranial shape. Landmark configurations of individuals within an age class are superimposed using the Procrustes generalized least squares (GLS) superimposition. The (gestational) age of each sample is given above the configuration. The 150-day-old sample is omitted.

Variance of cranial shape decreases rapidly and very early in postnatal growth as should be evident visually (Figure 12-4). That visual impression is again confirmed by the quantitative analysis (Figure 12-5). Variance drops substantially over the first week of sampling. The 36-day-old rats have only 66% of the variance of the 29-day-old rats. That decrease continues over the next week at a lower rate; the 43-day-olds have approximately 80% of the variance of the 36-day-olds. A transitory increase in variance occurs from 43 to 52 days (just after weaning); after 62 days, variance rises slightly but never appreciably exceeds that of the 43-day-old rats. Because the analyses of cotton rats, house mice, and *C. expulsus* are based on

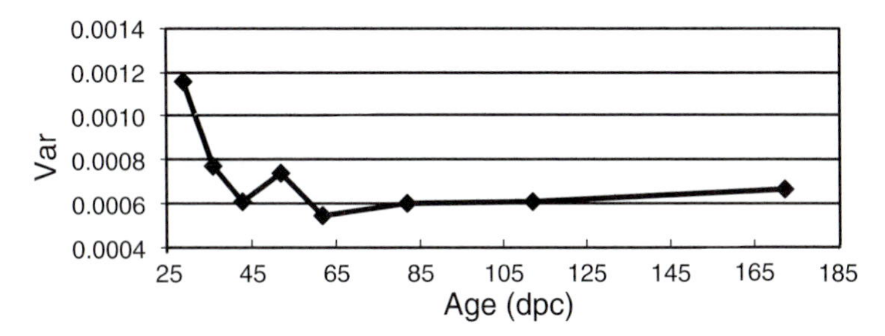

FIGURE 12-5. The ontogenetic dynamics of variance in cranial shape. Variance of cranial shape is plotted against gestational age.

a cross-sectional sampling design, inferences about changes in variance require statistical testing. Placing these results in that statistical framework, we could conclude that the decrease in variance over the first week is statistically significant ($p < 0.02$), but not the decrease over the second ($p = 0.282$). Neither is the transitory spike in the variance statistically significant ($p = 0.154$), but the subsequent decrease is ($p = 0.039$), presumably because variance drops slightly more than it rises.

III. BIOLOGICAL PATTERNS VERSUS ARTIFACTS

The ontogenetic dynamics of variance in cranial shape of male Norway rats resembles the pattern seen in cotton rats, house mice, and *C. expulsus* (palatal and lateral views). The two analyses of male Norway rats are virtually contradictory even if they both agree that variance decreases. At this point, it seems reasonable to propose that the dramatic, early postnatal drop in variance is indeed biologically meaningful, even if its magnitude might be inflated by preservational artifacts. The striking discrepancy between the two studies of Norway rats clearly needs an explanation, and so might the more subtle differences in the timing of variance reduction among cotton rats, house mice, *C. expulsus*, and Norway rats. Both differences among morphological regions sampled and differences in life history among species might explain much of the contrasts among studies.

A. Morphological Sampling

To this point, I have not mentioned the morphological features sampled because that would not be relevant were the results purely artifactual. However, once we begin to look for biological explanations for discrepant results, this is an obvious possibility. The cranial landmarks analyzed herein comprise a small subset of the data analyzed by Nonaka and Nakata. Theirs includes sagittal and lateral cranial landmarks, facial landmarks, and the mandible and tips of the upper and lower incisors as well. The dynamics of variance of cranial shape are very similar to those of skull shape in palatal view (Figure 12-2A, B, C) and lateral view absent the mandible and incisors (Figure 12-2C). The striking contrast between results arises when including the mandible and incisors. It is tempting to conclude that the striking decrease in variance immediately after weaning results from reducing the mismatch between the upper and lower jaws. However, it is still surprising that the mismatch would be reduced so late in development, i.e., after the coordination between jaws becomes critical to feeding.

B. Life History/Developmental Rate

Life history ought to be an important factor in the timing of variance reduction because skull morphology might be expected to normalize when (or before) the ability to function becomes critical for survival. Consequently, we would expect that precocial species such as cotton rats that see, hear, and walk the day they are born would normalize morphology shortly after birth (31 days postconception [dpc]), whereas those who are blind, deaf, and immobile for a week such as *C. expulsus*, would normalize their morphologies slightly later (27–28 dpc). Those that do not open their eyes for nearly two weeks after birth such as house mice (33 dpc) and Norway rats (28 days dpc) would normalize morphologies even later. Placing the species on a gestational age scale takes into account their different ages at birth, which explains many of their differences in degree of neonatal maturity. However, there are also differences in developmental rate, and these are reflected in both the degree of ossification of the skull and the timing of life-history milestones (Zelditch *et al.*, 2004).

Although I cannot estimate developmental age from age at eye opening, that milestone corresponds to a major transition in sensory and locomotory function. On those grounds, we might regard a 24-day-old *C. expulsus* as more comparable with a 29-day-old than with a 23-day-old house mouse, partly reconciling the subtle differences among temporal patterns shown in Figure 12-2. Life-history strategies certainly ought to matter to evolutionary explanations for canalization, but they are also relevant from a more strictly developmental perspective because of their consequences for developmental rate. Purely on first principles, we might anticipate that minor perturbations from normal development would have a large morphological impact when developmental rates are high. Errors introduced early in development are thus likely to have a larger impact than those introduced later. Compensating for those errors also involves perturbations of development because individuals have to return to the normal trajectory after deviating from it. However, until we have a better understanding of the actual processes that produce and reduce variation, it is difficult to do more than speculate about the impact of developmental rate.

IV. MECHANISMS GENERATING AND REGULATING CRANIOFACIAL SHAPE VARIANCE

Perturbations of a wide variety of developmental processes could generate craniofacial variation, but many are not likely to be reasonable candidates for sources of subtle variation because of their enormous impact on morphology. Moreover, the errors they produce are not likely to be compensated later in development. For example, the failure of the neural tube to close has a dramatic impact on cranial morphology, for example, no forebrain, cerebrum, or cranium are formed. However, infants born without those structures do not later develop them. The deviations that can be effectively compensated are likely to be modest. In the case of the rodents discussed in the previous section, the deviations from the norm are modest; for example, Figure 12-6 shows the deviation between the normal shape of a 7-day-old Norway rat cranium and the individual most deviant at that most variable age (Figure 12-6A), as well as its deviation from the norm 2 weeks later (Figure 12-6B).

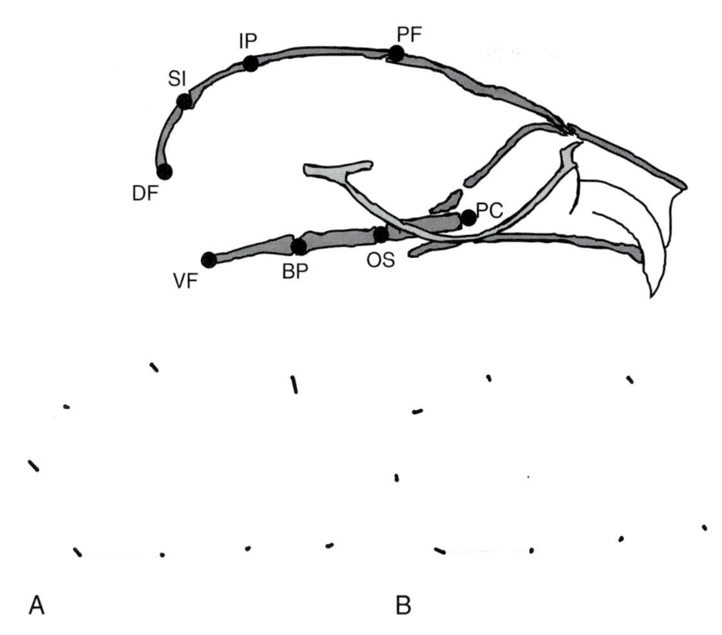

FIGURE 12-6. The difference between the average rat cranial shape and an individual: (A) The most deviant rat at 29 days relative to the 29-day-old mean and (B) the same rat relative to the 43-day-old mean.

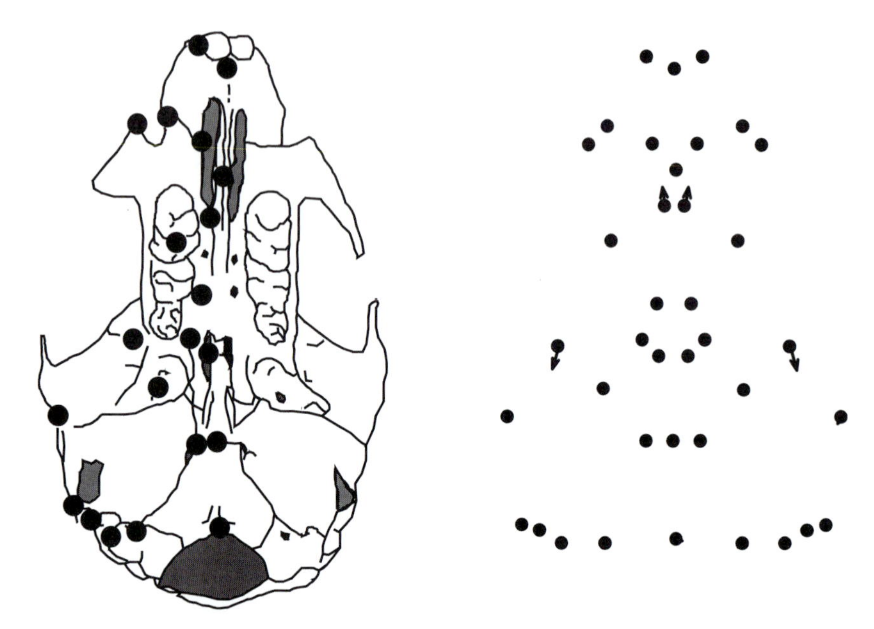

FIGURE 12-7. The difference between the average cotton rat skull and the two most deviant individuals at the most highly variable age (31 days). To ease visual interpretation, the landmarks for the half skull are reflected across the midline.

Considering that this is a laboratory strain, we might anticipate finding little variation, and indeed we find higher levels of variation in offspring of wild-caught mothers (i.e., samples of *S. fulviventer* and *C. expulsus*). However, even the most highly deviant cotton rats, at the most highly variable age, differ only subtly from the norm (Figure 12-7).

The mechanisms we are seeking are those producing and compensating for small deviations from the norm. Another characteristic of those mechanisms is that they have highly variable effects on morphology. Individuals do not closely resemble each other in their departures from the norm. If they did, we would expect to find a small number of dimensions accounting for the variation within age. However, a principal components analysis (PCA) of the 7-day-old Norway rats suggests that variation is only weakly structured; PC1 accounts for 21.2% of the variation, PC2 for 18.9%, and PC3 for 14.7%. Even though there are only 21 rats, in a 12-dimensional space, it takes six components to account for 81% of the variance. Similarly, it takes six components to account for 78% of the variability of 18 neonatal cotton rats, and seven to account for 80% of the variability of 28 10-day-old house mice. Either a few biological sources of variation produce different effects on different individuals, or different individuals are affected by different sources of variation.

We cannot go much further using purely exploratory methods. Given biological hypotheses, we can focus on what they predict for the dynamics and structure of variance. In the following text, I present four hypotheses regarding the causes of variance and canalization. They are not mutually exclusive because, as argued in the preceding text, multiple processes are likely to generate variation and also to canalize skull shape. Two of the hypotheses (targeted growth, organismal developmental timing) were tested in our analysis of cotton rats and house mice (Zelditch *et al.*, 2004). Another (neural regulation of musculoskeletal interactions) was suggested by our results and also by the general theory predicting patterns of morphological integration in mammalian crania. Herein, I introduce another (modular developmental timing), which formally resembles the other developmental timing hypothesis, except it pertains to skeletal modules rather than to the entire organism. The two models are formally similar even though the difference between them is biologically profound, so I discuss them in the following sequence.

V. TARGETED GROWTH

Targeted growth is named after Tanner's (1963) analogy between a growing child and a target-seeking rocket. The target sought by the growing child is the normal size for its age. Because growth is targeted, a child who is deflected from its normal growth trajectory will grow either unusually rapidly or for longer to catch up to the norm. As a result, the variance of body size is expected to decrease over ontogeny, and numerous studies support that hypothesis (e.g., Dickinson, 1960; Prader *et al.*, 1963; Monteiro and Falconer, 1966; Riska *et al.*, 1984). There are some rare exceptions (El Oksh *et al.*, 1967; Cheverud *et al.*, 1996), but it seems safe to say that mammalian growth is targeted (and avian growth also appears to be targeted, see e.g., Badayev and Martin, 2000; Kunz and Eckman, 2000). However, we cannot assume that every population exhibits targeted growth, much less that skull growth is invariably targeted. The hypothesis that variation in growth produces variance in shape, and that targeted growth reduces the variance in shape, is predicated on two assumptions: (1) skull growth, like body size, is targeted, and (2) variation in size produces much of the variance in shape, so removing the variance in size will reduce the variation in shape.

The first assumption is supported empirically by the findings that head length (Leamy and Cheverud, 1984) and skull size (e.g., Nonaka and Nokata, 1984) exhibit targeted growth. However, even in the absence of such empirical support, we would anticipate that skull size is targeted because of the systemic processes regulating growth. Adult body size is canalized by the regulation of (1) appetite (via hunger and satiety signals), (2) energy expenditure, (3) somatotrophic hormones, and (4) age at onset of sexual maturity (a limiting factor on growth). The system is complex, involving numerous interactions among these components, each of which interacts with other factors as well. The two principal signaling hormones are leptin, the product of the obesity gene (Zhang *et al.*, 1994), secreted primarily by adipocytes, and ghrelin, produced primarily by glands within the stomach (Sakata *et al.*, 2002). The role of leptin in regulating body size is evident by enormously obese leptin-deficient mice that never feel satiated (and are also sterile). Ghrelin is a more recent discovery and is a leptin antagonist with respect to appetite/energy regulation; ghrelin induces weight gain and reduces fat utilization (Gualillo *et al.*, 2003). Both hormones are involved in regulating growth. In fact, ghrelin was initially discovered when looking for the molecule binding to the receptor of a hormone that stimulates growth hormone (GH) secretion (Kojima *et al.*, 1999). The receptor was already known, and it was also known that its ligand stimulates GH secretion—the problem was to find that ligand. Using the orphan receptor strategy, Kojima and colleagues discovered ghrelin. Leptin also apparently influences GH secretion (Dixit *et al.*, 2003; Ghizzoni and Mastorakos, 2003) and may act more indirectly as well via regulation of insulinlike growth factor-1 (IGF-1) receptors (Maor *et al.*, 2002). In addition, leptin might be a skeletal growth factor (Nakajima *et al.*, 2003; Maor *et al.*, 2002), although the effects of leptin on bone growth are still under investigation. Leptin apparently stimulates bone growth directly, and leptin receptors are found in osteoblasts and chondrocytes (Steppan *et al.*, 2000). However, leptin also seems to inhibit growth via the sympathetic nervous system (Ducy *et al.*, 2000; Takeda *et al.*, 2002; Elefteriou *et al.*, 2004). Indirectly, leptin affects bone growth through its impact on puberty; leptin is a permissive factor for puberty, acting as a metabolic gate, perhaps signaling that energy reserves are sufficient to support the transition to the energetically demanding reproductive phase (Cunningham *et al.*, 1999). Very recently, leptin has also been determined to be a neurotrophic factor, when it was discovered that neural projection pathways from the hypothalamic region are permanently disrupted in the absence of leptin (Bouret *et al.*, 2004).

Although remarkably complex, this system ties together many previously recognized components of growth regulation, such as the long-known connection between weight (adiposity) and initiation of puberty (Kennedy and Mitra, 1963; Frisch and Revelle, 1971; Frisch, 1980), as well as the suspicion that growth is systemically regulated (Tanner, 1963). It also specifies the physiological mechanisms underlying ecological theories about trade-offs among energy allocated to maintenance, growth, and reproduction. The question is whether it also ties together targeted growth, and the canalization of shape.

For targeted growth to canalize skull shape, variation in shape must be causally related to variation in size. We might anticipate that it would be because growth does not simply enlarge animals; it also alters their proportions. The enormous literature on allometry focuses specifically on that relationship between size and shape; although that relationship between size and shape is often couched in terms of the biomechanical implications of body size, studies of ontogenetic allometries concentrate on morphogenetic consequences of (differential) growth. For example, transgenic mice that overproduce growth hormone and insulinlike growth factor-1 because of overexpression of the *MTrGH* gene are not just large; they are also shaped differently from their littermates, and those differences in shape are a direct result of the differences in size (Shea *et al.*, 1987, 1990; Corner and Shea, 1995). The shape of the giant transgenic mice can be predicted by extrapolating the allometric equation for normal mice to a larger body size.

Although intuitively plausible and seemingly well-supported empirically, our analysis of cotton rats and mice found that neither assumption was supported by our data (Zelditch *et al.*, 2004). We found no compelling evidence for targeted growth, which is troubling in light of the generality of targeted growth in mammals. More importantly, we found that variation in size accounts for extremely little of the variation in shape. That second finding suggests that targeted growth, when it

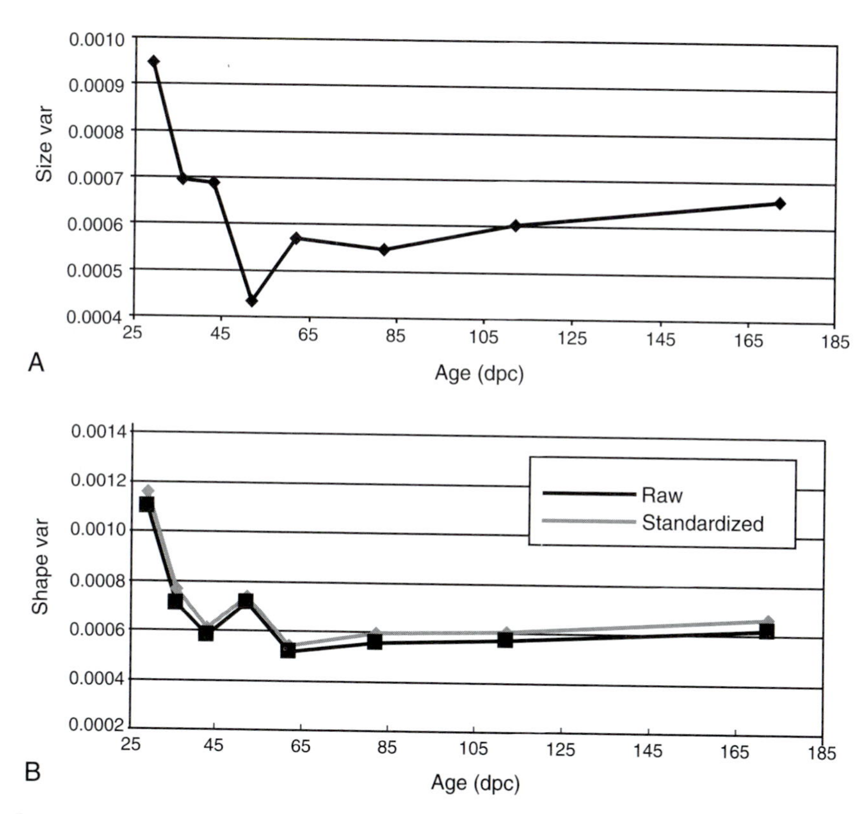

FIGURE 12-8. The ontogenetic dynamics of variance in cranial size and shape. (A) Variance in (ln) centroid size; (B) Variance in cranial shape including the variance in shape correlated with size (raw data) and excluding the variance in shape correlated with size (standardized data).

does occur, would not be an effective means of canalizing skull shape and that skull shape is canalized even when targeted growth does not occur. Additional support for this view is provided by the analysis of the Norway rat crania analyzed in the previous section. This population does exhibit targeted cranial growth, as evident from the decreasing variance in (ln-transformed) size (Figure 12-8A). Nevertheless, the variation in size never explains more than 8% of the variance in shape. As we might anticipate, statistically removing the variation in shape related to size has no impact on the dynamics of shape variance (Figure 12-8B). Of course, case studies of three populations cannot undermine a plausible hypothesis, but they do suggest that targeted growth is neither necessary nor sufficient to canalize shape.

VI. ORGANISMAL DEVELOPMENTAL TIMING

This hypothesis is based on the observation that individuals develop at different rates and that young fetuses seem to differ more than older ones in their degree of maturity at any given chronological age (Miyake *et al.*, 1996). Hallgrímsson (1999) proposed the hypothesis that variation in shape results from variation in maturity when looking for a plausible explanation to reconcile the observations that variation in shape decreases even as random deviations from bilateral asymmetry (fluctuating asymmetry) increase over postnatal growth of primates (Hallgrímsson, 1993, 1998). Interestingly, fluctuating asymmetry of mouse limbs decreases over prenatal growth, so it is possible that its increase is either restricted to postnatal growth or perhaps even to primates (Hallgrímsson *et al.*, 2003). Although the timing (and generality) of the increase in fluctuating asymmetry requires further investigation, it is still necessary to find an explanation for the decreasing variability in shape that accommodates these contradictory dynamics.

The hypothesis that variation in developmental timing produces much of the variation in shape does not require any mechanism that actively reduces that variation in timing. Even without any active regulation, variation in shape resulting from variation in developmental timing should naturally decrease over time. That is because very small differences in degree of maturity early in development have a far larger impact than they do later. For example, a mouse that is one day behind the norm for its age at 10 days past conception will still have a neural tube that is open anteriorly, but its littermates will have an anteriorly closed neural tube and a visibly enlarged brain. Sixty days later, that same individual would look indistinguishable from them in skull form, differing from them only in not yet having mature sperm in its epididymis. Thus, even without any active regulation of developmental timing, variation in shape because of it will naturally decline.

Developmental timing might be actively regulated in some cases, particularly in precocial species, which are characterized by their high degree of maturity at birth. The precocial life-history strategy maximizes neonatal maturity, subject to some constraints, including neonatal brain size. Precocial animals have relatively large brains compared with their adult brain sizes, which means the offspring's brains are large compared with their mothers' body size (Sacher and Staffeldt, 1974). The size of the birth canal may set an upper limit on neonatal brain size, limiting the degree of maturity that is optimal at birth. Precocial animals, however, might not actively regulate developmental timing because they are also typically older at birth compared with altricial animals because of their longer gestation lengths. By birth they may have already reached the point where variation in developmental timing has little morphological impact. The same arguments could be advanced for precocial organs, such as the brain. They too may have an optimal (intermediate) degree of maturity at birth and thus may under stabilizing selection. However, they also could be relatively "old" compared with other organs within the same organism and thus vary little in form despite variation in degree of maturity. We might thus expect precocial species (and organs) to be less variable in developmental timing than altricial species and later-forming organs.

The two assumptions of this hypothesis are (1) variation in developmental timing explains a large fraction of the variation at young ages; and (2) variation in developmental timing either decreases disproportionately (relative to variation resulting from other factors), or else its impact on shape diminishes disproportionately over time. The reason for emphasizing that the decrease is expected to be disproportionate is that the morphological variation from developmental timing is expected to be preferentially removed. If developmental timing accounts for 20% of the variation throughout ontogeny, for example, it is not preferentially lost and could not explain why overall variance decreases by 50%. I should note that this hypothesis partially overlaps the previous one because immature individuals may be small for their age. The idea of separating variation in growth from variation in development may even seem grossly unrealistic biologically; growth and development are sometimes viewed as nearly inextricable (e.g., Shea, 2002). However, even if there is an indisputable correlation between ontogenetic increases in size and increasing maturity, and between ontogenetic increases in size and changes in shape, that connection may be looser within individual cohorts. Small adults are not immature and do not necessarily look any more childlike than large adults. However, because the two hypotheses can partially overlap and therefore explain some of the same variation, we can test the hypothesis by removing all the shape variation correlated with size before estimating the proportion explained by developmental timing (see Zelditch *et al.*, 2004, for the method of estimating that proportion of variation).

Our empirical analysis of house mice and cotton rats found an interesting difference between these two species in the expected direction. A large fraction of the shape variation of altricial house mice results from variation in developmental timing; as much as 29.6% of the shape variation of the youngest sample is explained by developmental timing. However, even shortly after weaning, 27.8% of the shape variation is still explained by developmental timing. Thus variation in developmental timing accounts for a sizeable proportion of the variation in shape of house mice, but it does not diminish disproportionately—overall variation decreases, but the same proportion is explained by developmental timing until weaning. In contrast, very little of the shape variation of precocial cotton rats is explained by developmental timing (e.g., 10.9% of the variation of neonatal skull shape), and that proportion does not appreciably decrease. In the case of rat cranial shape, variation in developmental timing explains 15.3% of shape variation of 7-day-olds (29 dpc) and remains at nearly that level (11–13%) until sexual maturity, except for transiently dropping to 8.9% at 21 days (43 dpc). It does not appear that the diminishing impact of variation in developmental maturity explains the canalization of skull shape.

VII. VARIATION IN RELATIVE DEVELOPMENTAL TIMING OF MODULES

This hypothesis is based on the idea that individual modules can differ in maturity either because of differences in time at initiation of development or rate of development. In the case of craniofacial morphology, those modules would be mesenchymal condensations, considered the fundamental morphogenetic units of skeletal development (Atchley and Hall, 1991; Hall, 2003). According to the hypothesis, variation in shape is introduced by minor differences in timing or rate, such as in the time at which the condensation originally forms or in rate of cell proliferation of within it. Different modules will thus be at different degrees of maturity within an individual, and the same module will be at different degrees of maturity across individuals. For example, the molar alveolus might be immature both with respect to the incisor alveolus of that same animal and with respect to the molar alveolus of other individuals. To explain how that variation could be transitory (rather than produce intraspecific heterochrony), this hypothesis must assume a control over the final size of the condensation that is not related to the age at which the condensation forms. This is a reasonable assumption because the molecules or molecular cascades that initiate formation of condensations are not the same ones that determine the condensation's final size (see Hall, 2003). Another trio of critical assumptions is: (1) variation in timing can be introduced into a single condensation without affecting any others, (2) final size of a condensation is regulated locally so the size of each condensation is canalized independently of all others, and (3) the rate of development is a decreasing function of degree of maturity.

The first assumption seems reasonable for several reasons, including the definition of "modularity" itself. A more empirical basis for justifying the assumption is the prevalence of localized (dissociated) heterochrony. For example, empirical studies (of teleosts) show that bones vary in sequence of ossification across conspecific individuals (Mabee *et al.*, 2000). The second and third assumptions are strongly supported by an intriguing study of growth plates that are individually inhibited by experimentally administering a glucocorticoid to a single growth plate on one side of the animal. The treated side grows slowly and is stunted, but when released from inhibition; it grows rapidly and nearly catches up to the control (Baron *et al.*, 1994). To explain this highly local catch-up growth, Baron and colleagues drew upon the fact that growth rates of limbs decrease over time because of a decrease in rate of cell proliferation (Kember and Walker, 1971). They conjectured that growth plates senesce just like cells in culture (Hayflick, 1965) and that senescence is a function of the cumulative number of cell divisions. In effect, the final size of a growth plate would be regulated by a mechanism tracking the cumulative number of cell divisions, and because transiently inhibited growth plates have undergone fewer rounds of cell division, they are less senescent. They thus grow faster, and cells continue to proliferate until exhausting what Baron and colleagues call their "intrinsic growth potential." A follow-up study confirmed the histological immaturity of the inhibited growth plate and the higher proliferation rate of the transiently stunted limbs (Gafni *et al.*, 2001). It thus appears that regulation at a highly local level is possible. So long as final size or number of cells (or number of cell cycles) is controlled, the variation resulting from differences in degree of maturity will be transitory. Moreover, the variation in shape will naturally diminish over time just as it does when timing varies at the organismal level.

The assumption that any condensation is individually and freely variable is required to explain the diversity of deviations from the normal shape discussed in the preceding text. This assumption might seem difficult to justify if only because of the expectation that the proliferation of cells is stringently regulated in both space and time (e.g., Pratap *et al.*, 2003). In addition, systemic factors (such as IGF-1) might be expected to integrate the behavior of local cell populations. Of course, the existence of dissociated heterochrony argues otherwise, but it is possible that (evolutionary) heterochronic dissociations involve alterations in timing or rate of processes specific to one module, and subject to stringent regulation, rather than variation among individuals that share a common, more loosely controlled system. We clearly need more studies to determine the degree to which developmental timing of skeletogenic modules is controlled, the parameters that are most stringently controlled (e.g., final size versus timing of colony formation), and the extent to which such controls are systemic or local.

Variation introduced by differences in timing of modules should be compensated so long as the final size of the module is actively controlled. However, we might expect that variation introduced by this process would diminish over time and that minor discrepancies would eventually have little impact on shape. This hypothesis thus predicts that individual modules follow a normal ontogenetic trajectory, at a normal rate for their degree of development. Testing this hypothesis will likely require more than shape analysis because the key claim is about the behavior of cells and about the onset of their proliferation (which occurs long before there is any shape to measure). Labeling cells could provide the necessary "landmarks," but counting cells would probably provide a greater degree of resolution accuracy. Yet, before beginning a laborious study, it would be useful to determine that transiently retarded craniofacial bones do catch up to the norm and also that they do so late enough in development to explain the postnatal decrease in shape variance. Both could be determined by using locally administered growth inhibitors.

To date, little quantitative evidence supports the idea that the craniofacial variation (especially that of the mandible) is truly modular. However, that might be because of the difficulty of finding enough landmarks to test that hypothesis rigorously or because of to the age of the animals sampled. Some studies simplify the Atchley and Hall model, reducing it to just two modules: (1) ascending ramus and (2) alveolus, but even these do not appear to be independent in their variation (Klingenberg *et al.*, 2001, 2003). Another study finds that the Atchley and Hall mandibular model is consistent with the data, but the data are equally consistent with the hypothesis that variation arises from epigenetic interactions between muscle and bone and between bone and teeth (Duarte *et al.*, 2000). In that case, the modules are muscle–bone or muscle–teeth units, not individual mesenchymal condensations. In addition, the structure of those epigenetic interactions, which cross boundaries between modules, would argue that the skull (or mandible) does not behave as if its development is truly modular.

VIII. NEURAL REGULATION OF MUSCULOSKELETAL INTERACTIONS

This hypothesis proposes that epigenetic interactions between hard and soft tissues generate the (co)variation of craniofacial shape and that normalization of those interactions corrects for errors produced by earlier interactions. The basic idea is that bone responds to physical forces exerted by soft tissues, including muscle, brain, and eye. It is premised on the conventional explanation for the morphological integration of the mammalian cranium, namely, epigenetic interactions between bone and soft tissues (e.g., Cheverud, 1982, 1995; Zelditch, 1988; Zelditch and Carmichael, 1989). The impact of soft tissues on skeletal form has long been recognized and is the crux of the functional matrix model of cranial development (Moss and Salentijn, 1969). As summarized by Henderson and Carter (2002), the forces generated by muscle contraction and also by differential growth impose a time-dependent pattern of stresses (local force intensities) and strains (local deformations) throughout musculoskeletal tissues. These physical cues guide growth and differentiation of mesenchymal tissues and influence both rates and directions of growth. Despite uncertainties about some key events, several steps in the cascade have been identified, including an increase in intracellular Ca^{++} of osteoblasts, followed by an increase in protein kinase C activation, succeeded by expression of several genes (including transforming growth factor β [TGF-β] and IGF-1) which is followed by proliferation of osteoblasts, matrix synthesis, and eventually, mineralization (summarized in Skerry, 2000; Yu *et al.*, 2001; Fong *et al.*, 2003). Loading bone demonstrably affects the expression of some bone-specific genes, such as *CMF608*, which stimulates proliferation of osteoblasts and may contribute to differentiation of the cells (Segev *et al.*, 2004).

To explain how variance of craniofacial form might be regulated, this hypothesis emphasizes the strains resulting from muscle contraction and that musculoskeletal interactions are initially unpredictable and spatially disorganized because neural signals are transmitted by an immature central nervous system (CNS). As the CNS matures, the neural signals normalize, thereby normalizing the forces exerted by muscles on bone. Any deviations from the norm arising during the initial (disorganized) phase are later removed both because new bone will grow according to more predictable strains and also because bone is resorbed wherever it exceeds its (now predictable) needs.

The primary assumptions made by this hypothesis are (1) that bone growth responds to fetal and infant muscle activity; (2) fetal and infant muscle activity is relatively disorganized, and becomes more predictable as the CNS matures; and (3) that the normalization of neuromuscular control is sufficient to decrease the variance produced earlier when signals were disorganized. A secondary assumption is that functional transitions later in growth, for example, the transition from suckling to chewing, do not radically reorganize neuromuscular signals and generate more variance than can be removed with days.

There is a great deal of support for the assumption that fetal muscle activity has a morphogenetic impact on bone. During fetal growth, forces exerted by muscles are actually strong enough to fracture the fragile bones of fetuses with osteogenesis imperfecta (Rowe and Shapiro, 1998). In addition, those fetal movements are necessary for normal development of both face and cranium, as can be seen in the craniofacial abnormalities that result from defective muscle development (e.g., Pai, 1965; Herring and Lakars, 1981; Rodriguez *et al.*, 1988). Also, abnormal bone phenotypes can be produced by experimentally paralyzing fetal muscles (Rodriguez *et al.*, 1992). That fetal movements affect bone growth is not surprising because the response to stress is an increasing function of growth rate, and those rates are higher during fetal development than postnatally (Mosley and Lanyon, 2002). It might seem unlikely that muscle contractions affect the cranium, but the vault does experience strong tensile forces from muscle contractions (Herring and Teng, 2000; Sun *et al.*, 2004).

There is another source of strain, however, which is differential growth. The growth of the brain, not just muscle contractions, presumably affects the cranium, as suggested by the enlarged skulls sometimes found in patients with hydrocephaly or large brain tumors (Gladstone, 1905; Bognar, 1997; Medina *et al.*, 2001).

The expanding brain may not be able to generate large enough strains to stimulate growth at cranial sutures, but the dura mater might respond and modulate bone growth (Henderson *et al.*, 2004). Muscle growth, as well as muscle contraction, may also affect bone growth; although a hypothesis emphasizing epigenetic interactions among bone and other tissues might not differentiate among the causes of the stresses, this hypothesis does make that distinction because the explanation for canalization is specific to the forces exerted by muscle contraction.

The assumption that prenatal and perinatal muscle function is relatively disorganized and gradually fine-tunes is also supported. Ontogenetic studies of muscle function find that early movements are relatively disorganized compared with later ones (e.g., Wineski and Herring, 1984; Westneat and Hall, 1992; Green *et al.*, 1997), even to the point of being called "fidgety" (Forssberg, 1999). Some aspects of muscle function change ontogenetically beyond simple fine-tuning, but most taxa (including rodents) appear to undergo a relatively gradual change from suckling to weaning, allowing young animals to refine their chewing abilities before they depend on chewing for nutritive purposes (Herring, 1985; Lakars and Herring, 1980; Langenbach and van Eijen, 2001). Thus both a critical and a secondary assumption of the hypothesis seem to be met.

The increasing organization of neuromuscular control may direct bone growth along more predictable pathways, but that, by itself, might not be enough to correct for responses the bones made to the disorganized signals produced earlier. Actually removing prior "errors" may also be needed, which is possible because muscle *unloading* lead to resorption of bone by osteoclasts, a process of adaptive remodeling (see Lanyon, 1987). Spatial cues for resorption may be provided by programmed cell death (apoptosis) of bone cells (osteocytes) that initiate localized bone resorption (Noble *et al.*, 2003). Remodeling might thus be an important part of normalizing skull shape.

We proposed this hypothesis to explain the reduction of shape variance during early postnatal growth (Zelditch *et al.*, 2004). However, we did not explicitly test it. In light of the logistical difficulty of analyzing fetal neuromuscular interactions, testing this hypothesis may require knocking out critical features of the proposed regulatory system, such as fetal muscle contractions or neuromuscular coordination or the ability to resorb bone. In the absence of neuromuscular control, or indeed of any of the components of the regulatory system, variance should continually accumulate rather than decrease over time (Zelditch *et al.*, 2004).

IX. CANALIZED SHAPE AS AN EPIPHENOMENON

The four hypotheses proposed herein treat the canalization of shape as a developmental epiphenomenon; none supposes that shape itself is actively regulated. Viewing the canalization of shape as an outcome of regulating something else, such as neuromuscular coordination, might seem implausible if shape itself is seen as the target of stabilizing selection. No doubt that view is reasonable because there are obvious disadvantages in having teeth that cannot occlude, eyes that face posteriorly in an anteriorly directed face, or crania that fail to enclose the brain. It hardly seems necessary to argue at length that skull shape should be subject to stabilizing selection. Nevertheless, on developmental grounds, I doubt that organisms can monitor their shapes and detect minor deviations from the norm, much less correct for them.

Shape itself may be regulated early in development by the mechanisms controlling embryonic spatial patterning. When the embryo initially sets up its coordinate system, errors would be devastating and probably uncorrectable. Errors made so early in development might not be correctible postnatally, partly because processes of pattern formation are sensitive to the size of the field over which they operate. Errors are not likely to be corrected after the field enlarges by orders of magnitude. Moreover, those mechanisms do not operate over large distances. Finally, any perturbations arising early in development may be too severe to correct later—an infant born without a brain does not eventually compensate and develop one later. The stage in which shape itself is most likely to be regulated directly is also the one least likely to be canalized by a form of error correction.

There are many other possible mechanisms that could explain the canalization of craniofacial shape, including one briefly mentioned in the preceding text—the normalization of strains produced by differential tissue growth (Henderson and Carter, 2002). We clearly need additional models, especially for locally generated and locally canalized variation. In emphasizing local processes, I do not mean to dismiss the vital role that systemic regulators play in overall growth or bone development. Their impact can be substantial and their consequences profound even when they simply reduce growth rates and extend the duration of growth (see Miller and German, 1999; Reichling and German, 2000). Yet, such systemic factors, even though undeniably crucial to normal development, do not seem to produce or remove the local (and nearly random) shape variation seen in these rodents. Targeted growth is obviously an important part of the canalization of skull form because form encompasses size as well as shape. Still, understanding how shape is regulated will require pursuing hypotheses that explain locally generated and regulated variation. Even if shape itself is not directly regulated, as I suspect it is not, there *are* mechanisms that regulate it indirectly; variation in both size and shape is under organismal (developmental) control. This does not appear to be a peculiarity of mammalian skull growth; that same reduction of variance in shape has been found in a snail (Foote and Cowie, 1986). Perhaps the most striking finding of all these studies is that even if evolutionary biologists regard variation as a given, organisms do not.

ACKNOWLEDGMENTS

I wish to thank Benedikt Hallgrímsson and Brian Hall for inviting me to contribute. I am very grateful to Eladio Marquez and Donald Swiderski for critically reading earlier version(s) of this manuscript.

REFERENCES

Atchley, W. R., and Hall, B. K. (1991). A model for development and evolution of complex morphological structures. *Biological Reviews* 66, 101–157.

Badyaev, A. V., and Martin, T. E. (2000). Individual variation in growth trajectories: Phenotypic and genetic correlations in the ontogeny of the house finch (*Carpodacus mexicanus*). *Journal of Evolutionary Biology* 13, 290–301.

Baron, J., Klein, K. O., Colli, M. J., Yanovski, J. A., Novosad, J. A., Bacher, J. D., and Cutler, G. B. (1994). Catch-up growth after glucocorticoid excess: A mechanism intrinsic to the growth-plate. *Endocrinology* 135, 1367–1372.

Boersma, B., and Wit, J. M. (1997). Catch-up growth. *Endocrine Reviews* 18, 646–661.

Bognar, L. (1997). Brain tumors during the first year of life. *Annals of the New York Academy of Sciences* 824, 148–155.

Bookstein, F. L. (1991). *Morphometric Tools for Landmark Data: Geometry and Biology.* Cambridge, England: Cambridge University Press.

Bouret, S. G., Draper, S. J., and Simerly, R. B. (2004). Trophic action of leptin on hypothalamic neurons that regulate feeding. *Science* 304, 108–110.

Cane, W. P. (1993). The ontogeny of postcranial integration in the common tern, *Sterna hirundo*. *Evolution* 47, 1138–1151.

Cheverud, J. M. (1982). Phenotypic, genetic and environmental integration in the cranium. *Evolution* 36, 499–512.

Cheverud, J. M. (1984). Quantitative genetics and developmental constraints on evolution by selection. *Journal of Theoretical Biology* 110, 155–172.

Cheverud, J. M. (1995). Morphological integration in the saddle-back tamarin (*Saguinus fuscicollis*) cranium. *American Naturalist* 145, 63–89.

Cheverud, J. M., Routman, E. J., Duarte, F. A. M., van Swinderen, B., Cothran, K., and Perel, C. (1996). Quantitative trait loci for murine growth. *Genetics* 142, 1305–1319.

Corner, B. D., and Shea, B. T. (1995). Growth allometry of the mandibles of giant transgenic mice: An analysis based on the finite-element scaling method. *Journal of Craniofacial Genetics and Developmental Biology* 15, 125–139.

Cunningham, M. J., Clifton, D. K., and Steiner, R. A. (1999). Leptin's actions on the reproductive axis: Perspectives and mechanisms. *Biology of Reproduction* 60, 216–222.

Dickinson, A. G. (1960). Some genetic implications of maternal effects: An hypothesis of mammalian growth. *Journal of Agricultural Science* 54, 378–390.

Dixit, V. D., Mielenz, M., Taub, D. D., and Parvizi, N. (2003). Leptin induces growth hormone secretion from peripheral blood mononuclear cells via a protein kinase C- and nitric oxide-dependent mechanism. *Endocrinology* 144, 5595–5603.

Duarte, L. C., Monteiro, L. R., Zuben, F. J. V., and Reis, S. F. d. (2000). Variation in mandible shape in *Thrichomys apereoides* (Mammalia: Rodentia): Geometric analysis of a complex morphological structure. *Systematic Biology* 49, 563–578.

Ducy, P., Amling, M., Takeda, S., Priemel, M., Schilling, A. F., Bell, F. T., Shen, J., Vinson, C., Rueger, J. M., and Karsenty, G. (2000). Leptin inhibits bone formation through a hypothalamic relay: A central control of bone mass. *Cell* **100**, 197–207.

Elefteriou, F., Takeda, S., Ebihara, K., Magre, J., Patano, N., Kim, C. A., Ogawa, Y., Liu, X., Ware, S. M., Craigen, W. J., Robert, J. J., Vinson, C., Nakao, K., Capeau, J., and Karsenty, G. (2004). Serum leptin level is a regulator of bone mass. *Proceedings of the National Academy of Sciences USA* **101**, 3258–3263.

El Oksh, H. A., Sutherland, T. M., and Williams, J. S. (1967). Prenatal and postnatal maternal influence on growth in mice. *Genetics* **57**, 79–94.

Fong, K. D., Nacamuli, R. P., Loboa, E. G., Henderson, J. H., Fang, T. D., Song, H. M., Cowan, C. M., Warren, S. M., Carter, D. R., and Longaker, M. T. (2003). Equibiaxial tensile strain affects calvarial osteoblast biology. *Journal of Craniofacial Surgery* **14**, 348–355.

Foote, M., and Cowie, R. H. (1986). Developmental buffering as a mechanism for stasis. *Evolution* **42**, 396–399.

Forssberg, H. (1999). Neural control of human motor development. *Current Opinion in Neurobiology* **9**, 676–682.

Frisch, R. E. (1980). Pubertal adipose tissue: Is it necessary for normal sexual maturation? Evidence from the rat and human female. *Federation Proceedings* **39**, 2395–2400.

Frisch, R. E., and Revelle, R. (1971). Height and weight at menarche and a hypothesis of menarche. *Archives of Disease in Childhood* **46**, 695–710.

Gafni, R. I., Weise, M., Robrecht, D. T., Meyers, J. L., Barnes, K. M., De-Levi, S., and Baron, J. (2001). Catch-up growth is associated with delayed senescence of the growth plate in rabbits. *Pediatric Research* **50**, 618–623.

Ghizzoni, L., and Mastorakos, G. (2003). Interactions of leptin, GH, and cortisol in normal children. *Annals of the New York Academy of Sciences* **997**, 56–63.

Gladstone, R. J. (1905). A study of the relations of the brain to size of the head. *Biometrika* **4**, 105–123.

Green, J. R., Moore, C. A., Ruark, J. L., Rodda, P. R., Morvee, W. T., and VanWitzenburg, M. J. (1997). Development of chewing in children from 12 to 48 months: Longitudinal study of EMG patterns. *Journal of Neurophysiology* **77**, 2704–2716.

Gualillo, O., Lago, F., Gomez-Reino, J., Casanueva, F. F., and Dieguez, C. (2003). Ghrelin, a widespread hormone: Insights into molecular and cellular regulation of its expression and mechanism of action. *FEBS Letters* **552**, 105–109.

Hall, B. K. (2003). Unlocking the black box between genotype and phenotype: Cell condensations as morphogenetic (modular) units. *Biology and Philosophy* **18**, 219–247.

Hall, B. K., and Miyake, T. (1994). How do embryos measure time? In *Evolutionary Change and Heterochrony* (K. J. McNamara, ed.), pp. 3–20. New York: Wiley.

Hallgrímsson, B. (1993). Fluctuating asymmetry in *Macaca fascicularis*: A study of the etiology of developmental noise. *American Journal of Physical Anthropology* **14**, 421–443.

Hallgrímsson, B. (1998). Fluctuating asymmetry in the mammalian skeleton: Evolutionary and developmental implications. *Evolutionary Biology* **30**, 187–251.

Hallgrímsson, B. (1999). Ontogenetic patterning of skeletal fluctuating asymmetry in rhesus macaques and humans: Evolutionary and developmental implications. *International Journal of Primatology* **20**, 121–151.

Hallgrímsson, B., Miyake, T., Wilmore, K., and Hall, B. K. (2003). Embryological origins of developmental stability: Size, shape and fluctuating asymmetry in prenatal random bred mice. *Journal of Experimental Zoology* **296B**, 40–57.

Hayflick, L. (1965). The limited *in vitro* lifetime of human diploid strains. *Experimental Cell Research* **37**, 614–636.

Henderson, J. H., and Carter, D. R. (2002). Mechanical induction in limb morphogenesis: The role of growth-generated strains and pressures. *Bone* **31**, 645–653.

Henderson, J. H., Longaker, M. T., and Carter, D. R. (2004). Sutural bone deposition rate and strain magnitude during cranial development. *Bone* **34**, 271–280.

Herring, S. W. (1985). The ontogeny of mammalian mastication. *American Zoologist* **25**, 339–349.

Herring, S. W., and Lakars, T. C. (1981). Craniofacial development in the absence of muscle contraction. *Journal of Craniofacial Genetics and Developmental Biology* **1**, 341–357.

Herring, S. W., and Teng, S. (2000). Strain in the braincase and its sutures during function. *American Journal of Physical Anthropology* **112**, 575–593.

Hingst-Zaher, E., Marcus, L., and Cerqueira, R. (2000). Application of geometric morphometrics to postnatal size and shape changes in the skull of *Calomys expulsus*. *Hystrix, Italian Journal of Mammalogy* **11**, 99–113.

Kawecki, T. J. (2000). The evolution of genetic canalization under fluctuating selection. *Evolution* **54**, 1–12.

Kember, N. F., and Walker, K. V. R. (1971). Control of bone growth in rats. *Nature* **229**, 428–429.

Kennedy, G. C., and Mitra, J. (1963). Body weight and food intake as initiating factors for puberty in the rat. *Journal of Physiology* **166**, 408–418.

Klingenberg, C. P., Leamy, L. J., Routman, E. J., and Cheverud, J. M. (2001). Genetic architecture of mandible shape in mice: Effects of quantitative trait loci analyzed by geometric morphometrics. *Genetics* **157**, 785–802.

Klingenberg, C. P., Mebus, K., and Auffray, J.-C. (2003). Developmental integration in a complex morphological structure: How distinct are the modules in a mouse mandible. *Evolution and Development* **5**, 522–531.

Kojima, M., Hosoda, H., Date, Y., Nakazato, M., Matsuo, H., and Kangawa, K. (1999). Ghrelin is a growth-hormone-releasing acylated peptide from stomach. *Nature* **402**, 656–660.

Kunz, C., and Ekman, J. (2000). Genetics and environmental components of growth in nestling blue tits (*Parus caeruleus*). *Journal of Evolutionary Biology* **13**, 199–212.

Lakars, T. C., and Herring, S. W. (1980). Ontogeny of oral function in hamsters (*Mesocricetus auratus*). *Journal of Morphology* **165**, 237–254.

Langenbach, G. E. J., and van Eiden, T. M. G. J. (2001). Mammalian feeding motor patterns. *American Zoologist* **41**, 1338–1351.

Lanyon, L. E. (1987). Functional strain in bone as an objective and controlling stimulus for adaptive bone remodeling. *Journal of Biomechanics* **20**, 1083–1093.

Leamy, L., and Cheverud, J. M. (1984). Quantitative genetics and the evolution of ontogeny. II. Genetic and environmental correlations among age-specific characters in random bred mice. *Growth* **48**, 339–353.

Mabee, P., Olmstead, K. L., and Cubbage, C. C. (2000). An experimental study of intraspecific variation, developmental timing, and heterochrony in fishes. *Evolution* **54**, 2091–2106.

Maor, G., Rochwerger, M., Segev, Y., and Phillip, M. (2002). Leptin acts as a growth factor on the chondrocytes of skeletal growth centers. *Journal of Bone and Mineral Research* **17**, 1034–1043.

Medina, L. S., Frawley, K., Zurakowski, D., Buttros, D., DeGrauw, A. J. C., and Crone, K. R. (2001). Children with macrocrania: Clinical and imaging predictors of disorders requiring surgery. *American Journal of Neuroradiology* **22**, 564–570.

Miller, J. P., and German, R. Z. (1999). Protein malnutrition affects the growth trajectories of the craniofacial skeleton in rats. *Journal of Nutrition* **129**, 2061–2069.

Miyake, T., Cameron, A. M., and Hall, B. K. (1996). Detailed staging of inbred C57BL/6 mice between Theiler's [1972] stages 18 and 21 (11–13 days of gestation) based on craniofacial development. *Journal of Craniofacial Genetics and Developmental Biology* **16**, 1–31.

Monteiro, L. R. (1999). Multivariate regression models and geometric morphometrics: The search for causal factors in the analysis of shape. *Systematic Biology* **48**, 102–111.

Monteiro, L. S., and Falconer, D. S. (1966). Compensatory growth and sexual maturity in mice. *Animal Production* **8**, 179–192.

Mosley, J. R., and Lanyon, J. E. (2002). Growth rate rather than gender determines the size of the adaptive response of the growing skeleton to mechanical strain. *Bone* **30**, 314–319.

Moss, M. L., and Salentijn, L. (1969). The primary role of the functional matrix in facial growth. *American Journal of Orthodontics* **55**, 566–577.

Moss, M. L., and Vilmann, H. (1978). Studies on orthocephalization of the rat head. *Gegenbaurs Morphologisches Jahrbuch* **124**, 559–579.

Nakajima, R., Inada, H., Koike, T., and Yamano, T. (2003). Effect of leptin to cultured growth plate chondrocytes. *Hormone Research* **60**, 91–98.

Noble, B. S., Peet, N., Stevens, H. Y., Brabbs, A., Mosley, J. R., Reilly, G. C., Reeve, J., Skerry, T. M., and Lanyong, L. E. (2003). Mechanical loading: Biphasic osteocyte survival and targeting of osteoclasts for bone destruction in rat cortical bone. *American Journal of Physiology Cell Physiology* **284**, C934–C943.

Nonaka, K., and Nakata, M. (1984). Genetic variation and craniofacial growth in inbred rats. *Journal of Craniofacial Genetics and Development* **4**, 271–302.

Nonaka, K., and Nakata, M. (1988). Genetic and environmental factors in the longitudinal growth of rats. 3. Craniofacial shape change. *Journal of Craniofacial Genetics and Developmental Biology* **8**, 337–344.

Pai, A. C. (1965). Developmental genetics of a lethal mutation, muscular dysgenesis (mdg) in the mouse: I. Genetic analysis and gross morphology. *Developmental Biology* **11**, 82–92.

Prader, A., Tanner, J. M., and Harnack, G. A. V. (1963). Catch-up growth following illness or starvation: An example of developmental canalization in man. *Journal of Pediatrics* **62**, 646–659.

Pratap, J., Galindo, M., Zaidi, S. K., Vradii, D., Bhat, B. M., Robinson, J. A., Choi, J. Y., Komori, T., Stein, J. L., Lian, J. B., Stein, G. S., and v. Wijnen, A. J. (2003). Cell growth regulatory role of Runx2 during proliferative expansion of preosteoblasts. *Cancer Research* **63**, 5357–5362.

Reichling, T. D., and German, R. Z. (2000). Bones, muscles and visceral organs of protein malnourished rats (*Rattus norvegicus*) grow more slowly but for longer durations to reach normal final size. *Journal of Nutrition* **130**, 2326–2332.

Riska, B., Atchley, W. R., and Rutledge, J. J. (1984). A genetic analysis of targeted growth in mice. *Genetics* **107**, 79–101.

Rodriguez, J. I., Palacios, J., Garcia-Alix, A., Pastor, I., and Paniagua, R. (1988). Effects of immobilization on fetal bone development: A morphometric study in newborns with congenital neuromuscular diseases with intrauterine onset. *Calcified Tissue International* **43**, 335–339.

Rodriguez, J. I., Palacios, J., Ruiz, A., Sanchez., M., Alvarez, I., and Demiguel, E. (1992). Morphological changes in long bone development in fetal akinesia deformation sequence: An experimental study in curarized rat fetuses. *Teratology* **45**, 213–221.

Rowe, D. W., and Shapiro, J. R. (1998). Osteogenesis imperfecta. In *Metabolic Bone Disease and Clinically Related Disorders* (L. V. Avioli and S. M. Krane, eds.), pp. 651–695. San Diego, CA: Academic Press.

Sacher, G. A., and Staffeldt, E. F. (1974). Relation of gestation time to brain weight for placental mammals: Implications for the theory of vertebrate growth. *The American Naturalist* **108**, 593–615.

Sakata, I., Nakamura, K., Yamazaki, M., Matsubara, M., Hayashi, Y., Kangawa, K., and Sakai, T. (2002). Ghrelin-producing cells exist as two types of cells, closed- and opened-type cells, in the rat gastrointestinal tract. *Peptides* **23**, 531–536.

Scharloo, W., Hoogmoed, M. S., and Ter Kuile, A. (1967). Stabilizing and disruptive selection on a mutant character in *Drosophila*. I. The phenotypic variance and its components. *Genetics* **56**, 709–726.

Schmalhausen, I. I. (1949). *Factors of Evolution: The Theory of Stabilizing Selection*. Chicago: University of Chicago Press.

Segev, O., Samach, A., Faerman, A., Kalinski, H., Beiman, M., Gelfand, A., Turam, H., Boguslavsky, S., Moshayov, A., Gottleib, H., Kazanov, E., Nevo, Z., Robinson, D., Skaliter, R., Einat, P., Binderman, I., and Feinstein, E. (2004). CMF608: A novel mechanical strain-induced bone-specific protein expressed in early osteochondroprogenitor cells. *Bone* **34**, 246–260.

Shea, B. T. (2002). Are some heterochronic transformations likelier than others? In *Human Evolution Through Developmental Change* (N. Minugh-Purvis and K. J. McNamara, eds.), pp. 79–101. Baltimore: Johns Hopkins University Press.

Shea, B. T., Hammer, R. E., and Brinster, R. L. (1987). Growth allometry of the organs in giant transgenic mice. *Endocrinology* **121**, 1–7.

Shea, B. T., Hammer, R. E., Brinster, R. L., and Ravosa, M. R. (1990). Relative growth of the skull and postcranium in giant transgenic mice. *Genetical Research* **56**, 21–34.

Siegal, M. L., and Bergman, A. (2002). Waddington's canalization revisited: Developmental stability and evolution. *Proceedings of the National Academy of Sciences USA* **99**, 10528–10532.

Skerry, T. (2000). Biomechanical influences on skeletal growth and development. In *Development, Growth and Evolution: Implications for the Study of the Hominid Skeleton* (P. O'Higgins and M. J. Cohn, eds.), pp. 29–39. New York: Academic Press.

Steppan, C. M., Crawford, D. T., Chidsey-Frink, K. L., Ke, H., and Swick, A. G. (2000). Leptin is a potent stimulator of bone growth in *ob/ob* mice. *Regulatory Peptides* **92**, 73–78.

Sun, Z., Lee, E., and Herring, S. W. (2004). Cranial sutures and bones: Growth and fusion in relation to masticatory strain. *Anatomical Record A* **276**, 150–161.

Takeda, S., Elefteriou, F., Levasseur, R., Liu, X., Armstrong, D., Ducy, P., and Karsenty, G. (2002). Leptin regulates bone formation via the sympathetic nervous system. *Cell* **111**, 305–317.

Tanner, J. M. (1963). Regulation of growth in size in mammals. *Nature* **199**, 845–850.

Vilmann, H., and Moss, M. L. (1980). Studies on orthocephalization V. Peripheral positional stability of the rat cranial frame in the period between 14 and 150 days. *Acta Anatomica* **107**, 330–335.

Vilmann, H., and Moss, M. L. (1981). Studies on orthocephalization VII. Behavior of the rat cranial frame in the period between 1 day before birth to 14 days after birth. *Acta Anatomica* **109**, 157–160.

Waddington, C. H. (1952). Canalization of the development of a quantitative character. In *Quantitative Inheritance* (C. H. Waddington, ed.), pp. 43–46. London: Her Majesty's Stationary Office.

Waddington, C. H. (1957). *The Strategy of the Genes*. London: Allen & Unwin.

Wagner, A. (1996). Does evolutionary plasticity evolve? *Evolution* **50**, 1008–1023.

Westneat, M. W., and Hall, W. G. (1992). Ontogeny of feeding motor patterns in infant rats: An electromyographic analysis of suckling and chewing. *Behavioral Neuroscience* **106**, 539–554.

Wineski, L. E., and Herring, S. W. (1984). Postnatal development of masticatory muscle function. *American Zoologist* **24**, 1009.

Yu, J. C., Lucas, J. H., Fryberg, K., and Borke, J. L. (2001). Extrinsic tension results in FGF-2 release, membrane permeability change, and intracellular Ca^{++} increase in immature cranial sutures. *Journal of Craniofacial Surgery* **12**, 391–398.

Zelditch, M. L. (1988). Ontogenetic variation in patterns of phenotypic integration in the laboratory rat. *Evolution* **42**, 28–41.

Zelditch, M. L., and Carmichael, A. C. (1989). Ontogenetic variation in patterns of developmental and functional integration in skulls of *Sigmodon fulviventer*. *Evolution* **43**, 1738–1747.

Zelditch, M. L., Bookstein, F. L., and Lundrigan, B. L. (1992). Ontogeny of integrated skull growth in the cotton rat *Sigmodon fulviventer*. *Evolution* **46**, 1164–1180.

Zelditch, M. L., Bookstein, F. L., and Lundrigan, B. L. (1993). The ontogenetic complexity of developmental constraints. *Journal of Evolutionary Biology* **6**, 121–141.

Zelditch, M. L., Lundrigan, B. L., Sheets, H. D., and Garland, T. (2003). Do precocial mammals develop at a higher rate? A comparison of rates of skull development in *Sigmodon fulviventer* and *Mus musculus domesticus*. *Journal of Evolutionary Biology* **16**, 708–720.

Zelditch, M. L., Lundrigan, B. L., and Garland, T. (2004). Developmental regulation of skull morphology. I. Ontogenetic dynamics of variance. *Evolution and Development* **6**, 194–206.

Zhang, Y., Proenca, R., Maffei, M., Barone, M., Leopold, L., and Friedman, J. M. (1994). Positional cloning of the mouse *obese* gene and its human homologue. *Nature* **372**, 425–443.

CHAPTER 13

Role of Stress in Evolution: From Individual Adaptability to Evolutionary Adaptation

ALEXANDER V. BADYAEV
Department of Ecology and Evolutionary Biology, University of Arizona, Tucson, Arizona, USA

INTRODUCTION

Environments outside the range normally experienced by a population, and the associated changes in organisms' morphological, physiological, or behavioral homeostasis (stress), accompany most evolutionary changes (Bijlsma and Loeschcke, 1997; Hoffmann and Parsons, 1997; Hoffmann and Hercus, 2000). Depending on the intensity, predictability, and recurrence of stress, responses might range from stress tolerance and avoidance at organismal level to the rapid appearance of novel traits or extinction at population level. Yet, moderate stress is essential for normal growth and differentiation of metabolic, physiological, neurological, and anatomical systems of an organism (Huether, 1996; Clark and Fucito, 1998; Muller, 2003). For example, a large part of skeletal development is directed by exposure to tension and mechanical overloads in excess of those normally experienced by the organism (Hall, 1986; Carter 1987). Stress plays an important role in facilitating local adaptation by enabling better adjustments, synchronization, and functioning of many organismal systems (Simons and Johnston, 1997; Emlen *et al.*, 2003; Wingfield, 2003). Anyone who has experienced the invigorating effects of diving into icy-cold water after a sauna (both of which are extreme environments), the health benefits of rigorous exercise (which by definition exceeds the range of everyday environments), or analgesic and attention-sharpening effects accompanying stressful encounters (McEwen and Sapolsky, 1995; Shors and Servatious, 1997) will testify to these effects of stress. On the other hand, response to an acute and unfamiliar stressor precludes normal organismal functions (Sibly and Calow, 1989), and the high cost of stress tolerance or lack of evolved stress response strategies can lead to evolutionary stasis (Parsons, 1994).

Extreme environments not only disrupt normal development and induce large phenotypic changes in novel directions, but they also simultaneously exert strong phenotypic selection that favors changes in these directions (Waddington, 1941; Schmalhausen, 1949; Bradshaw and Hardwick, 1989; Jablonka *et al.*, 1995; Eshel and Matessi, 1998). Not surprisingly, evolutionary diversification, the appearance of phenotypic novelties, and mass extinction are all closely associated with extreme environmental changes (Howarth, 1993; Guex, 2001; Nicolakakis *et al.*, 2003). Yet, there exists a remarkable gap in our understanding of the mechanisms behind the evolutionary importance of stress. Whereas it is widely recognized, especially in physiological and neurological studies, that stress plays an important role in directing and organizing the adaptive adjustment of an organism to ever-changing environments, very little is known about the mechanisms that enable the organismal accommodation of stress-induced effects and the evolution of a response to stress.

Lack of a developmental perspective in evolutionary studies of stress has left us with several unresolved questions. First, how can organisms *prepare* for novel and extreme environmental change? The organismal ability to mount an appropriate reaction to a stressor requires recognition and evaluation of the extreme environment. How can this ability evolve in relation to stressors that are short and rare in relation to a species generation time? Second, numerous studies have documented an increase in phenotypic and genotypic variance under stress, and it is suggested that this variance is a source of novel adaptations under changed environments. Yet, for stress-induced modifications to have evolutionary importance they have to be inherited and persist in a sufficient number of individuals within a population. This requires an organism to survive stress and reproduce at least once; thus stress-induced variation has to be accommodated by an organism without reducing its functionality. How is such accommodation accomplished? Moreover, could existing organismal systems channel accumulation of stress-induced variance in some directions, but not others and thus direct evolutionary change in response to stress? The perspective outlined here, with specific focus on the effect of stress during development in animals, suggests that these questions are resolved by considering (1) the organization of developmental systems that enable accommodation and channeling of stress-induced variation without compromising organismal functionality; (2) the significance of phenotypic and genetic assimilation of neurological, physiological, morphological, and behavioral responses to stressors; as well as (3) multiple inheritance systems that transfer the wide array or developmental resources and conditions between the generations enabling long-term persistence and evolution of stress-induced adaptations.

I. EVOLUTION OF RESPONSE TO STRESS

A. Detection and Avoidance

Stress occurs when changes in the external or internal environment are interpreted by an organism as a threat to its homeostasis (e.g., Greenberg *et al.*, 2001; McEwen and Wingfield, 2003). The ability of an organism to mount an appropriate response to potentially stressful environmental changes requires correct recognition of environmental change and the activation of a stress response (e.g., Johnson *et al.*, 1992). The costs and benefits of stress detection and stress response implementation and the costs and benefits of maintaining stress resistance strategies vary among environments and individuals, favoring multiple solutions of dealing with stress. Crucial to these solutions is an organism's familiarity with the strength and types of stressors. This familiarity is determined, in turn, by the recurrence of a particular stressor in relation to a species' generation time (Lively, 1986; Lachmann and Jablonka, 1996; Meaney, 2001; Piersma and Drent, 2003). Yet, it is unclear how the ability to recognize and assess potentially stressful environments can evolve. How can organisms judge the appropriate reaction to a stressor, such as is required to select between stressor avoidance and stress tolerance? Are the mechanisms of assessment and avoidance specific to a particular stressor?

1. Familiarity with Stressor: Cognitive and Physiological Assimilation of a Rare Event

The response to stress depends crucially on prior experience and a "memory" of response to a stressor. Generally, repeated exposure to a particular stressor favors the evolution of mechanisms that suppress an organism-wide stress reaction and, instead, activate stress-specific responses (Johnson *et al.*, 1992; Veenema *et al.*, 2003).

For example, in higher vertebrates, stress-induced activation of the neuroendocrinological system increases its reactivity to internal and external stimuli, facilitates the processing of sensory information, and ultimately enables the formation of a behavioral or physiological strategy for dealing with a stressor. Furthermore, stress-induced activation of neuroendocrinological systems facilitates long-term retention of information about a stressful event and corresponding organismal response after the stressor is gone. Interestingly, once formed, the maintenance of such "memory" can be accomplished by periodic exposure to different stressors. For example, hormones associated with stress detection and avoidance also play a major role in modifications of neural circuits (Gold and McGaugh, 1978); once the stress-avoidance strategy is formed, exposure to even low concentration of these hormones maintains the strategy (McGaugh *et al.*, 1982).

Physiological studies of animals show that the repeated experience of successfully overcoming social stresses during ontogeny is a prerequisite for the acquisition of a normal repertoire of behavioral strategies (Huether, 1996; see also Gans, 1979). An insightful example comes from experiments that show that individuals exposed to repeatable but consistently unfamiliar (and thus "uncontrollable" by an animal) stressors develop "stressful helplessness"; i.e., they lose their ability to react to any stressor (Katz *et al.*, 1981; Johnson *et al.*, 1992; Avitsur *et al.*, 2001).* On the contrary, individuals that were allowed to develop a stress-avoidance strategy by exposure to a previously encountered stressor not only developed stress tolerance to a particular stressor but also actively sought out other mild stressors. In the absence of other stressors, their stress-avoidance abilities diminished (Katz *et al.*, 1981; Johnson *et al.*, 1992; Avitsur *et al.*, 2001). These results suggest that, once originated, a stress-response strategy can be maintained by other environments and that adaptation to one type of stressor, at least in "social" stresses, may facilitate adaptation to other stressors.

Phenotypic assimilation of the appropriate stress response is facilitated when neural circuits and hormones related to the stress response are also involved in other organismal functions (Aston-Jones *et al.*, 1986; Greenberg *et al.*, 2001). In such cases, even a single stressful experience during development is often enough to induce changes that, in the future, will prevent organism-wide stressful reactions and activate stress-specific behavioral and physiological responses (Levine *et al.*, 1967; 1989). Generally, stress-induced reorganization of developmental pathways and organismal function rather than the production of novel stress-specific pathways is thought to account for the ease with which individuals and populations lose and gain the ability to resist stress in laboratory populations (Chapin *et al.*, 1993).

*Organisms' "stress helplessness" from lack of opportunity to develop stress-specific avoidance strategy is conceptually analogous to "morphological stasis" of lineages that occur in environments with frequent acute and diverse stresses that prevent the evolution of stress-specific adaptations.

B. Stress-Avoidance Strategies

The ability to remove a stressor *actively* by either relocation or avoidance requires an evolved ability to detect or anticipate stressful changes and the "knowledge" or "memory" of stress-avoidance strategies or adjustments (Bradshaw and Hardwick, 1989; Jablonka *et al.*, 1995; Denver, 1999). Therefore, the evolution of stress avoidance is more likely when stressful events are predictable, prolonged, and frequent in relation to generation time (Ancel Meyers and Bull, 2002; Figure 13-1). Alternatively, the short-term avoidance of a frequent and mild stressor might be accommodated by behavioral or physiological plasticity of an organism (Figure 13-1; Schlichting and Smith, 2002; Nicolakakis *et al.*, 2003; Piersma and Drent, 2003; Wingfield and Sapolsky, 2003). For example, repeated challenges of an organism's immune system enable a more precise reaction to a specific pathogen, frequent and diverse stressors facilitate the formation of complex and robust metabolic

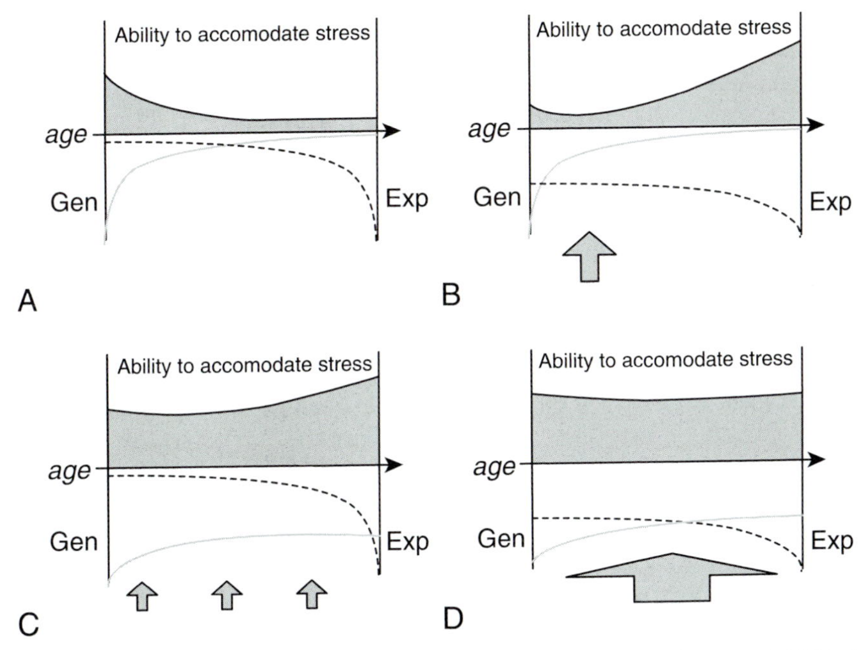

FIGURE 13-1. Conceptual outline of the acquisition of ability to accommodate stress (solid black line and gray area) across a life span of an individual under (A) normal (i.e., not stressful) environment; (B) novel strong stressor (examples: stress-enhanced learning, behavioral avoidance of a stressor, social stressors), (C) frequent mild stressor (examples: weather-induced migrations of arctic passerines, periodic torpor), (D) frequent, novel, and strong stressors (e.g., "living fossils," "stressful helplessness"). "Gen" (solid gray line) indicates genetic effects on acquisition of ability to accommodate stress; "exp" (dotted black line) indicates the effects of individual experience with stressor over the generation time; "age" is a duration of a single generation; gray arrows show the timing and strength of a stressor.

networks (Clark and Fucito, 1998), and challenges to skeletal tissues caused by mechanical overload during growth lower developmental errors (see also Simons and Johnston, 1997; Graham *et al.*, 2000). Wingfield (2003; Wingfield and Sapolsky, 2003) reviews the cases when selection favors stress avoidance and where suppression of organism-wide stress response is accomplished by a blockage of either neural system perceiving a stressor or sensitivity of individual organismal systems to stress-induced increase in circulating glucocorticosteroids. Generally, when the environment during growth is a good predictor of the environment to be experienced as an adult, developmental plasticity in morphology and behavior can enable the accommodation of internal and external environmental fluctuations (Levine *et al.*, 1967). Consequently, organisms activate stress reactions when there is a discordance between environments during their development and their current external and internal environments (Meaney, 2001; Bateson *et al.*, 2004, Weaver *et al.*, 2004).

On the longer time scale, avoidance of a predictable stressor can be accomplished by changes in an organism's life history, especially by altering the timing of reproduction or duration of development. Common cases include stress-induced modification of the timing of metamorphism in amphibians, changes in the duration of gestation in mammals, and the timing of flowering and seeding in plants (e.g., Bradshaw and Hardwick, 1989; Stanton *et al.*, 2000). For example, tadpoles of several species accelerate metamorphosis when environmental changes indicate a greater probability of desiccation; this sensitivity to stressor cues is regulated by the corticotropin-releasing hormone signaling system (Denver, 1999). Heil *et al.* (2004) describe evolutionary establishment of environmentally induced stress avoidance in *Acacia* plants.

In sum, initial behavioral accommodations of stress (e.g., hiding, relocation, lowering metabolism) may set the stage for the evolution of adaptive stress-avoidance strategies (e.g., periodic hibernation, migration, torpor). When a stressor is reliably preceded by other environmental changes, their mutual recurrence facilitates the establishment of stressor recognition, assessment and avoidance strategies, such that an evolved stress-specific strategy does not involve an activation of an organism-wide stress response. When individuals vary in their reaction to stress and when stress-induced strategies are favored by natural selection during and after stressful events, these strategies can become phenotypically and genetically assimilated in a population (Baldwin, 1896; Hinton and Nolan, 1980; Oyama, 2000; West-Eberhard, 2003; Figure 13-1).

II. EVOLUTIONARY CONSEQUENCES OF STRESS

A. Stress-Induced Variation

A stress-induced increase of phenotypic and genetic variance in a population has three main sources. First, directional selection imposed by a stressor can result in faster rates of mutation and recombination. Second, stress challenges to regulatory mechanisms can release and amplify previously accumulated, but unexpressed, genetic and phenotypic variation. Third, stressful environments can facilitate developmental expression of genetic variance that had accumulated, but was phenotypically neutral, under normal range of environments. These sources of variation can be adaptive under stressful conditions when they facilitate the population's persistence through a stressful event by the development of novel adaptations to changed environments.

1. Generated Variance

Organismal reaction to a stressor is often associated with generation of variation in a directional and locally adaptive manner (Jablonka and Lamb, 1995; West-Eberhard, 2003). In some cases, such directionality is attributed to the channeling effects of complex developmental networks (e.g., Walker, 1979; Roth and Wake, 1985). In other cases, it is associated directly with a stressful environment (e.g., Wills, 1983) or with stress effects on organismal fitness (Hadany and Beker, 2003). Some studies documented that an extreme environment increases genetic variation because of the increase in mutation and recombination rates (Imasheva, 1999, reviewed in Hoffmann and Parsons, 1997). When such mutations are directional (or "focused", *sensu* Caropale, 1999) in relation to a stressor—that is when the stressful environment both causes a mutation and favors phenotypic change associated with this mutation—such an increase in mutation rate results in greater similarity among individuals in response to a stressor facilitating evolutionary adaptation to novel environments (Shapiro, 1992; Jablonka and Lamb, 1995; Wright, 2000). For example, exposing *Chlamydomonadas* to a stressful ultraviolet irradiation increased mutation rates in traits affecting fitness (Goho and Bell, 2000). Similarly, stress induced directional and locally appropriate mutations in bacteria (Cairns *et al.*, 1988; Sniegowski *et al.*, 2000; Wright, 2000; Bjedov *et al.*, 2003). Exposure to acute stress was associated with rapid adaptive evolution of a gene family, primarily because of gene duplication, in cyanobacteria (Dvornyk *et al.*, 2002), with rapid amplification of a gene in humans (Prody *et al.*, 1989), and with greater frequency of sexual recombination in *Volvox* (Nedelcu and Michod, 2003). Other examples include long-term effects of stress on gene expression and DNA sequence, activation of previously unexpressed genes by stressful events, and stress-induced transposition in plants (Belyaev and Borodin, 1982; Ruvinsky *et al.*, 1983; McClintock, 1984; Wessler, 1996). At the level of phenotype, induction of a phenotypic trait by a stressor and concurrent selection on the induced trait are common (Jablonka *et al.*, 1995; Oyama, 2000; Nicolakakis *et al.*, 2003; Price *et al.*, 2003; West-Eberhard, 2003).

2. Hidden Variance

Stressful environments often reveal greater phenotypic and genetic variability than is seen under normal environments. It is commonly suggested that such hidden variation results from stress-induced challenge of preexisting genetic and developmental architecture of organismal homeostasis (Scharloo, 1991). In turn, an increase in variation in individual organismal systems and their subsequent reorganization is thought to enable the formation of novel adaptations (Bradshaw and Hardwick, 1989; Eshel and Matessi, 1998; Gibson and Wagner, 2000; Lipson *et al.*, 2002; Schlichting and Smith, 2002; Badyaev, 2004c). The idea that the extreme environment's challenge to previously canalized system is the source of such hidden variation is collaborated by observations of the stress-induced sudden appearance of primitive, ancestorlike forms in some lineages (Guex, 2001), by studies of phenotypic responses to stress that mimic the expression of mutation (Goldschmidt, 1940; Chow and Chan, 1999; Schlichting and Smith, 2002), by documentation that phenotypically neutral genetic variance in ancestral forms of cultivated plants becomes highly adaptive in the hybrid backgrounds of domesticated forms (Lauter and Doebley, 2002; Rieseberg *et al.*, 2003), and by numerous examples of environment dependency in expression of genetic variation (Kondrashov and Houle, 1994; Leips and MacKay, 2000; Badyaev and Qvarnström, 2002; Keller *et al.*, 2002; Badyaev, 2004b).

Yet, despite these examples, it is not clear how genetic and developmental systems accumulate and store phenotypically neutral genetic variance while not expressing it (Eshel and Matessi, 1998; Wagner and Mezey, 2000; Hermission *et al.*, 2003; Masel and Bergman, 2003). Specifically, the discussion has focused on the existence of "evolutionary capacitors" (Rutherford, 2000) and "adaptively inducible canalizers" (Meiklejohn and Hartl, 2002) which are *specific* mechanisms that buffer and accumulate developmental variation, producing "hidden reaction norms" of a phenotype. A debated question is whether "evolutionary capacitors" are stressor-specific regulatory systems or whether evolutionary capacity is a property of any complex and locally adapted organismal system. Rutherford and Lindquist (1998) described that mutations at the gene for the stress-induced chaperone proteins (Hsp90) harbor abundant but normally unexpressed genetic variation that when selected leads to the appearance and assimilation of novel phenotypes in the population (Ruden *et al.*, 2003). Thus Hsp90 might be a specialized evolutionary capacitor that buffers developmental variation but under stressful conditions facilitates adaptation (Meiklejohn and Hartl, 2002). However, recent studies suggested that "evolutionary capacity" is a property of most adapted developmental systems that when challenged by a novel environment (external or internal) reveal large genetic variation (Kirschner and Gerhart, 1998; Rutherford, 2000; Bergman and Siegal, 2003; Badyaev, 2004a). For example, Milton *et al.* (2003) showed experimentally that Hsp90 is involved in buffering of only some developmental pathways and not others. Similarly, Szafraniec *et al.* (2001) found that as long as mutant effects are not expressed, many complex and redundant developmental systems enable accumulation of mutational variance. Thus complex developmental processes and genetic networks can constrain variation in individual traits (Rice, 2004), and phenotypically neutral genetic variation can accumulate in such systems given sufficient time and population size (Hermission and Wagner, 2005).

Interestingly, in many complex social networks, a stress-induced decrease in integration accelerates acquisition of a new optimum phenotype. An interesting example is the stress-induced modification of foraging and nest site searches in social insects. In some ant species, a destruction of the nest site leads to the breaking of the strict hierarchical social structure and rapid proliferation of random individual nest search routes and patterns. When a few individuals find a new suitable site, their recruitment of other individuals to follow them to the site rapidly leads to crystallization of the relocation route and movement patterns and reinstatement of the social integration of the colony (Britton *et al.*, 1998; Couzin and Franks, 2003).

In sum, stress resistance might be a by-product of an organism's complexity, and accumulation of unexpressed variation by genetic and phenotypic developmental systems facilitate evolutionary change under extreme environments. Organismal homeostasis can be compromised by either novel directional selection on some organismal systems but not others, or by organism-wide effects of a stressor, resulting in weaker organismal homeostasis and greater phenotypic plasticity (Schlichting and Pigliucci, 1998; Newman and Muller, 2000). Under the former scenario, a more directional and faster response to a stressor at the population level is expected because stress-induced variation will be channeled and amplified by existing functional complexes. The latter scenario should produce a greater opportunity for the evolution of morphological novelty. Overall, the weakening of complex phenotypic regulatory systems and accumulation of neutral genetic variance provides a link between diversification, evolutionary change, and extreme environments.

B. Buffering, Accommodating, and Directing Stress-Induced Variation

Organisms can maintain functionality in stressful environments by channeling and accommodating stress-induced variation. This is accomplished by buffering some organismal functions while increasing the flexibility of others (Alberch, 1980; Nijhout, 2002). How can such organization evolve?

1. Stress Buffering: A By-Product of Complexity in Development or an Evolved Strategy?

Organismal functions most closely related to fitness are thought to be the most buffered against internal and external stressors (Waddington, 1941; Schmalhausen, 1949; Stearns and Kawecki, 1994). Yet, an organism's functioning in changing environments requires the ability to track and respond to these environments. Consequently, evolved systems that shield an organism from stressors restrict an organism's ability and capacity to adapt continuously to changing environments (Wagner *et al.*, 1997; Eshel and Matessi, 1998; Ancel, 1999; Schlichting and Smith, 2002). For example, suppression of stress-induced activation of the sensory systems limits an organism's ability to acquire and retain the sensory cues and behavioral strategies necessary for stress avoidance (see preceding text; Huether, 1996). On the one hand, a lack of phenotypic plasticity results in population extinction under stress (Gavrilets and Scheiner, 1993; Ancel, 1999). On the other hand, extensive phenotypic variability in organismal functions weakens the effects of directional selection imposed by stressful environments and thus lessens the opportunity for genetic assimilation and evolution of adaptations to stress (Fear and Price, 1998; Ancel, 2000; Huey *et al.*, 2003). Thus, for functioning of organismal systems that are most closely related to fitness, intermediate levels of phenotypic plasticity and environmental sensitivity should be the most optimal (Behera and Nanjundiah, 1995; Wagner *et al.*, 1997; Ancel, 2000; Price *et al.*, 2003). Yet, it is unclear how an optimal level of stress buffering can evolve. Specifically, is it shaped by natural selection exerted by extreme environments or by internal stabilizing selection for the cohesiveness of an organism?

Recent studies suggest that buffering is an emerging property of developmental complexity rather than an evolved stress-resistance mechanism (see preceding text); the increasing complexity of developmental pathways and networks leads directly to environmental and genetic stability and canalization (Baatz and Wagner, 1997; Clark and Fucito, 1998; Rice, 1998; Waxman and Peck, 1998; Meiklejohn and Hartl, 2002; Siegal and Bergman, 2002; Ruden *et al.*, 2003; Rice, 2004). Complex genetic and developmental networks can accommodate the effects of stressful perturbations without the loss of function or structure, while building up neutral genetic variation (Rutherford, 2000; Bergman and Siegal, 2003; Masel, 2004).

An organism's resistance to extreme environments depends on the historical recurrence of stressors as well as the ability of existing developmental processes to accommodate stress-induced changes (Gans, 1979; Lively, 1986; Jablonka and Lamb, 1995; Chipman, 2001; Arthur, 2002; Emlen *et al.*, 2003). Thus differences among organisms and organismal systems in response to stress may reflect different histories of past selection. Some traits (such as foraging or sexual traits) may experience recurrent and fluctuating directional selection that favors rapid transformations in response to changing environments, whereas other parts of a phenotype might be under concurrent stabilizing selection favoring canalization (Olson and Miller, 1958; Wagner, 2001). A combination of long-term stabilizing selection on the entire organism with strong and variable directional selection imposed by a stressor on a few organismal components should favor the evolution of modular organization where stress-induced modifications of traits can be accomplished with minimum interference with the rest of the phenotype (Simpson, 1953; Berg, 1960; Kirschner and Gerhart, 1998; Wagner and Mezey, 2004; Wagner *et al.*, 2005; Figures 13-2 and 13-3). Persistence of such modular organization under fluctuating selection pressures is enabled by developmental complexity of its components (Badyaev, 2004a,c); such organization channels stress-induced variation while buffering

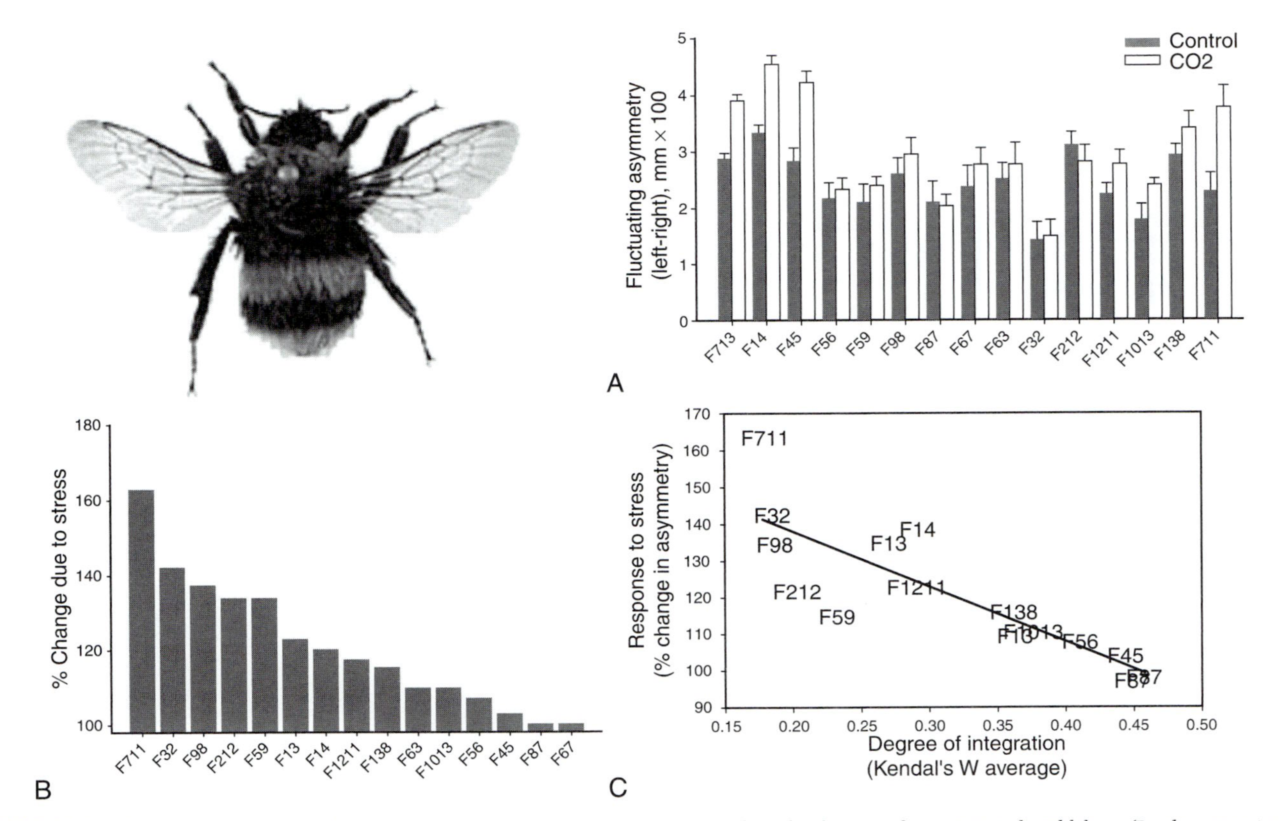

FIGURE 13-2. Degree of developmental integration in relation to accumulation of random developmental variation in bumblebees (*Bombus empatiens*). Bumblebees were raised under normal (control) and elevated (stress) concentration of CO_2. (A, B) Individual wing traits (intervein distances, shown on X axis) varied in reaction to the stressor, i.e., accumulated different amounts of developmental noise (measured as fluctuating asymmetry between left and right side). (C) Traits that were most closely developmentally integrated with other traits had lower response to the stressor and accumulated lower amount of random developmental variation (Sowry and Badyaev, 1999).

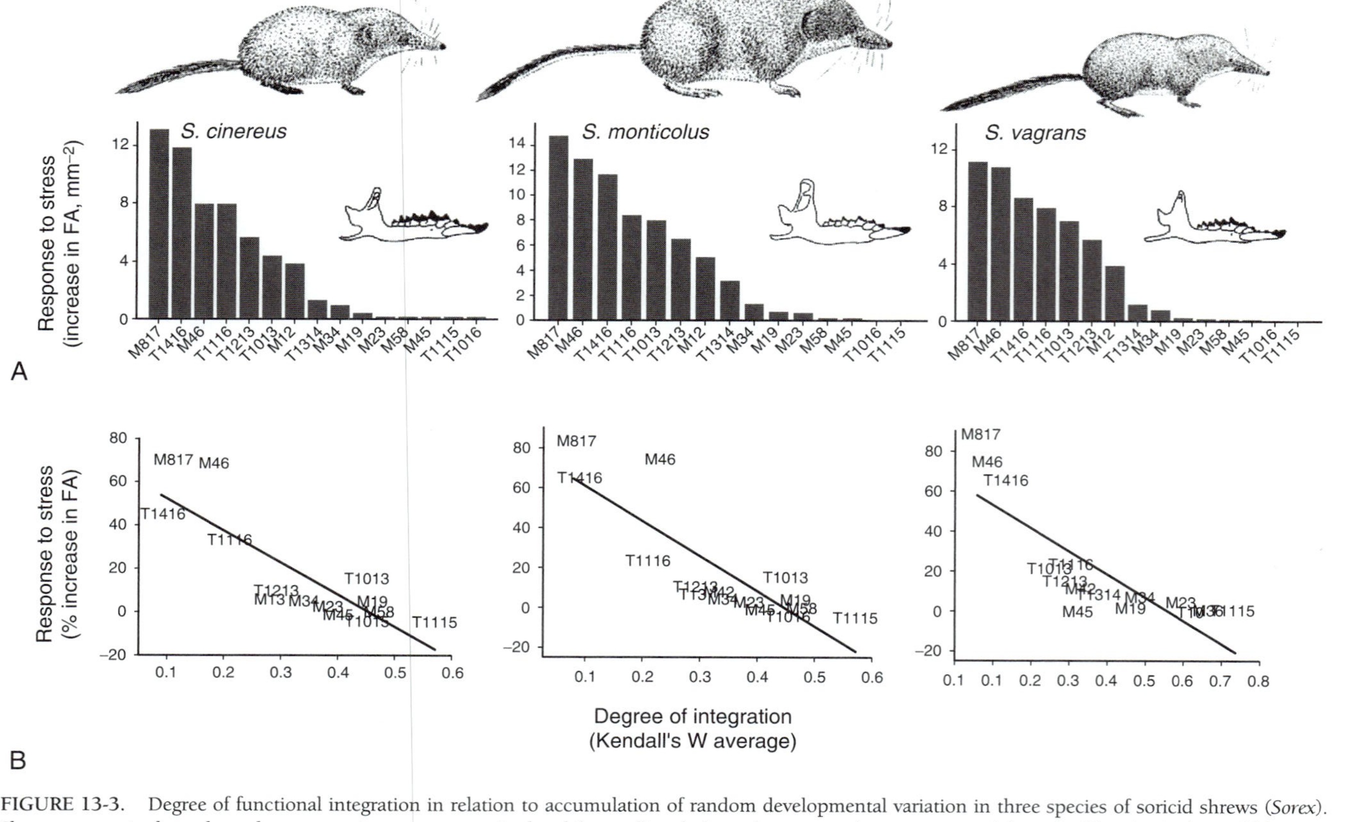

FIGURE 13-3. Degree of functional integration in relation to accumulation of random developmental variation in three species of soricid shrews (*Sorex*). Shrews were raised on plots where vegetation was not manipulated (control) and plots where vegetation was removed (stress). (A) Individual mandible traits (distances between muscle attachments) varied in response to the stressor as measured by an increase in fluctuating asymmetry (FA). (B) Traits that were most closely integrated with other traits had lower response to the stressor and accumulated less random developmental variation (Fiumara and Badyaev, 1998; Badyaev and Foresman, 2004).

organismal components and thus enables a greater and more similar response to a stressful environment among individuals.

2. Stress Accommodation by Changes in an Organism's Integration

Organisms might accommodate stress-induced variation without the loss of function by lessening homeostasis of individual systems. Such a decrease in an organism's integration under stress allows exploration of new environmental niches and novel solutions to adapt to these environments (Holloway *et al.*, 1990, 1997; Raberg *et al.*, 1998; Badyaev and Foresman, 2000; Hoffmann and Woods, 2001). For example, individual hormonal systems have a greater potential range of performances and can remain functional under a wider range of environments than is allowed by homeostasis under normal range of environments (Dickinson, 1988; Johnson *et al.*, 1992; Greenberg *et al.*, 2001). In other words, organism-wide homeostasis is accomplished at the expense of the potential of individual systems and components (e.g., Smith-Gill, 1983), and organisms might react to a stressor by actively weakening homeostasis. For example, frequently documented suppression of immunocompetence under stressful conditions might facilitate novel adaptations to a stressor by realizing full capabilities of individual immune systems (Raberg *et al.*, 1998; Avitsur *et al.*, 2001). When stress is associated with damage of tissues and accumulation of heat shock proteins, as is the case with hypertension and greater activity, suppression of immunological functions enables individual organismal systems to respond to a stressor without activation of organism-wide autoimmunological response (Dickinson, 1988; Raberg *et al.*, 1998; Avitsur *et al.*, 2001).

However, there are examples of stress-induced increases in organismal integration and corresponding suppression of random genetic and developmental variation under stress (e.g., Siegel and Doyle, 1975; Bennington and McGraw, 1996; Badyaev and Foresman, 2004). For example, exposure to stress prevented the expression of deleterious mutations in *Escherichia coli* (Kishony and Leibler, 2003). Similarly, fluctuating asymmetries of developmentally independent forewings and hindwings became integrated in bumblebees raised under stressful, but not under control conditions, apparently as a result of a greater resource exchange between different tissues under stress (Klingenberg *et al.*, 2001). An increase in overall integration accounted for lesser phenotypic variation in the foraging structures of several mammalian species raised under stressful conditions (Badyaev, 1998; Badyaev *et al.*, 2000). Similarly, when breeding opportunities are limited or when the benefit of the current breeding attempt exceeds the costs of stress response, organisms can "buffer" reproductive systems by blocking or reducing their sensitivity to stress or by increasing compensatory interactions within reproductive systems to counteract the stress effects on the organism (Wingfield and Sapolsky, 2003).

When stressors are mild and occur during ontogeny, individual organismal systems often accommodate stress-induced variation without the reduction in functionality

(Bradshaw and Hardwick, 1989; Clarke and McKenzie, 1992; Huether, 1996; Schandorff, 1997). The degree of phenotypic plasticity is usually the highest, and ability to accommodate stress-induced variation (and to be shaped by stress) is the greatest during early stages of development (Figure 13-1). The importance of the timing of stress for directing the evolution of morphological traits is well documented. For example, when components of foraging structures differ in patterns of ossification, morphological variation in later ossified components is directed by stress-induced modifications of earlier ossified components (Figure 13-1; Badyaev, 1998; Mabee *et al.*, 2000; Badyaev and Foresman, 2004; Badyaev *et al.*, 2005). Neyfakh and Hartl (1993) documented that prior exposure to a stressor makes the ontogeny of morphological structures more amenable to subsequent modification. Moreover, when stress occurs early in ontogeny, accommodation and channeling of stress-induced variation by existing organismal structures causes similar reorganization in many individuals simultaneously (Roth and Wake, 1985; Chapin *et al.*, 1993; Figure 13-2), which facilitates adaptive evolution (Goldschmidt, 1940; West-Eberhard, 2003). Our studies of shrew mandibles (Fiumara and Badyaev, 1998; Badyaev and Foresman, 2000, 2004; Foresman and Badyaev, 2003; Badyaev *et al.*, 2005) and bumblebee wings (Sowry and Badyaev, 1999; Klingenberg *et al.*, 2001), as well as other studies (Leamy, 1993; Badyaev, 1998; Klingenberg and McIntyre, 1998; Klingenberg and Zaklan, 2000; Badyaev *et al.*, 2001; Juste *et al.*, 2001; Badyaev and Young, 2004), support these ideas; patterns of expression of stress-induced developmental variation were similar among individuals because of the similar effects of integration on channeling stress-induced variation and because variation from different sources is expressed by the same developmental pathways (Cheverud, 1982; Meiklejohn and Hartl, 2002, Figures 13-2 and 13-3).

Gans (1979) suggested that because extensive reorganization of the organismal phenotype is needed to deal with extreme environments, only a small portion of the population survives the stressful environment. This increases the probability of appearance of extreme phenotypes. Population size fluctuations under stress also affect the ability of genetic and developmental systems to accumulate and retain solutions to rare environmental events (Wagner, 2003; Hermission and Wagner, 2005). However, the effect of population size on the probability of establishing and retaining a novel phenotype differs between normal and extreme environments. Under normal conditions, modifications are likely to be established in larger populations that are more buffered from stochastic fluctuations and are able to accumulate larger amounts of neutral genetic variance (Gavrilets, 2004). Under stressful environments, that not only introduce greater variability, but also select toward a new phenotypic optimum, smaller populations should allow greater evolutionary change (Barton and Charlesworth, 1984; Gavrilets, 2004). Moreover, when a stressor is associated with abrupt changes in population composition—as is the case with extensive mortality or dispersal—it can lead to the modification of genetic and phenotypic interrelationships among the traits (Bryant and Meffert, 1988; Cheverud *et al.*, 1999; Badyaev and Foresman, 2000).

3. Accommodation of Stressor by Channeling Stress-Induced Variation

Natural selection favors organismal homeostasis that maintains some developmental variation for adjustment of the organism to its external and internal environments (Simons and Johnston, 1997; Wagner *et al.*, 1997; Eshel and Matessi, 1998; Emlen *et al.*, 2003; Nanjundiah, 2003). Stressful conditions can increase this variation and differences among organismal systems in their reaction to a stressor (and the corresponding channeling of stress-induced variation) might bias the introduction and expression of variation available for selection and thus bias evolutionary change (Bonner, 1965; Roth and Wake, 1985; Jablonka and Lamb, 1995; West-Eberhard, 2003).

Empirical studies show that the coordinated development of morphological traits leads to their similarity in expression of stress-induced developmental variation (Leamy, 1993; Smits *et al.*, 1996; Badyaev and Foresman, 2000; Klingenberg and Zaklan, 2000; Klingenberg *et al.*, 2001, 2004; Badyaev *et al.*, 2005). Our studies of four species of shrews showed that stress-induced variation was largely confined to the directions delimited by groups of traits involved in the same function (muscle attachments) (Badyaev *et al.*, 2000; Badyaev and Foresman, 2004). Interestingly, this channeling was concordant with the direction of species divergence—species differed most in the same traits that were most sensitive to stress within each species (Badyaev *et al.*, 2000). These results not only confirm a strong effect of functional complexes on directing and incorporating stress-induced variation during development, but also might explain the historical persistence of complex groups of traits despite the effects of stressful environments.

C. Inheritance

For a stress-induced modification to be preserved in a lineage, it needs to be accommodated by an organism, and if conditions favoring this modification recur, transmitted between generations, i.e., inherited. This presents two problems. First, can environmentally induced effects become inherited? Second, if each organism accommodates a stressor by different adjustments, then how can this diversity enable directional evolution of a stress-response strategy?

Stress-induced phenotypic changes commonly persist across several generations; such across-generations carry-over effects (*sensu* Jablonka *et al.*, 1995) can be caused by the transfer of physical substances, inheritance and developmental incorporation of a stressor, hormonal effects that influence expression of genetic variance in subsequent generations, epigenetic inheritance of stress-induced variation and structures, as well as behavioral effects (Jablonka and Lamb, 1995; Oyama, 2000; West-Eberhard, 2003). For example, inheritance of dominant-subdominant relationships in groups of many social mammals is accomplished by mechanisms different from original stressful encounters that established the dominance structure (Creel *et al.*, 1996; Goymann and Wingfield, 2004). Similarly, maternal care often sets the stage for a lifelong reaction to stressors, by modifying the expression of genes that regulate behavioral, physiological and endocrinological responses to stressors (Mousseau and Fox, 1998; Meaney, 2001; Badyaev, 2002; Weaver *et al.*, 2004). Stress-induced changes in neuroendocrinological systems often occur with significant delay after the exposure to stress and persist for a long time. This led to the suggestion that the primary function of such delayed changes is integration of past stress-induced responses and sensitization of the organism to future occurrences of similar stressors (Huether, 1996). In turn, within-generation and between-generation maintenance of stress-induced changes in neural and physiological systems is accomplished by similar hormonal mechanisms (McGaugh *et al.*, 1982; Meaney, 2001). Poststress fluctuating environments are often different from both the environment before the stressor and the stressful environment itself and have few predictable cues to organisms that survived stressful event. Under such conditions, a short-term inheritance of developmental resources is highly advantageous (Jablonka *et al.*, 1995). More generally, short-term and nongenetic inheritance is beneficial when the frequency of stress recurrence is greater than the generation time, but shorter than is necessary for the spread and fixation of adaptive mutation (i.e., the evolution of genetic adaptation) (Levins, 1963; Ancel Meyers and Bull, 2002).

In sum, accommodation of stress-induced variance by an organism can be facilitated by recurrent developmental stressors; genetic assimilation replaces stress-induced developmental modification if this modification has a fitness advantage in both stressful and poststress environments (Schmalhausen, 1949; Waddington, 1952). Even when the short-term organismal responses to a stressor are not genetically heritable, differences among organisms in the ability to survive stress and the recurrence of stressful environments will canalize stress-induced responses developmentally (Baldwin, 1896; Schlichting and Pigliucci, 1998; Ancel, 1999; West-Eberhard, 2003).

III. EVOLUTIONARY ADAPTATION

Close association between extreme environments and the pattern and rate of adaptive evolution is one of the best-documented patterns in evolutionary biology; stressful environments uncover, generate, and amplify phenotypic and genetic variation among individuals in the population and facilitate population divergence (Hoffmann and Parsons, 1997). Unlike environmental fluctuations within a range normally experienced by a population, stressful environments modify and reorganize integrated developmental and genetic networks simultaneously in a large group of individuals; directional change produced by these networks in combination with strong and novel directional selection by stressful environment facilitates rapid evolution and diversification (Jablonka and Lamb, 1995; West-Eberhard, 2003). Moreover, when a stressor compromises an organismal trait, releasing accumulated and unexpressed genetic variation associated with the trait's function, such variance enhances the organismal response to selection acting on this trait (Zakharov, 1993; Robinson and Dukas, 1999; Bergman and Siegal, 2003). Example include stress-induced cartilage changes during development of bird skeletons that lead to the formation of novel structures (Muller, 2003), and stress-induced modifications in integration of foraging structures that facilitate diversification of cichlid jaw morphology (Chapman *et al.*, 2000; Albertson *et al.*, 2003). Moreover, extreme environments cause evolutionary change by modifying population dynamic processes such as immigration, population size, inbreeding, and competition (Kawata, 2002; Gavrilets, 2004). For example, in shrews, periods of environmental stress are accompanied by increased food competition and extensive mortality (Zakharov *et al.*, 1991; Badyaev *et al.*, 2000). In turn, greater interspecific competition for food amplified and extended the effects of stress exposure on the ontogeny of morphological structures (Foresman and Badyaev, 2003).

A. Stress-Induced Evolution versus Stress-Induced Stasis

Stress specificity, intensity, and recurrence are of fundamental importance for its evolutionary consequences (Bradshaw and Hardwick, 1989; Parsons, 1994; Ancel Meyers and Bull, 2002). Parsons (1994) suggested that only some subsets of stressful environments—narrowly fluctuating and slowly changing in relation to generation time—are associated with a rapid evolutionary change, whereas extreme and rapidly changing environments promote morphological stasis because of the costs associated with stress tolerance (see also Hoffmann *et al.*, 2003). Furthermore, only stressors specific to an organismal system are expected to enable assimilation and evolutionary persistence of stress-induced adaptations, because more general stressors favor stress tolerance by increasing homeostasis, in turn leading to a reduction in organismal metabolism and fitness. Thus, among the array of organismal responses to stressful environments, only accommodation of stress-induced variation and stress avoidance leads to evolutionary change (Parsons, 1993). In turn, because of its association with lower metabolism and stronger regulatory systems, stress tolerance is unlikely to be associated with greater organismal plasticity, thus leading to stasis under extreme environments, which is observed in "living fossils" (Parsons, 1993, 1994; Figure 13-1).

IV. CONCLUSIONS

Several themes and approaches in recent studies significantly further our understanding of the relationship between stressful environments and evolution. First, stressful environments modify (most often reduce) the integration of neurological, endocrinological, morphological, and behavioral regulatory systems. Second, such reduced integration and subsequent accommodation of stress-induced effects by complex developmental systems enables organismal "memory" of a stressful event as well as phenotypic and genetic assimilation of the response to a stressor. Third, the widely held assumption of randomly generated variance under stressful conditions is not correct. In complex functional systems, a stress-induced increase in phenotypic and genetic variance is often directional, channeled and amplified by the existing developmental system, which accounts for similarity among individuals in stress-induced change and thus significantly facilitates the rate of adaptive evolution. Fourth, accumulation of phenotypically neutral genetic variance might be a property of any locally adapted and complex developmental system; novel or extreme environments facilitate the phenotypic expression of this variance. Fifth, stress-induced effects and stress-resistance strategies can persist for several generations. In animals, such carry-over effects are enabled by hormonal effects on learning and gene expression and are facilitated by maternal inheritance of either a stressor or a stress-induced response. These transgenerational effects along with the complexity of developmental systems and stressor recurrence might lead to genetic assimilation of stress-induced effects. Accumulation of neutral genetic variance by developmental systems and phenotypic accommodation of stress-induced effects, together with the inheritance of stress-induced modifications, ensures the evolutionary persistence of stress-response strategies and provides a link between individual adaptability and evolutionary adaptation.

ACKNOWLEDGMENTS

I thank Renee Duckworth, Dana Seaman, Kevin Oh, and Rebecca Young for comments that helped improve this manuscript, Joachim Hermission and Gunter Wagner for discussions, Susanna Sowry for collaborations on the bumblebee projects, Kerry Foresman and Celeste Fuimara for collaborations on the shrew projects, the National Science Foundation (DEB-0075388, DEB-0077804, and IBN-0218313) and University of Arizona for partially funding this work, and Benedikt Hallgrimsson and Brian Hall for the invitation to contribute to this book.

REFERENCES

Alberch, P. (1980). Ontogenesis and morphological diversification. *American Zoologist* **20**, 653–667.

Albertson, R. C., Steelman, J. T., and Kocher, T. D. (2003). Directional selection has shaped the oral jaws of Lake Malawi cichlid fishes. *Proceedings of the National Academy of Sciences of the United States of America* **100**, 5252–5257.

Ancel, L. W. (1999). A quantitative model of the Simpson-Baldwin effect. *Journal of Theoretical Biology* **196**, 197–209.

Ancel, L. W. (2000). Undermining the Baldwin expediting effect: Does phenotypic plasticity accelerate evolution. *Theoretical Population Biology* **58**, 307–319.

Ancel Meyers, L., and Bull, J. J. (2002). Fighting change with change: Adaptive variation in an uncertain world. *Trends in Ecology and Evolution* **17**, 551–557.

Arthur, W. (2002). The emerging conceptual framework of evolutionary developmental biology. *Nature* **415**, 757–764.

Aston-Jones, G., Ennis, M., Pieribone, V. A, Nickel, W. T., and Shipley, M. T. (1986). The brain nucleus locus coeruleus: Restricted afferent control of a broad efferent network. *Science* **234**, 734–737.

Avitsur, R., Stark, J., and Sheridan, J. (2001). Social stress induces glucocorticoid resistance in subordinate animals. *Hormones and Behavior* **39**, 247–257.

Baatz, M., and Wagner, G. P. (1997). Adaptive inertia caused by hidden pleiotropic effects. *Theoretical Population Biology* **51**, 49–66.

Badyaev, A. V. (1998). Environmental stress and developmental stability in dentition of the Yellowstone grizzly bears. *Behavioral Ecology* **9**, 339–344.

Badyaev, A. V. (2002). Growing apart: An ontogenetic perspective on the evolution of sexual size dimorphism. *Trends in Ecology and Evolution* **17**, 369–378.

Badyaev, A. V. (2004a). Colorful phenotypes of colorless genotypes: Towards a new evolutionary synthesis of animal color displays. In *Animal Coloration: Proximate and Ultimate Mechanisms* (G. E. Hill and K. J. McGraw, eds.). Cambridge, MA: Harvard University Press.

Badyaev, A. V. (2004b). Developmental perspective on the evolution of sexual displays. *Evolutionary Ecology Research* **6**, 975–991.

Badyaev, A. V. (2004c). Integration and modularity in the evolution of sexual ornaments: An overlooked perspective. In *Phenotypic Integration: The Evolutionary Biology of Complex Phenotypes* (M. Pigliucci and K. Preston, eds.), pp. 50–79. Oxford, England: Oxford University Press.

Badyaev, A. V., and Foresman, K. R. (2000). Extreme environmental change and evolution: Stress-induced morphological variation is strongly concordant with patterns of evolutionary divergence in shrew mandibles. *Proceedings of the Royal Society Biological Sciences Series B* **267**, 371–377.

Badyaev, A. V., and Foresman, K. R. (2004). Evolution of morphological integration: I. Functional units channel stress-induced variation in shrew mandibles. *American Naturalist* **163**, 868–879.

Badyaev, A. V., and Qvarnström, A. (2002). Putting sexual traits into the context of an organism: A life-history perspective in studies of sexual selection. *Auk* **119**, 301–310.

Badyaev, A. V., and Young, R. L. (2004). Complexity and integration in sexual ornamentation: An example with carotenoid and melanin plumage pigmentation. *Journal of Evolutionary Biology* **17**, 1355–1366.

Badyaev, A. V., Foresman, K. R., and Fernandes, M. V. (2000). Stress and developmental stability: Vegetation removal causes increased fluctuating asymmetry in shrews. *Ecology* **81**, 336–345.

Badyaev, A. V., Hill, G. E., Dunn, P. O., and Glen, J. C. (2001). Plumage color as a composite trait: Developmental and functional integration of sexual ornamentation. *American Naturalist* **158**, 221–235.

Badyaev, A. V., Foresman, K. R., and Young, R. L. (2005). Evolution of morphological integration: II. Developmental accommodation of stress-induced variation. *American Naturalist*, In press.

Baldwin, J. M. (1896). A new factor in evolution. *American Naturalist* **30**, 441–451.

Barton, N. H., and Charlesworth, B. (1984). Genetic revolutions, founder effects, and speciation. *Annual Review of Ecology and Systematics* **15**, 133–164.

Bateson, P., Barker, B., Clutton-Brock, T., Deb, D., D'Udine, B., Foley, R. A., Gluckman, P., Godfrey, K., Kirkwood, T. B., Lahr, M. M., McNamara, J., Metcalfe, N. B., Monaghan, P., Spencer, H. G., and Sultan, S. E. (2004). Developmental plasticity and human health. *Nature* **430**, 419–421.

Behera, N., and Nanjundiah, V. (1995). An investigation into the role of phenotypic plasticity in evolution. *Journal of Theoretical Biology* **172**, 225–232.

Belyaev, D. K., and Borodin, P. M. (1982). The influence of stress on variation and its role in evolution. *Biologisches Zentrablatt* **100**, 705–714.

Bennington, C. C., and McGraw, J. B. (1996). Environment-dependence of quantitative genetic parameters in *Impatiens pallida*. *Evolution* **50**, 1083–1097.

Berg, R. L. (1960). The ecological significance of correlation pleiades. *Evolution* **14**, 171–180.

Bergman, A., and Siegal, M. L. (2003). Evolutionary capacitance as a general feature of complex gene networks. *Nature* 424, 549–552.

Bijlsma, R., and Loeschcke, V. (1997). *Environmental Stress, Adaptation and Evolution*. Basel, Switzerland: Birkhauser Verlag.

Bjedov, I., Tenaillon, O., Gerard, B., Souza, V., Denamur, E., Radman, M., Taddei, F., and Matic, I. (2003). Stress-induced mutagenesis in bacteria. *Science* 300, 1404–1409.

Bonner, J. T. (1965). *Size and Cycle*. Princeton, NJ: Princeton University Press.

Bradshaw, A. D., and Hardwick, K. (1989). Evolution and stress: Genotypic and phenotypic components. *Biological Journal of Linnean Society* 37, 137–155.

Britton, N. F., Stickland, T. R., and Franks, N. R. (1998). Analysis of ant foraging algorithms. *Journal of Biological Systems* 6, 315–336.

Bryant, E. H., and Meffert, L. M. (1988). Effects of experimental bottleneck on morphological integration in the housefly. *Evolution* 42, 698–707.

Cairns, J., Overbaugh, J., and Miller, S. (1988). The origin of mutants. *Nature* 335, 142–145.

Caropale, L. H. 1999. Chance favors the prepared genome. *Annals New York Academy of Sciences* 870, 1–21.

Carter, D. R. (1987). Mechanical loading history and skeletal biology. *Journal of Biomechanics* 20, 1095–1109.

Chapin, F. S., III., Autumn, K., and Pugnaire, F. (1993). Evolution of suites of traits in response to environmental stress. *American Naturalist* 142, S78–S92.

Chapman, L. J., Galis, F., and Shinn, J. (2000). Phenotypic plasticity and the possible role of genetic assimilation: Hypoxia-induced trade-offs in the morphological traits of an African cichlid. *Ecology Letters* 3, 387–393.

Cheverud, J. M. (1982). Phenotypic, genetic, and environmental morphological integration in the cranium. *Evolution* 36, 499–516.

Cheverud, J. M., Vaughn, T. T., Pletscher, L. S., King-Ellison, K., Bailiff, J., Adams, E., Erickson, C., and Bonislawski, A. (1999). Epistasis and the evolution of additive genetic variance in populations that pass through a bottleneck. *Evolution* 53, 1009–1018.

Chipman, A. (2001). Developmental exaptation and evolutionary change. *Evolution and Development* 3, 299–301.

Chow, K. L., and Chan, K. W. (1999). Stress-induced phenocopy of C. elegans defines functional steps of sensory organ differentiation. *Developmental Growth and Differentiation* 41, 629–637.

Clark, A. G., and Fucito, C. D. (1998). Stress tolerance and metabolic response to stress in *Drosophila melanogaster*. *Heredity* 81, 514–527.

Clarke, G. M., and McKenzie, J. A. (1992). Coadaptation, developmental stability, and fitness of insecticide resistance genotypes in the Australian sheep blowfly, *Lucilia cuprina*: A review. *Acta Zoologica Fennica* 191, 107–110.

Couzin, I. D., and Franks, N. R. (2003). Self-organized lane formation and optimized traffic flow in army ants. *Proceedings of Royal Society London: Biological Sciences* 270, 139–146.

Creel, S., Creel, N. M., and Monfort, S. L. (1996). Social stress and dominance. *Nature* 379, 212.

Denver, R. J. (1999). Evolution of the corticotropin-releasing hormone signaling system and its role in stress-induced phenotypic plasticity. *Annals of New York Academy of Sciences* 897, 46–53.

Dickinson, W. J. (1988). On the architecture of regulatory systems: Evolutionary insights and implications. *Bioessays* 8, 204–208.

Dvornyk, V., Vinogradova, O., and Nevo, E. (2002). Long-term microclimatic stress cases rapid adaptive radiation of kaiABC clock gene family in a cyanobacterium, *Nostoc linckia*, from "Evolution Canyons" I and II, Israel. *Proceedings of the National Academy of Sciences of the United States of America* 99, 2082–2087.

Emlen, J. M., Freeman, D. C., and Graham, J. H. (2003). The adaptive basis of developmental instability: A hypothesis and its implication. In *Developmental Instability: Causes and Consequences* (M. Polak, ed.), pp. 51–61. Oxford, England: Oxford University Press.

Eshel, I., and Matessi, C. (1998). Canalization, genetic assimilation, and preadaptation: A quantitative genetic model. *Genetics* **149**, 2119–2133.

Fear, K. K., and Price, T. (1998). The adaptive surface in ecology. *Oikos* **82**, 440–448.

Fiumara, C., and Badyaev, A. V. (1998). Ecological and historical patterns of morphological integration: Stress-induced variation and evolutionary divergence in shrew mandibles. *Honour Thesis, Organismal Biology Program*. Missoula, MT: University of Montana.

Foresman, K. R., and Badyaev, A. V. (2003). Developmental instability and the environment: Why are some species better indicators of stress than others? In *Advances in the Biology of Shrews* (J. F. Meritt, J. Kirkland, G. L. Kirkland, and R. K. Rose, eds.). Pittsburgh, PA: Carnegie Museum of Natural History.

Gans, C. (1979). Momentary excessive construction as the basis for protoadaptation. *Evolution* **33**, 227–233.

Gavrilets, S. (2004). *Fitness Landscapes and the Origin of Species*. Princeton, NJ: Princeton.

Gavrilets, S., and Scheiner, S. (1993). The genetics of phenotypic plasticity. V. Evolution of reaction norm shape. *Journal of Evolutionary Biology* **6**, 31–48.

Gibson, G., and Wagner, G. P. (2000). Canalization in evolutionary genetics: A stabilizing theory? *Bioessays* **22**, 372–380.

Goho, S., and Bell, G. (2000). Mild environmental stress elicits mutations affecting fitness in Chlamydomonas. *Proceedings of Royal Society London: Biological Sciences* **267**, 123–129.

Gold, P. E., and McGaugh, J. L. (1978). Neurobiology and memory: Modulators correlates and assumptions. In *Brain and Learning* (T. Teilor, ed.), pp. 93–104. Stanford, CT: Greylock.

Goldschmidt, R. B. (1940). *The Material Basis of Evolution*. New Haven, CT: Yale University Press.

Goymann, W., and Wingfield, J. C. (2004). Allostatic load, social status and stress hormones: The costs of social status matter. *Animal Behaviour* **67**, 591–602.

Graham, J. H., Fletcher, C., Tigue, J., and McDonald, M. (2000). Growth and developmental stability of *Drosophila melanogaster* in low frequency magnetic fields. *Bioelectromagnetics* **21**, 465–472.

Greenberg, N., Carr, J. A., and Summers, C. H. (2001). Ethological causes and consequences of the stress response. *Integrative and Comparative Biology* **42**, 508–516.

Guex, J. (2001). Environmental stress and atavism in ammonoid evolution. *Eclogae Geologicae Helvetiae* **94**, 321–328.

Hadany, L., and Beker, T. (2003). Fitness-associated recombination on rugged adaptive landscapes. *Journal of Evolutionary Biology* **16**, 862–870.

Hall, B. K. (1986). The role of movement and tissue interactions in the development and growth of bone and secondary cartilage in the clavicle of the embryonic chick. *Journal of Embryology and Experimental Morphology* **93**, 133–152.

Heil, M., Greiner, S., Meimberg, H., Kruger, R., Noyer, J. -L., Heubl, G., Linsenmair, K. E., and Boland, W. (2004). Evolutionary change from induced to constitutive expression of an indirect plant resistance. *Nature* **430**, 205–208.

Hermission, J., and Wagner, G. P. (2004). The population genetic theory of hidden variation and genetic robustness. *Genetics*, **168**, 2271–2284.

Hermission, J., Hansen, T. F., and Wagner, G. P. (2003). Epistasis in polygenic traits and the evolution of genetic architecture under stabilizing selection. *American Naturalist* **161**, 708–734.

Hinton, S. J., and Nolan, G. E. (1980). How learning can guide evolution. *Complex Systems* **1**, 495–502.

Hoffmann, A. A., and Hercus, M. J. (2000). Environmental stress as an evolutionary force. *Bioscience* **50**, 217–226.

Hoffmann, A. A., and Parsons, P. A. (1997). *Extreme Environmental Change and Evolution*. Cambridge, England: Cambridge University Press.

Hoffmann, A. A., and Woods, R. (2001). Trait variability and stress: Canalization, developmental stability and the need for a broad approach. *Ecology Letters* **4**, 97–101.

Hoffmann, A. A., Hallas, R. J., Dean, J. A., and Schiffer, M. (2003). Low potential for climatic stress adaptation in a rainforest *Drosophila* species. *Science* **231**, 100–102.

Holloway, G. J., Povey, S. R., and Sibly, R. M. (1990). The effect of new environment on adapted genetic architecture. *Heredity* **64**, 323–330.

Holloway, G. J., Crocker, H. J., and Callaghan, A. (1997). The effects of novel and stressful environments on trait distribution. *Functional Ecology* **11**, 579–584.

Howarth, F. G. (1993). High-stress subterranean habitats and evolutionary change in cave-inhabiting arthropods. *American Naturalist* **142**, s65–s77.

Huether, G. (1996). The central adaptation syndrome: Psychosocial stress as a trigger for adaptive modifications of brain structure and brain function. *Progress in Neurobiology* **48**, 569–612.

Huey, R. B., Hertz, P. E., and Sinervo, B. (2003). Behavioral drive versus behavioral inertia in evolution: A null model approach. *American Naturalist* **161**, 357–366.

Imasheva, A. G. (1999). Environmental stress and genetic variation in animal populations. *Russian Journal of Genetics* **35**, 343–351.

Jablonka, E., and Lamb, M. J. (1995). *Epigenetic Inheritance and Evolution: The Lamarckian Dimension*. Oxford, England: Oxford University Press.

Jablonka, E., Oborny, B., Molnar, I., Kisdi, E., Hofbauer, J., and Czaran, T. (1995). The adaptive advantage of phenotypic memory in changing environments. *Philosophical Transactions of the Royal Society of London B Biological Sciences* **350**, 133–141.

Johnson, E. O., Kamilaris, T. C., Chrousos, G. P., and Gold, P. W. (1992). Mechanisms of stress: A dynamic overview of hormonal and behavioral homeostasis. *Neuroscience and Biobehavioral Reviews* **16**, 115–130.

Juste, J., Lopez-Gonzalez, C., and Strauss, R. E. (2001). Analysis of asymmetries in the African fruit bats *Eidolon helvum* and *Rousettus egyptiacus* (Mammalis: Magachiroptera) from the islands of the Gulf of Guinea. II. Integration and levels of multivariate fluctuating asymmetry across a geographical range. *Journal of Evolutionary Biology* **14**, 672–680.

Katz, R. J., Roth, K. A., and Carrol, B. J. (1981). Acute and chronic stress effects on open field activity in the rats: Implication for a model of depression. *Neuroscience and Biobehavioral Reviews* **5**, 247–251.

Kawata, M. (2002). Invasion of vacant niches and subsequent sympatric speciation. *Proceedings of Royal Society London: Biological Sciences* **269**, 55–63.

Keller, L. F., Grant, P. R., Grant, B. R., and Petren, K. (2002). Environmental conditions affect the magnitude of inbreeding depression in survival of Darwin's finches. *Evolution* **56**, 1229–1239.

Kirschner, M., and Gerhart, J. (1998). Evolvability. *Proceedings of the National Academy of Sciences of the United States of America* **95**, 8420–8427.

Kishony, R., and Leibler, S. (2003). Environmental stresses can alleviate the average deleterious effect of mutations. *Journal of Biology* **2**, 14–19.

Klingenberg, C. P., and McIntyre, G. S. (1998). Geometric morphometrics of developmental instability: Analyzing patterns of fluctuating asymmetry with Procrustes methods. *Evolution* **53**, 1363–1375.

Klingenberg, C. P., and Zaklan, S. D. (2000). Morphological integration among developmental compartments in the *Drosophila* wing. *Evolution* **54**, 1273–1285.

Klingenberg, C. P., Badyaev, A. V., Sowry, S. M., and Beckwith, N. J. (2001). Inferring developmental modularity from morphological integration: Analysis of individual variation and asymmetry in bumblebee wings. *American Naturalist* **157**, 11–23.

Kondrashov, A. S., and Houle, D. (1994). Genotype-environment interactions and the estimation of the genomic mutation rate in *Drosophila melanogaster*. *Proceedings of the Royal Society of London Series B* **258**, 221–227.

Lachmann, M., and Jablonka, E. (1996). The inheritance of phenotypes: An adaptation to fluctuating environments. *Journal of Theoretical Biology* **181**, 1–9.

Lauter, N., and Doebley, J. (2002). Genetic variation for phenotypically invariant traits detected in teosine: Implications for the evolution of novel forms. *Genetics* **160**, 333–342.

Leamy, L. J. (1993). Morphological integration of fluctuating asymmetry in the mouse mandible. *Genetica* **89**, 139–153.

Leips, J., and MacKay, T. F. C. (2000). Quantitative trait loci for life span in *Drosophila melanogaster*: Interactions with genetic background and larval density. *Genetics* **160**, 333–342.

Levine, S., Haltmeyer, G. C., Karas, G. G., and Denenberg, V. H. (1967). Physiological and behavioral effects of infantile stimulation. *Physiology and Behavior* **2**, 55–63.

Levine, S., Coe, C. H., and Wiener, S. (1989). Psychoneuroendocrinology of stress: A psychobiological perspective. In *Psychoendocrinology* (F. R. Brush and S. Levine, eds.), pp. 341–380. San Diego, CA: Academic Press.

Levins, R. (1963). Theory of fitness in a heterogeneous environment. II. Developmental flexibility and niche selection. *American Naturalist* **97**, 75–90.

Lipson, H., Pollack, J. B., and Suh, N. P. (2002). On the origin of modular variation. *Evolution* **56**, 1549–1556.

Lively, C. M. (1986). Canalization versus developmental conversion in a spatially variable environment. *American Naturalist* **128**, 561–572.

Mabee, P. M., Olmstead, K. L., and Cubbage, C. C. (2000). An experimental study of intraspecific variation, developmental timing, and heterochrony in fishes. *Evolution* **54**, 2091–2106.

Masel, J. (2004). Genetic assimilation can occur in the absence of selection for the assimilating phenotype, suggesting a role for the canalization heuristic. *Journal of Evolutionary Biology* **17**, 1106–1110.

Masel, J., and Bergman, A. (2003). The evolution of evolvability properties of the yeast prion [*PSI+*]. *Evolution* **57**, 1498–1512.

McClintock, B. (1984). The significance of responses of the genome to challenge. *Science* **226**, 792–801.

McEwen, B. S., and Sapolsky, R. M. (1995). Stress and cognitive function. *Current Opinion in Neurobiology* **5**, 205–216.

McEwen, B. S., and Wingfield, J. C. (2003). The concept of allostasis in biology and biomedicine. *Hormones and Behavior* **43**, 2–15.

McGaugh, J. L., Martinez, J. L., Jensen, R. A., Hannan, T. J., Vasquez, B. J., Messing, R. B., Liang, K. C., Brewton, K. C., and Spiehier, V. (1982). Modulation of memory storage by treatments affecting peripheral catecholamines. In *Neuronal Plasticity and Memory Formation* (C. Ajmone-Marsan and H. Matthies, eds.), pp. 129–141. New York: Raven Press.

Meaney, M. J. (2001). Maternal care, gene expression, and the transmission of individual differences in stress reactivity across generations. *Annual Reviews of Neurosciences* **24**, 1161–1192.

Meiklejohn, C. D., and Hartl, D. L. (2002). A single mode of canalization. *Trends in Ecology and Evolution* **17**, 468–473.

Milton, C. C., Huynh, B., Batterham, P., Rutherford, S. L., and Hoffmann, A. A. (2003). Quantitative trait symmetry independent of Hsp90 buffering: Distinct modes of genetic canalization and developmental stability. *Proceedings of the National Academy of Science USA* **100**, 13396–13401.

Mousseau, T. A., and Fox, C. W. (1998). The adaptive significance of maternal effects. *Trends in Ecology and Evolution* **13**, 403–407.

Muller, G. B. (2003). Embryonic motility: Environmental influences and evolutionary innovation. *Evolution and Development* **5**, 56–60.

Nanjundiah, V. (2003). Phenotypic plasticity and evolution by genetic assimilation. In *Origination of Organismal Form: Beyond the Gene in Developmental and Evolutionary Biology* (G. B. Muller and S. Newman, eds.), pp. 245–263. Cambridge, MA: The MIT Press.

Nedelcu, A. M., and Michod, R. E. (2003). Sex as a response to oxidative stress: The effect of antioxidants on sexual induction in a facultatively sexual lineage. *Proceedings of Royal Society London: Biological Sciences* **270**, S136–S139.

Newman, S. A., and Muller, G. B. (2000). Epigenetic mechanisms of character origination. *Journal of Experimental Zoology* **288**, 304–314.

Neyfakh, A. A., and Hartl, D. L. (1993). Genetic control of the rate of embryonic development: Selection for faster development at elevated temperatures. *Evolution* **47**, 1625–1631.

Nicolakakis, N., Sol, D., and Lefebvre, L. (2003). Behavioural flexibility predicts species richness in birds, but not extinction risk. *Animal Behaviour* **65**, 445–452.

Nijhout, H. F. (2002). The nature of robustness in development. *Bioessays* **24**, 553–563.

Olson, E. C., and Miller, R. L. (1958). *Morphological Integration*. Chicago: University of Chicago Press.

Oyama, S. (2000). *The Ontogeny of Information: Developmental Systems and Evolution*. Durham, NC: Duke University Press.

Parsons, P. A. (1993). The importance and consequences of stress in living and fossil populations: From life-history variation to evolutionary change. *American Naturalist* **142**, s5–s20.

Parsons, P. A. (1994). Morphological stasis: An energetic and ecological perspective incorporating stress. *Journal of Theoretical Biology* **171**, 409–414.

Piersma, T., and Drent, J. (2003). Phenotypic flexibility and the evolution of organismal design. *Trends in Ecology and Evolution* **18**, 228–233.

Price, T. D., Qvarnstrom, A., and Irwin, D. E. (2003). The role of phenotypic plasticity in driving genetic. *Evolution* **270**, I1433–1440.

Prody, C. A., Dreyfus, P., Zamir, R., Zakut, H., and Soreq, H. (1989). *De novo* amplification within a "silent" human cholinesterase gene in a family subjected to prolonged exposure to organophosphorus insecticides. *Proceedings of the National Academy of Science USA* **86**, 690–694.

Raberg, L., Grahn, M., Hasselquist, D., and Svensson, E. (1998). On the adaptive significance of stress-induced immunosuppression. *Proceedings of Royal Society of London Series B* **265**, 1637–1641.

Rice, S. H. (1998). Evolution of canalization and the breaking of von Baer's laws: Modeling the evolution of development with epistasis. *Evolution* **52**, 647–656.

Rice, S. H. (2004). Developmental associations between traits: Covariance and beyond. *Genetics* **166**, 513–526.

Rieseberg, L. H., Raymond, O., Rosenthal, D. M., Lai, Z., Livingstone, K., Nakazato, T., Durphy, J. L., Schwarzbach, A. E., Donovan, L. A., and Lexer, C. (2003). Major ecological transitions in wild sunflowers facilitated by hybridization. *Science* **301**, 1211–1216.

Robinson, B. W., and Dukas, R. (1999). The influence of phenotypic modifications on evolution: The Baldwin effect and modern perspectives. *Oikos* **85**, 582–589.

Roth, G., and Wake, D. B. (1985). Trends in the functional morphology and sensorimotor control of feeding behavior in salamanders: an example of the role of internal dynamics in evolution. *Acta Biotheoretica* **34**, 175–192.

Ruden, D. M., Garfinkel, M. D., Sollars, V. E., and Lu, X. (2003). Waddington's widget: Hsp90 and the inheritance of acquired characters. *Seminars in Cell and Developmental Biology* **14**, 301–310.

Rutherford, S. L. (2000). From genotype to phenotype: Buffering mechanisms and the storage of genetic information. *Bioessays* **22**, 1095–1105.

Rutherford, S. L., and Lindquist, S. (1998). Hsp90 as a capacitator for morphological evolution. *Nature* **396**, 336–343.

Ruvinsky, A. O., Lobkiv, Y. I., and Belyaev, D. K. (1983). Spontaneous and induced activation of genes affecting the phenotypic expression of glucose 6-phosphate dehydrogenase in *Daphnia pulex* II. Glucose induced changes in the electrophoretic mobility of G6PD. *Molecular and General Genetics* **189**, 490–494.

Schandorff, S. (1997). Developmental stability and skull lesions in the harbour seal (*Phoca vitulina*) in the 19th and 20th centuries. *Annales Zoologi Fennici* **34**, 151–166.

Scharloo, W. (1991). Canalization: genetic and developmental aspects. *Annual Reviews in Ecology and Systematics* **22**, 65–93.

Schlichting, C. D., and Pigliucci, M. (1998). *Phenotypic Evolution: A Reaction Norm Perspective*. Sunderland, MA: Sinauer Associates.

Schlichting, C. D., and Smith, H. (2002). Phenotypic plasticity: Linking mechanisms with evolutionary outcomes. *Evolutionary Ecology* **16**, 189–211.

Schmalhausen, I. I. (1949). *Factors of Evolution*. Philadelphia: Blakiston.

Shapiro, J. A. (1992). Natural genetic engineering in evolution. *Genetica* **86**, 99–111.

Shors, T. J., and Servatious, R. J. (1997). The contribution of stressor intensity, duration, and context to the stress-induced facilitation of associative learning. *Neurobiology of Learning and Memory* **67**, 92–96.

Sibly, R. M., and Calow, P. (1989). A life-cycle theory of responses to stress. *Biological Journal of Linnean Society* **37**, 101–116.

Siegal, M. L., and Bergman, A. (2002). Waddington's canalization revisited: Developmental stability and evolution. *Proceedings of the National Academy of Sciences of the United States of America* **99**, 10528–10532.

Siegel, M. I., and Doyle, W. J. (1975). The differential effects of prenatal and postnatal audiogenic stress on fluctuating dental asymmetry. *Journal of Experimental Zoology* **191**, 211–214.

Simons, A. M., and Johnston, M. O. (1997). Developmental instability as a bet-hedging strategy. *Oikos* **80**, 401–405.

Simpson, G. G. (1953). *The Major Features of Evolution*. New York: Simon and Schuster.

Smith-Gill, S. J. (1983). Developmental plasticity: Developmental conversion versus phenotypic modulation. *American Zoologist* **23**, 47–55.

Smits, J. D., Witte, F., and Van Veen, F. G. (1996). Functional changes in the anatomy of the pharyngeal jaw apparatus of *Astatoreochromis alluaudi* (Pisces, Cichlidae), and their effects on adjacent structures. *Biological Journal of Linnean Society* **59**, 389–409.

Sniegowski, P., Gerrish, P., Johnson, T., and Shever, A. (2000). The evolution of mutation rates: Separating causes from consequences. *Bioessays* **22**, 1057–1066.

Sowry, S. M., and Badyaev, A. V. (1999). Does developmental integration mediate trait response to environmental stress? Experiment with bumblebee (*Bombus empatiens*) wing venation pattern and CO2 exposure. *Honour Thesis. Organismal Biology Program*. Missoula, MT: University of Montana.

Stanton, M. L., Roy, B. A., and Thiede, D. A. (2000). Evolution in stressful environments. I. Phenotypic variability, phenotypic selection, and response to selection in five distinct environmental stresses. *Evolution* **54**(1), 93–111.

Stearns, S. C., and Kawecki, T. J. (1994). Fitness sensitivity and the canalization of life-history traits. *Evolution* **48**, 1438–1450.

Szafraniec, K., Borts, R. H., and Korona, R. (2001). Environmental stress and mutational load in diploid strains of the yeast *Saccharomyces cerevisiae*. *Proceedings of the National Academy of Sciences USA* **98**, 1107–1112.

Veenema, A. H., Meijer, O. C., de Kloet, E. R., Koolhaas, J. M., and Bohus, B. G. (2003). Differences in basal and stress-induced HPA regulation of wild house mice selected for high and low aggression. *Hormones and Behavior* **43**, 197–204.

Waddington, C. H. (1941). Evolution of developmental systems. *Nature* **147**, 108–110.

Waddington, C. H. (1953). The genetic assimilation of an acquired character. *Evolution* 7, 118–126.

Wagner, A. (2003). Risk management in biological evolution. *Journal of Theoretical Biology* **225**, 45–57.

Wagner, G. P. (2001). *The Character Concept in Evolutionary Biology*. Academic Press.

Wagner, G. P., and Mezey, J. G. (2000). Modeling the evolution of genetic architecture: A continuum of alleles model with pairwise AxA epistasis. *Journal of Theoretical Biology* **203**, 163–175.

Wagner, G. P., Booth, G., and Bacheri-Chaichian, H. (1997). A population genetic theory of canalization. *Evolution* **51**, 329–347.

Wagner, G. P., and Mezey, J. G. (2004). The role of genetic architecture constraints for the origin of variational modularity. In *Modularity in Development and Evolution* (G. Schlosser and G. P. Wagner, eds.), pp. 338–358. Chicago: Chicago University Press.

Wagner, G. P., Mezey, J. G., and Callabretta, R. (2005). Natural selection and the origin of modules. In *Modularity: Understanding the Development and Evolution of Complex Systems*. Cambridge, MA: MIT Press.

Walker, I. (1979). The mechanical properties of proteins determine the laws of evolutionary change. *Acta Biotheoretica* **28**, 239–282.

Waxman, D., and Peck, J. R. (1998). Pleiotropy and the preservation of perfection. *Science* **279**, 1210–1213.

Weaver, I. C. G., *et al.* (2004). Epigenetic programming by maternal behavior. *Nature Neuroscience* **7**, 847–854.

Wessler, S. R. (1996). Plant retrotransposons: Turned on by stress. *Current Biology* **6**, 959–961.

West-Eberhard, M. J. (2003). *Developmental Plasticity and Evolution*. Oxford, England: Oxford University Press.

Wills, C. (1983). The possibility of stress-triggered evolution. *Lecture Notes in Biomathematics* **53**, 299–312.

Wingfield, J. C. (2003). Control of behavioural strategies for capricious environments. *Animal Behaviour* **66**, 807–816.

Wingfield, J. C., and Sapolsky, R. M. (2003). Reproduction and resistance to stress: When and how. *Journal of Neuroendocrinology* **15**, 711–724.

Wright, B. E. (2000). Minireview: A biochemical mechanism for nonrandom mutations and evolution. *Journal of Bacteriology* **182**, 2993–3002.

Zakharov, V. M. (1993). Appearance, fixation and stabilization of environmentally induced phenotypic changes as a microevolutionary event. *Genetica* **89**, 227–234.

Zakharov, V. M., Pankakoski, E., Sheftel, B. I., Prltonen, A., and Hanski, I. (1991). Developmental stability and population dynamics in the common shrew, *Sorex araneus*. *American Naturalist* **138**, 797–810.

CHAPTER 14

Environmentally Contingent Variation: Phenotypic Plasticity and Norms of Reaction

SONIA E. SULTAN* AND STEPHEN C. STEARNS†

*Biology Department, Wesleyan University, Middletown, Connecticut, USA
†Department of Ecology and Evolutionary Biology, Yale University, New Haven, Connecticut, USA

INTRODUCTION

Environmentally influenced variation in phenotypic expression or *phenotypic plasticity* is a fundamental property of organisms with consequences for developmental and ecological genetics, evolutionary biology, population and community ecology, conservation biology, and medicine (Lewontin, 1985; Stearns, 1989; Scheiner, 1993; Schlichting and Pigliucci, 1998; Tollrian and Harvell, 1999; Gilbert, 2001; Pigliucci, 2001; Lummaa, 2003; Sultan, 2003; and references therein). Phenotypic plasticity is "the rule rather than the exception" (Gilbert and Bolker, 2003), because genetic and environmental information interact to shape virtually all aspects of the organism's development and function. As our understanding of DNA enters its "mid-life crisis" (Angier, 2003), it is appropriate to restore the genetic material to its context: in an organism with integrative transduction systems that construct the phenotype. With this recognition comes an altered understanding of the genotype as coding for a set of phenotypic potentialities rather than a single, specific outcome. (Cases in which phenotypic expression is insensitive to either genetic or environmental variation are known as "canalization" [see previous chapter by Gibson]; here the phenotypic outcomes are narrowly constrained). To understand phenotypic diversity and its evolution, it helps to see organisms as systems of genotype–environment integration rather than to focus solely on DNA sequences or phenotypic trait states. In addition, to understand natural selection, it is logical to focus on the key phenotypic product of genotype–environment interaction—fitness—in the natural environment.

Although phenotypic plasticity is the general case, it is tricky to study empirically, and it is currently described by an imprecise and often inconsistent terminology. In this chapter we begin with definitions and distinctions that identify ways to conceive of the environment and of plasticity. We then discuss these issues: (1) What causes plasticity? What types of genetic architecture and signal transduction mechanisms underpin plastic responses to environment? (2) What is the nature of genetic variation for these responses? (3) How do plastic responses, expressed within a single generation, interact with the slowly changing developmental frameworks characteristic of entire clades? (4) What are the consequences of plasticity for populations and communities? We have aimed to write a constructive guide to key issues of phenotypic variation, rather than a comprehensive review of a vast field. We also hope that this chapter can serve as a research agenda for this rapidly unfolding area.

I. PLASTICITY CONCEPTS

The concept of phenotypic plasticity encompasses many phenomena usefully subdivided by distinctions to which every researcher should be sensitive. Plasticity refers broadly to all aspects of the phenotype in which expression varies as a result of variation in the environment. This concept includes virtually all traits—morphological, behavioral, physiological, reproductive, epidemiological—and virtually all environmental factors, biotic and abiotic. The organism's expressed phenotype itself can shape its biotic and abiotic environment, creating an iterative, dynamic feedback (Lewontin, 2001; Gray, 1992). To understand the plastic response, we also need to consider the physical and biochemical state of the organism at the cell, tissue, and whole-body levels, including hormonal integration, cell–cell interactions, and the mechanisms underpinning dynamic coadaptive adjustment among traits. Thus plastic responses to the environment depend on the state of the organism, itself the product of prior genotype and environment interactions. Importantly, the internal environment often reflects events in the previous generation as well as earlier in the life of the organism.

A. Specific Types of Plasticity

One can, and some have, distinguished subsets of phenotypic plasticity based on ecological categories (e.g., plasticity for shade responses in plants), on qualitative trait categories (life-history plasticity, physiological plasticity), or on the timescale, specific mechanism, or occurrence pattern of plastic responses (Slobodkin, 1968; Piersma and Drent, 2003). Such distinctions can help to point to shared mechanisms or interpretations and are best drawn in the context of specific research questions rather than *a priori*. For instance, one might choose to distinguish irreversible developmental plasticity from rapid metabolic or physiological responses that occur on a shorter timescale, but the precise boundaries will necessarily be system specific. A case in point is that of animal behavior. Some see behavior as continuous with plasticity and doubt the utility of distinguishing between the two. They might regard behavior as a particular type of plasticity consisting of iterated, dynamic reaction norms for short-term responses, mediated by a central nervous system. Acclimation, a term from physiological ecology, is another aspect of environmental response that one could include within a broad concept of plasticity. Others prefer to limit the concept of plasticity to unique, irreversible developmental events in the lifetime of an individual occurring on a timescale that is an appreciable fraction of that lifetime.

We think it wise not to make a decision on this issue until one is confronted with a particular problem in need of solution. The problems should shape the categories used, categories judged for their utility in helping to solve the problem. The categories should not shape the problems. In some cases, for example, distinguishing plasticity from behavior may help. In others, viewing them as elements of a continuum may help. Rather than waste time on arguing about whether it is a good idea to adopt this distinction, we prefer to remain aware that it is a possible distinction and to reserve judgment on its utility until confronted with a concrete research problem. When we have more experience on this issue, it may be possible to recommend classes of problems in which one or the other stance is more productive. However, that is not yet the case.

B. Reaction Norms

Phenotypic plasticity can be studied by means of several types of phenotypic response data or "norms of reaction." Here we build up an array of plasticity concepts starting with the most elemental, the "genotypic reaction norm" (Woltereck, 1909; Schmalhausen, 1949; examples in Sultan and Bazzaz, 1993), then passing through a series of classes of reaction norms that integrate increasingly complex effects, and ending with a comment on the relationship of reaction norms to behavior.

The reaction norm is most precisely conceptualized as *the phenotypic expression of a given genotype for a single trait at several specified levels of a particular environmental factor*, for example, the reaction norm of a particular clone of *Daphnia pulex* for head spine length as a function of a defined range of ambient temperatures. As with all reaction norms, both the trait of interest and the environmental variable must be precisely defined, so that comparable measurements can be made on the different phenotypes produced by genetic replicates in different experimental environments. For instance, because developmental rates often vary with environment, traits may be defined with respect to either absolute age or specified ontogenetic stages (Gedroc *et al.*, 1996). In the preceding example, head spine would be measured in *Daphnia* individuals at a particular instar at each temperature. Environmental factors should be carefully controlled so as to avoid covarying changes that can confound interpretation.

From this concept we can build up to several others that are experimentally useful, particularly in organisms for which genotypic replicates cannot be readily obtained. If the study organism has many offspring, we can estimate "family mean reaction norms" by measuring several sibling offspring at each environmental level (Gebhardt and Stearns, 1992). Such family mean data provide the best estimate of reaction norms for organisms that cannot be cloned and are particularly robust when siblings are inbred full sibs (e.g., Gupta and Lewontin, 1992).

When familial relationship cannot be determined in the sample, one may still wish to compare the plastic reactions of two populations based on a random sample of individuals from each population split into subgroups and measured at a series of environmental levels. Such data represent "undefined population mean reaction norms." It is also possible to build up population mean reaction norms from a collection of genotypic reaction norms, then termed "genotype-based population mean reaction norms," or from a collection of family mean reaction norms, then termed "family-based population mean reaction norms."

Such distinctions, though cumbersome, are critical, because the essential aim of reaction norm analysis is to establish clarity about the genotype–phenotype relationship. Another approach taken in the literature is to reserve the term *reaction norm* for genotypic data and refer to "plasticity patterns" when the sample is genetically undefined. Whatever approach is taken, it is essential to make clear the type of data being presented.

C. Parental Effect Reaction Norms (Cross-Generational Plasticity)

"Parental effect reaction norms" refer to effects of environmental variation on the parental generation as measured in the offspring. In many systems, including flowering plants and mammals, these cross-generational effects reflect the fact that the maternal body constitutes the offspring's early developmental environment. However, this type of plasticity may also arise from environmental effects on the maternal or paternal individuals or (in self-fertilizing organisms) on both simultaneously. In estimating "family mean" parental effect reaction norms in outcrossing organisms, maternal and paternal effects can be separately estimated with a full-sib/half-sib mating design, in which case the resulting data form a sheaf of surfaces in two dimensions. With the reaction norm measured over two generations, the offspring can be measured in a single environment, or alternatively at each level of environmental factor used to define the reaction norm in the parents, thus estimating a reaction surface. The evolutionary implications of these patterns depend on offspring dispersal and the consequent distribution of environmental states across as well as within generations, information that is rarely collected. "Grandparental effect reaction norms" are known in some systems (Reznick, 1981; Wulff, 1986); their importance depends on the relationship of the generation time of the study organism to the rate of environmental change. If the effect of the environment encountered by the grandparent on the grandchild depends on the environment encountered by the parent, as is probably the case, then one needs a design complicated and powerful enough to measure the grandparent–parent interaction effects.

D. Imprinted Reaction Norms

Many organisms have a sensitive period early in life during which basic patterns are determined irreversibly for critical traits. These early environmental effects on development condition the organism's physiological and behavioral responses to subsequent environmental inputs. For instance, the fetal nutritional environment in humans determines not just birth weight but the state of the circulatory and endocrine systems, affecting the individual's metabolic response to subsequent nutritional states (Barker, 1998; Lummaa, 2003). Thus the interaction of environmental events at different points in the life cycle may be an important aspect of plasticity and a key step toward understanding the real-life complexity of developmental trajectories. To study these effects, one could expose organisms to a series of levels of environmental factors for a defined period early in life and then measure their reactions to those environmental factors later in life. As with maternal-effect reaction norms, such measurements will result in a sheaf of surfaces defined by three axes: level of environmental factor (1) early in life, (2) during the rearing period, and (3) later in life. We call such responses "imprinted reaction norms." Depending on the genetic structure of the experiment, they may be elemental, family mean, or undefined population mean reaction norms. Examples include birth weight of a particular population of mosquitofish as a function of the salinity of the environment in the first week of the life of the mother (Stearns, 1980) and insulin resistance, hypertension, cardiovascular disease, and obesity late in life as a function of fetal nutrition in humans (Barker, 1998; Lummaa, 2003) and rats (Vickers *et al.*, 2000).

E. Iterated Reaction Norms

Reaction norms typically describe environmental reactions for a single trait that is expressed once in the life of an individual or that can be measured at a single meaningful point. However, some traits, such as clutch size in iteroparous organisms, are expressed repeatedly. Their repeated expression raises the possibility that the second expression depends on the first, as is known to be the case with clutch size in many birds—a type of cumulative parental effect. The third expression could in principle depend on both the first and on the second expression, and so forth. Reactions that depend on events occurring repeatedly in the life of the same organism are termed "iterated parental-effect reaction norms."

The gentle reader may wish to consider the insights that might be gained by attempting to measure the multidimensional objects carrying the label of "family mean, parental effect, imprinted, iterated reaction norms." The difficulty of measuring them detracts neither from the logic that indicates the plausibility of their existence nor from their importance as complicated elements of the genotype–phenotype map. It is interesting to contemplate how far out into the web of causation suggested by that label adaptation and identifiable genetic and environmental effects could be tested and applied. In addition, it is worth noting that simply uttering the label suggests the kinds of experimental controls needed to isolate particular effects.

F. Dynamic Reaction Norms

In some cases, reaction norms may be of interest for inherently dynamic traits, such as growth rates, or for traits that comprise continuous responses to temporally variable environments. Indeed, in indeterminately growing organisms such as plants and some animals, development itself is such a trait, and the norm of reaction must be conceived as a dynamically and continuously remodeled phenotype. This concept describes the continuous readjustment of root morphology as growing roots encounter different soil environments, the continuous replacement of surface proteins in the malaria pathogen, and the continuous updating of the status of the vertebrate immune system as organisms encounter, respond to, and "remember" pathogens.

II. THE GENETIC AND DEVELOPMENTAL BASIS OF PHENOTYPIC PLASTICITY

To understand phenotypic plasticity as variation subject to evolutionary change, we must trace the complex causal chains that link ecologically meaningful patterns of plastic response to underlying developmental pathways through specific genetic and environmental components. This is of course a demanding research program and one that is likely to require collaborative efforts across biological disciplines (Callahan *et al.*, 1997). Our success will depend both on careful studies of genetic cascades and signal transduction pathways and on a developmental paradigm that accommodates the real-world complexity of phenotypic expression (Gilbert, 2001). To some extent, the study of evolutionary development provides a model for comparisons of regulatory cascades and their genetic components. However, thus far this kind of "evo-devo" study has focused on the few highly conserved genes involved in major structural traits such as generation of body axes and location of appendages and sense organs (Carroll *et al.*, 2001; Wagner, 2000). The mechanisms of subtler aspects of development such as plasticity involve more transient, interacting regulatory events that incorporate the whole dynamic apparatus of environmental signal transduction (Carroll *et al.*, 2001).

At present, the precise developmental pathways underlying patterns of phenotypic plasticity are known in only a few cases (Pigliucci and Schmitt, 1999; Schlichting and Smith, 2002; Nijhout, 2003). Here we review two of the best-documented: morphogenetic response to shade in green plants and plasticity for metamorphic rate in amphibians (a third elegant example, butterfly wing spots, is discussed in a later section). These cases are particularly compelling because the adaptive consequences of the response are well understood. In choosing these two examples, we emphasize a common paradigm for studies of plasticity in diverse organisms. In our view, shaping such a general paradigm (and a correspondingly inclusive terminology) is an important step toward incorporating plasticity into both evolutionary and developmental biology. As mentioned previously, this is one reason for avoiding categorical distinctions in types of plasticity that are unlikely to apply well to all organisms (Schlichting and Smith, 2002).

Rather than add to parallel literatures on plant and animal plasticity that are necessarily both redundant and incongruent, one can specify within a general framework how developmental differences between these groups are likely to affect their expression of plasticity. For instance, animals have highly differentiated cells in relatively rigid developmental trajectories. In "higher" (bilaterian) animals, few tissues are expected to retain plasticity in adulthood, and the body plan is ordinarily fixed (Walbot, 1996). In contrast, plants express plasticity continually at both the cell and organ levels and continue to form and extend body parts throughout life. Indeed, the fate of differentiated cells in the adult plant body can change, for instance, in response to a wound (Walbot, 1996). In addition, each meristem on a plant body is capable of independent response to environmental signals, because there is no single central tissue analogous to the nervous system (Gilroy and Trewavas, 2001).

These differences lead to three distinct predictions: First, in contrast with plants, the kinds of plasticity expressed by animals will change during the life cycle (e.g., from developmental to physiological and behavioral). Second, plastic responses in animals may be more highly integrated at the whole-organism level because of joint nervous and endocrine control. Third, many structural traits will be highly canalized in animals but not in plants. Colonial organisms, fungi, and microorganisms will also have particular modes of plastic response that reflect both cellular constraints and emergent developmental repertoires, for instance, in lichens and biofilms. Recognizing these kinds of differences among organisms can inform our research questions and lead to a more sophisticated understanding of the nature of plasticity in general.

A. Photomorphogenetic Plasticity in Plants

Plants perceive impending shade by neighbors as a change in light spectral quality (Ballaré *et al.*, 1987). Perception of these spectral signals initiates "shade avoidance" responses such as internode elongation and suppressed branching (Casal and Smith, 1989) that have been shown to enhance fitness in dense plant stands (Schmitt *et al.*, 1995; Dudley and Schmitt, 1996). In contrast to other aspects of the plant environment, the perception mechanism for these adaptive plastic responses is both well understood and straightforward (Schmitt *et al.*, 1999; Gilroy and Trewavas, 2001). Most plants perceive and transduce this subtle cue with phytochrome photoreceptor molecules (Smith, 1995), which are sensitive detectors of small changes in the red to far-red (R to FR) ratio of incident light. Such changes directly modify the ratio of two photoconvertible forms of the phytochrome molecule (Pfr to P) in vegetative tissues. The morphogenetic impact of this chemical signal is remarkably rapid: Low Pfr/P in a growing stem induces growth promotion with a lag time of only 10 minutes and a reversal lag of 16 minutes. The magnitude of the stem elongation response depends on how low the R to FR light ratio is and for how long, as well as when in the diurnal photoperiod the low ratio occurs (Casal and Smith, 1989).

Specific loci that encode several phytochrome proteins have been sequenced in *Arabidopsis thaliana*, and photoreceptor-deficient mutants have been isolated that show disabled plastic responses to light (Callahan *et al.*, 1999; Pigliucci and Schmitt 1999). Studies of these mutants and of transgenic constructs have demonstrated the distinct though partly overlapping light-sensing and regulatory functions of these phytochrome genes, which constitute a "gene family" with both evolutionarily conserved and more rapidly evolving, variable functional domains (Schlichting and Smith, 2002). The phytochrome genes interact throughout development to regulate a host of ecologically important morphological and life-history traits (Schmitt *et al.*, 1999, and references therein). Similarly, two *Arabidopsis* blue light receptors encoded by paralogous genes initiate signaling pathways that are partly redundant, but because of differential sensitivities, act to cue phototropic response in different light habitats (Galen *et al.*, 2003). Gene duplication and regulatory diversification can thus enable the origin of novel plastic responses that, if adaptive, can be reinforced by natural selection. These fascinating studies show how genetic variation at the molecular level contributes to the epigenetic regulation of adaptive phenotypic response to environmental variability.

B. Adaptive Plasticity for Timing of Amphibian Metamorphosis

Many amphibians breed in temporary ponds and face the challenge of completing metamorphosis to the terrestrial form before their aquatic larval environment dries up. Because the rate of pond drying is unpredictably variable, plasticity for rate of development provides an adaptive solution to the fitness trade-off between risk of larval mortality and maximal size at metamorphosis (Newman, 1992). Several species, particularly those living in dry habitats, are able to accelerate metamorphosis in response to the onset of pond desiccation. Such species can show a continuous, graded acceleration in developmental rate that correlates to the rate of water loss in drying ponds (Denver 1997a,b, 1998; Denver *et al.*, 1998; Boorse and Denver, 2004).

How do individual animals recognize reduced pond volume, and how do they subsequently change their metamorphic rate? In the spadefoot toad *Spea hammondii*, a carefully controlled study eliminated several potential covariates of pond drying as cues for the plastic response: water temperature, chemical concentration, and physical interactions among more crowded tadpoles. Instead, the tadpoles sensed their imminent risk by perceiving both a reduction in swimming volume signaled by their own reduced movement and closer proximity to the water surface perceived either visually or through changes in water pressure (Denver *et al.*, 1998). These environmental signals initiate an increase in corticotropin-releasing hormone (CRH) in the brain, which in turn rapidly activates two distinct endocrine systems that control metamorphosis, the thyroid and the interrenal (Denver, 1997a, 1998). Thyroid hormone and other hormones released by these systems regulate gene expression to shape the suite of morphogenetic changes that comprise metamorphosis, from limb development to dramatic remodeling of the gut (Shi, 1994).

Since CRH is considered the primary stress neurohormone in vertebrates, its role as a mediator of complex environmental signals may be phylogenetically linked to various types of developmental plasticity, including for instance early parturition in mammals resulting from fetal stress (Denver, 1997b). Despite such profound evolutionary conservation, congeneric amphibian species can evolve to use different environmental cues to initiate accelerated metamorphosis, because diverse sensory cues may be transduced through the neuroendocrine system to elicit similar hormonal events (Denver, 1997a). Subtle changes in sensitivity to various environmental signals, shaped by selection depending on cue reliability and perception, would act upstream of CRH to create diverse cues for a similar adaptive plastic response. In another spadefoot toad, *Scaphiopus couchii,* the increased density of tadpoles in a drying pond functions as the metamorphic stimulus, probably through increased physical interactions among the more crowded individuals (Newman, 1994). In *Scaphiopus multiplicatus*, the environmental signal is directly hormonal: As pond volume declines, the density of brine shrimp increases, and tadpoles ingest more of these prey items with their high constituent levels of thyroid hormone (Pfennig, 1992). Although the molecular variation that underlies these cases remains to be fully understood, these studies demonstrate the interplay of conserved and variable transduction events that enable plastic responses to environmental challenges.

C. Mediation of Phenotypic Expression

Let us now summarize the picture of the developmental process that emerges from these and a host of molecular developmental studies (see also discussions by Nijhout, 1990, 2003; Stern, 2000; Stearns, 2003; Sultan, 2003). Development integrates both genetic and environmental information through a complex series of regulatory steps. External environmental signals are transduced into internal signals by the organism in specific ways that depend on its sensory and metabolic systems, typically by means of hormones and other signals. (In some cases, environmental conditions can directly affect the action of transcription factors, as in the case of heat shock proteins.) The effects of a given hormone or other signal vary depending on the type of cell, tissue, or organ as well as on developmental stage and on other environmental conditions (Voesenek and Blom, 1996; Gilroy and Trewavas, 2001); cells can differ in the threshold at which they initiate a response as well as in the response duration and magnitude (Gilroy and Trewavas, 2001). That response is accomplished in many cases by the cell signal (either directly or through a secondary messenger or "transducer") binding to nuclear DNA. (Some plastic responses may also result from phosphorylation and protein activation cascades rather than novel gene expression, something we suspect in principle without yet knowing in fact; H. F. Nijhout, personal communication, 2004.) This event elicits a regulatory cascade that consists of the expression of transcription factors that bind to the control sites of many genes, inducing expression patterns that upregulate or downregulate networks of subsequent gene activity with phenotypic consequences. These transcription factors can be either "promiscuous" or targeted in the genomic regions they affect, and their effects on gene expression can be cell-specific, tissue-specific, and stage-specific (Carroll *et al.*, 2001). The effects of mediating cell signals on gene expression are thus both phenotypically specific and highly pleiotropic (Ketterson and Nolan, 1999). Furthermore, many developmental processes are conditioned by multiple rather than single hormonal signals. Clearly, the regulatory systems that integrate these many internal and external signals involve substantial interaction and "cross-talk."

In sum, then, phenotypic expression is mediated by a layered network of regulatory events, from environmental signal transduction to coordinated effects on gene expression with specific morphogenetic and physiological results. One critical implication of our increasingly sophisticated understanding of developmental mechanisms is that the underlying pathways of plastic response are no different in nature from those of other developmental processes (see Schlichting and Smith, 2002). Plasticity may differ in the range of possible phenotypic outcomes for a given trait and organism, but not in the nature of its genetic mechanisms. As developmental genetics has matured, its initial goal of identifying genes with pronounced effects on certain developmental events has shifted to the goal of understanding gene regulation—that is, the complex interplay of signal transduction, epigenetic interactions, and physical and biochemical factors that regulates gene expression and consequently underlies the process of development (Carroll *et al.*, 2001). Because development in general depends on regulatory cascades that include signals from both external and internal environments, there is no reason to posit that phenotypic plasticity requires a distinct genetic architecture or regulatory process. Indeed, it has become increasingly clear that plasticity per se is not regulated by genes separate (and separately evolving) from those that otherwise affect traits (Scheiner and Lyman, 1991; Kliebenstein and Mitchell-Olds, 2002).

Several studies have concluded that plastic responses even to such a relatively straightforward environmental factor as temperature point to a complex epistatic system of gene regulation (Scheiner, 2002). Published data have not readily distinguished alternative hypotheses about underlying genetic architecture (Scheiner and Lyman, 1991; Karan *et al.*, 2000; see also Wu, 1998). In general, the lack of simple correspondence between genotypic and phenotypic variation (Stern, 2000) suggests complex regulatory systems. Even the quantitative trait locus (QTL) approach, which identifies entire regions involved in multilocus traits, may not encompass the genetic complexity of these systems, for this approach cannot resolve many small regulatory effects (Vitzhum, 2003), and different QTL alleles may be expressed in different multifactorial environments such as growth season (Weinig *et al.*, 2002).

Those cases of phenotypic expression that we characterize as "plastic" may simply entail a relatively broad diversity of outcomes, compared with "canalized" regulatory systems, in which environmental effects are buffered such that phenotypes are more uniform (see previous chapter). According to Nijhout (2003), the inevitable and likely maladaptive sensitivity of individual development to environmental factors such as temperature and ionic concentration that directly affect chemical and metabolic processes has been shaped by evolution of regulatory pathways in one of two directions: either to refine this environmental sensitivity into adaptive norms of reaction (i.e., plasticity) or to buffer it through developmental canalization and physiological homeostasis (similar to developmental buffering of genetic variation). Thus canalization and plasticity can be understood as two sides of the same evolving developmental coin; rather than define them both as special kinds of gene regulation, both can be seen *as* gene regulation (Stearns, 2003). Note that what may appear to the researcher as distinct alternative phenotypes may just be those discrete points on a continuous norm of reaction that are expressed because of (natural or experimental) environmental discontinuities, not necessarily the result of some exceptional kind of major developmental switch (Stearns and Hoekstra, 2000; Karan *et al.*, 2000; Nijhout, 2003). At this point, few studies have included enough environments to assess an entire norm of reaction (Karan *et al.*, 2000; see Windig, 1994, for a well-analyzed case), so these distinctions have perhaps been overly emphasized.

Recent theoretical results on networks of gene expression also call into question the idea that canalization—and, by inference, plasticity—is a distinct developmental phenomenon demanding special explanation. Model genetic regulatory networks that are allowed to evolve to maintain function despite knock-out mutations in randomly chosen genes have, as a by-product, the buffering of the expression of the genes in the network (Siegal and Bergman, 2002). Knock-out mutations in yeast appear to occur in a real-world context that has similar properties (Bergman and Siegal, 2003). After Waddington (1960) suggested that canalizing mechanisms evolved to buffer phenotypes against genetic and environmental perturbations, canalization came to be seen as a real phenomenon caused by distinct genes and regulatory mechanisms. However, the property of canalization may simply be a by-product, and not necessarily a well-localized by-product but a diffuse property of the entire control network rather than any single gene. The same insight may well apply to plasticity.

The fact that plasticity does not call for a distinct developmental genetic paradigm does not at all lessen the importance of plastic expression as an aspect of phenotypic variation, but it does suggest some changes to our research questions. It remains fundamentally important to evaluate the ecological and selective consequences of environmentally contingent patterns of phenotypic expression versus buffered, canalized expression, but we believe our understanding of the underlying mechanisms and their evolution should be integrated into broader investigations. For instance, possible "costs of plasticity" should be studied in the general context of gene regulation, rather than assumed to be inevitable results of a uniquely cumbersome genetic architecture. If plasticity does depend on the same kind of regulatory pathways as do more uniform phenotypic outcomes, this would account for the fact that despite considerable effort the evidence for the costs of plasticity is scant at best (Dorn *et al.*, 2000). Constraints on the evolution of plasticity may reside more in phylogenetic or environmental limits to accurate cue perception and transduction than in hypothetical costs (Sultan and Spencer, 2002). Most importantly, this insight changes our approach to development in general to one in which both environmental context and genetic constituents interact in a dynamic integrative process (Lewontin, 2001; Sultan, 2003).

D. Genetic Variation and the Evolution of Plasticity

One challenge is to incorporate this view of development as both dynamically complex and context dependent into our notion of heredity. It has become clear that a linear model of causation running from gene to phenotype is rarely true, that the phenotype emerges from an epigenetic system that integrates interacting genes and gene products with both internal and external signals (Trewavas and Malho, 1997). Increasing knowledge of molecular mechanisms, combined with advances in evolutionary development, have made clear that it is these complex epigenetic systems that "descend with modification," not specific genetic variants or alleles. Starting with Waddington (1960), several artificial selection experiments have shown clearly that norms of reaction are heritable and can evolve (Hillesheim and Stearns, 1991; Scheiner and Lyman, 1991; Brakefield, 2003, and references in these papers; see Scheiner, 2002, for full discussion). Indeed, artificial selection can shape plasticity patterns (the slope of reaction norms to post-hoc response-ordered environments) in as few as five generations (Falconer, 1990; Hillesheim and Stearns, 1991). However, there have been relatively few studies of selection directly for plasticity (Fischer *et al.*, 2000); more often artificial selection on other traits has produced correlated effects on plasticity patterns (Scheiner, 2002).

One important goal of these selection experiments has been to clarify the genetic architecture of plastic response, which will of course affect evolutionary predictions (Via *et al.*, 1995). Unfortunately in this area "theory has far outstripped data" (Scheiner, 2002). The complex genetic basis of these response systems—and the view that plasticity is a difference in outcome and not mechanism—is confirmed by findings that selective change in the mean and plasticity of a trait are interrelated (Scheiner and Lyman, 1991; Kliebenstein and Mitchell-Olds, 2002). Although there is clearly widespread genetic variation for reaction norms in natural populations (revealed in quantitative genetic studies as significant genotype by environment interaction or "$g \times e$"; Falconer and Mackay, 1996), the molecular basis of this variation is still largely unknown (Scheiner, 1993; Wu, 1998; Nager *et al.*, 2000; Kliebenstein *et al.*, 2002; Nijhout, 2003). This is in part because of the complexity both of the phenotypic traits studied and of the environmental factors that elicit variation in those traits (Kliebenstein *et al.*, 2002). One approach is to use QTL to map major loci that influence differential gene expression in different environments (Jansen *et al.*, 1995). For instance, Lukens and Doebley (1999) identified a difference at a single QTL locus between maize and its wild ancestor teosinte that contributes to reduced architectural plasticity in response to density, evidently through a change in regulation rather than in protein coding. The QTL approach will no doubt remain very useful, despite its limited resolution. However, in cases where $g \times e$ interaction is influenced by many genes of small effect, it may be impossible to resolve in mapping experiments (Stratton, 1998; Vitzthum, 2003).

Like other aspects of phenotypic variation, potential for norm of reaction evolution will vary among traits and taxa depending on available genetic variation and on phylogenetic, developmental, and other constraints. Consequently, even direct selection for plasticity may produce no response (van Kleunen *et al.*, 2002). Artificial selection for high versus low plasticity for temperature-induced adaptive variation in butterfly wing patterns revealed genetic constraints to change in plasticity, possibly because of positive genetic correlations across temperatures (Wijngaarden and Brakefield, 2001). Fischer *et al.* (2000) artificially selected for change in the spatial plasticity of clonal spread in buttercups, an important aspect of competitive response in the field. Despite substantial genetic variation for the trait ($g \times e$ interaction), no response was found after two generations of selection, suggesting that additive genetic variation contributing to plasticity may have been depleted in these natural populations because of previous selection. Like other adaptive traits, plasticity for ecologically important responses may thus show relatively low response to selection (Scheiner, 1993; Fischer *et al.*, 2000). One interesting aspect of evolutionary constraint is specific to plasticity: When some traits are plastic, genetic correlations among traits and therefore response to selection will vary from one environment to another (Newman, 1988; Hillesheim and Stearns, 1991).

Evolutionary change in plastic response—and phylogenetic constraint to such change—can occur at any link in the intricate causal chain that regulates phenotypic expression, from the initial perception of an environmental cue, to its transduction as an internal (often hormonal) signal, to cellular reception mechanisms, to transcription factors, to sequence variation in DNA-binding sites, to gene expression. Clearly many points of regulation can evolve (Nijhout, 2003); "genes for plasticity" can refer to heritable change at any of these levels. Indeed, cryptic variation within and among populations can occur from variation at any of a number of points in the regulatory cascade (A. C. Burke, personal communication, 2004). Mutational changes to upstream receptors may alter many aspects of gene regulation and thus have broad pleiotropic effects, while allelic changes to the promotor regions of downstream genes are likely to show more specific effects on the phenotype (Stern, 2000; Kliebensbeck *et al.*, 2002, and references). For instance, a QTL "for" plasticity might be a region of a chromosome that codes for a transcription factor. In a carefully focused study on the production of a plant defense compound with a known genetic basis and hormonal signal, Kliebenstein and Mitchell-Olds (2002) showed that the same QTL regulated both the plasticity of defense biosynthesis and its mean level. Interestingly, they found substantial among-population variation in the signal transduction pathway even for this biosynthetically simple secondary metabolite. Genetic variation for regulatory proteins is probably a key aspect of variation for plasticity. Conversely, similar plastic response patterns may result from different evolutionary changes in the underlying regulatory cascade. At a higher level in the regulatory chain, genetic changes in hormone physiology can affect many seemingly unrelated developmental traits (Voesenek and Blom, 1996). At its end, changes to the sensory and behavioral apparatus for environmental perception will alter the initial steps in signal transduction that initiate the entire cascade. Future research on the evolution of plasticity must be aimed at understanding both genetic variation at these diverse points in the process and the evolutionarily conserved aspects of environmental perception, hormonal control, transcription, and expression that constrain the evolution of plastic responses systems (deWitt *et al.*, 1998; Ketterson and van Nolan, 1999).

III. HOW PLASTICITY INTERACTS WITH CONSERVED DEVELOPMENTAL PATTERNS

One of the major biological discoveries of the last 20 years is the existence of deeply conserved developmental patterns controlled by equally conserved genes. Examples include *HOX* control of the development of the body axis of arthropods and vertebrates, *HOX-d* control of vertebrate limb development, and the *MADS* genes that interact in the ABC model of angiosperm flower development (Carroll *et al.*, 2001; Weigel and Meyerowitz, 1994). One possible inference is that with the exposure of these developmental mechanisms we have discovered the origin of phylogenetic constraints and the explanation of Baupläne. Whether that is the case or not, this much appears to be true: Some aspects of developmental regulation change very slowly and are nearly invariant within fairly large clades. If some developmental pathways change slowly, with rates measured in phylogenetic time on a scale of many millions of years, then how do those pathways interact with the variation in regulatory steps involved in plasticity, which occurs within a single generation?

The butterfly wing has been developed beautifully as a model system that would appear to be ideal to answer this question. It is a flat sheet of cells with many natural markers, such as wing veins and compartment boundaries. The possible locations of eyespots and stripes within that system of natural markers—the so-called Nymphalid Ground Plan—appears to be fairly invariant within a large clade of butterflies (Nijhout, 1991). The development of the butterfly wing can be manipulated through surgery on wing discs that lie on the surface of metamorphosing pupae. For example, eyespot primordia can be transplanted to new locations, producing eyespots where none occur in nature (Brakefield *et al.*, 1996). The genetic control over the development of some of its major features, including eyespots, is now quite well understood and has been shown to involve developmental control genes with known functions in *Drosophila* and other insects. For example, *distal-less* is expressed at the distal tips of *Drosophila* appendages and plays a role in determining whether the appendage turns into an antenna or a leg, and if a leg, which leg. It thus provides positional information; its expression signals to the surrounding cells that they are at the distal end of an appendage. In the butterfly wing, this is the gene whose expression triggers the development of eyespots, which are in the middle of the wing blad—not anywhere near its distal end—a good example of cooption of an existing developmental regulatory mechanism to a new function (Brakefield *et al.*, 1996).

These eyespots undergo adaptive seasonal change, a form of plasticity known as seasonal polyphenism (Shapiro, 1976). In the best-studied case, *Bicyclus anynana,* a tropical African butterfly, the selective reasons for the seasonal change have been investigated in the field. The eyespots are large in the wet season, small in the dry season. When the dry season forms are released in the wet season, they have lower fitness than wet season controls. The selective pressure is predator avoidance—what is cryptic in the dry season is not so cryptic in the wet season. Laboratory experiments have established that the seasonal changes in the size of the eyespots are in fact not a discontinuous polyphenism but a continuous reaction norm that appears to be discontinuous in the field simply because the temperatures that induce the change in development differ between seasons. All intermediate forms can be reared by using a range of temperatures in the laboratory (Windig, 1994). This impressive and actively growing work can be accessed through Brakefield (2000), Brunetti *et al.* (2001), Beldade and Brakefield (2002), Beldade *et al.* (2002c), and Monteiro *et al.* (2003). It builds on insights that can be traced through Nijhout (1991), who played a key role in the development of the system, back to Schwanwitsch (1924), who first recognized the Nymphalid Ground Plan.

At first, this system appears to offer a straightforward answer to the question addressed here: How does plasticity interact with conserved developmental patterns? Plasticity would appear to be a short-term response embedded in an ancient framework. The shared ancestors of butterflies and flies had developmental control genes that shaped appendage structure. In the fly they controlled the development of legs and antennae; in the butterfly, of wing spots and stripes. The positional information was phylogenetically conserved in developmental control genes whose expression was constrained. They could only induce the production of spots and stripes in certain parts of the wing and not in others. Once, however, it had been determined that an eyespot was going to develop in a certain place, its size and color could be altered during the development of a single individual by temperature. This gives a fairly classical picture of a phylogenetically conserved, rigid developmental pattern expression early in development that laid down structures that were then subject to "fine-tuning" by plasticity later in development. The developmental control genes, in particular *distal-less*, created a vase, the eyespot, in which we could conceptually place a bundle of reaction norms for size and color like a spray of flowers representing the phenotypic plasticity of the butterfly population in seasonal Africa. Just as a vase holds a bouquet of flowers on a table, so did the phylogenetically conserved developmental control over the eyespot hold the reaction norms in position on the wing.

A. Genetic Causation and the Butterfly Wing: A More Complicated Picture

We are no longer convinced that the simple metaphor of a vase with flowers is a helpful way to think about the system, because we believe this view is based on problematic assumptions about the nature of genetic causation. Our uncertainty arises from remarkable new experimental results on butterfly wings (Beldade *et al.*, 2002a,b; Monteiro *et al.*, 2003), from the new understanding of complex regulatory control on gene expression discussed in the preceding text, from the history of explanations of the segmentation of the *Drosophila* embryo, and from questions about the impact of solid geometry on the expression of developmental control genes. Bear with us while we sketch these key elements of a more complicated and indeed more interesting picture.

First, recent experimental results make clear that the simple conserved genetic architecture of eyespot expression is in fact neither simple nor conserved.

- Artificial selection can uncouple the size of the anterior and posterior eyespots, essentially removing one while retaining the other (Beldade *et al.*, 2002b). Thus it is fairly straightforward to eliminate one eyespot module while retaining another.
- One of the genes responding to selection for eyespot size is *distal-less* itself (Beldade *et al.*, 2002a). Thus the gene thought to represent a deeply conserved phylogenetic constraint itself responds rapidly to microevolutionary selection pressures with allelic change.
- In *B. anynana* there are normally seven eyespots. X-ray mutagenesis yields mutants that remove some eyespots but not others—three and four either reduced or completely absent; one, two, three, and four absent; one, three, four, and seven absent; and all absent (Monteiro *et al.*, 2003). These results suggest that the differentiation of eyespot foci in each wing cell is controlled by one or more focus regulator genes that are under the control of regional regulator genes. Mutations to the local regulators have local effects; mutations to the regional regulators have regional effects. Some mutants yield wing patterns not found in any of the 80 existing species of *Bicyclus*, including some thought to be forbidden by the Nymphalid Ground Plan.

To summarize, not all conceivable changes in butterfly wing patterns can be selected, nor does mutagenesis produce all conceivable variants. Thus the concept of a basic butterfly wing pattern remains intact, as does the idea that an eyespot is a module. However, questions have been raised about how one should think about genetic determination: Clearly the same gene is involved in both macroevolutionary patterns and microevolutionary change. Developmental control genes are not hands-off managers that initiate development and then leave the scene; they are micromanagers that remain involved in the details of the process until it is complete (Akam, 1998). Artificial selection readily modifies eyespots separately, so each eyespot must have independent genetic control as well as general modular control. In addition, x-ray mutagenesis produces completely novel variants, including those outside the accepted possibilities of the classical model.

The second source of uncertainty is the emerging picture of transcriptional control in eukaryotes, a picture that is becoming increasingly detailed (Stern, 2000). As discussed in the previous section, eukaryotic genes are *cis*-regulated by the binding of transcription factors to upstream control regions. For example, in *Drosophila* whether or not a gene will be expressed, and the level at which it will be expressed, is determined by the binding of from 1 to 20 transcription factors to the control region. Some transcription factors bind only to the control regions of only a few genes, but many bind to those of hundreds of genes. Some may even bind to a considerable fraction of the whole genome.

Now take that picture, and apply it to the control over the production of a module, the eyespot, in the butterfly wing. There may well be a cascade of regional and local effects, as the mutagenesis experiments suggest, but they are effects felt in a network of hundreds of genes. In addition, some of the genes thought to be triggering the cascade, such as *distal-less*, are themselves part of the fine-tuning of the module.

The third source of uncertainty is the history of explanations of segmentation in the *Drosophila* embryo, a history that addresses questions about the genotype–phenotype map rather similar, in an abstract way, to those addressed by canalization and plasticity. Turing (1952) suggested that segmentation could arise as a result of gradients of interacting morphogens producing stripes of concentration as they diffused in a cylinder. Meinhardt (1982) developed this idea into a mathematical model that appeared to produce all the essential features seen in the early segmentation of the *Drosophila* embryo. However, when the molecular mechanisms were exposed (Nüsslein-Volhard and Wieschaus, 1980; St. Johnston and Nüsslein-Volhard, 1992), we learned that stripes in embryos could be produced by entirely different mechanisms relying on local interactions of gene expression rather than global patterns of reaction-diffusion. Thus Turing's elegant insight became another "beautiful idea killed by an ugly fact," reminding us of the wisdom of agnosticism until beautiful ideas have survived rigorous tests. There are many possible alternative explanations lurking in the complexities of the genotype–phenotype map, and some of them doubtless apply to the links between phylogenetic constraints and phenotypic plasticity.

The fourth source of our uncertainty arises simply from the contemplation of three-dimensional geometry. When *distal-less* is expressed in *Drosophila*, it is near the distal tip of a narrow cylinder. When *distal-less* is expressed in *Bicyclus*, it is well within a flat, nearly two-dimensional sheet of cells. How much of what we attribute to the "control" of *distal-less* over the production of eyespots results from a targeted evolutionary modification of its activity, and how much results from more global events that restructured the appendage from a cylinder to a sheet? We cannot know without being able to make a fly turn a developing leg into the flat sheet of a wing or being able to make a butterfly turn a wing into the thin cylinder of a leg and then studying what *distal-less* elicits in the changed geometry. We might be surprised by how much results from the geometry and its consequent cell–cell interactions and not to this particular gene.

B. The Same Networks May Give Rise to Both Plasticity and Constraint

Let us now return to the original idea, the notion of a phylogenetic constraint holding a bundle of reaction norms like an old and durable vase holding an ephemeral bouquet of flowers. In some general sense that may be an appropriate analogy, but an unknown part of it may be a naive reification. For one thing, the components of the vase may themselves be part of the bouquet. We may thus think we see plasticity fine-tuning the phenotype within a long-established framework of phylogenetic and developmental constraint, when in fact one network of interactions may be causing the entire pattern, interactions of effects that cannot be cleanly assigned on the one hand to phylogeny and constraint and on the other hand to plasticity, interactions that produce both the appearance of constraint in one and the appearance of plasticity in another part of the same system, both as by-products. Thus we see research on the mechanisms that produce plasticity merging completely with research on the mechanisms of development in general, with the new element being systematic study of the interactions of developmental mechanisms with environmental inputs.

IV. WHAT EFFECTS DOES PLASTICITY HAVE ON POPULATIONS AND COMMUNITIES?

Although the mechanisms that cause plasticity are best approached at the level of the individual organism, many of the most important consequences of plasticity are realized at the levels of populations and communities. Population dynamics can be viewed as systematic change in the population density encountered by individual organisms, and their reactions to that change in density are a very important type of phenotypic plasticity: As population density declines, the surviving individuals each have more to eat, and they respond by growing faster to larger sizes and producing more offspring. As density increases, individual growth rates decline, fecundity drops, and mortality increases. Thus phenotypic plasticity lies at the center of density dependence. The magnitude of such effects is illustrated by the reactions of California sardines to the sharp decline in density they experienced when their population crashed in the 1950s (Murphy, 1967): near the end of the crash, only a few adult sardines were being caught, but some of them were 50-cm long and had enormous fecundities.

Ecologists have traditionally translated such responses into models that implicitly assume population mean reaction norms and ignore genetic variation in reaction norms. The loss of insight caused by making that simplifying assumption could be significant (cf. Barbault and Stearns, 1991). To see why, we build up a picture of what is going on from the reaction norm perspective. First, consider a fitness measure, b/d (b = per capita birthrate, d = per capita death rate) as a function of its own population density and that of a competitor (Figure 14-1). As its own density increases, the fitness measure declines from above 1.0 to below 1.0. As the density of its competitor increases, it does the same with different details. The population-mean reaction norm is thus a reaction surface describing the response to changes in density of both species. Species 1 replaces itself at all

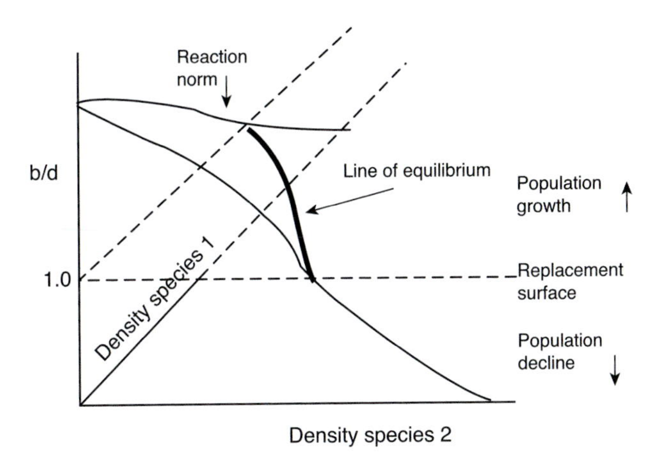

FIGURE 14-1. Density-dependent population-mean plasticity in fitness defined as *b/d* response to the density of one's own species (species 1) and a competitor (species 2). (Reprinted from Acta Oecologica, Vol. 12, Barbault & Stearns, "Towards an evolutionary ecology linking species," pp. 3–10, Copyright [1991], with permission from Elsevier.)

combinations of densities of its own population and that of its competitor where b/d = 1.0. This happens along the line of equilibrium depicted, which is nothing more than the Lotka-Volterra or Tilman zero-growth isocline (ZGI). When we add the population mean plastic response of the second species (Figure 14-2), we see that the system can have an equilibrium point at which the two species coexist. This happens when the two lines of equilibrium cross each other on the replacement surface where b/d = 1.0.

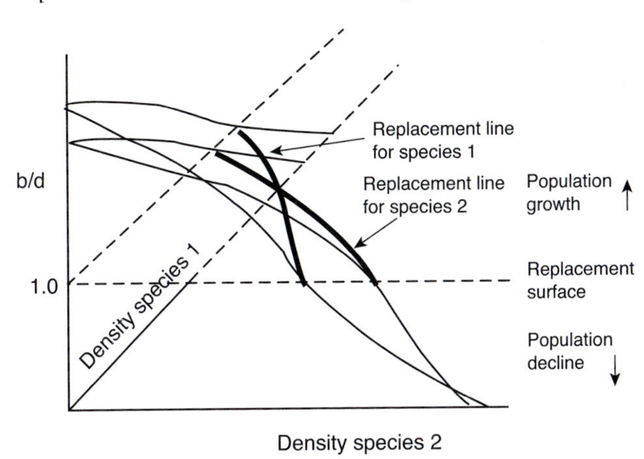

FIGURE 14-2. Density-dependent population-mean plasticity in fitness plotted for two species for the case where the species coexist where the replacement lines cross on the replacement surface. (Reprinted from Acta Oecologica, Vol. 12, Barbault & Stearns, "Towards an evolutionary ecology linking species," pp. 3–10, Copyright [1991], with permission from Elsevier.)

Thus far we are in well-explored territory first mapped in the 1920s and revisited in detail in the 1970s and 1980s. Recently Takimoto (2003) has extended this approach, using population-mean plastic responses, to the problem of ontogenetic niche shifts. Ontogenetic niche shifts are changes in what organisms eat or where they live during their life cycle. For example, tadpoles that eat algae in ponds turn into frogs that eat insects on land. Such life cycles consist of stages with each stage interacting with a different community. Takimoto asked whether adaptive plasticity in the timing of niche shifts stabilizes consumer-resource dynamics. It would do so by accelerating the shift to the next niche if resources are scarce or delaying that shift if resources are abundant in the first niche. That plasticity will cause scarce resources to increase and abundant resources to decrease in the first niche, stabilizing the interaction, by comparison with nonplastic shifts. Takimoto's theoretical comparison of the plastic strategy with two nonplastic alternatives (shift at fixed size or fixed age) confirmed the logic of the idea: Only the plastic strategy had a locally stable equilibrium, and that equilibrium was the result of density-dependent negative feedback in resource dynamics, as postulated. The analysis was elegant, but it did not include the next step—considering genetic variation for the plastic response.

When we take that variation into account, we must put into the picture the fact that each population is a bundle of reaction norms (Figure 14-3). These reaction norms intersect the replacement surface not along a line of equilibrium, but within a certain area. That area defines the set of densities of both species within which some genotype is at its replacement density with $b/d = 1$. This is true for both species. Thus there is an area on the replacement surface defined by the intersection of the replacement sets of both species. Within that area some genotype in each species is at density equilibrium.

Figure 14-3 is a static description of a dynamic system. When we consider the dynamics, several questions arise, including these:

- Does the ecological interaction itself maintain a certain level of genetic variation? The answer will depend on whether the reaction surfaces of the genotypes of both species cross or do not cross over the ranges of densities encountered, not just on the equilibrium surface but within the volume of fluctuations about that surface.
- If reaction norms cross and recross in the space defined by the normal fluctuation of a population in response to changes in all of its biotic and abiotic interactions, then the fitness of any given genotype is always shifting up and down, and the fitness of many genotypes is varying from positive to negative. Genotype × environment interactions—crossing reaction norms—can make selection diffuse, slowing the rate of response even to directional selection (see Mitchell-Olds and Rutledge, 1986; Barton and Turelli, 1989; Gillespie and Turelli, 1989; further discussion and references in Sultan, 1987, 2003).

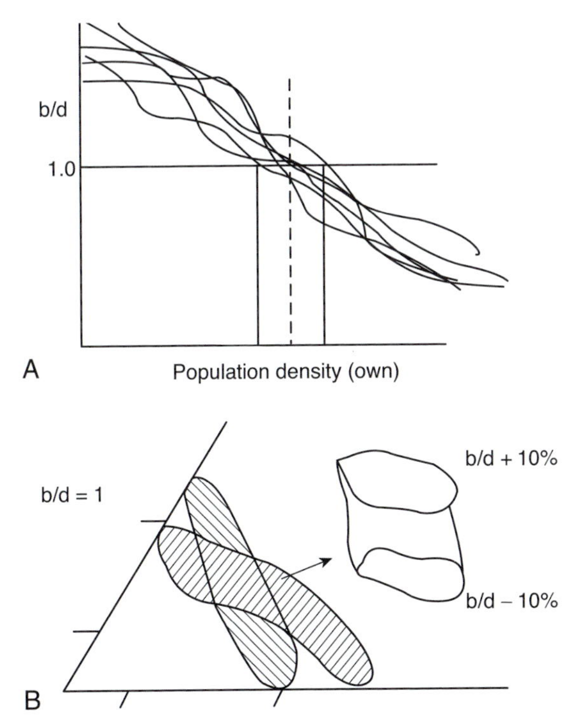

FIGURE 14-3. (A) Each population is a bundle of reaction norms, not a population mean reaction norm. The genotypes cross the replacement surface across a range of densities. At any point within that range, indicated by the dashed line, various genotypes are increasing, decreasing, or replacing themselves. (B) Replacement surfaces for both species; they replace themselves not along zero-growth isoclines (ZGIs) but within areas defined by genetic variation in their reaction norms. Both populations fluctuate above and below the replacement line in response to forces, such as resource dynamics and variation in abiotic factors, not depicted here. The volume depicted indicates where the system might be found if those fluctuations cause increases and decreases of 10%. (Reprinted from Acta Oecologica, Vol. 12, Barbault & Stearns, "Towards an evolutionary ecology linking species," pp. 3–10, Copyright [1991], with permission from Elsevier.)

- Does the existence of genetic variation for plastic response stabilize both the population dynamics of each species and their ecological interaction? It would appear to, because each species is now interacting with the other, not at a point or along a line but within a space.
- If the answer to the previous question is "yes," then we will be led to ask, how much of the stability of the natural world results from genetic variation for plastic responses to population density or to any other important ecological factor?

Answering that question requires simultaneous consideration of the ecological and the evolutionary dynamics, because the ecological interactions cause selection to change the properties of the interacting bundles of reaction norms. The evolutionary changes in plastic response then change the ecological interactions, and so forth. Equilibrium analysis is not appropriate for such systems: only analysis of the full temporal dynamics can tell us what will happen. The questions just listed are ripe for the application of adaptive dynamics (Dieckmann, 1997), which has precisely as its goal a full temporal analysis of the dynamic interaction of evolutionary and ecological change.

V. RESEARCH AGENDA

We seek a fresh conceptualization of the genotype–phenotype map unencumbered by the historical baggage of causal models formulated decades ago. Our reflections on recent discoveries regarding the intricacies of genetic and environmental regulation has made us question the usefulness of distinguishing the mechanisms of plasticity and canalization from the those of development in general. Yet these terms remain useful categories of phenotypic expression. The day may come when the structure of the genotype–phenotype map will be laid out in enough detail to justify a new vocabulary to describe its essential features and their consequences, but that time has not yet arrived.

In the meantime, what sorts of research will help that day arrive sooner rather than later? We offer a few opinions on this very open question, in the spirit of stimulating a discussion to help others more precisely formulate their own, perhaps quite different, points of view.

Since the ubiquity of plasticity is now well established, further descriptions of plasticity will only advance knowledge if they are done in comparative, environmental, or experimental contexts that are so well structured that they produce new insights into mechanisms and consequences. That said, in many organisms the natural history of plasticity is not yet well described. In many cases we do not yet know to which factors of natural environments organisms are responding with plasticity, nor do we have comprehensive descriptions of the diversity of patterns of reaction norms within and among populations and species. As novel environmental challenges arise in many disturbed and contaminated habitats, it is particularly critical to understand the limits of plasticity and its evolutionary potential in natural populations (Gilbert, 2000).

The potential macroevolutionary role of plasticity as a starting point for adaptive divergence remains controversial (see discussions in Weber and Depew, 2003; West-Eberhard, 2003). To what degree is plasticity important in speciation and in major transitions, such as the move from water to land? Does selection on bundles of reaction norms expressed in novel environments result in genetic assimilation of novel traits? Answering such questions will probably depend on integrating norm of reaction experiments with both phylogenetic comparisons and insights from developmental studies.

If plasticity is a by-product of environmental interactions with genetic regulatory networks, then asking what is the cost of plasticity is really not very different from asking what is the cost of development. We suspect that we will learn more by concentrating on the mechanisms of development and how they yield specific plastic responses than by focusing on the putatively unique costs of plasticity.

One of the earliest speculations about plasticity (Wright, 1931) was that plasticity in some sense substituted for or made unnecessary genetic variation. Although that claim may have resulted from a view of selection acting on populations rather than individuals, the issue cannot simply be discarded as a red herring thrown into our path by group selection. Is plasticity only involved in fine-tuning the phenotype, or is it a central and essential part of local adaptation? Are there any generalities about the relationship between the adaptive plasticity of traits and their genetic variation and evolvability?

The twenty-first century promises to witness the resolution of many open questions about the relationship of genotype to phenotype. Among those questions, those concerning the nature of plasticity and its evolution are among the most important. The nature of plasticity is essentially the nature of development; when the connection of development to environmental signals has been properly understood, we will have understood the nature of plasticity. When the consequences of plasticity for variation in reproductive success in natural populations are better understood, we will know more about how plasticity evolves and what consequences it has for population and community ecology. The major message of this chapter is that, to achieve those goals, we may have to fully integrate development into our understanding of biological variation.

ACKNOWLEDGMENTS

We thank Fred Nijhout, Bob Denver, and Annie Burke for constructive feedback.

REFERENCES

Akam, M. (1998). Hox genes: From master genes to micromanagers. *Current Biology* **8**, R676–678.

Angier, N. (2003). Not just genes: Moving beyond nature vs. nurture. *New York Time: Science Times*, Feb. 25, pp. 1, 10.

Ballarè, C. L., Sanchez, R. A., Scopel, A. L., Casal, J. J., and Ghersa, C. M. (1987). Early detection of neighbour plants by phytochrome perception of spectral changes in reflected sunlight. *Plant Cell and Environment* **10**, 551–557.

Barbault, R., and Stearns, S. (1991). Towards an evolutionary ecology linking species interactions, life-history strategies and community dynamics: An introduction. *Acta Oecologia* **12**, 3–10.

Barker, D. J. P. (1998). The fetal origins of coronary heart disease and stroke: Evolutionary implications. In *Evolution in Health and Disease* (S. C. Stearns, ed.), pp. 246–250. Cambridge, England: Oxford University Press.

Barton, N. H., and Turelli, M. (1989). Evolutionary quantitative genetics–how little do we know? *Annual Review of Genetics* **23**, 337–370.

Beldade, P., and Brakefield, P. M. (2002). The genetics and evo-devo of butterfly wing patterns. *Nature Reviews Genetics* **3**, 442–452.

Beldade, P., Brakefield, P. M., and Long, A. D. (2002a). Contribution of distal-less to quantitative variation in butterfly eyespots. *Nature* **415**, 315–318.

Beldade, P., Koops, K., and Brakefield, P. M. (2002b). Developmental constraints versus flexibility in morphological evolution. *Nature* **416**, 844–847.

Beldade, P., Koops, K., and Brakefield, P. M. (2002c). Modularity, individuality, and evo-devo in butterfly wings. *Proceedings of the National Academy of Science* **99**, 14262–14267.

Bergman, A., and Seigal, M. L. (2003). Evolutionary capacitance as a general feature of complex gene networks. *Nature* **424**, 549–552.

Boorse, G. C., and Denver, R. J. (2004). Expression and hypophysiotropic actions of corticotropin releasing factor in *Xenopus laevis*. *General and Comparative Endocrinology* **137**, 272–282.

Brakefield, P. M. (2000). Structure of a character and the evolution of butterfly eyespot patterns. In *The Character Concept in Evolutionary Biology* (G. P. Wagner, ed.), San Diego, CA: Academic Press. Reprinted in *Journal of Experimental Zoology (Molecular and Developmental Evolution)* **291**, 93–104.

Brakefield, P. M. (2003). Artificial selection and the development of ecologically relevant phenotypes. *Ecology* **84**, 1661–1671.

Brakefield, P. M., Gates, J., Keys, D., Kesbeke, F., Wijngaarden, P. J., Monteiro, A., French, V., and Carroll, S. B. (1996). Development, plasticity and evolution of butterfly eyespot patterns. *Nature* **383**, 236–242.

Brunetti, C. R., Selegue, J. E., Monteiro, A., French, V., Brakefield, P. M., and Carroll, S. B. (2001). The generation and diversification of butterfly eyespot color patterns. *Current Biology* **11**, 1578–1585.

Callahan, H. S., Pigliucci, M., and Schlichting, C. D. (1997). Developmental phenotypic plasticity: Where ecology and evolution meet molecular biology. *Bioessays* **19**, 519–525.

Callahan, H. S., Wells, C. L., and Pigliucci, M. (1999). Light-sensitive plasticity genes in *Arabidopsis thaliana:* Mutant analysis and ecological genetics. *Evolutionary Ecology Research* **1**(6), 731–751.

Carroll, S. D., Grenier, J. K., and Weatherbee, S. D. (2001). *From DNA to Diversity: Molecular Genetics and the Evolution of Animal Design*, p. 214. Malden, MA: Blackwell Science.

Casal, J. J., and Smith, H. (1989). The function, action and adaptive significance of phytochrome in light-grown plants. *Plant, Cell and Environment* **12**, 855–862.

Denver, R. J. (1997a). Proximate mechanisms of phenotypic plasticity in amphibian metamorphosis. *American Zoology* **37**, 172–184.

Denver, R. J. (1997b). Environmental stress as a developmental cue: Corticotropin-releasing hormone is a proximate mediator of adaptive phenotypic plasticity in amphibian metamorphosis. *Hormones and Behavior* **31**, 2.

Denver, R. J. (1998). Hormonal correlates of environmentally induced metamorphosis in the Western Spadefoot Toad, *Scaphiopus hammondii*. *General and Comparative Endocrinology* **110**, 326–336.

Denver, R. J., Mirhadi, N., and Phillips, M. (1998). Adaptive plasticity in amphibian metamorphosis: Response *Scaphiopus hammondii* tadpoles to habitat desiccation. *Ecology* **79**, 1859–1873.

DeWitt, T. J., Sih, A., and Wilson D. S. (1998). Costs and limits to benefits as constraints on the evolution of phenotypic plasticity. *Trends in Ecology and Evolution* **13**, 77–81.

Dieckmann, U. (1997). Can adaptive dynamics invade? *Trends in Ecology and Evolution.* **12**, 128–131.

Dorn, L. A., Pyle, E. H., and Schmitt, J. (2000). Plasticity to light cues and resources in *Arabidopsis thaliana*: Testing for adaptive value and costs. *Evolution* **54**, 1982–1994.

Dudley, S. A., and Schmitt, J. (1996). Testing the adaptive plasticity hypothesis: Density-dependent selection on manipulated stem length in *Impatiens capensis*. *American Naturalist* **147**, 445–465.

Falconer, D. S. (1990). Selection in different environments: Effects on environmental sensitivity (reaction norm) and on mean performance. *Genetical Research Cambridge* **56**, 57–70.

Falconer, D. S., and McKay, T. F. C. (1996). *Introduction to quantitative genetics*. Harlow, England: Prentice Hall.

Fischer, M., van Kleunen, M., and Schmid, B. (2000). Genetic allele effects on performance, plasticity and developmental stability in a clonal plant. *Ecology Letters* **3**, 530–539.

Galen, C., Huddle, J., and Liscum, E. (2004). An experimental test of the adaptive evolution of phototropins: Blue-light photoreceptors controlling phototropism in *Arabidopsis thaliana. Evolution* **58**, 515–523.

Gebhardt, M. D., and S. C. Stearns. (1992). Phenotypic plasticity for life history traits in *Drosophila melanogaster.* III. Effect of the environment on genetic parameters. *Genetical Research* **60**, 87–101.

Gedroc, J. J., McConnaughay, K. D. M., and Coleman, J. S. (1996). Plasticity in root/shoot partitioning: Optimal, ontogenetic, or both? *Functional Ecology* **10**, 44–50.

Gilbert, S. F. (2001). Ecological developmental biology: Developmental biology meets the real world. *Developmental Biology* **233**, 1–12.

Gilbert, S. F., and Bolker, J. A. (2003). Ecological development biology: Preface to the symposium. *Evolution and Development*. **5**(1), 3–8.

Gillespie, J. H., and Turelli, M. (1989). Genotype-environment interactions and the maintenance of polygenic variation. *Genetics* **121**, 129–138.

Gilroy, S., and Trewavas, A. (2001). Signal processing and transduction in plant cells: The end of the beginning? *Nature Reviews (Molecular and Cell Biology)* **2**, 307–314.

Gray, R. D. (1992). Death of the gene: Developmental systems strike back. In *Trees of Life: Essays in the Philosophy of Biology* (P. Griffiths, ed.), pp. 165–210. Dordrecht, Netherlands: Kluwer.

Gupta, A. B., and Lewontin, R. C. (1982). A study of reaction norms in natural populations of *Drosophila pseudoobscura*. *Evolution* **36**, 934–948.

Hillesheim, E., and Stearns, S. C. (1991). The responses of *Drosophila melanogaster* to artificial selection on body weight and its phenotypic plasticity in two larval food environments. *Evolution* **45**(8), 1909–1923.

Hillesheim, E., and Stearns, S. C. (1992). Correlated responses in life-history traits to artificial selection for body weight in *Drosophila melanogaster*. *Evolution* **46**(3), 745–752.

Jansen, R. C., Van Ooijen, J. W., Stam, P., Lister, C., and Dean, C. (1995). Genotype-by-environment interaction in genetic mapping of multiple quantitative trait loci. *Theoretical and Applied Genetics* **91**, 33–37.

Karan, D., Morin, J. P., Gibert, P., Moreteau, B., Scheiner, S. M., and David, J. R. (2000). The genetics of phenotypic plasticity. IX. Genetic architecture, temperature, and sex differences in *Drosophila melanogaster*. *Evolution* **54**(3), 1035–1040.

Ketterson, E. D., and Nolan, V. (1999). Adaptation, exaptation, and constraint: A hormonal perspective. *The American Naturalist* **54**, S4–S25.

Kliebenstein, D. J., Figuth, A., and Mitchell-Olds, T. (2002). Genetic architecture of plastic methyl jasmonate responses in *Arabidopsis thaliana*. *Genetics* **161**(4), 1685–1696.

Lewontin, R. C. (1985). Population genetics. *Annual Review of Genetics* **19**, 81–102.

Lewontin, R. C. (2001). Gene, organism and environment: A new introduction. In *Cycles of Contingency* (S. G. Oyama, P. E. Griffiths, and R. D. Gray, eds.), p. 55–66. Cambridge, MA: The MIT Press.

Lukens, L. N., and Doebley, J. (1999). Epistatic and environmental interactions for quantitative trait loci involved in maize evolution. *Genetical Research* **74**(3), 291–302.

Lummaa, V. (2003). Early developmental conditions and reproductive success in humans: Downstream effects of prenatal famine, birthweight, and timing of birth. *American Journal of Human Biology* **15**, 370–379.

Meinhardt, H. (1982). *Models of Biological Pattern Formation.* San Diego, CA: Academic Press.

Mitchell-Olds, T., and Rutledge, J. J. (1986). Quantitative genetics in natural plant populations. A review of the theory. *American Naturalist* **127**, 379–402.

Monteiro, A., Prijs, J., Bax, M., Hakkaart, T., and Brakefield, P. M. (2003). Mutants highlight the modular control of butterfly eyespot patterns. *Evolution and Development* **5**, 180–187.

Murphy, G. I. (1967). Vital statistics of the Pacific Sardine (*Sardinops caerulea*) and the population consequences. *Ecology* **48**, 731–736.

Nager, R. G., Keller, L. F., and Van Noordwijk, A. J. (2000). Understanding natural selection on traits that are influenced by environmental conditions. In *Adaptive Genetic Variation in the Wild* (T. Mousseau, B. Sinervo, J. A. Endler, eds.), pp. 95–115. New York: Oxford University Press.

Newman, R. A. (1988). Genetic variation for larval anuran (*Scaphiopus couchii*) developmental time in an uncertain environment. *Evolution* **42**, 763–773.

Newman, R. A. (1992). Adaptive plasticity in amphibian metamorphosis. *Bioscience* **42** (9), 671–678.

Newman, R. A. (1994). Effects of changing density and food level on metamorphosis of a desert amphibian, *Scaphiopus couchii*. *Ecology* **75**, 1085–1096.

Nijhout, H. F. (1990). Metaphors and the role of genes in development. *Bioessays* **12**, 441–446.

Nijhout, H. F. (1991). *The Development and Evolution of Butterfly Wing Patterns*. Washington, DC: Smithsonian.

Nijhout, H. F. (2003). Development and evolution of adaptive polyphenisms. *Evolution and Development* **5**(1), 9–18.

Nüsslein-Volhard, C., and Wieschaus, E. (1980). Mutations affecting segment number and polarity in *Drosophila*. *Nature* **287**, 795–801.

Pfennig, D. W. (1992). Proximate and functional causes of polyphenism in an anuran tadpole. *Functional Ecology* **6**, 167–174.

Piersma, T., and Drent, J. (2003). Phenotypic flexibility and the evolution of organismal design. *Trends in Ecology and Evolution*. **18**(5), 228–233.

Pigliucci, M. (2001). *Phenotypic Plasticity: Beyond Nature and Nurture*. Baltimore: Johns Hopkins University Press.

Pigliucci, M., and Schmitt, J. (1999). Genes affecting phenotypic plasticity in *Arabidopsis*: Pleiotropic effects and reproductive fitness of photomorphogenic mutants. *Journal of Evolutionary Biology* **12**, 551–562.

Reznick, D. (1981). Grandfather effects: The genetics of inter-population differences in offspring size in the mosquito fish. *Evolution* **35**, 941–953.

St. Johnston, R., and Nüsslein-Volhard, C. (1992). The origin of pattern and polarity in the *Drosophila* embryo. *Cell* **68**, 201–219.

Scheiner, S. M. (1993). Genetics and evolution of phenotypic plasticity. *Annual Review of Ecology and Systematics* **24**, 35–68.

Scheiner, S. M. (2002). Selection experiments and the study of phenotypic plasticity. *Journal of Evolutionary Biology* **15**, 889–898.

Scheiner, S. M., and Lyman, R. F. (1991). The genetics of phenotypic plasticity. II. Response to selection. *Journal of Evolutionary Biology* **4**, 23–50.

Schlichting, C. D., and Pigliucci, M. (1998). *Phenotypic Evolution: A Reaction Norm Perspective*. Sunderland, MA: Sinauer Associates.

Schlichting, C. D., and Smith, H. (2002). Phenotypic plasticity: linking molecular mechanisms with evolutionary outcomes. *Evolutionary Ecology* **16**, 189–211.

Schmalhausen, I. I. (1949). *Factors of Evolution: The Theory of Stabilizing Selection*. Philadelphia: Blakiston. (Reprinted 1986) Chicago: University of Chicago Press.

Schmitt, J., Dudley, S. A., and Pigliucci, M. (1999). Manipulative approaches to testing adaptive plasticity: Phytochrome-mediated shade-avoidance responses in plants. *The American Naturalist* **154**, S43–S54.

Schmitt, J. A., McCormac, A. C., and Smith, H. (1995). A test of the adaptive plasticity hypothesis using transgenic and mutant plants disabled in phytochrome-mediated elongation responses to neighbors. *American Naturalist* **147**, 937–953.

Schwanwitsch, B. N. (1924). On the groundplan of wing-pattern in nymphalids and certain other families of rhopalocerous Lepidoptera. *Proceedings of the Zoological Society of London B* **34**, 509–528.

Shapiro, A. M. (1976). Seasonal polyphenism. *Evolutionary Biology* **9**, 259–333.

Shi, Y. B. (1994). Molecular biology of amphibian metamorphosis: A new approach to an old problem. *Trends in Endocrinology and Metabolism* **5**, 14–20.

Siegal, M. L., and Bergman, A. (2002). Waddington's canalization revisited: Developmental stability and evolution. *Proceedings of the National Academy of Science USA* **99**, 10528–10532.

Slobodkin, L. B. (1968). Toward a predictive theory of evolution. In *Population Biology and Evolution* (R. C. Lewontin, ed.), pp. 187–205. Syracuse, NY: Syracuse University Press.

Smith, H. (1995). Physiological and ecological function within the phytochrome family. *Annual Review of Plant Physiology and Plant Molecular Biology* **46**, 289–315.

Stearns, S. C. (1980). A new view of life-history evolution. *Oikos* **35**, 266–281.

Stearns, S. C. (1989). The evolutionary significance of phenotypic plasticity. *Bioscience* **39**, 436–445.

Stearns, S. C. (2003). Safeguards and spurs. *Nature*. **424**, 501–504.

Stearns, S. C., and Hoekstra, R. F. (2000). *Evolution: An Introduction*. Oxford, England: Oxford University Press.

Stern, D. L. (2000). Perspective: Evolutionary developmental biology and the problem of variation. *Evolution* **54**, 1079–1091.

Stratton, D. A. (1998). Reaction norm functions and QTL-environment interactions for flowering time in *Arabidopsis thaliana*. *Heredity*. **81**,144–155.

Sultan, S. E. (1987). Evolutionary implications of phenotypic plasticity in plants. *Evolutionary Biology* **21**, 127–178.

Sultan, S. E. (2003). Commentary: The promise of ecological developmental biology. *Journal of Experimental Zoology (Molecular and Developmental Evolution)* **296B**, 1–7.

Sultan, S. E., and Bazzaz, F. A. (1993). Phenotypic plasticity in *Polygonum persicaria*. I. Diversity and uniformity in genotypic norms of reaction to light. II. Norms of reaction to soil moisture and the maintenance of genetic diversity. III. The evolution of ecological breadth for nutrient environment. *Evolution* **47**, 1009–1071.

Sultan, S. E., and Spencer, H. G. (2002). Metapopulation structure favors plasticity over local adaptation. *American Naturalist* **160**, 271–283.

Takimoto, G. (2003). Adaptive plasticity in ontogenetic niche shifts stabilizes consumer-resource dynamics. *American Naturalist* **162**, 93–109.

Tollrian, R., and Harvell, C. D. (1999). *The Ecology and Evolution of Inducible Defenses*. Princeton, NJ: Princeton University Press.

Trewavas, A. J., and Malho, R. (1997). Signal perception and transduction: The origin of the phenotype. *The Plant Cell*. **9**, 1181–1195.

Turing, A. M. (1952). The chemical basis of morphogenesis. *Philosophical Transactions of the Royal Society of London B* **237**, 37–72.

Van Kleunen, M., Fischer, M., and Schmid B. (2002). Experimental life-history evolution: Selection on the allocation to sexual reproduction and its plasticity in a clonal plant. *Evolution* **56**(11), 2168–2177.

Via, S., Gomulkiewicz, R., De Jong, G., Scheiner, S. M., Schlichting, C. D., and Van Tienderen, P. H. (1995). Adaptive phenotypic plasticity—consensus and controversy. *Trends in Ecology and Evolution* **10**, 212–217.

Vickers, M. H., Breier, B. H., Cutfield, W. S., Hofman, P. L., and Gluckman, P. D. (2000). Fetal origins of hyperphagia, obesity, and hypertension and postnatal amplification by hypercaloric nutrition. *American Journal of Physiology-Endocrinology and Metabolism* **279**, E83–E87.

Vitzthum, V. J. (2003). A number no greater than the sum of its parts: The use and abuse of heritability. *Human Biology* **75** (4), 539–558.

Voesenek, L. A. C. J., and Blom, C. W. P. M. (1996). Plants and hormones: An ecophysiological view on timing and plasticity. *Journal of Ecology* **84**, 111–119.

Waddington, C. H. (1960). Experiments on canalizing selection. *Genetical Research* **1**, 140–150.

Wagner, G. P. (2000). What is the promise of developmental evolution? Part I. Why is developmental biology necessary to explain evolutionary innovations? *Journal of Experimental Zoology* **288**, 95–98.

Walbot, V. (1996). Sources and consequences of phenotypic and genotypic plasticity in flowering plants. *Trends in Plant Science: Perspectives* **1**(1), 27–32.

Weber, B. H., and Depew, D. J. (2003). *Evolution and Learning: The Baldwin Effect Reconsidered.* Cambridge, MA: MIT Press.

Weigel, D., and Meyerowitz, E. M. (1994). The ABCs of floral homeotic genes. *Cell* **78**, 203–209.

Weinig, C., Ungerer, M. C., Dorn, L. A., Kane, N. C., Toyonaga, Y., Halldorsdottir, S. S., Mackay, T. F. C., Purugganan, M. D., and Schmitt, J. (2002). Novel loci control variation in reproductive timing in *Arabidopsis thaliana* in natural environments. *Genetics* **162**, 1875–1884.

West-Eberhard, M. J. (2003). *Developmental Plasticity and Evolution.* Oxford, England: Oxford University Press.

Wijngaarden, P. J., and Brakefield, P. M. (2001). Lack of response to artificial selection on the slope of reaction norms for seasonal polyphenism in the butterfly *Bicyclus anynana*. *Heredity* **87**(4), 410–420.

Windig, J. J. (1994). Reaction norms and the genetic basis of phenotypic plasticity in the wing pattern of the butterfly *Bicyclus anynana. Journal of Evolutionary Biology* **7**, 665–695.

Woltereck, R. (1909). Weitere experimentelle Untersuchungen über Artveränderung, speziell über das Wesen quantitativer Artunterschiede bei Daphniden. *Verhandlungen der Deutschen Zoologischen Gesellschaft* **1909**, 110–72.

Wright, S. (1931). Evolution in mendelian populations. *Genetics* **16**, 97–159.

Wu, R. (1998). The detection of plasticity genes in heterogeneous environments. *Evolution*. **52**(4), 967–977.

Wulff, R. (1986). Seed size variation in *Desmodium paniculatum*. II. Effects on seedling growth and physiological performance. *Journal of Ecology* **74**, 99–114.

CHAPTER 15

Variation and Life-History Evolution

DEREK A. ROFF
Department of Biology, University of California, Riverside, California, USA

INTRODUCTION

Life histories are variable at many different levels, among species, among populations of a species, and within populations of a species. In this chapter, I shall focus upon the last category of variation. What factors lead to variation in life-history traits within a population? This is potentially an immense topic, and I take as my task to give a relatively brief overview of the multiplicity of circumstances leading to variation. I divide the discussion into three sections according to the type of environment, namely a constant environment, a stochastic environment, and a predictable environment. Variation can occur in each type of environment, but each has unique features. There are two sources of variation, genetic and phenotypic. Genetic variation, in most cases, will be manifested also as phenotypic variation, whereas phenotypic variation may be the result of a single genotype showing genotype by environment interaction. Ideally this discussion would be divided according to both environment and type of variation (genetic, phenotypic): However, research does not always proceed in such neat packages, and thus it will be necessary to shift emphasis as available information dictates.

I. PHENOTYPIC VARIATION IN A CONSTANT ENVIRONMENT

At first sight we might expect that in a constant environment there would be a rapid erosion of phenotypic variation. However, there are mechanisms that even in a constant environment ensure the maintenance of phenotypic variation. Broadly speaking, these can be divided into four categories: (1) mutation–selection balance, (2) heterozygous advantage, (3) antagonistic pleiotropy, and (4) frequency-dependent selection. The first three processes maintain both genetic and phenotypic variability, whereas the last can maintain phenotypic variation in the absence of genetic variation.

A. Mutation–Selection Balance

For reviews of the mathematical basis of this concept, see Bulmer (1989) and Burger (1998). There are two models, differences being a consequence of the assumption concerning the effect of new mutations. These two models are known as the continuum-of-alleles model and the house-of-cards model.

The continuum-of-alleles model assumes that there are n loci, at which mutation can produce an infinite series of alleles, the average effect being zero (i.e., there is symmetry of effects). Assuming, for example, a normal distribution of effects, the probability that an allele at locus i with effect x_i mutates to an allele with effect y_i is

$$\exp\left(-\frac{1}{2}\left[\frac{y_i - x_i}{\alpha}\right]^2\right) \Big/ \left(\alpha\sqrt{2\pi}\right),$$

where α^2 is the variance of the mutational effect. There is no exact solution for the continuum-of-alleles model, but if it can be assumed that the variance resulting from a new mutation is much smaller than the existing genetic variance at the locus, then the standing additive genetic variance is

$$\sigma_A^2 = \sigma_m(\gamma + \sigma_E^2)\sqrt{2n} \quad (1)$$

where σ_m^2 is the average amount of new genetic variance introduced per zygote per generation by mutation, σ_E^2 is the environmental variance, and γ is the parameter of the Gaussian stabilizing selection function (fitness of trait value $x = \exp(-(x-\mu)^2/2\gamma)$, μ is the optimum trait value).

Turelli (1984) argued that the variance associated with new mutations should generally overwhelm existing genetic variation at a locus (the house-of-cards model). As a consequence each locus will comprise one common wild-type allele in high frequency and rare mutant alleles with large effect. Altering the continuum-of-alleles model for this assumption leads to

$$\sigma_A^2 = 2(\gamma + \sigma_E^2)\alpha^2\sigma_m^2 \qquad (2)$$

Note that the amount of additive genetic variance depends very much upon the assumption of mutational effects. Further, it strongly depends upon parameter values. This issue cannot be resolved by theory, only by empirical estimation.

The range in estimates for the mutational variances is large and can support per generation increases in heritability from 0.1% to 1% (Lynch *et al.*, 1998), which under some conditions are sufficient to account for the observed heritabilities (see Figure 4-13 in Roff, 1997). On the other hand, most mutations are deleterious and hence will be being continuously purged from the populations. To investigate this process, Lynch *et al.*, (1998) studied the accumulation of mutations in the cladoceran *Daphnia pulex*. The experimental protocol was to start with 100 lines of *D. pulex* and for each line initiate the next generation with a single individual: Under this scheme selection is essentially nonexistent and mutations can accumulate. The mutational heritability, defined as the ratio of the mutational variance to the environmental variance, is quite high for all the traits measured and averages 7% of the observed heritability (Table 15-1). The mean persistence times of mutations was estimated as the ratio of the genetic variance to the mutational variance and ranged from 22 to 72 generations (Table 15-1). These data also indicate that starting from zero variance the equilibrium standing level of variance could be achieved

Table 15-1. Estimated Heritabilities, h^2, Mutational Heritabilities, h^2_m, and Persistence Times of Mutations, t_p, of Morphological and Life-History Traits in *Daphnia pulex*[a]

Character	h^2	h^2_m	$100\frac{h^2_m}{h^2}$	t_p
Size at maturity	0.3412	0.0097	2.84	46.6
Age at maturity	0.0183	0.0046	25.10	57.1
Duration of instar	0.1434	0.0057	3.97	35.6
Size at birth	0.7500	0.0048	0.64	24.1
1st clutch size	0.1135	0.0102	8.99	21.9
2nd clutch size	0.1923	0.0175	9.10	23.8
3rd clutch size	0.1524	0.0031	2.03	59.1
4th clutch size	0.1578	0.0096	6.08	33.8
5th clutch size	0.1684	0.0038	2.26	71.6

[a]Data from Lynch *et al.* (1998).

within 100 generations. Based on these data, Lynch *et al.* (1998, p. 731) concluded "that standing levels of genetic variance for life-history traits are largely a reflection of the recurrent introduction of mildly deleterious alleles by mutation."

Houle (1998) examined the question in a slightly different manner, arguing that if mutational variance were a significant contributor to the standing additive genetic variance then there would be a positive correlation between the additive coefficient of variation, CV_A, and the coefficient of mutational variance, CV_M. To test this hypothesis, Houle examined data for *Drosophila melanogaster*, finding indeed a significant correlation between the two parameters (Figure 15-1). A similar pattern is also evident in the *Daphnia* data (Figure 15-1).

In summary, whereas theory is of relatively little help at present in estimating the contribution of mutation to standing additive genetic variance, the empirical evidence points toward mutation being a significant contributor.

B. Heterozygous Advantage

For a single locus with two alleles, it is well established that a polymorphism can be maintained if the heterozygote has a higher fitness than either homozygote. However, extension of the two-locus model to multiple loci and hence life-history traits is not immediate. As in the previous models, the theoretical answer changes with the model used (Roff, 1997).

There have been a large number of studies that have related heterozygosity measured at electrophoretically detectable loci with quantitative characters. Three principle hypotheses have been advanced for a correlation between a metric trait and heterozygosity: (1) true overdominance hypothesis—the allozyme loci have a direct effect on the quantitative trait; (2) associative overdominance hypothesis—allozyme loci are themselves neutral but are linked to loci that are themselves determinants of the quantitative trait, the linked loci under selection may show overdominance or be subject to an input of deleterious mutations (the partial dominance hypothesis for inbreeding depression (Charlesworth, 1991); (3) inbreeding depression hypothesis—heterozygosity at the allozyme loci are general indicators of the overall level of heterozygosity (Mitton and Pierce, 1980), and individuals with high levels of homozygosity suffer inbreeding depression.

Heterozygosity at loci coding for enzymes might be favored because of the metabolic buffering it confers, but direct evidence for this appears to be lacking. Pogson and Zouros (1994) tested for a causal relationship between heterozygosity and growth rate in the deep-sea scallop by examining the pattern between restriction fragment length polymorphism (RFLP) heterozygosity and growth rate. They argued that if the observed pattern with electrophoretic loci resulted from associative overdominance then the same pattern should be observed with other assumed neutral markers. No relationship was observed between RFLP heterozygosity and

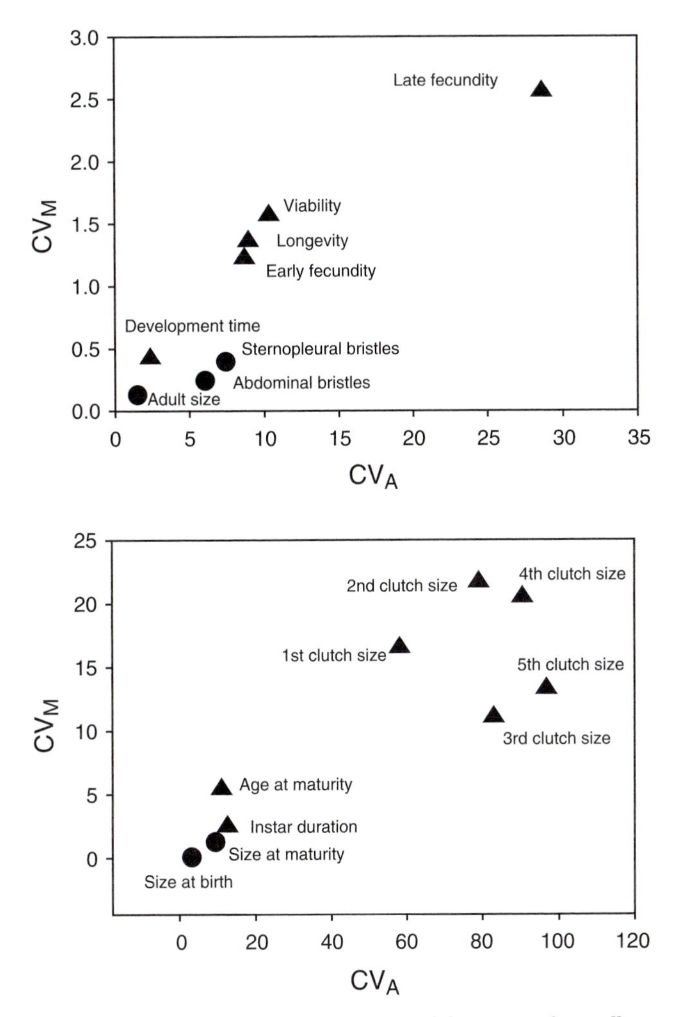

FIGURE 15-1. Plots of the coefficient of mutational variability versus the coefficient of additive genetic variance for *Drosophila melanogaster* (upper panel) and *Daphne pulex* (lower panel). Data from Houle (1998) and Lynch *et al.* (1998).

growth rate providing support for a causative relationship between the electrophoretic loci and growth rate. Only further studies can determine whether this is or is not a common phenomenon.

The true overdominance hypothesis could also be distinguished from the other two hypotheses by examining the relationship between heterozygosity, the quantitative trait, and population size: There should be no effect of population size if the true overdominance hypothesis is correct, but a loss of the correlation between heterozygosity and the quantitative trait in very large populations where linkage disequilibrium and inbreeding depression should be absent. Houle (1989) tested this by examining the correlation between heterozygosity and three metric traits—wing area, growth, and development rate—in a large outbred population of *D. melanogaster*. Houle found no significant correlations and so rejected the true overdominance hypothesis.

If the variance in a trait is entirely additive, then the question of heterozygous advantage does not arise (fitness itself is excluded since pleiotropic effects can confer heterozygote advantage even when the genetic control is strictly additive). Because most genetic variance in morphologic traits appears to be additive, while a considerable portion results from dominance variance in life-history traits (see next section), it follows that heterozygous advantage, if important, will be found in fitness-related traits. The available evidence, though relatively slight, favors the partial dominance hypothesis (Crow, 2000; Roff, 2002a) arguing against heterozygote superiority.

C. Antagonistic Pleiotropy

Consider a locus that influences two life-history traits: Mutations that increase the fitness contribution of each trait will be quickly fixed in the population, whereas mutations that are deleterious to both traits will be quickly lost from the population leaving, segregating, at least temporarily, in the population those mutations that increase fitness from one trait while decreasing it from the other trait (Prout, 1980). This is a nonequilibrium hypothesis, pleiotropic effects resulting from a constant flux of mutations that are eventually removed. This is a component of the set of alleles maintained by mutation–selection balance. An alternative mechanism is that it is the antagonistic pleiotropy itself that maintains the genetic variation. L. N. Hazel (1943) argued that alleles affecting two traits that had a positive effect on fitness would be quickly fixed, whereas alleles that had a negative effect would be lost, leaving alleles that had a positive effect via one trait and a negative effect via the other. Rose (1982, 1985) and Curtsinger *et al.* (1994) have shown that this scenario is insufficient to ensure genetic variation at equilibrium except by mutation–selection balance. Equilibrium can occur if there is dominance variance of sufficient magnitude.

As in the case of heterozygous advantage, the conditions for a polymorphism maintained by antagonistic pleiotropy for multiple loci are more stringent than with a single locus but, to date, the exact requirements have not been worked out (Curtsinger *et al.*, 1994). Approximately, at least, the probability that genetic variation in trade-offs is maintained by antagonistic pleiotopy increases as the ratio of dominance to additive genetic variance increases, with a minimum requirement of approximately 0.25. More than 40% of morphologic traits and more than 65% of life-history traits satisfy this requirement (Figure 15-2) with 54% of life-history

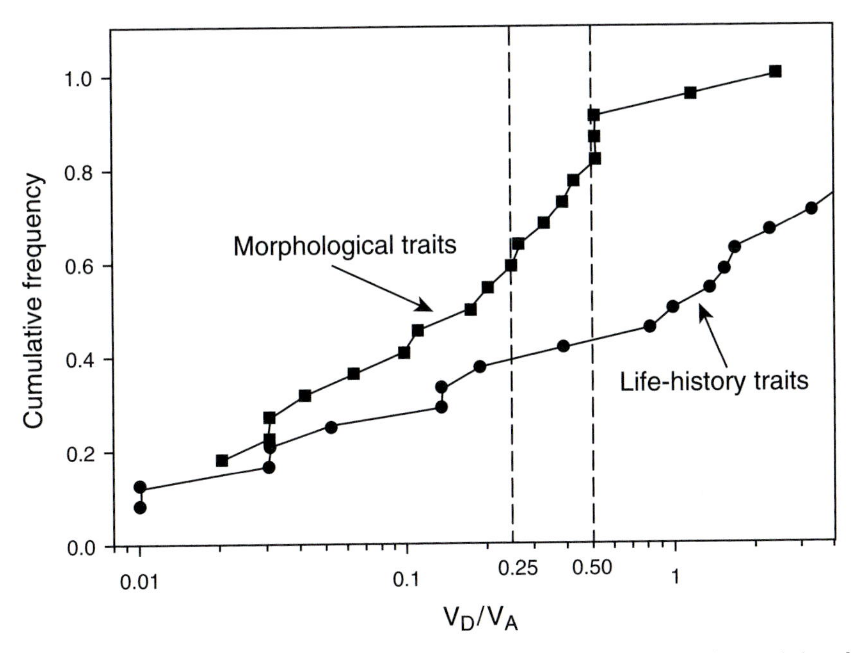

FIGURE 15-2. Cumulative plots of the ratio of dominance to additive genetic variance for morphological and life-history traits. Redrawn from Roff (2002b).

traits having ratios greater than 0.5 and 35% with ratios greater than 2. These data suggest that genetic variation in trade-offs, which will thus lead to correlations less than −1, can very well be sustained by antagonistic pleiotropy.

D. Frequency-Dependent Selection

Frequency-dependent selection can be defined in a variety of ways, which are not identical (Heino *et al.*, 1998), but all carry the sense that the fitness of a phenotype or genotype varies with the phenotypic or genotypic composition of the population. Such selection can readily maintain phenotypic variation and genetic variation at a single locus (Roff, 1997). The maintenance of polygenic variation with frequency-dependent selection are somewhat more restrictive than in the single locus case. Using a simulation model, Mani *et al.* (1990) explored the combined effect of mutation, stabilizing-dependent selection, and frequency-dependent selection on a genetic system in which there are n (≤ 12) loci, each with up to 32 alleles that act additively, the ith allele contributing an amount i to the genotypic value. With the exception of no frequency dependence, the number of alleles maintained at equilibrium was independent of the strength of the frequency-dependent selection and only weakly related to the strength of stabilizing selection. Despite beginning with 32 alleles per locus, at equilibrium the number of alleles per locus was only 1.5–3.0. Nevertheless, these simulations show very clearly that frequency-dependent selection can play a major role in the maintenance of genetic variation.

For life-history theory, the maintenance of phenotypic variation is the central factor, whether or not this variation is underlain by genetic variation. Frequency-dependent selection can be generated both by intraspecific and interspecific interactions. The latter category includes visually oriented predation on color morphs (Clarke, 1979) and selection by parasites and disease (Brunet and Mundt, 2000). Most typically, evidence for frequency-dependent selection is found for discrete traits, which is not surprising given the relative ease with which the frequency of such morphs can be monitored. In several cases the variation in the morph can be linked to variation in some life-history trait. For example, Cosmidis *et al.* (1999) observed frequency-dependent selection at the *Adh* locus in the olive fruit fly (*Bactrocera oleae*), which was linked to variation in development time of the different genotypes.

One of the most common sources of frequency-dependence is alternate male morphs that differ both morphologically and behaviorally. Although in some cases these behaviors are determined by very simple Mendelian genetics (single locus), in the majority of cases, the genetic determination is polygenic (Roff and Fairbairn, 1991) and can be understood using the threshold model of quantitative genetics (Roff, 1986, 1997). According to this model, the dimorphic trait is determined by a continuously distributed underlying trait, called the liability, and a threshold of sensitivity: If the liability lies below the threshold, one morph is produced, whereas the alternate morph is produced when the liability is above the threshold. The liability can be viewed as a quantitative trait in the same manner as, say, body size, and hence its genetic determination calculated in the same manner (although the statistical protocols have to be modified to take into account that only two phenotypes are observed: See Roff [1997] for a description of the methodology). When the dimorphism is polygenic, there appears to be a relatively large heritability under specific conditions, but there are strong environmental influences such that conditions during development modulate the expression of the trait (Roff, 1996). It appears that, in general, the dimorphism, although undoubtedly being influenced by frequency-dependent factors, is also a condition-dependent response, such response being an evolutionary consequence of environmental variation. An interesting case of a color polymorphism that is determined in a simple Mendelian manner is the triple color morphs of the side-blotched lizard, *Uta stansburiana*. The three morphs are distinguished by differing throat color, which is determined by a single locus with three alleles (*o*, *b*, *y*), and mating behavior (Sinervo and Clobert, 2003). Orange males (*oo*, *bo*, *yo*) usurp territories, blue males (*bb*) are mate-guarders, and yellow males (*by*, *yy*) are sneakers. The interaction of the three behaviors creates cyclical frequency shifts, because no single morph is more fit than both of the other two: Sneakers can invade a population of usurpers, mate-guarders are more fit than sneakers, and usurpers can beat mate-guarders (Sinervo and DeNardo, 1996; Sinervo, 2001). Females also show variation in throat color but with only two morphs, orange and yellow, which are correlated with fecundity patterns: Orange females lay large clutches of small eggs, whereas yellow females lay smaller clutches of larger eggs. At low density orange females are favored, and at high density yellow females have higher fitness (Sinervo *et al.*, 2001; Svensson *et al.*, 2001).

II. PHENOTYPIC VARIATION IN A STOCHASTIC ENVIRONMENT

The opposite extreme of a constant environment is one in which there is stochastic variation between generations and/or among sites. No cues are available to indicate what the environment might be in the following generation or next site. Because the approaches are somewhat different, I shall consider temporal and spatial variation separately and then a single example of how the two components can interact to preserve variation.

A. Temporal Variation

Temporal variation acting on a trait under stabilizing selection will not by itself maintain genetic variation, as can be illustrated with the following intuitive argument: In each generation there is selection against extremes at one tail of the distribution, which tail being selected against varying over time, and hence a steady erosion of genetic variation. Alternatively, we may think of temporal variation as stabilizing selection in which the optimal value fluctuates: Because stabilizing selection erodes genetic variance, there is no reason to suppose that fluctuation in the optimal value will be itself sufficient to prevent this erosion. With overlapping generations there are conditions under which genetic variance is maintained (Ellner, 1996). A necessary condition for preservation of genetic variation is that there does not exist a single genotype that produces a range of phenotypes; in the presence of such a genotype, genetic variance will be eroded (Hairston *et al.*, 1996).

Although fluctuating selection may not by itself favor the maintenance of genetic variation, it can favor the maintenance of phenotypic variance. To examine this question, we need to define the appropriate fitness measure for a temporally fluctuating environment. Thoday (1953) suggested that an appropriate measure of fitness is the probability of persistence. If clone A has a persistence time of 100 generations and clone B a persistence time of 200 generations, then on average, after a sufficiently long time, only clone B will still be extant. Thus persistence time is an index of fitness, but a better one (or at least easier to work with) suggested by Cohen (1966), is the geometric mean of the finite rate of increase.

The preceding concept and the finding that temporal stochasticity can favor variation is well illustrated by Cohen's (1966) analysis of the evolution of germination rate in an annual plant. Each year some fraction, G, of the seeds germinate, the rest remaining dormant. Failure to emerge is not without cost, with some fraction D dying each year. The recursion equation for population (clone) size, N, is

$$N_{t+1} = F_t G N_t + (1 - G)(1 - D)N_t \quad (3)$$

where F_t is the number of seeds per parent. The first term on the right represents the contribution from the organisms that emerge in the present year, and the second term is the "seed bank" (=proportion not emerging/germinating times proportion not dying). For any given year, the fitness is measured by the per year rate of increase $W = N_{t+1}/N_t = F_t G + (1-G)(1-D)$. Variation in the rate of increase is generated by temporal variation in the expected fecundity, F_t, of each emergent. For purposes of illustration, let us assume that fecundity can take two values, 0 and F, with the probability of F being P. Taking the geometric mean to be the correct measure of fitness the fitness, W_G for emergence fraction, G, is

$$W_G = [G0 + (1 - G)(1 - D)]^{(1 - P)}[GF + (1 - G)(1 - D)]^{P} \quad (4)$$

To find the optimum emergence fraction, G_{opt}, we first take logs and then differentiate with respect to G

$$\frac{\partial lnW_G}{\partial G} = \frac{(1-P)(1-D)(-1)}{(1-G)(1-D)} + \frac{P(1-D-F)(-1)}{(1-G)(1-D)+GF} \quad (5)$$

The optimum emergence fraction occurs when the derivative equals zero. After algebraic manipulation of the left hand side, we obtain

$$G_{opt} = \frac{PF-(1-D)}{F-(1-D)} \quad (6)$$

Thus provided P is less than 1 (and greater than 0, which is obvious since the population would collapse), the optimal emergence fraction is less than 1 and selection favors phenotypic variation. Most typically, fecundity will not have only two outcomes but many, and the fitness function is written as

$W_G = \prod_i [GF_i + (1 - G)(1 - D)]^{P_i}$, where P_i is the probability of fecundity F_i. Following the same procedure as previously, we find (Roff, 2002b, p. 296) that an intermediate fraction will be favored when

$$1 - D > \left(\sum_i \frac{P_i}{F_i} \right)^{-1}.$$

The expression on the right hand side is the harmonic mean of F_i, which means that an intermediate germination fraction is favored whenever the proportion surviving the dormancy phase is greater than the harmonic mean of the fecundity of an emergent. Because the harmonic mean is less than the geometric mean, it follows that a sufficient condition for G_{opt} to be less than 1 is that $1 - D >$ geometric mean of F_i.

The idea that in a temporally variable world selection may favor variation in offspring phenotype was suggested independently at least three times. The idea is implicit in Cohen's analysis of optimal germination rate in a randomly varying environment, was explicitly put forward as a verbal argument by Den Boer (1968) who termed the phenomenon "spreading the risk," and finally developed by Gillespie (1974, 1977) in the context of variation in offspring number. In reviewing Gillespie's analysis, Slatkin (1974) labeled it as "bet-hedging," a term that has stuck, although den Boer's term actually has priority. Cooper and Kaplan (1982) used extinction probability as an index of fitness and suggested that environmental variation would favor random variation in phenotypic values, calling this phenomenon "adaptive coin-flipping." Again this idea is really implicit in those of Cohen, den Boer, and Gillespie, nonrandom variation falling under the umbrella of phenotypic plasticity.

Bet-hedging may underlie numerous life-history patterns, such as dimorphic variation, iteroparity, brood size reduction, and propagule size (Roff, 2002). Cohen (1966) considered a particular type of dimorphic variation (dormancy). There are many other examples in which two distinct morphs are produced: Body size (e.g., major/minor males), trophic morphology, wing morphology, and cyclomorphosis (Roff, 1996). As an example, consider the case of a protective dimorphism, as is found in numerous zooplankton species. Suppose that one morph is resistant to attack from a particular predator while another is not. However, the unprotected morph has a higher fecundity (e.g., various invertebrates such as *Daphnia*). First, consider the relative fitness of two phenotypes, one that produces only the unprotected morph and one that produces only the protected morph. In the "predator-free" environment, let the fitness of the unprotected morph be 1, and in the "predator-present" environment, let it be w_2. For the protected morph, let the respective fitness be w_2 and w_3. Given a temporal frequency of predator-free environments of f, the fitness of the unprotected morph will exceed that of the protected morph whenever

$$f > \frac{ln(w_3) - ln(w_1)}{ln(w_3) - ln(w_1) - ln(w_2)} \tag{7}$$

To examine the likelihood that a mutant that produces both morphs with frequency P can invade the population, I (Roff, 1996) varied all five parameters (three fitnesses, f, P) within the range 0.1 to 0.9 in increments of 0.1. Of the 13,359 combinations examined, the unprotected morph had the highest fitness in 46.2% of cases, the protected morph in 45.6% of case, and the dimorphic phenotype in 8.2% of cases. Thus, although in this arbitrary parameter space phenotypic variation can be favored, the greatest majority of combinations resulted in a population without phenotypic variation. Although a general theoretical argument can be made for bet-hedging in this example, it is quite possible that over most of parameter space it is unlikely. In an extensive review of the literature on bet-hedging in insects, Hopper (1999) was unable to find any definitive tests that it has been of consequence in the evolution of insect life histories.

B. Spatial Variation

In contrast to temporal variation, spatial variation can readily preserve genetic and hence phenotypic variation. The reason for this is that different selection regimens are operating at the same time on different portions of the populations; consequently, genetic variation lost in one patch may be restored by migration of individuals from another patch. This intuitive argument also applies to the case of overlapping generations, patches being replaced by generations. The first analysis of spatial variation was that of Levene (1953), who assumed a single-locus, two-allele model. A sufficient, but not necessary, condition for a stable equilibrium to occur is that the weighted harmonic means of the fitnesses are less than 1. Numerous variants of Levene's model have been analyzed. The results of these investigations have verified the qualitative conditions for a polymorphism to be maintained in a spatial environment. Three important models, fundamentally different from that of Levene, are those of Bulmer (1971), Gillespie (1976), and Gillespie and Turelli (1989).

Bulmer (1971) investigated the stability of an additive genetic model in a two-patch universe. Within each patch there is stabilizing selection with the optimal value differing between patches and migration between patches. Significant genetic variation can be maintained by migration, but the phenotypic variance must be significantly smaller than the difference between the optima and the stabilizing selection coefficient.

Gillespie (1976, 1978) introduced a model in which enzyme activity is an additive function of alleles, and fitness is a concave function of enzyme activity.

As a consequence, the heterozygote is intermediate in fitness between the two homozygotes. The fitnesses of the homozygotes are assumed to vary across patches: In the two patch case, one homozygote, say *AA*, is the most fit in the first habitat, while the second homozygote, *aa*, is dominant in the second habitat. This reversal in dominance is a critical element of the model. Because the function relating fitness to enzyme activity is concave, the heterozygote is fitter than the arithmetic mean of the two homozygotes. The parameter space over which this model produces a stable polymorphism far exceeds that of Levene's model (Maynard Smith and Hoekstra, 1980).

Gillespie and Turelli (1989) proposed an alternative model in which there cannot be a single genotype that is most fit in all environments. The phenotype of an individual is the sum of three components, G, E and Z, where G is the average phenotype produced by a given genotype averaged over all environments, E is an environmental effect that is independent of the genotype, and Z is a genotype x environment effect. Assuming neither dominance nor epistasis, $G + Z$ is simply the sum of the individual contributions of the alleles. A consequence of this particular formulation is that the variance of the average phenotype produced by a given genotype across all environments is a decreasing function of the number of heterozygous loci. Gillespie and Turelli then assumed that there is a single phenotype that is optimal in all environments (i.e., there is a single stabilizing selection function). With this assumption it can be shown that the mean fitness of a genotype is an increasing function of the number of heterozygous loci. Because of the overdominance averaged across all environments, selection will tend to preserve genetic variation. The important assumption of this model is that increasing heterozygosity buffers the organism against environmental perturbations.

All of the preceding analyses assume that there is migration between sites, but in the absence of any cues as to the state of an environment, selection would, in most cases, simply select for the absence of migration and hence evolution to the optimum trait combination for each separate habitat. An exception to this would be if all available resources were taken up and hence density dependence was operating. In this case migration may be selected for because it is evolutionarily better to compete against nonrelatives than relatives. This proposition does suppose that there is genetic variation, which would lead to inbreeding depression. Whereas there is considerable evidence that the requisite type of genetic variation (directional dominance) exists for life-history traits, there is no evidence that such variation would itself remain given only spatial variation. It may be mathematically convenient to consider a model in which there is spatial variation and migration, but it is probably unrealistic. A more reasonable scenario is one in which there is both spatial and temporal variation.

C. Spatial and Temporal Variation

Southwood (1962) advanced the hypothesis that the proportion of migrants in a population would be inversely related to the persistence time of habitat patches.

An operational difficulty in testing this hypothesis is that in many species it is not possible to morphologically distinguish migrants from nonmigrants. However, there are a group of taxa both animal and plant in which two morphs occur, one that is incapable of migration and another that is capable of migration but may not choose to do so. For example, in paedomorphosis in salamanders or wing dimorphism in insects, there is a morph that clearly cannot migrate to the same extent as the other morph (thus the neotenic form of the salamander cannot leave its natal pond, and the flightless insect morph is restricted, generally, to migration by walking, which is clearly more limiting than flight). In these cases we can call one morph the "nonmigrant" and the other morph a "potential migrant." Using the presence of this dimorphism in planthoppers, Denno *et al.* (1991) showed a negative correlation between the proportion of the macropterous (long-winged) morph and habitat persistence, confirming Southwood's hypothesis.

To illustrate how a dimorphism for wing morph (or some other morphologic trait that enhances/prevents migration) can evolve and be maintained in an environment that is spatially and temporally variable, consider an environment consisting of discrete patches, with each patch persisting for some finite time period and the intervening habitat being such that the "nonmigrant" form cannot move from one patch to another. Patches are coming into and out of existence, and hence a population consisting solely of nonmigrants will not persist. Initially, therefore, we suppose that the population consists of only winged individuals, which is the ancestral state of all wing-dimorphic insects. Now a mutation arises that produces a flightless morph. Because of the initially low frequency of the mutant allele, it will occur overwhelmingly in the heterozygous state. If a mutant allele is recessive, it will not be expressed in the heterozygous condition, and hence such mutations will only increase initially by genetic drift. However, if the allele is dominant, the flightless (nonmigrant) morphology will be immediately expressed, and any advantage it confers will be realized. In this example, therefore, I shall assume that the mutant allele is dominant, which is consistent with the observation that where wing dimorphism is determined by a single-locus, two-allele mechanism almost without exception the flightless allele is dominant (Roff, 1986; Roff and Fairbairn, 1991).

As the flightless morph is not diverting energy into the maintenance of flight muscles and associated structures, the flightless form will be assumed to have a higher fecundity than the winged morph, evidence for which is plentiful (Roff and Fairbairn, 1991). Therefore, the flightless allele will increase in frequency in the patch in which it arises. This increase will be enhanced if, as is likely, the emigration of the winged morph exceeds immigration. Because the flightless morph cannot move between patches and the flightless allele is dominant, the flightless allele can only be passed to another patch if a flightless male mates with a winged female prior to migration. If mating always occurs after migration, the flightless allele will never move to another patch and will be eliminated from the population when the patch in which it arose ceases to exist. Assuming random mating within a patch, the frequency of mating between a volant (flight capable = winged) female that subsequently migrates and a flightless male increases with the frequency of the flightless allele (which itself must increase with the persistence time of the patch). Because of its fecundity advantage and relatively greater survival rate within a patch (i.e., no emigration), the flightless allele will spread though the population. However, it cannot replace the winged allele, because a population homozygous for the flightless allele will never colonize new patches. Hence it will become extinct as each patch eventually ceases to exist. Thus an equilibrium frequency will be established that is dependent upon the persistence time of the patch, the cost of being winged, the relative survival rate of the winged morph, the probability that a winged female migrates, and possibly the genetic determination of the trait (single locus versus polygenic). Simulation of this model shows that two morphs can be maintained in the population and that it can correctly predict the observed relationship in the planthopper data discussed in the preceding text (Roff, 1994).

III. PREDICTABLE ENVIRONMENTS

In many circumstances there are cues that provide information on the future state of conditions an individual or its progeny will encounter. For example, the time available for growth and reproduction in a seasonal environment can be predicted with varying degrees of accuracy by the day length and current temperature. A particularly fine example is found in the pitcher-plant mosquito, *Wyeomyia smithii*. The cold winter months are passed over by diapause in the third or fourth larval instar (Bradshaw and Holzapfel, 1990). Diapause is induced when the larvae encounter a critical photoperiod. Changes in latitude and/or altitude reduce the season length, and hence there should be a corresponding shift in the critical photoperiod inducing diapause. An extensive investigation of diapause induction over a wide range of latitudes and altitudes have confirmed this (Bradshaw, 1976; Bradshaw and Lounibos, 1977, Figure 15-3). The northward expansion of the mosquito has required adaptation in diapause induction, for which experiments have demonstrated abundant additive genetic variation—$h_2 \approx 0.5$ (Hard *et al.*, 1993). What is maintaining the genetic variation at any given site has not been studied.

Selection favors interactions with the environment that increase fitness. Suppose there are two types of habitats, designated E_1 and E_2, in which the optimal trait values are X_{1*} and X_{2*}, respectively. The most fit phenotype/genotype is that which is able to perceive the environmental value and react in such a manner that its trait values in habitats 1 and 2 are the optimal values (X_{1*} and X_{2*}, respectively), that is, selection will favor the evolution of some response $f(E)$ such that $X_i = f(E_i) = X_{i*}$. Thus, for example, in a two-patch spatial model, the optimal ages at maturity if the organism were able to perceive what type of patch it was in would differ between patches, whereas in a stochastic environment only a single age would evolve.

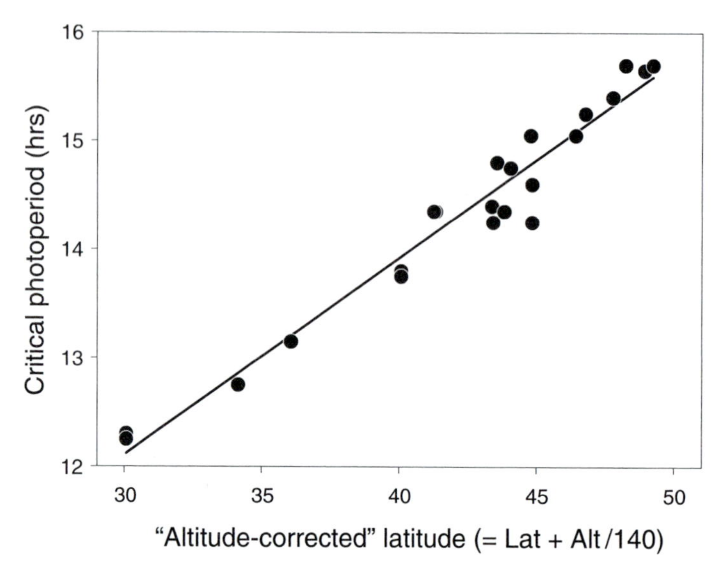

FIGURE 15-3. The correspondence between latitude and altitude and the critical photoperiod in the pitcher-plant mosquitoe, *Wyeomyia smithii*. After Bradshaw and Lounibos (1977).

The production of different phenotypes according to the environmental conditions is known as phenotypic plasticity and the function, $f(\cdot)$, is termed the norm of reaction. An implicit assumption in the preceding description is that there is a single genotype (i.e., the different phenotypes are not a consequence of selective mortality of particular genotypes). A general definition of phenotypic plasticity is "a change in the average phenotype expressed by a genotype in different macro-environments" (Via, 1987, p. 47). In a similar manner, the norm of reaction can be defined as the following: "A reaction norm as coded for by a genotype is the systematic change in mean expression of a phenotypic character that occurs in response to a systematic change in an environmental variable" (De Jong, 1990, p. 448). This definition does not exclude discrete environments because they can be subsumed under the definition by the statistical approach of dummy variables. In some instances the same phenotype can be produced by several different genotypes or in the face of environmental variation: This phenomenon is termed "canalization" (Waddington, 1942).

Phenotypic plasticity is extensively discussed in the chapter by Stearns. Here I shall present a single example of how phenotypic variation can persist when the trait is phenotypically plastic. The example I discuss is the evolution of inducible defenses.

If there is a cost to a defensive trait, we might expect it to be phenotypically plastic, appearing when some environmental stimulus indicates a need. Many traits do indeed show such plasticity, this plasticity consisting of either a primarily dimorphic, irreversible condition or a reversible, largely quantitative response. In the first group are many animal examples, such as helmets and neckteeth (Cladocera), spines (rotifers, protozoans), and crypic morphs (Lepidoptera). Plants seem to fall mainly in the second group, though there are also animal examples (Table 15-2). The induction of the defensive structures may require only a chemical cue released by the potential predator or be a response to a physical assault (e.g., consumption of leaves by insect herbivores induces the production of defensive compounds in many plant species). As predicted in the preceding text, the overall finding supports the hypothesis that inducible defenses are costly and reduce fitness in the absence of the predator (Table 15-2). In general, defended morphs show a reduced rate of growth and a reduced fecundity and in some cases an elevated mortality rate in the absence of the predator.

The fundamental assumption of models for the evolution of an inducible defense is that survival probability can be assessed by some environmental stimulus. Now, suppose that there are two morphs, an unprotected morph and a protected

TABLE 15-2. **Examples of Defensive Polymorphisms with Assessment, where Available, of the Costs and Benefits of the Two Morphologies**[a]

Taxon	Defensive morph induced	Inducing factor produced by	Costs[b] α, $l(x)$, $m(x)$
Daphnia spp., Cladoceran	Helmeted	Invertebrate and vertebrate[c] predators	+ , − , −
Daphnia pulex, Cladoceran	Necktoothed	Invertebrate predators[c]	+ , − , ND
Brachionous calyciflorus, *Keratella* spp., Rotifers	Spined	Invertebrate predators[c]	ND, − , −
Asplanchna spp., Rotifer	(a) Cruciform (b) Giant	(a) α-tocopherol released by algal cells (b) Dietary α-tocopherol + large prey	ND, ND, −
Protozoa, *Onychodromus quadricornutus*, *Euplotes* spp.,	Spined	Giant morph, Predatory ciliates	Reduced growth
Membranipora membranacea, Bryozoan	Spined	Grazing by nudibranch	Reduced colony growth
Chthmalus anisopoma, Gastropod	Bent	Predatious gastropod	+, ND, −
Thais lamellosa, Gastropod	Larger apertural teeth	Predatory crab	+, ND, ND
Littorina obtusata, Gastropod	Thicker shell	Predatory crab	Possibly reduced growth

Continued

TABLE 15-2. Examples of Defensive Polymorphisms with Assessment, where Available, of the Costs and Benefits of the Two Morphologies[a]—*Cont'd*

Taxon	Defensive morph induced	Inducing factor produced by	Costs[b] α, $l(x)$, $m(x)$
Corals and sea anemones	Catch tentacles	Proximity of competitors	Reduced number of feeding tentacles
Papilionidae, swallowtail butterflies	Pupal color background	Photoperiod, substrate color, foodplant odor	Increased mortality on wrong background
Acyrthosiphon pisum, aphid	Macroptery	Predatory beetle	Reduced fecundity, increased development time
Nemoria arizonaria, Caterpillars	Twig mimic	Tannin concentration	+, − , −[d]
Hyla chrysoscelis, Tadpoles	Inactive, larger more brightly colored tailfin	Dragonfly larvaed[e]	Lowered survival in absence of predator
Carassius carassius, carp	Deep bodied	Predatory fish (pike)	+, ND, ND
Harmonia axyridis, ladybird beetle	Reflex bleeding	Predators	+, ND, ND
Gossypium thurberi, Plant	Resistance	Natural damage by leaf miners early in season	None (survival, growth, reproduction)
Hordeum vulgare, Plant	Resistance(?)	Avirulent strain of powdery mildew[f]	Grain yield, Kernel weight, grain protein
Lycopersicon esculentum, Plant	Resistance	Chitin injection[j] (simulating caterpillar attack)	None (survival, growth, reproduction)
Nicotiana sylvestris, Plant	Resistance	Artificial damage (simulating caterpillar attack)	Total plant mass, number of fruits
Brassica rapa, Plant	Cyanide compounds	Moth larvae and fungal pathogen	Decreased seed production
Pastinaca sativa, Plant	Furanocoumarins	Attack by herbivores or pathogens	Decreased fruit production[g]
Phytoplankton spp.	Colony form, spines	Grazing by small zooplankters	Higher sinking rate

[a]Adapted from Roff (2002b).
[b]Columns indicate: development time (in some cases growth was slowed but time to maturity not presented), mortality in absence of predator, fecundity. +; increased. −; decreased. 0; no effect. ND; no data.
[c]Induction is initiated by the presence of chemicals released by the predator.
[d]The dimorphism is seasonal with catkin mimics being produced early in the season and twig mimics later in the season, when catkins are no longer present.
[e]Dragonfly larvae kept in cage inside tadpole container.
[f]Costs assessed from lines genetically altered to increase production of defensive compounds.
[g]Negative genetic correlation between production of compound and fruit set.

morph, and that the strength of the stimulus indicates the probability of being attacked and eaten. As shown in Figure 15-4, the probability for the protected morph will lie below that of the unprotected morph. As shown by many empirical studies, the protected morph suffers a cost in another fitness component, say fecundity. Taking lifetime reproductive success as our fitness measure (for

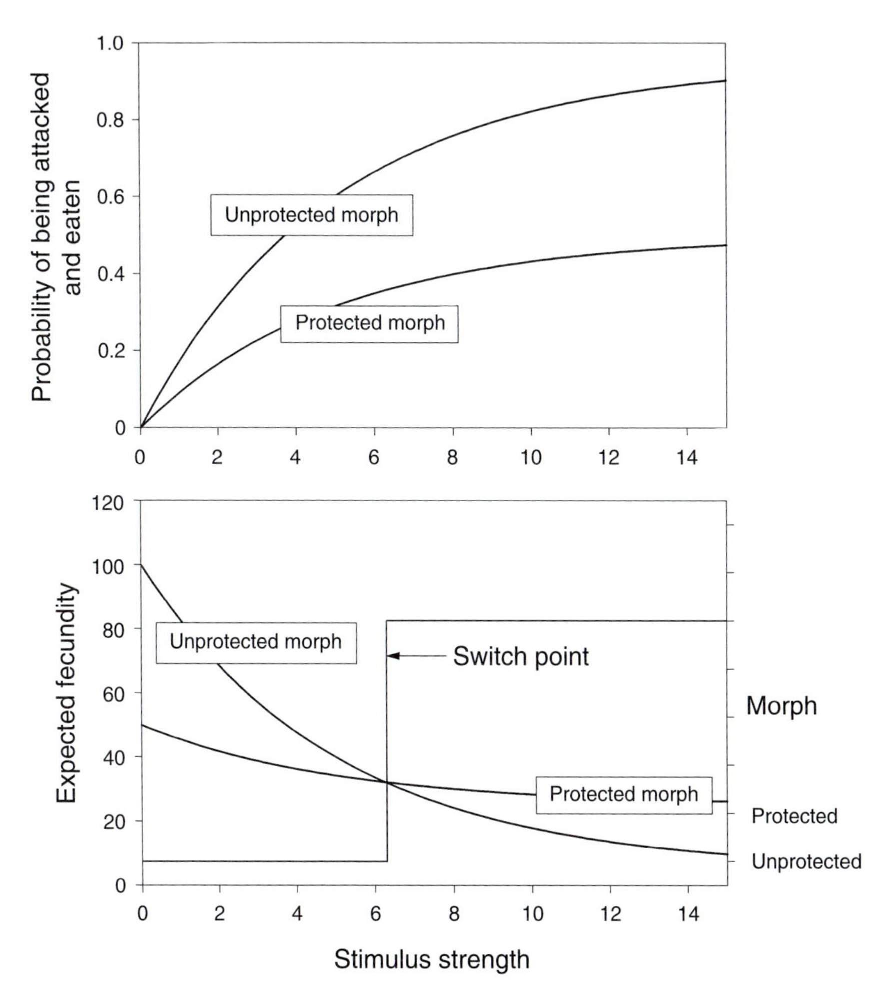

FIGURE 15-4. A hypothetical example of an inducible trait. Some cue (=stimulus strength) is present in the environment that indicates the probability of being attacked and eaten (upper panel). This probability is greater for the unprotected morph, but its higher fecundity gives it a higher expected fitness (= survival × fecundity) for low levels of the stimulus (lower panel). However, at some value the expected fecundity of the protected morph exceeds that of the unprotected morph, and the organism should switch to the protected morph. Redrawn from Roff (2002).

simplicity only) then, at low stimulus strength, the fitness of the unprotected morph will be higher than that of the protected morph, but eventually the curves cross and the fitness ranks are reversed (Figure 15-4). The immediate prediction from this model is that the optimal reaction norm is a step function (Figure 15-4). If there is genetic variation for the threshold of induction, we have the standard threshold model of quantitative genetics and a population reaction norm that is a cumulative normal (Myers and Hutchings, 1986; Hazel *et al.*, 1990). As in the case of a stochastic environment, we can consider temporal and spatial variation separately.

A. Temporal Variation

There are four possible life histories: monomorphically unprotected, monomorphically protected, dimorphic with some given probability (i.e., the bet-hedging response), and dimorphic according to the environmental conditions (phenotypic plasticity). Given that there is an environmental cue that predicts fitness, will a reaction norm evolve? The simple answer is "not necessarily." The evolution of phenotypic plasticity will depend upon how well the cue predicts the environment and what are the costs of making the incorrect developmental decision.

To examine how frequent the four life-history responses (monomorphic protected, monomorphic unprotected, bet-hedging, reaction norm) are the most fit, I ran the following simulation for a two-state universe (Roff, 2002, Table 15-3). Within-state fitness values were generated at random subject to the constraint $0 < w_1 < w_3 < w_2 < 1$. The probability of adopting the protected morph in the "many predator" state, P_M (which is a measure of the reliability of the signal), was also selected at random from the interval 0–1 and P_f selected at random from the interval 0–P_m. For each combination of parameters and given patch frequency, f, the most fit life-history response was determined. One thousand replicates were generated for each value of f (Figure 15-5). For the bet-hedging life history,

TABLE 15-3. Parameter Definitions for the Evolution of an Inducible Defense in a Two-State/Patch Universe[a]

	State/Patch type	
	Few predators	Many predators
Frequency of states/patches	f	$1-f$
Fitness of the unprotected morph	1	W_1
Fitness of the protected morph	W_2	W_3
Probability of developing into the protected morph	P_f	P_m

[a]The maximum fitness is 1, achieved by the unprotected morph in the environment with few predators.

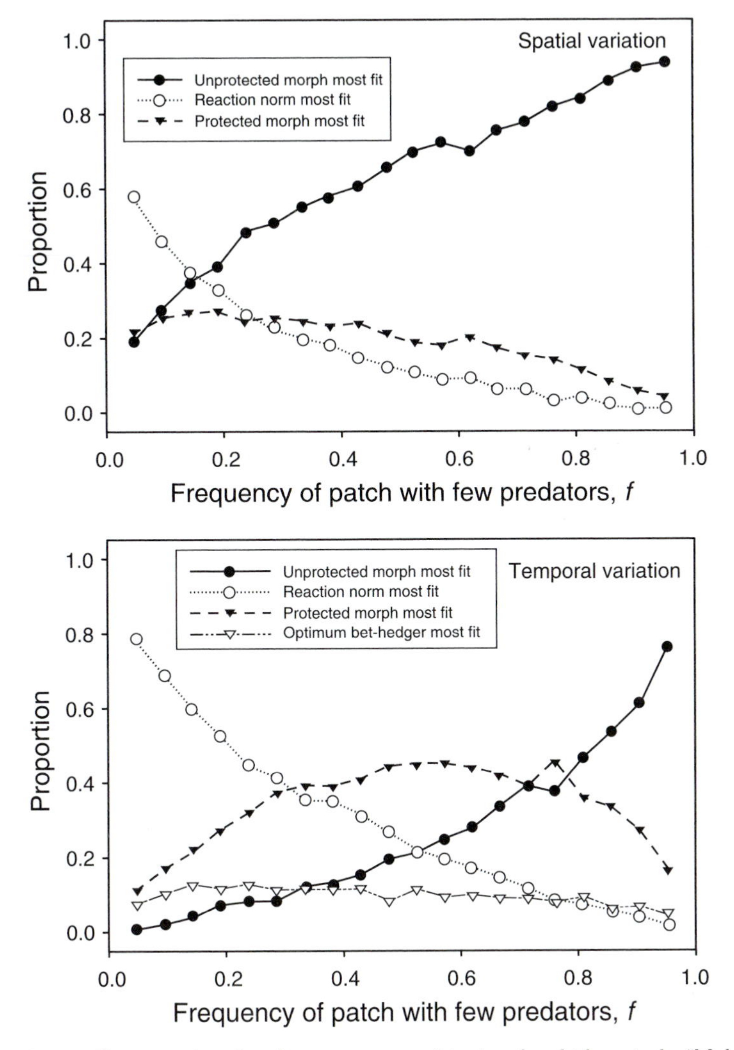

FIGURE 15-5. The proportion of random parameter combinations for which particular "life-history" responses are the most fit in a two-patch, spatially variable (top panel) or temporally variable (bottom panel) environment. Fitnesses are as defined in Table 15-3. See Roff (2002) for details of the calculation of the fitnesses of the three (spatial) or four (temporal) types of responses.

I calculated the optimum P for each combination. When the frequency of few predator states was low (approx $0 < f < 0.2$), the reaction norm life history was most frequently favored (Figure 15-5). At intermediate frequencies (approx $0.3 < f < 0.7$), the protected morph had the highest fitness most often, and at high frequencies ($f > 0.8$), the unprotected morph dominated most often. Optimal bet-hedging had the highest fitness in approximately 10% of the combinations, regardless of state frequency, but never had the highest overall probability of being the most fit. These results demonstrate (1) that for particular parameter values any of the four life histories might be the most fit (i.e., all possible life histories were observed) and (2) that a reaction norm is much more likely that an optimal bet-hedging response.

B. Spatial Variation

As in the preceding text, I shall assume a two-state universe, in this case the states being two patches. An animal lives throughout its life in a single patch, but there is random migration between generations. The first question we can ask is "Can a general bet-hedging response have a higher fitness than either of the monomorphic types?" The answer is "no" (p. 448, Roff, 2002): Spatial heterogeneity will by itself never select for a bet-hedging response (Lloyd, 1984; Lively, 1986; Moran, 1992).

Given that there is an environmental cue that predicts fitness, will a reaction norm evolve? The obvious answer to this question is: Yes, depending on parameter values. Superiority of the reaction norm response depends on the frequency of the two patches relative to the difference in fitness between the two morphs within patches and the probability of developing into the protected morph. An important point is that the most fit morph is not determined by the patch with the highest frequency, and thus it can happen that the most fit morph is actually determined by a highly infrequent patch that has a high fitness cost for making the wrong decision. Therefore, it could appear to the casual observer who fails to record these infrequent patches that the organism is maladapted to its environment. To examine how often each of the three possible life histories is favored, I generated parameter combinations as in the case of temporal variation. When most patches contain many predators (approximately, $0 < f < 0.1$), a reaction norm is favored most often, but when the frequency of patches with few predators exceeds approximately 0.2, the unprotected morph most frequently has the highest fitness (Figure 15-5).

IV. CONCLUDING COMMENTS

Variation is the norm for virtually all traits, particularly life-history traits. In part the variation is a consequence of genetic mechanisms—mutation–selection balance, antagonistic pleiotropy, heterozygous advantage—and in part a result of external factors that lead to, for example, frequency-dependent selection, bet-hedging, and the evolution of phenotypic plasticity. The challenge is to determine the role each of these phenomena plays in generating and maintaining variation. It most likely varies both among traits and among species. There is no doubt that phenotypic plasticity is important, and present evidence suggests that mutation may be contributing significantly to standing variation. The role of bet-hedging is still uncertain, as is the importance of antagonistic pleiotropy (the latter is certainly important in determining the optimum combination of traits but not necessarily in preserving variation). Frequency-dependent selection is important for discrete traits, but its impact on quantitative traits remains to be demonstrated, except in the cases in which the quantitative variation is coupled to discrete variation. All in all there is still much work to be done.

REFERENCES

Bradshaw, W. E. (1976). Geography of photoperiodic response in a diapausing mosquito. *Nature* **262**, 384–386.

Bradshaw, W. E., and Lounibos, L. P. (1977). Evolution of dormancy and its photoperiodic control in pitcher-plant mosquitoes. *Evolution* **31**, 546–547.

Bradshaw, W. E., and Holzapfel, C. M. (1990). Evolution of phenology and demography in the pitcher plant mosquito, *Wyeomyia smithii*. In *Insect Life Cycles: Genetics, Evolution, and Co-ordination* (F. Gilbert, F. ed), pp. 47–67. London Springer-Verlag.

Brunet, J., and Mundt, C. C. (2000). Disease, frequency-dependent selection, and genetic polymorphisms: Experiments with stripe rust and wheat. *Evolution* **54**, 406–415.

Bulmer, M. G. (1971). Stable equilibria under the two island model. *Heredity* **27**, 321–330.

Bulmer, M. G. (1989). Maintenance of genetic variability by mutation–selection balance: A child's guide through the jungle. *Genome* **31**, 761–767.

Burger, R. (1998). Mathematical properties of mutation–selection models. *Genetica (Dordrecht)* **102–103**, 279–298.

Charlesworth, D. (1991). The apparent selection on neutral marker loci in partially inbreeding populations. *Genetic Research* **57**, 159–175.

Clarke, B. C. (1979). The evolution of genetic diversity. *Proceedings of the Royal Society of London B* **205**, 453–474.

Cohen, D. (1966). Optimizing reproduction in a randomly varying environment. *Journal of Theoretical Biology* **12**, 119–129.

Cooper, W. S., and Kaplan, R. H. (1982). Adaptive 'coin-flipping': A decision-theoretic examination of natural selection for random individual variation. *Journal of Theoretical Biology* **94**, 135–151.

Cosmidis, N., Loukas, M., and Zouros, E. (1999). Rarer need not be better if commoner is worse: Frequency-dependent selection for developmental time at the alcohol dehydrogenase locus of the olive fruit fly, *Bactrocera oleae*. *Evolution* **53**, 518–526.

Crow, J. F. (2000). The rise and fall of overdominance. *Plant Breeding Reviews* **17**, 225–257.

Curtsinger, J. W., Service, P. M., and Prout, T. (1994). Antagonistic pleiotropy reversal of dominance and genetic polymorphism. *American Naturalist* **144**, 210–228.

De Jong, G. (1990). Quantitative genetics of reaction norms. *Journal of Evolutionary Biology* **3**, 447–468.

Den Boer, P. J. (1968). Spreading of risk and stabilization of animal numbers. *Acta Biotheoretica* **18**, 165–194.

Denno, R. F., Roderick, G. K., Olmstead, K. L., and Dobel, H. G. (1991). Density-related migration in planthoppers (Homoptera: Delphacidae): The role of habitat persistence. *American Naturalist* **138**, 1513–1541.

Ellner, S. (1996). Environmental fluctuations and the maintenance of genetic diversity in age or stage-structured populations. *Bulletin of Mathematical Biology* **58**, 103–127.

Gillespie, J. H. (1974). Natural selection for within-generation variance in offspring number. *Genetics* **76**, 601–606.

Gillespie, J. H. (1976). A general model to account for enzyme variation in natural populations II. Characterization of the fitness function. *American Naturalist* **110**, 809–821.

Gillespie, J. H. (1977). Natural selection for variance in offspring numbers: A new evolutionary principle. *American Naturalist* **111**, 1010–1014.

Gillespie, J. H. (1978). A general model to account for enzyme variation in natural populations. V. The SAS-CFF model. *Theoretical Population Biology* **14**, 1–45.

Gillespie, J. H., and Turelli, M. (1989). Genotype-environment interactions and the maintenance of polygenic variation. *Genetics* **121**, 129–138.

Hairston, J. N. G., Ellner, S., and Kearns, C. M. (1996). Overlapping generations: The storage effect and the maintenance of biotic diversity. In *Population Dynamics in Ecological Space and Time* (O. E. Rhodes, J. R. K. Chesser, and M. H. Smith, eds.), pp. 109–145. Chicago: The University of Chicago Press.

Hard, J. J., Bradshaw, W. E., and Holzapfel, C. M. 1993. The genetic basis of photoperiodism and its evolutionary divergence among populations of the pitcher-plant mosquito, *Wyeomyia smithii*. *American Naturalist* **142**, 457–473.

Hazel, L. N. (1943). The genetic basis for constructing selection indices. *Genetics* **28**, 476–490.

Hazel, W. N., Mock, R. S., and Johnson, M. D. (1990). A polygenic model for the evolution and maintenance of conditional strategies. *Proceedings of the Royal Society of London B* **242**, 181–187.

Heino, M., Metz Johan, A. J., and Kaitala, V. (1998). The enigma of frequency-dependent selection. *Trends in Ecology & Evolution* **13**, 367–370.

Hopper, K. R. (1999). Risk spreading and bet-hedging in insect population biology. *Annual Review of Entomology* **44**, 535–560.

Houle, D. (1989). Allozyme-associated heterosis in *Drosophila melanogaster*. *Genetics* **123**, 789–801.

Houle, D. (1998). How should we explain variation in the genetic variance of traits? *Genetica (Dordrecht)* **102–103**, 241–253.

Levene, H. (1953). Genetic equilibrium when more than one ecological niche is available. *American Naturalist* **87**, 331–333.

Lively, C. M. (1986). Canalization versus developmental conversion in a spatially variable environment. *American Naturalist* **128**, 561–572.

Lloyd, D. G. (1984). Variation strategies of plants in heterogeneous environments. *Biological Journal of the Linnean Society* **21**, 357–385.

Lynch, M., Latta, L., Hicks, J., and Giorgianni, M. (1998). Mutation, selection, and the maintenance of life-history variation in a natural population. *Evolution* **52**, 727–733.

Mani, G. S., Clarke, B. C., and Sheltom, P. R. (1990). A model of quantitative traits under frequency-dependent balancing selection. *Proceedings of the Royal Society of London B* **240**, 15–28.

Maynard Smith, J., and Hoekstra, R. (1980). Polymorphism in a varied environment: How robust are the models? *Genetic Research* **35**, 45–57.

Mitton, J. B., and Pierce, B. A. (1980). The distribution of individual heterozygosity in natural populations. *Genetics* **95**, 1043–1054.

Moran, N. (1992). The evolutionary maintenance of alternative phenotypes. *American Naturalist* **139**, 971–989.

Myers, R. A., and Hutchings, J. A. (1986). Selection against parr maturation in Atlantic Salmon. *Aquaculture* **53**, 313–320.

Pogson, G. H., and Zouros, E. (1994). Allozyme and RFLP heterozygosities as correlates of growth rate in the scallop *Placopecten magellanicus*: A test of the associative overdominance hypothesis. *Genetics* **137**, 221–231.

Prout, T. (1980). Some relationships between density-dependent selection and density dependent population growth. *Evolutionary Biology* **13**, 1–68.

Roff, D. A. (1986). The evolution of wing dimorphism in insects. *Evolution* **40**, 1009–1020.

Roff, D. A. (1994). Habitat persistence and the evolution of wing dimorphism in insects. *American Naturalist* **144**, 772–798.
Roff, D. A. (1996). The evolution of threshold traits in animals. *Quarterly Review of Biology* **71**, 3–35.
Roff, D. A. (1997). *Evolutionary Quantitative Genetics*. New York: Chapman and Hall.
Roff, D. A. (2002a). Inbreeding depression: Tests of the overdominance and partial dominance hypotheses. *Evolution* **56**, 768–775.
Roff, D. A. (2002b). *Life History Evolution*. Sinauer Associates, Sunderland, MA.
Roff, D. A., and Fairbairn, D. J. (1991). Wing dimorphisms and the evolution of migratory polymorphisms among the insecta. *American Zoologist* **31**, 243–251.
Rose, M. R. (1982). Antagonistic pleitropy, dominance, and genetic variation. *Heredity* **48**, 63–78.
Rose, M. R. (1985). Life history evolution with antagonistic pleiotropy and overlapping generations. *Theoretical Population Biology* **28**, 342–358.
Sinervo, B. (2001). Runaway social games, genetic cycles driven by alternative male and female strategies, and the origin of morphs. *Genetica (Dordrecht)* **112–113**, 417–434.
Sinervo, B., Bleay, C., and Adamopoulou, C. (2001). Social causes of correlational selection and the resolution of a heritable throat color polymorphism in a lizard. *Evolution* [print] **55**, 2040–2052.
Sinervo, B., and Clobert, C. (2003). Morphs, dispersal behavior, genetic similarity, and the evolution of cooperation. *Science* **300**, 1949–1951.
Sinervo, B., and DeNardo, D. F. (1996). Costs of reproduction in the wild: Path analysis of natural selection and experimental tests of causation. *Evolution* **50**, 1299–1313.
Slatkin, M. (1974). Hedging one's evolutionary bets. *Nature* **250**, 704–705.
Southwood, T. R. E. (1962). Migration of terrestrial arthropods in relation to habitat. *Biological Reviews* **37**, 171–214.
Svensson, E., Sinervo, B., and Comendant, T. (2001). Condition, genotype-by-environment interaction, and correlational selection in lizard life-history morphs. *Evolution* [print] **55**, 2053–2069.
Thoday, J. M. (1953). Components of fitness. *Symposium of the Society for Experimental Biology* **7**, 96–113.
Turelli, M. (1984). Heritable genetic variation via mutation-selection balance: Lerch's zeta meets the abdominal bristle. *Theoretical Population Biology* **25**, 138–193.
Via, S. (1987). Genetic constraints on the evolution of phenotypic plasticity. In *Genetic Constraints on Adaptive Evolution* (V. Loeschcke, ed.), pp. 47–71. Berlin: Springer-Verlag.
Waddington, C. H. (1942). Canalization of development and the inheritance of acquired characters. *Nature* **150**, 563–565.

CHAPTER 16

Antisymmetry

A. Richard Palmer

Systematics and Evolution Group, Department of Biological Sciences,
University of Alberta, Edmonton, Alberta, and Bamfield Marine Sciences Centre,
Bamfield, British Columbia, Canada

INTRODUCTION

The notion of antisymmetry likely strikes most people as bizarre. How can any variation exist that is "anti-" something else? To dismiss antisymmetry as mere intellectual catnip of academic snoots would seem easy. To dismiss it too hastily would be a big mistake.

Antisymmetry is a peculiar kind of variation whose evolutionary significance is surprisingly unappreciated, no doubt in part *because* the term seems odd and foreboding. However, the phenomenon, with its particularly apt moniker, is actually widespread and offers the promise of valuable insights into a century-old debate about the interplay between development and evolution.

At its simplest, antisymmetry refers to the condition where right-sided and left-sided—or dextral and sinistral—forms are equally common within a species (Palmer, 1996a), as seen in the major claws of lobsters and male fiddler crabs, the side to which the upper mandible crosses in most crossbill finches, or the spiral orientation of palm-tree trunks (Neville, 1976). In other words, each individual within a species is conspicuously asymmetric, but the *direction* of that asymmetry is random. Antisymmetry can therefore only be confirmed by examining multiple individuals or multiple parts on an individual.

Antisymmetry is a particularly important kind of phenotypic variation because, with very few exceptions, the direction of asymmetry is not inherited (see Inheritance of Direction in Antisymmetric Species, which follows). In other words, although the phenotype "asymmetric" is clearly heritable, the conspicuous and readily observable phenotype "right-handed" is not. Such a claim can rarely be made for other kinds of variation.

This lack of a genetic basis to a readily observable phenotype lends antisymmetry its great significance. Because direction of asymmetry is not inherited, each evolutionary transition from an antisymmetric ancestor to a directionally asymmetric descendent represents an example of the seemingly heretical, neo-Lamarckian phenomenon of genetic assimilation (Waddington, 1953): a conspicuous phenotype (e.g., right-handed) with no heritable basis arising evolutionarily before that phenotype comes under genetic control. It is a clear and compelling case where the phenotype leads and the genotype follows (West-Eberhard, 2003).

Because evolutionary transitions between antisymmetry and directional asymmetry, and between symmetry and directional asymmetry, may be enumerated using standard phylogenetic methods, a rare quantitative estimate is possible of the relative contribution of genetic assimilation and conventional evolution to the origin of novel forms (Palmer, 2004). This claim depends critically on an understanding of the biology of antisymmetry.

I. ASYMMETRY TERMINOLOGY

The significance of antisymmetry can only be appreciated when viewed against the entire spectrum of asymmetry variation. Regrettably, the terminology used to describe this variation is cumbersome and confusing (Table 16-1). Complications arise on four fronts.

First, departures from symmetry may be either subtle—on the order of 1% of trait size—or quite conspicuous (Palmer, 1996b). Some terms applied to subtle asymmetries also apply to conspicuous ones, even though their precise meanings differ. Second, conspicuous asymmetries come in a bewildering variety of forms that seem to defy a straightforward and comprehensive nomenclature (e.g., see Taxomonic distribution and functional significance of antisymmetry, which follows). For example, although many asymmetries occur between antimeres of paired structures on otherwise bilaterally symmetric organisms, some take the form of deviations of a single medial structure—or the whole body—toward one side, and yet others are spirals or helices (Ludwig, 1932). Terms typically applied to spiral asymmetries often sound odd when applied to bilateral ones.

Third, unlike a bilaterally paired structure—where the term *right-handed*, by convention, has a precise meaning (i.e., the right side is larger or used preferentially)—the "direction" of a whole-body, spiral, or helical asymmetry is defined arbitrarily, and experts themselves sometimes disagree (Edwards, 1966; Kihara, 1972; Galloway, 1989; Robertson, 1993; Fujinaga, 1997). Fourth, even where a convention exists for defining asymmetries unambiguously in an individual—as right or left, dextral or sinistral—such terms lose their precision when referring to populations or species because groups of individuals may include both enantiomorphs in different frequencies. So additional terms are required to capture the spectrum of asymmetry variation exhibited by groups of individuals.

A. Terms for Subtle Asymmetries

Terms for subtle asymmetries refer to the form of the frequency distribution of differences—on the order of 1% of trait size—between the right and left sides in a sample of individuals (Palmer, 1994). They include *fluctuating asymmetry*

TABLE 16-1. Asymmetry Terminology[a]

Description	Term for the phenomenon	Descriptive noun or adjective
A form is not superimposable on its mirror image.	Asymmetry Handedness Chirality	Asymmetric(al) Handed Chiral
a. Terms referring to a particular individual or structure (orientation not specified)		
One member of a pair of forms having a mirror image relationship		Enantiomer(ic) Enantiomorph(ic)
The homologous structure on the opposite side of an individual		Antimer -e, -ic
b. Terms referring to orientation (individual or population)		
Situated to the right side of the body or coiling in a clockwise direction[b]	Right-handedness Dextrality	Right-handed Dextral
Situated to the left side of the body or coiling in an anticlockwise direction[b]	Left-handedness Sinistrality	Left-handed Sinistral
c. Terms referring to populations or species (i.e., a group of individuals)		
Only one of two possible mirror-image forms occurs, *orientation not specified*	Directional asymmetry Homochirality Monostrophy Fixed asymmetry Handed asymmetry Monomorphic asymmetry	Directionally asymmetric(al) Homochiral Monostrophic
A mixture of two mirror-image forms, where the two forms *may or may not be equally frequent* (i.e., relative abundance not specified)	Enantiomorphy Heterochirality Heterostrophy Amphidromy Dimorphic asymmetry	Enantiomorph -ic,-ous Heterochiral Heterostrophic Amphidrom -ic, -ous Ambidextrous
A mixture of two mirror-image forms, where the two forms *are explicitly equally frequent*	(Pure) antisymmetry Random asymmetry Indifferent asymmetry Equal heterochirality	Antisymmetric(al) Randomly asymmetric(al) Indifferently asymmetric(al) Racemate, racemic
A mixture of two mirror-image forms, where the two forms *are explicitly not equally frequent*	Biased antisymmetry Unequal heterochirality	Biased antisymmetric(al) Unequally heterochiral

[a]Compiled from various sources.
[b]Clock direction traced by a point moving from the near end to the far end of a spiral or helix as viewed from the near end.

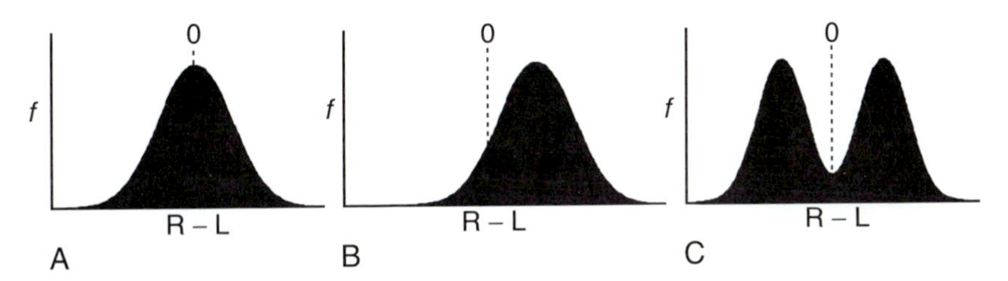

FIGURE 16-1. Three patterns of variation in subtle asymmetries (those around 1% of trait size): (A) *fluctuating asymmetry* (Ludwig, 1932), mean $(R-L)=0$, variation = normally distributed; (B) *directional asymmetry* (Van Valen, 1962), mean $(R-L)\neq 0$, variation = normally distributed; and (C) *antisymmetry* (Timoféeff-Ressovsky, 1934), mean $(R-L)=0$, variation = platykurtic or bimodal. R, measurement or count on the right side; L, measurement or count on the left side; *f*, frequency of the difference between measurements (counts) on the right and left sides. In a perfectly symmetric individual $R-L=0$.

(mean = 0, variation normally distributed; Figure 16-1A), *directional asymmetry* (mean ≠ 0, variation normally distributed; Figure 16-2B) and *antisymmetry* (mean = 0, variation bimodal or nonnormal in the direction of *platykurtosis*; Figure 16-1C).

B. Terms for Conspicuous Asymmetry in an Individual

Two convenient terms allow reference to a particular individual or a trait that is asymmetric without implying anything about orientation (Table 16-1a): *enantiomorph* (one member of a mirror-image pair of forms) and *antimere* (the homologous structure on the opposite side). A third term—enantiomer—is more widely used in the biochemical literature but occasionally applied to organismal asymmetry.

C. Terms for the Orientation of Bilateral or Spiral Asymmetries

Other terms refer explicitly to orientation or chirality (Table 16-1b). These include the familiar *right-handed* and *left-handed* and the less familiar *dextral* and *sinistral*. Although essentially synonymous, dextral and sinistral are applied more commonly to spiral asymmetries whereas right-handed and left-handed typically apply to bilateral ones. However, dextral and sinistral are more useful for a wide-ranging discussion of asymmetry variation because (1) they are shorter, (2) they apply more comfortably to both spiral and bilateral asymmetries, and (3) they permit less clumsy adverbs (e.g., dextrally or sinistrally).

Implied in these terms is a common frame of reference. For bilateral structures, by convention, the term dextral (right-handed) means "larger on the right *as viewed from the dorsal side.*"

For spiral or helical asymmetries, things are not so simple, and care must be taken to specify the convention being followed. In his delightful review of plant asymmetries, Kihara (1972) traces the history of confusion regarding twining or spiraling direction in plants all the way back to pre-Linnean times and enumerates those botanists falling into *orthodox* and *anti-orthodox* camps. Galloway (1989) and Robertson (1993) note similar confusion among malacologists regarding shell coiling. Fortunately, most zoologists, biochemists, physicists, and engineers follow the same convention, as did the *anti-orthodox* botanists (Kihara, 1972; Galloway, 1989). A spiral or helix is called dextral (right-handed) if the clock direction traced by a point moving from the near end to the far end is clockwise *as viewed from the near end*. Fortunately, choice of viewing end does not matter, only the convention of tracing the helix starting at the near end. So, for example, the spiral orientation of a snail's shell will be clockwise (dextral) regardless of whether the shell is viewed from the apex or the abapical end, so long as the spiral is traced starting at the end nearest the observer. Happily, for snails, and twining plants, this is also the direction of spiral growth as viewed from the oldest part.

D. Terms for Conspicuous Asymmetries in a Population or Species

A veritable cornucopia of terms have been coined to capture the many possible patterns of asymmetry variation exhibited by a group of individuals (Duncker, 1904; Ludwig, 1932; Timoféeff-Ressovsky, 1934; Grüneberg, 1935; Van Valen, 1962; Brown and Wolpert, 1990; Palmer *et al.*, 1993). These include terms for cases where (Table 16-1c): (1) all individuals are asymmetric in the same direction, (2) a mixture exists but nothing is implied about prevalence, (3) a mixture exists with equal frequencies of two enantiomorphs, and (4) a mixture exists but the two enantiomorphs are not equally common. Rather few receive regular use.

The problem is a simple one. Although an individual (or a trait in an individual) may be described unambiguously as symmetric, asymmetric, dextral, or sinistral, this precision evaporates when referring to populations or species, because not all individuals necessarily have the same chirality. Even in species of flatfish and snails, where virtually all individuals are asymmetric in the same direction, occasional reversals or *sports* occur (Hubbs and Hubbs, 1945; Welch, 2001). So frequencies of the rarer enantiomorph can vary anywhere from <0.01% to nearly 50% (Ludwig, 1932).

Although acutely aware of previous failed attempts, I suggest a simplified convention for referring to observable patterns of conspicuous asymmetry variation (Table 16-2). Another set of terms could be erected based on presumed causes of asymmetry, as Grüneberg (1935) did, but these would likely be more contentious, as causes would have to be identified in each case, and similar patterns of asymmetry variation may have different underlying causes.

Clearly, when referring to a sample of individuals, a continuum exists between two extreme states: (1) two enantiomorphs equally frequent (not statistically

TABLE 16-2. Terms Referring to Mixtures of Enantiomorphs in a Population or Species

Prevalence of common form	Duncker (1904)[d]	Ludwig (1932)	Timoféeff-Ressovsky (1934)[e]	Van Valen(1962)[f]	Brown and Wolpert (1990)	Palmer (this chapter)
50%[a]	Complete right- vs. left-handed asymmetry	Racemic	Antisymmetry	Pure antisymmetry	Random asymmetry	Antisymmetry
>50[b]–90%	Incomplete asymmetry	Amphidromic–nonracemic	?	Antisymmetry	?	Biased antisymmetry
>90%	Incomplete asymmetry	Monostrophic	?	?	?	Biased antisymmetry
>95%[c]	Incomplete asymmetry		?	?	?	Directional asymmetry
95–99%	?	Weakly monostrophic	?	?	?	Directional asymmetry
99–99.9%	?	Strongly monostrophic	?	?	?	Directional asymmetry
>99.9%	Complete asymmetry	Extremely monostrophic	?	Directional asymmetry	Handed asymmetry	Directional asymmetry

[a]Not significantly different from 50%.
[b]Significantly greater than 50%.
[c]Significantly greater than 95%.
[d]Duncker (1904) does not define the threshold between "incomplete asymmetry" and "complete asymmetry"; his definitions are *complete symmetry*—"exists in a group of individuals in a trait pair if the difference series of this trait varies uniformly around zero," *complete asymmetry*—"all individuals of a group behave asymmetrically in various degree but always in the same sense (either right- or left-handed)," *incomplete asymmetry*—"[lies] between these two extreme limits" (p. 634), *complete right-versus-left-handed asymmetry*—refers to the pattern observed in male fiddler crabs where right-sided and left-sided major claws are equally frequent (p. 584).
[e]Timoféeff-Ressovsky (1934) defines several terms to describe the correlation between subtle wing-vein variants on opposite sides (*symmetrical*—"absolute or very strong positive right-left-correlation," *dyssymmetrical*—"[weak] positive right-left correlation," *asymmetrical*—"characters in which the manifestation probability is actually equal for right and left, but varies independently from each other… [so] no right-left-correlation is present," *dysantisymmetrical*—"a negative right-left-correlation is present"), but clearly states that for *antisymmetry* "the manifestation of the trait on one side of the body precludes its manifestation on the other side" and in a figure illustrates right and left variants as equally frequent (p. 80).
[f]Van Valen's definition of antisymmetry is "where asymmetry is normally present but it is variable which side has greater development"; he also does not define a clear threshold between antisymmetry and directional asymmetry, but cites human handedness as an example of antisymmetry (p. 126), so he would consider a common form up to 90% as antisymmetry; similarly, he cites the human heart as example of directional asymmetry (p. 125), so he must accept samples with up to 0.01% reversal (Belmonte, 1999) as directional asymmetry.
? indicates the author did not state an explicit term for this situation.

different from 50 : 50) and (2) only one enantiomorph present. Only Ludwig (1932) advanced an array of terms to capture this full range of possibilities (Table 16-2), but none became mainstream.

Rather than trying to subdivide such a continuum into many arbitrary divisions, perhaps we should merely recognize four fundamental patterns of variation in a collection of individuals (Table 16-2). Although many descriptors can apply to these cases (Table 16-1), the following four seem most attractive based on familiarity, length, etymologic simplicity, and versatility.

1. Directional Asymmetry

The sample is for all practical purposes not a mixture; all individuals have the same chirality except for occasional reversals. Unfortunately, this term has a different meaning when applied to subtle asymmetries (Figure 16-1B), but the context usually makes the meaning clear. In practice and in keeping with the familiar—though still arbitrary—convention in statistics, directional asymmetry should apply to any sample where the rarer enantiomorph does not make up significantly more than 5%.

2. Enantiomorphy

The sample is a mixture of two enantiomorphs (i.e., the rarer form makes up significantly more than 5%), but nothing is implied about the prevalence of dextral or sinistral forms.

3. Antisymmetry

The sample is a mixture of two enantiomorphs, and both are equally frequent (i.e., do not depart significantly from a frequency of 50%). Antisymmetry is therefore a special case of enantiomorphy. This term has a somewhat different meaning when applied to subtle asymmetries (Figure 16-1C), but fortunately the most common pattern is two modes of equal height (Palmer and Strobeck, 2003), so even in this context it captures the essence of the term: Dextral and sinistral variants are equally common.

4. Biased Antisymmetry

The sample is a mixture of two enantiomorphs, and one is significantly more common than the other. In practice, and in keeping with the familiar convention in statistics, this would apply to any sample where the rarer enantiomorph makes up significantly less than 50% but significantly more than 5% of the sample. If needed, this could be further refined to *right-biased antisymmetry* (dextral form predominates) or *left-biased antisymmetry* (sinistral form predominates) to indicate the direction of imbalance.

II. THE HISTORY OF ANTISYMMETRY

Because antisymmetry—an equal mix of dextral and sinistral forms—is such an important phenomenon both developmentally and evolutionarily, its history is of more than passing interest.

Duncker (1904) may have been the first to name the general phenomenon (Table 16-2). When referring to the random occurrence of large claws on either side of male fiddler crabs, he called it *complete right- vs. left-handed asymmetry* (*"vollkommene rechts-resp. linksseitige Asymmetrie"*), as distinct from *complete asymmetry* (*"volkommene Asymmetrie"*—all individuals the same) and *incomplete asymmetry* (*"unvolkommene Asymmetrie"*—an unequal mix of dextral and sinistral forms). However, these descriptors never gained favor.

Although Astauroff (1930) was mainly concerned with subtle departures from symmetry, he drew attention to the random sidedness of colored spots on the elytra of female *Bruchus* beetles "as a case of *absolute negative correlation between the expression of a feature on different sides* [that] parallels some well-known cases of heterochely in higher crustaceans where in several species one claw is hyperdeveloped compared to its antimere (in 50% on one side, and in 50% on the other side)" (p. 258, italics added). This hardly qualifies as a useful descriptor, but Astauroff clearly meant "absolute negative correlation" to refer to an equal mix of dextrals and sinistrals.

In the most comprehensive review of asymmetry variation ever written, Ludwig (1932) advanced an elaborate set of terms to describe different proportions of dextral and sinistral forms (Table 16-2). Rather than invent a term to refer specifically to an equal mix of both forms, he borrowed one from chemistry, *racemic*. Although elegant, short, and well established among chemists, few biologists seem to have used the term. Shortly afterwards, Grüneberg (1935) coined an independent set of terms to refer to causes of asymmetries, but he largely followed Ludwig's terminology when referring to patterns of variation.

Timoféeff-Ressovsky (1934) appears to have coined *antisymmetry*. Like Astauroff, he too was concerned primarily with subtle asymmetries, and he repeated Astauroff's phrase "absolute negative correlation" to define it. Significantly, he also drew attention to a figure (his Figure 16) where right and left forms are shown as equally frequent and noted "for antisymmetry the manifestation of the trait on one side of the body precludes its manifestation on the other side" (p. 80).

Van Valen (1962) popularized *antisymmetry*, and his highly influential paper lead to its widespread use in studies of subtle asymmetries (Palmer and Strobeck, 2003). Unfortunately, the example he used is misleading. By citing human handedness, where nearly 90% are right-handed (Perelle and Ehrman, 1994), he departs from Timoféeff-Ressovsky's original use of the term to refer to an equal mix of dextral and sinistral forms. He even refers to the "pure situation, with equal frequencies of each handedness," implying that antisymmetry can refer to equal or unequal frequencies of two forms (Table 16-2). He does, however, acknowledge that "the present use of 'antisymmetry' is slightly expanded from that of Timoféeff-Ressovsky" (p. 126). For this reason, I recommend a return to Timoféeff-Ressovsky's more restricted use of antisymmetry, without any qualifiers, to refer explicitly to an equal mix of dextral and sinistral forms.

Brown and Wolpert (1990) used yet another pair of terms to distinguish between a sample of only one enantiomorph, *handed asymmetry*, and a sample of two enantiomorphs in equal frequency, *random asymmetry*. Unfortunately, the subtle deviations that give rise to fluctuating asymmetry are also random, which renders this descriptor potentially confusing.

Etymologically, antisymmetry arises from an unexpectedly attractive visual metaphor. When viewed as a frequency distribution of differences between the right and left sides, the relation between symmetry and antisymmetry is immediately obvious (Figures 16-1A, C). For bilaterally symmetric traits, the difference between the sides is zero on average, and what little variation remains is normally distributed about zero. For traits that exhibit antisymmetry, the frequency distribution of right–left differences shows a distinct valley between two equal-sized peaks that are equidistant from zero. This valley of *missing* observations—the "anti-" in antisymmetry—is centered on zero the same way the peak is centered on zero for symmetric traits.

Finally, a peculiar situation arises in plants that have flowers where the style bends toward one side or the other (Barrett *et al.*, 2000). Because some individual plants may have more than one flower, and the chirality of flowers on a single plant may either be fixed or variable, Barrett *et al.* distinguish "two forms of enantiostyly depending on whether both style orientations occur on the same plant (*monomorphic enantiostyly*) or on different plants (*dimorphic enantiostyly*)." These terms parallel the terminology for gender in plants where monomorphic refers to species in which all individual plants possess both male and female flowers and dimorphic refers to species where individual plants have different genders and where a population is a mixture of gender types (S. Barrett, personal communication, 2004). Unfortunately, the use of monomorphic and dimorphic to refer to variation at the population level conflicts with the pattern seen in an individual plant. So a species that exhibits monomorphic enantiostyly actually has dimorphic flowers (both right-styled and left-styled) on each individual, whereas one that exhibits dimorphic enantiostyly has monomorphic flowers (either all right-styled or all left-styled) on each individual. Terms such as intraplant enantiostyly and interplant enantiostyly might have been less confusing.

III. TAXONOMIC DISTRIBUTION AND FUNCTIONAL SIGNIFICANCE OF ANTISYMMETRY

More than 450 species from 67 families in eight phyla exhibit antisymmetry (Appendix 16-1). Its wide taxonomic distribution and its cooccurrence with symmetric taxa in most groups suggests that antisymmetry has evolved independently multiple times (Palmer, 1996a). Beyond its evolutionary significance, which is great (see penultimate section), the natural history of antisymmetry is fascinating. A few examples deserve attention.

A. Plants

I have not yet attempted a comprehensive survey of antisymmetry in plants, but many interesting cases exist (Kihara, 1972; Hudson, 2000; Welch, 2001; Klar, 2002) (see Table 16-3 for additional references). Some plants exhibit spiral growth of the stem or trunk, sometimes referred to as phyllotaxis, and quite often dextral and sinistral spiraling are equally common within a species. Both the arrangement of the earliest leaves in seedlings and the orientation of the paired margins of individual leaves that unroll laterally during development (i.e., one side is rolled about the other as it develops) exhibit antisymmetry in some species. The overlap of petals about the margins of a flower and the twining of stems and tendrils may also occur in two equally frequent chiral forms. Finally, among flowers with an overt plane of bilateral symmetry, many have evolved a style polymorphism where left-leaning and right-leaning styles are equally common within a species (Jesson and Barrett, 2003), although proportions may vary among local populations (S. Barrett, personal communication, 2004).

B. Cnidaria

Two large and familiar colonial hydrozoans exhibit antisymmetry: the By-the-Wind Sailor *Velella velella* (Figure 16-2A), and the Portuguese Man o' War *Physalia physalis*. Both spend their adult lives floating at the air-water interface of northern-hemisphere and southern-hemisphere oceans feeding on subsurface fish or invertebrates (Savilov, 1961). In both groups an upright, saillike structure deflects either clockwise (by convention, right-sailing or dextral) or anticlockwise (left-sailing or sinistral) from the colony's long axis when viewed dorsally (Edwards, 1966), and when blown by the wind the two forms actually do tack in different directions (Francis, 1991). Samples of individuals stranded on the shore typically depart substantially from 50 : 50, probably a result of the wind direction that brought them ashore (Shannon and Chapman, 1983). However, extensive open-ocean samples more closely approximate equal frequencies of dextral and sinistral forms (Savilov, 1961). The prevalence of dextral and sinistral forms in different ocean regions appears to result from a combination of prevailing winds and the orientation of oceanic gyres (Savilov, 1961).

C. Mollusca

Among mollusks, pure antisymmetry is most common in bivalves (Appendix 16-1). Although many gastropod species are polymorphic for coiling direction (Davis, 1987a), only occasionally are dextral and sinistral forms equally common and then likely just by chance, except possibly in some species of the pulmonate *Amphidromus* (Davis, 1978) (Figure 16-2B). Furthermore, because snails of opposite coil have a harder time mating (Asami *et al.*, 1998), such a polymorphism should be intrinsically unstable evolutionarily.

→

FIGURE 16-2. Examples of animals that exhibit antisymmetry: equal frequencies of dextral and sinistral forms within a species. (A) *Velella velella*—a sinistral form (sail twisted counterclockwise from long body axis) of the "By-the-Wind-Sailor." (B) *Amphidromus heerianus*—a snail that produces both dextrally coiled (left image) and sinistrally coiled (right image) shells. (C) *Chama lazarus*—free upper valve (right in this image) of a bivalve mollusk that cements one valve—the left or the right—to the substratum. (D) *Pandora inaequivalvis*—right valve (upper image) and hinge view (lower image) of a bivalve mollusk that lies in a horizontal orientation near the sediment–water interface. (E) *Psettodes erumei*—right-eyed individual of the most primitive living flatfish family (Psettodidae). (F) *Circeis amoricana*—a sinistrally coiled species of tube-building polychaete (Spirorbidae); the tubes attach to the substratum along the ventral side (the worm's body bends to the right or left); only species in one genus of this family (*Neomicrorbis*) exhibit pure antisymmetry. (G) *Tidarren fordum*—a male spider with only the right pedipalp remaining; either the right or left one is torn off at random just after the penultimate molt. (H) *Dinalloptes anisopus*—a male feather mite (ventral view); the grotesque asymmetry apparently reduces the likelihood that males will be dislodged during grooming while grasping females prior to mating. (I) *Neotrypaea californiensis*—a large male thalassinid mud shrimp with a hypertrophied right claw; these claws are not used to feed but may aid in agonistic encounters or mating (Labadie and Palmer, 1996). (J) *Cardisoma guanhumi*—a male land crab; males of most species have one enlarged claw indifferently on the right or left. (K) *Triaenodes basis*—the aquatic larvae of one of several species of caddis flies that build conical tubes from precisely cut pieces of plant stems arranged in a spiral, in this case dextral. (L) *Archimedes*—an extinct genus of bryozoan famous for its helically coiled colony form; in this example both helical forms originate from a common origin (sinistral on the right side of the image and dextral on the left). (M) *Macrorhamphosodes uradoi*—a peculiar, scale-eating triacanthodid fish (one of at least three groups in which asymmetric mouths for scale eating have evolved independently, Appendix 16-1); the mouth bends to the right or left to facilitate attack from the rear of the host. (N) *Verruca* sp.—barnacles where one pair of the normally movable opercular plates, either the right or left, is incorporated into the ring of rigid lateral wall plates, leaving only the other pair (the left pair in this picture) free to open and close to allow feeding legs to be extended and retracted. (O) *Torquirhynchia*—a brachiopod where the dorsal (upper) valve is elevated on one side (right side of valve in this image) and depressed on the other, at random. (P) *Alpheus hippothoe*—one of many species from the diverse family that includes snapping shrimp (Alpheidae); the hypertrophied claw can produce a loud snap that can stun prey, or deter predators or conspecifics. (Q) *Bugula*—one of several species that produce spiral colonies (the second frond from the left is clearly dextral). (R) *Loxia leucoptera*—a crossbill finch where the upper mandible crosses indifferently to the left or right of the lower one (to the left in this image); this crossing allows them to better pry apart conifer cones to extract the seeds with their tongue. (*A*, Photo by Monty Graham. *D*, Photo from Femorale.com.br. *H*, © Royal Museum for Central Africa; Terveren, Belgium, Dinalloptes Anisopus [feather mite]. *J*, Photo by Todd Zimmerman. *K*, Photo by Jens Schou. *L*, From F.K. McKinney, D.W. Brudick, Palaeontology 44, 855–859, Blackwell Publishing. *O*, From Leslie Harris, Courtesy, Natural History Museum of Los Angeles County. *P*, Reprinted from Commissural asymmetry in brachiopods by F.T. Fursich and T. Palmer from Lethaia 17:251–265, by permission of Taylor & Francis AS. *Q*, Photo by R.S. Santos ImagDOP. *R*, Photo by Frode Falkenberg.)

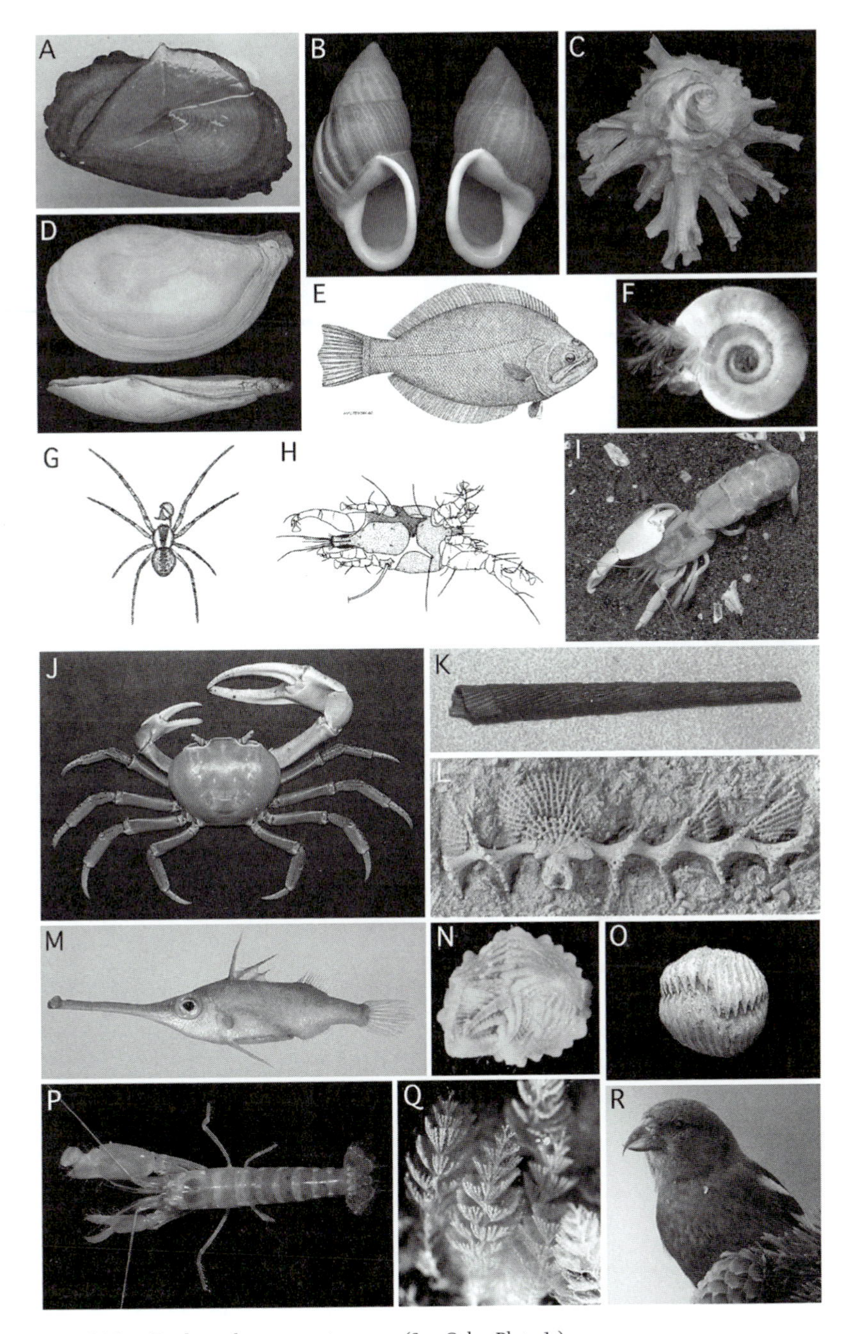

FIGURE 16-2. For legend see opposite page. (See Color Plate 1.)

Inequivalve (unequal) shells are most well developed in bivalve taxa where one valve is fixed to the substratum and the other is free to move (Figure 16-2C). Harper (1991) estimates that attached or cemented shells have evolved independently up to 20 times among 16 families. Which valve is attached—the right or the left—depends on the group (Bogan and Bouchet, 1998). Some families consistently attach by the right valve (Pectinidae, Terquemiidae, Myochamidae), others attach by the left valve (Ostreidae and Chondrodontidae), and some include species that attach by either valve (Chamidae, Diceratidae, Etheriidae, Corbulidae). Because the attached valve differs from the unattached one, species composed of indifferently attached individuals exhibit antisymmetry. Antisymmetry has arisen independently in at least eight families, including both cemented (Figure 16-2C) and bysally attached groups, as well as some free-living groups (Figure 16-2D) (Appendix 16-1). Functionally, the valves differ in form either because of constraints imposed by asymmetric attachment or, in some cases, because this improves filter-feeding efficiency (Savazzi, 1984).

D. Annelida

Two kinds of antisymmetry occur in polychaetes (Appendix 16-1). First, in calcareous tube-dwelling serpulid polychaetes, one of a bilateral pair of anterior gill filaments transforms into an operculum that plugs the tube when the animal withdraws (Zeleny, 1905). In adult worms, the plug may occur on either the right or left side with equal frequency. Although opercular plugs occur throughout the Serpulidae, I am aware of too few reports of the incidence of dextral and sinistral forms to say how widespread antisymmetry is.

Some spirorbid polychaetes—minute, cosmopolitan worms that inhabit spiral calcareous tubes that inhabit spiral calcareous tubes—exhibit another type of antisymmetry. The anatomically ventral margin of the tube is cemented to the bottom, so their pinwheel-shaped tubes coil either dextrally (clockwise in the direction of growth when viewed from above the substratum) or sinistrally (Figure 16-2F). Although most species exhibit directional asymmetry, some are polymorphic, and at least one—*Neomicrorbis,* the sister taxon to the remaining spirorbids (Macdonald, 2003)—exhibits clear antisymmetry (Fauchald, 1977). The adaptive significance of coiling is not clear and that of coiling direction even less so.

E. Arthropoda: Chelicerata

The most bizarre example of antisymmetry I have encountered occurs in theridiid spiders (Knoflach and van Harten, 2003, and references therein). In two genera (*Tidarren* and *Echinotheridion*), adult males possess a single highly modified pedipalp, which may be either the right or left at random. Shortly after the penultimate molt, subadult males entangle one of their two palps in threads and rotate repeatedly until the palp detaches, after which it is eaten. Throughout their remaining life, males possess only this single palp (Figure 16-2G). During mating with a much larger female, this palp transfers sperm to the female's contralateral receptacula, breaks off, and remains attached to her epigynum as a sperm plug. The female then eats the male. A virgin female can mate twice, receiving one insemination in each of her two receptacula. The second male deposits his palp in the remaining receptacula, regardless of his handedness. Why males sacrifice one of their palps prior to mating, when two could be used to inseminate both receptacula of a single female, remains mysterious.

Species in several families of feather mites (Acari, Astigmata) also exhibit antisymmetry (Figure 16-2H). The peculiarly twisted bodies of males appear to allow them to hold on simultaneously to the midrib of the feather and to a female with which they are mating (Gaud and Atyeo, 1996). Presumably, dextral and sinistral males occur preferentially on the right and left wings of their avian host, but I know of no data to support this.

F. Arthropoda: Crustacea

Antisymmetries are particularly widespread among crustaceans (Appendix 16-1). Some are whole-body asymmetries. The peculiar gooseneck barnacle *Koleolepas avis* lives symbiotically on the column of sea anemones (Yusa *et al.*, 2001). The asymmetry appears to facilitate feeding on the host's tentacles. Verrucomorph barnacles exhibit a most peculiar asymmetry (Figure 16-2N). Normal acorn barnacles are surrounded by a fused ring of from four to eight bilaterally paired lateral plates and have two pairs of symmetric movable plates that open and close like trapdoors. However, in the weirdly deformed verrucomorph barnacles, one pair of opercular plates replaces a pair of lateral plates and is cemented into the ring of plates surrounding the body, leaving only two moveable opercular plates. This deflection leaves barnacles tilting toward one side and occurs to the right or left at random (Newman and Hessler, 1989). Although the orientation ensures their feeding fans face into the prevailing (directional) current (Newman, 1989), the actual functional significance of this asymmetry remains decidedly obscure. Finally, some bopyrid isopods, which are parasitic on larger crustaceans, exhibit whole-body asymmetries correlated with the curvature of the gill cavity on the side of the body in which they live (Markham, 1985).

A species of the looking-glass copepod *Pleuromamma* exhibits an unusual concordant asymmetry of several bilateral traits, including a pigment spot and multiple limbs (Ferrari, 1984). Males possess more asymmetric characters than females, but right-sided and left-sided individuals are equally frequent in both sexes. Here too, the functional significance is not clear.

Finally, in virtually all decapod crustaceans in which conspicuous claws have evolved, some taxa show antisymmetric claws (Appendix 16-1). This includes several caridean shrimp groups (e.g., Figure 16-2P), most asymmetric ghost shrimp

(Thalassinidea; e.g., Figure 16-2I), at least two genera of lobsters, some anomuran crabs, and a host of brachyuran crabs (e.g., Figure 16-2J). Functionally, such asymmetries presumably provide an advantage because of increased versatility (Vermeij, 1973): Different claw forms can assume different functions for feeding, agonistic interactions, signaling, or mating (Mariappan *et al.*, 2000).

G. Arthropoda: Insecta

I have encountered only a few examples of antisymmetry in insects (Appendix 16-1). Undoubtedly, many more exist. In mantids (Mantodea) and true bugs (Heteroptera) that cross one wing over the other when folded, antisymmetry appears to be the norm (Skapec and Stys, 1980). Either the right or the left wing ends up on top at random, and sometimes this is associated with pigment differences. Functionally, larger wings may be held more closely against the back if one lies on top of the other.

Two types of antisymmetry occur in caddisflies (Trichoptera). In at least one genus, several species exhibit antisymmetry in the paired genital lobes (Morse, 2001). In several other genera, even though the tubes in which the larvae live are relatively straight cones, the walls are assembled from many short fragments of plant stems glued together in a spiral fashion (Wiggins, 1996). These spiral walls may coil either dextrally or sinistrally, and in most species both enantiomorphs occur with equal frequency (A. R. Palmer, unpublished, 2004). Spirally arranged stem fragments (Figure 16-2K) may reduce the chance that tubes fracture at the regularly spaced sutures that would exist if stem fragments were arranged as a successive stack of rings, as in the tubes of other caddisfly larvae (Wiggins, 1996). Whether spiral asymmetries in the tube wall correlate with anatomic asymmetries in the larvae remains unclear.

Finally, many insect groups have evolved twisted abdomens to facilitate mating (Richards 1927; Boudreaux, 1979). Rotated abdomens permit males to orient in the same direction as females during copulation, thus making flight while mating easier. Although the direction of twisting is fixed in most species, it is random in at least one chironomid fly (Dordel, 1973).

H. Brachiopoda

Antisymmetry occurs in several species from two extinct orders of brachiopods (Appendix 16-1). The asymmetry is a curious one. Unlike bivalves, where the paired valves are right and left, the paired valves of brachiopods are dorsal and ventral. In antisymmetric species, the gape between the valves deflects upward on one side yielding brachiopods with a twisted smile (Figure 16-2O). The functional significance is unknown, but it may facilitate filter feeding by increasing access to unfiltered water, decreasing mixing of excurrent and incurrent flow, or enhancing induced flow through the mantle cavity.

I. Bryozoa

Many bryozoans have independently evolved spiral colonial forms (McKinney, 1980). Perhaps the most famous example is the extinct fenestrate genus *Archimedes* (Figure 16-2L). The coiling direction of all known species is random (Condra and Elias, 1944), and some evidence suggests that branches of opposite chirality can arise from a single larva (McKinney and Burdick, 2001) (Figure 16-2L). Spiral colonial forms occur in several other fossil and living bryozoan groups, including *Spiralaria*, *Retiflustra*, and *Bugula* (Figure 16-2Q), and may improve filter-feeding efficiency by directing water flow through the colony (McKinney *et al.*, 1986).

J. Echinodermata

The pentaradial symmetry of most echinoderms might seem to preclude antisymmetry, but two examples occur in extinct, stalked crinoids (Appendix 16-1). In one group, the basal circlet of plates that lies at the transition between the stalk and crown overlaps in a chiral way. Although most species exhibit directional asymmetry or biased antisymmetry (Rozhnov, 1998), at least three exhibit pure antisymmetry. In another species, the stalk itself has a spiral form, and dextral and sinistral forms are equally common (Rozhnov, 1998).

K. Chordata

Among vertebrates, antisymmetry is most common in fish (Appendix 16-1). Scale eating has evolved independently in several groups, and in four of these (Characidae, Triacanthodidae [Figure 16-2M], and two genera of Cichlidae), examples of antisymmetric mouth deflections occur. The deflected mouth allows scale eaters to remove scales from the flanks of their prey by approaching them more from behind, where they are less likely to be seen (Hori, 1993). In the wholly asymmetric flatfishes (Pleuronectiformes), the eyed side in most species is fixed, but in some groups it is to the right and others to the left (Hubbs and Hubbs, 1945). However, two species in the Psettodidae (Figure 16-2E)—the living sister group to all remaining pleuronectiform fishes—exhibit eye-side antisymmetry (Hubbs and Hubbs, 1945). Asymmetric gonopodia, clasping structures, or genital openings, all of which likely increase fertilization success in side-to-side matings, have also evolved independently in at least four groups (Phallostethidae, Anablepidae, Poeciliidae, and Hemiramphidae). Examples of antisymmetric species occur in all four.

Antisymmetry is less common among tetrapods, but a few examples do occur (Appendix 16-1). In urodele amphibians, lower jaw bones may interdigitate or cartilaginous elements of the pectoral girdle may overlap at the midline and do so in a handed fashion. In at least one bat with an asymmetric dental formula, the missing incisor is from either the right or left side at random. The small jaw in this species appears unable to accommodate a full tooth complement, so one tooth is dropped from the arcade (Juste and Ibanez, 1993). Perhaps the most dramatic antisymmetries occur in the bills of some fringillid finches. In most crossbills (e.g., Figure 16-2R) and in the Hawaiian akepa, the upper mandible deflects to the left or right of the lower one at random. Combined with the ability to abduct their mandibles laterally, the crossed bill tips increase the efficiency with which these birds remove seeds from conifer cones (Benkman, 1988) or from between other closely paired structures (Hatch, 1985).

IV. DEVELOPMENT AND REGENERATION OF ASYMMETRY IN ANTISYMMETRIC SPECIES

Although an asymmetric individual from an antisymmetric species may not appear to differ from the same enantiomorph in a directionally asymmetric species, the causes of symmetry breaking are likely quite different (Grüneberg, 1935). Significantly, several studies of antisymmetric species confirm that external environmental stimuli determine direction of asymmetry during ontogeny. In other words, random external stimuli—sometimes from the right and sometimes from the left—provide the symmetry-breaking information that yields the equal frequencies of dextral and sinistral forms characteristic of antisymmetry.

A. Ontogeny

In most antisymmetric species, asymmetry first appears late in ontogeny, typically after hatching, settlement, or metamorphosis (Palmer, 1996a). By this time, precursors to most adult body features are already present.

Although data are sparse, lateral inhibition of one side by the other appears to ensure that only one antimere overdevelops in any one individual. In both lobsters (Govind, 1989) and snapping shrimp (Mellon, 1981)—where one claw is considerably larger than its antimere—experimental studies confirm that the first claw to hypertrophy inhibits the development of its antimere. Inhibition appears to be under nervous control. Significantly, differential use of one claw during the fifth intermolt of juvenile lobsters actually determines which side develops a crusher claw. Most surprising of all, in the absence of suitable stimuli, no crusher claw ever develops (Govind, 1989)! So the developmental program that generates a crusher claw depends on environmental stimuli to initiate it. The paired, antisymmetric opercular plug in serpulid polychaetes also develops via inhibition of one side by the other (Zeleny, 1905), although the nervous system is not involved (Okada, 1933).

In verrucomorph barnacles and several attached bivalves, the premetamorphic larvae and early postmetamorphic juveniles are initially symmetric. In both groups, the developing juveniles effectively tilt toward one side at random and then begin to develop asymmetrically (Kriz, 2001; Newman, 1989). The settled cyprids of verrucomorph barnacles tend to orient into the prevailing current (Newman, 1989), so they may be responding directly to environmental cues.

In at least two groups, antisymmetry is achieved by a bizarre mechanism: random autotomy of one member of a symmetric pair. Postmetamorphic fiddler crabs (*Uca*) have small, symmetric claws. Between 4-mm and 4.5-mm carapace width, a high incidence of *U. lactea* males exhibit two large-type claws (Yamaguchi, 1978). However, by 5-mm carapace width, the frequency of such males declines, and virtually all retain only a single master claw, accompanied by a minor claw that regenerates in place of the lost large claw. Morgan (1923) describes a similar pattern for *U. pugilator*. As noted in the preceding text (see Arthropoda: Chelicerata), autotomy creates asymmetry in species from two genera of theridiid spiders (*Tidarren*, *Echinotheridion*). Following the penultimate molt, males actively twist off one palp at random, but the missing palp never regenerates (Knoflach and van Harten, 2003).

The ontogeny of the unilateral operculum is well known in the serpulid polychaete *Hydroides dianthus* (Zeleny 1905; Schochet 1973). Curiously, a functional operculum first differentiates consistently from the left member of a bilateral pair of gill filaments. However, opercula are regularly lost, and direction of asymmetry reverses upon regeneration. This ultimately yields an equal mix of dextral and sinistral adult worms.

B. Regeneration of Missing Antimeres

Rules regarding regeneration of a missing antimere of an antisymmetric pair are puzzling. In all heterochelous crustaceans examined, a minor claw regenerates following loss of the minor claw (e.g., see Goss 1969). In some antisymmetric taxa, such as fiddler crabs (*Uca*) and lobsters (*Homarus*), a major claw regenerates after loss of the major claw. In other words, direction of asymmetry is maintained.

Curiously, direction of asymmetry reverses following loss of the major claw in many other decapod crustaceans, including snapping shrimps (*Alpheus heterochaelis*, Darby, 1934), which exhibit equal antisymmetry, and shore crabs (*Carcinus maenas*, Abby-Kalio and Warner, 1989) and stone crabs (*Menippe mercenaria*, Simonson, 1985), both of which exhibit a strong right-side bias. As noted in the preceding section, the direction of asymmetry also reverses when the unilateral operculum is lost in serpulid polychaetes.

These limited observations preclude any general conclusions, but they do hint at one intriguing pattern. In the two antisymmetric decapods where differential use (lobsters) or autotomy (fiddler crabs) determine the direction of asymmetry, asymmetry does not reverse following major claw loss. In contrast, the two directionally asymmetric crabs do reverse asymmetry following major claw loss. Are environmentally triggered asymmetries somehow more refractory to reversal once established?

Perhaps they are not, because asymmetry does reverse following loss of the major claw in antisymmetric snapping shrimp. Clearly, more work is needed.

V. INHERITANCE OF DIRECTION IN ANTISYMMETRIC SPECIES

The most surprising observation about antisymmetry relates to inheritance. With only one compelling exception, the direction of asymmetry in antisymmetric species is not inherited.

In 15 of 16 studies in plants, dextral and sinistral offspring were equally common regardless of whether both parents were dextral or both sinistral (Table 16-3a). Many different antisymmetric traits were examined, including direction

TABLE 16-3. Inheritance of Direction of Asymmetry in Plants and Animals Exhibiting Different Types of Asymmetry Variation[a]

Organism and trait	Direction of asymmetry inherited?	Source
a. Antisymmetry		
Plants		
Direction of phyllotaxy in palm tree trunks	No	(Davis, 1962; Toar *et al.*, 1979)
Direction of phyllotaxy in tobacco stems	No	(Allard, 1946)
Direction of phyllotaxy in Egyptian cotton stems	No	(Lugard, 1931)
Direction of phyllotaxy in wild teasel stems	No	(de Vries, 1911)
Direction of twining in morning-glory stems	No	(Imai, 1927)
Seedling handedness in two-rowed barley	No	(Compton, 1912)
Seedling handedness in triticale	No	(Rama Swamy and Bahadur, 1999)
Seedling handedness in pigeon pea	No	(Rao *et al.*, 1983)
Direction of leaf rolling in cocoyams	No	(Venkateswarlu, 1982)
Direction of leaf rolling in *Begonia*	No	(Ringe, 1971)
Direction of leaf torsion in *Plantago major*	No	(Ikeno, 1923)
Direction of leaf spiral in *Corchorus capsularis*	No	(Kundu and Sarma, 1965)
Frond handedness in duckweed	No	(Kasinov, 1973)
Direction of floral spiral in *Spiranthes*	No	(Koriba, 1914)
Direction of style bend in monomorphic enantiostyly	No	(Jesson and Barrett, 2003)
Direction of style bend in *Heteranthera* flowers	Yes (M)	(Jesson and Barrett, 2002a)
Animals		
Side of operculum in serpulid polychaetes	No	(Zeleny, 1905)
Side of red spots on elytra of female *Bruchus* beetles	No	(Breitenbecher, 1925)
Side of major claw in American lobsters	No	(Govind and Pearce, 1986)
Side of major claw in snapping shrimp	No	(Darby, 1934)
Side of major claw in male fiddler crabs	No	(Yamaguchi, 1977)
Top wing when folded in fruit flies	No	(Purnell and Thompson, 1973)

Continued

Color Plate 1. Examples of animals that exhibit antisymmetry: equal frequencies of dextral and sinistral forms within a species. (A) *Velella velella*—a sinistral form (sail twisted counterclockwise from long body axis) of the "By-the-Wind-Sailor." (B) *Amphidromus heerianus*—a snail that produces both dextrally coiled (left image) and sinistrally coiled (right image) shells. (C) *Chama lazarus*—free upper valve (right in this image) of a bivalve mollusk that cements one valve—the left or the right—to the substratum. (D) *Pandora inaequivalvis*—right valve (upper image) and hinge view (lower image) of a bivalve mollusk that lies in a horizontal orientation near the sediment–water interface. (E) *Psettodes erumei*—right-eyed individual of the most primitive living flatfish family (Psettodidae). (F) *Circeis amoricana*—a sinistrally coiled species of tube-building polychaete (Spirorbidae); the tubes attach to the substratum along the ventral side (the worm's body bends to the right or left); only species in one genus of this family (*Neomicrorbis*) exhibit pure antisymmetry. (G) *Tidarren fordum*—a male spider with only the right pedipalp remaining; either the right or left one is torn off at random just after the penultimate molt. (H) *Dinalloptes anisopus*—a male feather mite (ventral view); the grotesque asymmetry apparently reduces the likelihood that males will be dislodged during grooming while grasping females prior to mating. (I) *Neotrypaea californiensis*—a large male thalassinid mud shrimp with a hypertrophied right claw; these claws are not used to feed but may aid in agonistic encounters or mating (Labadie and Palmer, 1996).

Continued

Color Plate 1—cont'd. (J) *Cardisoma guanhumi*—a male land crab; males of most species have one enlarged claw indifferently on the right or left. (K) *Triaenodes basis*—the aquatic larvae of one of several species of caddis flies that build conical tubes from precisely cut pieces of plant stems arranged in a spiral, in this case dextral. (L) *Archimedes*—an extinct genus of bryozoan famous for its helically coiled colony form; in this example both helical forms originate from a common origin (sinistral on the right side of the image and dextral on the left). (M) *Macrorhamphosodes uradoi*—a peculiar, scale-eating triacanthodid fish (one of at least three groups in which asymmetric mouths for scale eating have evolved independently, Appendix 16-1); the mouth bends to the right or left to facilitate attack from the rear of the host. (N) *Verruca* sp.—barnacles where one pair of the normally movable opercular plates, either the right or left, is incorporated into the ring of rigid lateral wall plates, leaving only the other pair (the left pair in this picture) free to open and close to allow feeding legs to be extended and retracted. (O) *Torquirhynchia*—a brachiopod where the dorsal (upper) valve is elevated on one side (right side of valve in this image) and depressed on the other, at random. (P) *Alpheus hippothoe*—one of many species from the diverse family that includes snapping shrimp (Alpheidae); the hypertrophied claw can produce a loud snap that can stun prey, or deter predators or conspecifics. (Q) *Bugula*—one of several species that produce spiral colonies (the second frond from the left is clearly dextral). (R) *Loxia leucoptera*—a crossbill finch where the upper mandible crosses indifferently to the left or right of the lower one (to the left in this image); this crossing allows them to better pry apart conifer cones to extract the seeds with their tongue.

TABLE 16-3. Inheritance of Direction of Asymmetry in Plants and Animals Exhibiting Different Types of Asymmetry Variation[a]—Cont'd

Organism and trait	Direction of asymmetry inherited?	Source
Direction of bill crossing in crossbill finches	No	(Edelaar and Knops, 2003)
Direction of upper-beak crossing in cross-beak chickens	No[b]	(Landauer, 1938)
Preferred paw (right or left) in mice	No	(Biddle and Eales, 1996)
Visceral situs in *iv* mutant mouse line	No	(Layton, 1976)
Direction of crossing of optic chiasma in trout	No	(Larrabee, 1906)
Direction of crossing of optic chiasma in cod	No	(Larrabee, 1906)
Side of mouth opening in scale-eating cichlids	Yes (M)?[c]	(Hori, 1993)
b. Both wild-type and mutant exhibit directional asymmetry (§) or biased antisymmetry (†).		
Plants		
Direction of phyllotaxy in *Medicago*[§]	Yes (M)	(Lilienfeld, 1959)
Direction of helical growth in roots, hypocotyls petioles and petals in mutant *Arabidopsis*[§]	Yes (M)	(Hashimoto, 2002)
Animals		
Fiber chirality in the cuticle of mutant nematodes[§]	Yes (M)	(Bergmann *et al.*, 1998)
Shell coiling direction in the snail *Lymnaea*[†]	Yes (M)	(Freeman and Lundelius, 1982)
Shell coiling direction in the snail *Partula*[†]	Yes (M)	(Murray and Clarke, 1966)
Shell coiling direction in the snail *Laciniaria*[†]	Yes (M)	(Degner, 1952)
Eye side in polymorphic starry flounder[†]	Yes (P)?	(Policansky, 1982) (Hashimoto *et al.*, 2002)
Preferred hand (right or left) in humans[†]	Yes (weak)	(McManus, 2002)
Direction of heart asymmetry in *inv* mutant mice[c,§]	Yes (M)	(Yokoyama *et al.*, 1993)
c. Wild-type exhibits directional asymmetry; mutant exhibits antisymmetry.		
Randomized heart asymmetry in mutant mice:		
ActRIIb, *Cryptic (EGF-CFC)*, *Foxj1*, *Gdf1*, *iv(lrd)*, *Kif3a,b*, *Mgat1*, *Nodal*[d], *Polaris*, *Sil* and *Smad5* genes; no turning		Reviewed in Hamada *et al.* (2002)
Delta-like1 (Dll1)		(Krebs *et al.*, 2003)
Fused toes (ft)		(Heymer *et al.*, 1997)
Pkd2		(Pennekamp *et al.*, 2002)
RBPjk		(Raya *et al.*, 2003)
Randomized heart asymmetry in mutant zebrafish:		
Cyclops (nodal-related2)		(Rebagliati *et al.*, 1998)
no tail (Brachyury), *floating head (xnot)*		(Danos and Yost, 1996)
Notch		(Raya *et al.*, 2003)
Inversin (invs) knock-down		(Otto *et al.*, 2003)
Randomized heart jogging in majority of mutant zebrafish embryos: *curly up*, *dino*, *locke*, *schmalhans*, *tj2a*, *tm243b*, *tm317b* and *tw29b*		(Table 4 of Chen *et al.*, 1997)

Continued

TABLE 16-3. Inheritance of Direction of Asymmetry in Plants and Animals Exhibiting Different Types of Asymmetry Variation[a]—Cont'd

Organism and trait	Direction of asymmetry inherited?	Source
Randomized diencephalic asymmetry in zebrafish *oep* mutant		(Concha *et al.*, 2000)
Randomized parapineal asymmetry in mutant zebrafish: *cas, cyc=ndr2, flh, ntl, oep* (EGF-CFC), *sqt, sur*		Reviewed in Halpern *et al.*, (2003)
Randomized liver and pancreas asymmetry in zebrafish: *chordino* mutant		(Tiso *et al.*, 2002)
Randomized rotation of P11/P12 nerve cell homologues in *C. elegans unc-40* and *dpy-19* mutants		(Hobert *et al.*, 2002)
d. Wild-type exhibits directional asymmetry; mutant exhibits symmetry.		
Symmetric heart tube in mutant mice:		
Fgf8, Foxa2 (Hnf3β) Furin, Lefty1[e], and *Lefty2*[e]		Reviewed in Hamada *et al.* (2002)
Symmetric lungs in mutant mice:		
ActRIIb, Gdf1, Pitx2[e]; smoothened		Reviewed in Hamada *et al.* (2002)
Lefty1[e]		(Meno *et al.*, 1998)
Pkd2		(Pennekamp *et al.*, 2002)
Shh		(Tsukui *et al.*, 1999)
Zic3		(Purandare *et al.*, 2002)
No heart jogging in mutant zebrafish embryos: *floating head, lost-a-fin, no tail, snailhouse, spadetail*		(Table 4 of Chen *et al.*, 1997)
Symmetric heart tube in mutant zebrafish: *lost-a-fin (laf)*		(Chen *et al.*, 2001)

[a]Modified from (Palmer, 2004). M, inheritance is Mendelian, single-locus, two-allele. P, inheritance is polygenic.
[b]Upper beaks crossed to the right are weakly but consistently more common than left-crossed ones, but the proportions of dextral and sinistral offspring did not vary among parents of different orientation.
[c]Data open to alternate interpretations as breeding was not controlled (see text).
[d]Approximately 10% of *inv* homozygotes still exhibit normal visceral situs (*situs solitus*).
[e]Genes known to be expressed asymmetrically in wild-type individuals.
?-actual mode of inheritance unclear, see text.

of stem or trunk phyllotaxy (five cases), seedling handedness (three cases), leaf rolling or asymmetry (five cases), and floral asymmetry (two cases). In the only exception, careful breeding studies confirmed that direction of style bending in flowers of *Heteranthera multiflora* (Jesson and Barrett, 2002a)—a species that exhibits dimorphic enantiostyly (styles of all flowers on an individual plant bend in the same directions)—is controlled by two alleles at a single locus, with right-bending dominant. Roughly equal frequencies of dextral and sinistral plants are maintained in natural populations by negative frequency-dependent selection

(Jesson and Barrett, 2002b) because outcrossing rates are higher when pollinators move between flowers of opposite handedness. This pattern of floral asymmetry contrasts with the much more widespread phenomenon of monomorphic enantiostyly (both right-bending and left-bending flowers occur on an individual plant), where direction of style bending is clearly not inherited (Jesson and Barrett, 2003).

Antisymmetric animals exhibit the same pattern of inheritance (Table 16-3a). In 12 of 13 cases, dextral and sinistral offspring were equally frequent regardless of whether both parents were dextral or both sinistral. Here, too, many traits were examined, including side of the operculum (one case), side of colored spots (one case), side of the major claw (three cases), direction of bill or wing crossing (three cases), preferred paw use (one case), and two internal asymmetries: direction of visceral situs (one case) and orientation of crossed optic chiasma (two cases). Presumably, direction of asymmetry would not be inherited in pure lines of the 31 mutations that transform directional asymmetry to antisymmetry (Table 16-3c), as shown conclusively for the *iv* mutation, which randomizes visceral situs in mice (Layton, 1976). Only scale-eating cichlid fish appear to differ. Based on field-collected offspring from broods guarded by a pair of parents of known asymmetry, Hori (1993) concluded that mouth deflection was controlled by two alleles at a single locus, with dextral dominant. Unfortunately, as Hori noted, individuals of this species "have the unusual habit of farming out their fry to other breeding pairs" (p. 217), so until controlled laboratory crosses are conducted, these results must be considered tentative.

The evidence seems overwhelming: With virtually no exceptions, direction of asymmetry is not inherited in traits that exhibit antisymmetry.

VI. INHERITANCE OF DIRECTION IN DIRECTIONALLY ASYMMETRIC SPECIES

The results for antisymmetric species contrast sharply with studies of directionally asymmetric species, where three qualitatively different kinds of genetic effects are possible (Table 16-3b–d).

First, genetic variation may cause direction of asymmetry to be reversed. In these cases, variation in the direction of asymmetry is almost always inherited (Table 16-3b). In seven of nine cases—two in plants and five in animals—direction of asymmetry is controlled by two alleles at a single locus. In the remaining two cases, direction was weakly heritable and presumably polygenic. The *Inversin* (*inv*) gene, where heart asymmetry is reversed in 90% of mutant mice (Yokoyama *et al.*, 1993), is also associated with laterality defects in human viscera (Otto *et al.*, 2003).

Second, genetic variation may cause direction of asymmetry to be randomized (i.e., mutations yield antisymmetry). All the evidence here derives from studies of identified mutants affecting visceral or nervous-system asymmetry in model organisms (Table 16-3c). Many mutations randomize heart asymmetry in mice (at least

17 genes) and zebrafish (at least 13 genes). In zebrafish, as many as seven genes randomize habenular or parapineal asymmetries in the brain, and one randomizes liver and pancreas asymmetry. Finally, one gene randomizes nervous-system asymmetries in the nematode *Caenorhabditis elegans*. Among these genes, mutations in *CFC1* are also associated with laterality defects in humans (Bamford *et al.*, 2000).

Third, genetic variation may also cause directional asymmetries to revert to bilateral symmetry (Table 16-3d). Several mutations yield a symmetric heart tube in both mice (five genes) and zebrafish (one gene) or eliminate jogging of the embryonic zebrafish heart (six genes). Similarly, directional asymmetry is lost in eight mutations that yield symmetric lungs in mice.

Collectively, even though the mutant phenotypes vary, these observations all suggest that direction of asymmetry in traits exhibiting directional asymmetry depends on the action of one or more genes.

VII. EVOLUTIONARY SIGNIFICANCE OF ANTISYMMETRY

The virtual absence of heritable variation for direction of asymmetry in antisymmetric species, coupled with a nearly universal inheritance of direction in directionally asymmetric species (previous two sections), provides a unique opportunity to examine the prevalence of two fundamentally different modes of evolution (West-Eberhard, 2003):

- *conventional evolution*—where novel phenotypic variation arises initially from random mutations of small effect (i.e., genetic variation *precedes* observable phenotypic variation), and
- *genetic assimilation*—where novel phenotypic variation arises initially from environmental during development and where mutations that canalize these novel phenotypes arise later evolutionarily (i.e., genetic variation *follows* observable phenotypic variation).

Although acknowledged as a hypothetical possibility by many evolutionary biologists, rather few believe genetic assimilation is common or widespread (Hall, 1999; Schlichting and Pigliucci, 1999; West-Eberhard, 2003). In addition, resistance to this view remains strong (e.g., see de Jong and Crozier, 2003), primarily because compelling examples are hard to find.

Phylogenetic changes among asymmetry states permit a powerful test of the prevalence of these two modes of evolution (Palmer, 2004), because two alternative evolutionary routes to directional asymmetry are possible. First, directionally asymmetric taxa could arise directly from symmetric ancestors. Because direction of asymmetry is heritable in directionally asymmetric species (Table 16-3b), this suggests that asymmetry arose initially as mutations that induced right-handedness or left-handedness. Therefore this route to directional asymmetry qualifies as conventional evolution.

The second, alternative route is far more interesting (Palmer, 2004). If the earliest lineage derived from a symmetric ancestor was antisymmetric, and directionally asymmetric taxa derived from it, then antisymmetry (direction not inherited, Table 16-3a) preceded directional asymmetry (direction inherited, Table 16-3b) evolutionarily. Therefore, each evolutionary transition from antisymmetry to directional asymmetry qualifies as a case of genetic assimilation because phenotypic variation for direction of asymmetry arose before mutations that influence direction of asymmetry.

Data from an earlier wide-ranging survey (Palmer, 1996a) suggest that directional asymmetry arose via genetic assimilation (i.e., by way of antisymmetric ancestors) almost as frequently as by conventional evolution from symmetric ancestors (Table 16-4). One test, of course, does not a generalization make, but the results are tantalizing. Perhaps genetic assimilation really is a common mode of evolutionary change, as West-Eberhard (2003) has argued so passionately. It is the peculiar nature of antisymmetry—where the conspicuous alternative variants right-handed and left-handed are not heritable—that permits such a robust inference.

One final observation deserves attention because it is an exception that dramatically proves the rule regarding the evolutionary origin of conspicuous asymmetries. Direction of asymmetry was convincingly heritable in only one case of antisymmetry: direction of style bending in a plant that exhibits dimorphic enantiostyly (Jesson and Barrett, 2002a). This could imply that heritability of direction is more common among antisymmetric species than implied by Table 16-3a. However, a closer examination of this one exception is particularly illuminating (Jesson and Barrett, 2003). Dimorphic enantiostyly—where styles on all flowers of an individual plant bend in

TABLE 16-4. **Independent Phylogenetic Transitions among Asymmetry States**[a]

	Number of independent clades exhibiting derived state[b]		
Ancestral state	Symmetry	Antisymmetry	Directional asymmetry[c]
Symmetry	—	27 (33)	28 (35)
Antisymmetry	0 (0)	—	16 (28)
Directional asymmetry[d]	4 (4)	6 (10)[c]	30 (30)

[a]Condensed from Table 16-3 of Palmer (1996a).

[b]Early-arising and late-arising asymmetries combined (L and P in Table 3 of Palmer, 1996a). Numbers outside parentheses include only the most reliably inferred phylogenetic transitions. Numbers inside parentheses include all phylogenetic transitions regardless of reliability of inference (see Palmer 1996a for details).

[c]These cases are not pure antisymmetry; they are all cases of biased antisymmetry (one enantiomorph more numerous than the other) and are exclusively flatfish and snails.

[d]Both right/dextral and left/sinistral forms pooled; entries under directional-to-directional indicate dextral-to-sinistral pooled with sinistral-to-dextral transitions.

the same direction and where direction of asymmetry is inherited—is actually rather rare among enantiostylous plants. It occurs in only five species from three families. In sharp contrast, monomorphic enantiostyly—where styles on flowers of an individual plant bend in either direction and where direction of style bending is not inherited—is widespread. It occurs in many species from three dicot and five monocot families. Significantly, all cases of dimorphic enantiostyly (direction inherited) appear to be derived evolutionarily from ancestors that exhibited monomorphic enantiostyly (direction not inherited) (Jesson and Barrett, 2003). So even here, heritable variation evolved after phenotypic variation for style bending.

VIII. WHAT NEXT?

Antisymmetry is a particularly intriguing type of variation because of its simplicity: Two discrete, easily recognizable mirror-image phenotypes that imply a simple *binary switch* controls whether organisms "go right" or "go left" ontogenetically (Palmer, 2004). However, although widespread among animals and plants (Appendix 16-1), we still know little about its inheritance, development, and evolution. Is direction of asymmetry generally not inherited in antisymmetric species, as implied by the preliminary results in Table 16-3a? If true, then antisymmetry offers an attractive opportunity to study how a nonheritable phenotype—direction of asymmetry—both develops and evolves. For example, can direction of asymmetry be biased toward one side by manipulating conditions during development, as it can in lobsters where differential use of one claw induces its transformation into a crusher claw (Govind, 1989)? If generally true, then direction of asymmetry in antisymmetric species would legitimately be considered an environmentally induced state. Finally, does antisymmetry consistently precede directional asymmetry evolutionarily within clades where both occur? Each such clade offers an independent test for genetic assimilation and therefore further insight into a century-old debate about prevailing modes of evolution (West-Eberhard, 2003; Palmer, 2004).

ACKNOWLEDGMENTS

I thank Carolyn Bergstrom, Lois Hammond, and Spencer Barrett for valuable comments on an earlier draft, Arthur Anker for translating some of the German literature, and Kate Witkowska and Jennifer Mersereau (Bamfield Marine Sciences Centre librarian) for help obtaining references. I am particularly grateful to the Natural Sciences and Engineering Research Council of Canada for their generous, sustained, and unfettered support of my research.

APPENDIX 16-1. Taxa and Traits of Animals that Exhibit Conspicuous External Antisymmetry

Taxon[a]	Trait	Source
CNIDARIA		
Hydrozoa		
Hydroida		
Anthomedusae		
Porpitidae		
Velella velella [2A]	Angle of "sail" relative to long axis	(Edwards, 1966)
Siphonophora		
Cystonectae		
Physaliidae (Portuguese man of war)		
Physalia physalis	Angle of "sail" relative to long axis	(Totten and Mackie, 1956)
MOLLUSCA		
Gastropoda		
Amphidromidae		
Amphidromus (>1 spp.) [2B]	Direction of shell coiling	(Davis, 1978)
Bivalvia		
Filibranchia (= Pteriomorphia)		
Mytiloida (mussels)		
Mytiloidea		
Modiolus americanus	Direction of commissure twisting	(Savazzi, 1984)
Pterioida (wing/purse oysters, penshells)		
Lunulacardiidae[b]		
Maminka comata	Side of attached valve	(Kriz, 2001)
Mila	Side of attached valve	(Kriz, 2001)
Antipleuridae[b]		
Stolidotus (2 spp.)	Side of attached valve	(Kriz, 2001)
Dualina (2 spp.)	Side of attached valve	(Kriz, 2001)
Antipleura bohemica	Side of attached valve	(Kriz, 2001)
Silurina	Side of attached valve	(Kriz, 2001)
Hercynella	Side of attached valve	(Kriz, 2001)
Eulamellibranchia (= Heterodonta)		
Paleoheterodonta		
Unionoida (freshwater clams and mussels)		
Etheriidae		
Acostaea rivoli	Side of attached valve	(Yonge, 1978)
Etheria elliptica	Side of attached valve	(Yonge, 1978)
Pseudomulleria dalyi	Side of attached valve	(Yonge, 1978)
Unionidae		
Anconia lanceolata	Direction of commissure twisting	(Savazzi and Yao, 1992)
Veneroida (marine clams)		
Corbiculoidea		
Corbiculidae		
Posostrea anomioides	Side of attached valve	(Bogan and Bouchet, 1998)

Continued

APPENDIX 16-1. Taxa and Traits of Animals that Exhibit Conspicuous External Antisymmetry—Cont'd

Taxon[a]	Trait	Source
Chamoidea		
Chamidae (jewel-box shells)		
Carditochama mindorensis	Side of attached valve	(Matsukuma, 1996)
Eopseuma palaeodontica	Side of attached valve	(Matsukuma, 1996)
Amphichama (2 spp.)	Side of attached valve	(Matsukuma, 1996)
Chama (2 spp.) [2C]	Side of attached valve	(Campbell *et al.*, 2003)
Anomalodesmata		
Pholadomyoida		
Pandoridae		
Pandora inaequivalvis [2D]	Side of curved (lower) valve	(Allen and Allen, 1955)
ANNELIDA		
Polychaeta		
Palpata		
Canalipalpata		
Serpulidae		
Hydroides	Side of operculum	(Okada, 1933)
Spirorbidae		
Neomicrorbis (1 spp.) [2F]	Tube coiling direction	(Fauchald, 1977)
ARTHROPODA		
Chelicerata		
Arachnida		
Araneae (spiders)		
Araneoidea		
Theridiidae		
Tidarren (10 spp.) [2G]	Side of single palpus in male	(Knoflach and van Harten, 2003)
Echinotheridion (6 spp.)	Side of single palpus in male	
Acari (mites)		
Astigmata		
Alloptidae (2 spp.) [2H]	Side of elongated legs	(Gaud and Atyeo, 1996)
Freyanidae (3 spp.)	Side of elongated legs	(Gaud and Atyeo, 1996)
Pterolichidae (1 spp.)	Side of elongated legs	(Gaud and Atyeo, 1996)
Crustacea		
Cirripedia (barnacles)		
Pedunculata (stalked barnacles)		
Lepadomorpha		
Lepadidae		
Koleolepas avis	Side to which body bends	(Yusa *et al.*, 2001)
Sessilia (attached barnacles)		
Verrucomorpha (all spp.) [2N]	Side of attached opercular plates	(Newman and Hessler, 1989)
Copepoda		
Calanoida (calanoid copepods)		
Metridinidae		
Pleuromamma indica	Side of "black organ" + 3 other char.	(Ferrari, 1984)

Continued

APPENDIX 16-1. Taxa and Traits of Animals that Exhibit Conspicuous External Antisymmetry—Cont'd

Taxon[a]	Trait	Source
Malacostraca		
Peracarida		
Isopoda (isopods)		
Bopyridae (most asymmetrical spp.)	Direction of whole-body asymmetry	(Markham, 1985)
Decapoda		
Caridea		
Alpheidae (snapping shrimp)		
Most asymmetrical spp. [2P]	Side of master claw	(A. Anker, personal communication, 2004)
Palaemonidae		
Macrobrachium australe	Side of master claw	(Davis, 1987b)
Thalassinidea (ghost shrimp)		
Axioidae (most spp.)	Side of master claw	(Poore, 1994)
Callianassoidea (most spp.) [2I]	Side of master claw	(Poore, 1994)
Astacidea (lobsters and crayfish)		
Nephropidae (clawed lobsters)		
Nephrops norvegicus	Side of master claw	(Mori *et al.*, 1994)
Homarus (2 spp.)	Side of master claw	(Mariappan *et al.*, 2000)
Anomura		
Porcellanidae (porcelain crabs)		
Most asymmetrical spp.	Side of master claw	(e.g., Imafuku, 1993)
Polyonyx cometes	Side of master claw	(Johnson, 1967)
Brachyura (true crabs)		
Gecarcinidae (land crabs) [2J]		
Gecarcinus (3 spp.)	Side of master claw	(Rathbun, 1918)
Cardisoma (American spp.)	Side of master claw	(Rathbun, 1918)
Ocypodidae (ghost and fiddler crabs)		
Oxypode quadrata	Side of master claw	(Mariappan *et al.*, 2000)
Uca (all spp. but *Thalassuca*)	Side of master claw	
Grapsidae (shore crabs)		
Glyptograpsus	Side of master claw	(Rathbun, 1918)
Goniopsis	Side of master claw	(Rathbun, 1918)
Atelecyclidae		
Acanthocyclus (all spp.)	Side of master claw	(Rathbun, 1930)
Xanthidae		
Chlorodopsis melanochira	Side of master claw	(Mariappan *et al.*, 2000)
Leptodius exaratus	Side of master claw	(Imafuku, 1993)
Portunidae (swimming crabs)		
Trachycarcinus (1 spp.)	Side of master claw	(Rathbun, 1930)

Continued

APPENDIX 16-1. Taxa and Traits of Animals that Exhibit Conspicuous External Antisymmetry—Cont'd

Taxon[a]	Trait	Source
Insecta		
Pterygota		
Mantodea (mantids)		
Mantidae		
Tithrone reseipennis	Side and color of uppermost wing	(Barabas and Hancock, 1999)
Heteroptera (true bugs)		
Most spp. with crossed wings	Side of uppermost wing or wing cover	(Skapec and Stys, 1980)
Trichoptera (caddisflies)		
Mystacides (8 spp.)	Side of genital lobe	(Morse, 2001)
Triaenodes [2k]	Chirality of larval tube wall	(Glover, 1996)
Ylodes	Chirality of larval tube wall	(Glover, 1996)
Diptera (true flies)		
Chironomidae		
Clunio marinus	Direction of abdominal twisting	(Dordel, 1973)
BRACHIOPODA		
Orthida		
Streptis[h] (1 spp.)	Dorsal-most side of twisted gape	(Fürsich and Palmer, 1984)
Rhynchonellida[b]		
Stolomorhynchia (1 spp.)	Dorsal-most side of twisted gape	(Fürsich and Palmer, 1984)
Sphenorhynchia asymmetrica	Dorsal-most side of twisted gape	(Ghosh, 1988)
Rhactochynchia pseudoinconstans	Dorsal-most side of twisted gape	(Ghosh, 1988)
Rhynchonella (2 spp.)	Dorsal-most side of twisted gape	(Fürsich and Palmer, 1984)
Torquirhynchia (5 spp.) [2O]	Dorsal-most side of twisted gape	(Fürsich and Palmer, 1984)
BRYOZOA		
Fenestrata		
Archimedes[b] (all spp.) [2L]	Coiling direction of colony	(Condra and Elias, 1944)
Glymnolaemata		
Cheilostomata		
Bugula (3 spp.) [2Q]	Coiling direction of colony	(McGhee and McKinney, 2000)
Stenolaemata		
Cyclostomata		
Crisidmonea archimediformis[b]	Coiling direction of colony	(Taylor and McKinney, 1996)
Heterocrisina candelabrum[b]	Coiling direction of colony	(Voigt, 1987)
ECHINODERMATA		
Crinoidea		
Pisocrinidae		
Pisocrinus[b] (3 spp.)	Alignment of basal circlet plates	(Rozhnov, 1998)
Platycrinidae[b] (1 spp.)	Coiling direction of stalk	(Rozhnov, 1998)

Continued

APPENDIX 16-1. Taxa and Traits of Animals that Exhibit Conspicuous External Antisymmetry—Cont'd

Taxon[a]	Trait	Source
CHORDATA		
Vertebrata		
Actinopterygii (ray-finned fishes)		
Teleostei		
Characiformes		
Characidae (characins)		
Probolodus heterostomus	Deflection of mouth for scale eating	(Roberts, 1970)
Pleuronectiformes (flatfishes)		
Psettodidae		
Psettodes (2 spp.) [2E]	Eye-side of body	(Hubbs and Hubbs, 1945)
Atheriniformes		
Phallostethidae (>10 spp.)	Side of pectoral clasper, males	(Parenti, 1989)
Tetraodontiformes		
Triacanthodidae		
Macrorhamphosodes uradoi [2M]	Deflection of mouth for scale eating	(Nakae and Sasaki, 2002)
Cyprinodontiformes		
Anablepidae		
Anableps anableps	Side of genital opening	(Hubbs and Hubbs, 1945)
Poeciliidae (livebearers)		
Gambusia affinis	Concavity on side of gonopodium	(Hubbs and Hubbs, 1945)
Xiphophorini (7 spp.)	Concavity on side of gonopodium	(Hubbs and Hubbs, 1945)
Beloniformes		
Hemiramphidae (halfbeaks)		
Dermogenys pusillus	Concavity on side of gonopodium	(Hubbs and Hubbs, 1945)
Perciformes		
Cichlidae (cichlids)		
Perissodus microlepis	Deflection of mouth for scale eating	(Hori, 1993)
Telmatochromis temporalis	Deflection of mouth for scale eating	(Mboki *et al.*, 1998)
Amphibia		
Lissamphibia		
Albanerpetontidae[b]		
Albanerpeton[b] (7 spp.)	Interdigitation of mandibles at midline	(Gardner, 2001)
Celtedens[b] (2 spp.)	Interdigitation of mandibles at midline	(Gardner, 2001)
Urodela (salamanders)		
Triturus (2 spp.)	Ventral epicoracoid cartilage of pectoral girdle	(Greer and Mills, 1997)

Continued

APPENDIX 16-1. Taxa and Traits of Animals that Exhibit Conspicuous External Antisymmetry—Cont'd

Taxon[a]	Trait	Source
Aves		
Passeriformes		
Fringillidae		
Loxia (2 spp.) [2R]	Side to which upper mandible crosses	(Benkman, 1996)
Loxops coccinea	Side to which upper mandible crosses	(Hatch, 1985)
Mammalia		
Chiroptera		
Pteropodidae		
Myonycteris brachycephala	Side of missing lower internal incisor	(Juste and Ibanez, 1993)

[a]Numbers insides square brackets refer to pictures of that species or a related species in Figure 16-2.
[b]Extinct fossil taxa.

REFERENCES

Abby-Kalio, N. J., and Warner, G. F. (1989). Heterochely and handedness in the shore crab *Carcinus maenas* (L.) (Crustacea: Brachyura). *Zoological Journal of the Linnean Society* **96**, 19–26.

Allard, H. A. (1946). Clockwise and counterclockwise spirality in the phyllotaxy of tobacco. *Journal of Agricultural Research* 73, 237–242.

Allen, M. F., and Allen, J. A. (1955). On habits of *Pandora inaequivalvis* (Linné). *Proceedings of the Malacological Society of London* **31**, 175–185.

Asami, T., Cowie, R. H., and Ohbayashi, K. (1998). Evolution of mirror images by sexually asymmetric mating behavior in hermaphroditic snails. *American Naturalist* **152**, 225–236.

Astauroff, B. L. (1930). Analyse der erblichen Störungsfälle der bilaterelen Symmetrie in Zusammenhang mit der selbstandingen Variabilitat ahnlicher Strukturen (Analysis of the hereditary disturbances of biological asymmetry related to the autonomous variability of similar structures). *Zeitschrift für induktive Abstammungs- und Vererbungslehre* **55**, 183–262.

Bamford, R. N., Roessler, E., Burdine, R. D., Saplakoglu, U., de la Cruz, J., Splitt, M., Towbin, J., Bowers, P., Marino, B., Schier, A. F., Shen, M. M., Muenke, M., and Casey, B. (2000). Loss-of-function mutations in the EGF-CFC gene CFC1 are associated with human left-right laterality defects. *Nature Genetics* **26**, 365–369.

Barabas, S. P., and Hancock, E. G. (1999). Asymmetrical colour and wing-folding in *Tithrone roseipennis* (Saussure 1870) a neotropical praying mantis (Mantodea Hymenopodidae). *Tropical Zoology* **12**, 325–334.

Barrett, S. C. H., Jesson, L. K., and Baker, A. M. (2000). The evolution and function of stylar polymorphisms in flowering plants. *Annals of Botany* **85**(Suppl. A), 253–265.

Belmonte, J. C. I. (1999). How the body tells left from right. *Scientific American* **280**, 46–51.

Benkman, C. W. (1988). On the advantages of crossed mandibles: An experimental approach. *Ibis* **130**, 288–293.

Benkman, C. W. (1996). Are the ratios of bill crossing morphs in crossbills the result of frequency dependent selection. *Evolutionary Ecology* **10**, 119–126.

Bergmann, D. C., Crew, J. R., Kramer, J. M., and Wood, W. B. (1998). Cuticle chirality and body handedness in *Caenorhabditis elegans*. *Developmental Genetics* **23**, 164–174.

Biddle, F. G., and Eales, B. A. (1996). The degree of lateralization of paw usage (handedness) in the mouse is defined by three major phenotypes. *Behavior Genetics* **26**, 391–406.

Bogan, A., and Bouchet, P. (1998). Cementation in the freshwater bivalve family Corbiculidae (Mollusca: Bivalvia): A new genus and species from Lake Poso, Indonesia. *Hydrobiologia* **389**, 131–139.

Boudreaux, H. B. (1979). *Arthropod Phylogeny with Special Reference to Insects*. New York: Wiley-Interscience.

Breitenbecher, J. K. (1925). The inheritance of sex-limited bilateral asymmetry in *Bruchus*. *Genetics* **10**, 261–277.

Brown, N. A., and Wolpert, L. (1990). The development of handedness in left-right asymmetry. *Development* **109**, 1–9.

Campbell, M. R., Steiner, G., and Campbell, L. D. (2003). Recent Chamidae of the western Atlantic: Jewel box or Pandora's box? Abstracts. Ann Arbor, MI: *American Malacological Society Annual Meeting*.

Chen, J. N., van Bebber, F., Goldstein, A. M., Serluca, F. C., Jackson, D., Childs, S., Serbedzija, G., Warren, K. S., Mably, J. D., Lindah, P., Mayer, A., Haffter, P., and Fishman, M. C. (2001). Genetic steps to organ laterality in zebrafish. *Comparative and Functional Genomics* **2**, 60–68.

Chen, J. N., van Eeden, F. J. M., Warren, K. S., Chin, A., Nusslein-Volhard, C., Haffter, P., and Fishman, M. C. (1997). Left-right pattern of cardiac BMP4 may drive asymmetry of the heart in zebrafish. *Development* **124**, 4373–4382.

Compton, R. H. (1912). A further contribution to the study of right- and left-handedness. *Journal of Genetics* **2**, 53–70.

Concha, M. L., Burdine, R. D., Russell, C., Schier, A. F., and Wilson, S. W. (2000). A nodal signaling pathway regulates the laterality of neuroanatomical asymmetries in the zebrafish forebrain. *Neuron* **28**, 399–409.

Condra, G. E., and Elias, M. K. (1944). Study and revision of *Archimedes* (Hall). *Geological Society of America Special Papers* **53**, 1–243.

Danos, M. C., and Yost, J. H. (1996). Role of notochord in specification of cardiac left-right orientation in Zebrafish and *Xenopus*. *Developmental Biology* **177**, 96–103.

Darby, H. (1934). The mechanism of asymmetry in the Alpheidae. Carnegie Institute of Washington. *Papers from the Tortugas Laboratory* **28**, 349–361.

Davis, T. A. (1962). The non-inheritance of asymmetry in *Cocos nucifer*. *Journal of Genetics* **58**, 42–50.

Davis, T. A. (1978). Reversible and irreversible lateralities in some animals. *Behavioral and Brain Sciences* **2**, 291–293.

Davis, T. A. (1987a). Dextral and sinistral coiling in gastropod mollusks. *Proceedings of the Indian National Science Academy Part B Biological Sciences* **53**, 323–327.

Davis, T. A. (1987b). Laterality in Crustacea. *Proceedings of the Indian National Academy of Science (B)* **52**, 47–60.

de Jong, G., and Crozier, R. H. (2003). A flexible theory of evolution. *Nature* **424**, 16–17.

de Vries, H. (1911). *The Mutation Theory*. London: Kegan, Paul, Trench, Trübner.

Degner, E. (1952). Der Erbgang der Inversion bei *Laciniaria biplicata* MTG. (Gastropoda, Pulmonata) nebst Bemerkungen zur Biologie dieser Ar. Mitteilungen aus dem Hamburgerischen *Zoologischen Museums und Institut* **51**, 3–61.

Dordel, H.-J. (1973). Functional anatomical investigations of the abdominal torsion in the male imago of *Clunio marinus* Haliday (Diptera, Chironomidae). *Zeitschrift fur Morphologie der Tiere* **75**, 165–221.

Duncker, G. (1904). Symmetrie und Asymmetrie bei bilateralen Thieren. *Roux's Archiv fur Entwickelungsmechanik* **17**, 533–682.

Edelaar, P., and Knops, P. (2005). On the genetic system determining the direction of mandible crossing in crossbills (*Loxia* spp.). *Auk*, in review.

Edwards, C. (1966). *Velella velella* (L.): The distribution of its dimorphic forms in the Atlantic Ocean and the Mediterranean, with comments on its nature and affinities. In *Some Contemporary Studies in Marine Science* (H. Barnes, ed.), pp. 283–296. London: Allen & Unwin.

Fauchald, C. (1977). The polychaete worms. definitions and keys to the orders, families and genera. *Natural History Museum of Los Angeles County, Science Series* **28**, 1–190.
Ferrari, F. D. (1984). Pleiotropy and *Pleuromamma*, the looking-glass copopods (Calanoida). *Crustaceana* 7(Suppl.), 166–181.
Francis, L. (1991). Sailing downwind: Aerodynamic performance of the *Velella* sail. *Journal of Experimental Biology* **158**, 117–132.
Freeman, G., and Lundelius, J. W. (1982). The developmental genetics of dextrality and sinistrality in the gastropod *Lymnaea peregra*. *Roux's Archives of Developmental Biology* **191**, 69–83.
Fujinaga, M. (1997). Development of sidedness of asymmetric body structures in vertebrates. *International Journal of Developmental Biology* **41**, 153–186.
Fürsich, F. T., and Palmer, T. (1984). Commissural asymmetry in brachiopods. *Lethaia* **17**, 251–265.
Galloway, J. W. (1989). Reflections on the ambivalent helix. *Experientia* **45**, 859–872.
Gardner, J. D. (2001). Monophyly and affinities of albanerpetontid amphibians (Temnospondyli; Lissamphibia). *Zoological Journal of the Linnean Society* **131**, 309–352.
Gaud, J., and Atyeo, W. T. (1996). Feather mites of the world (Acarina, Astigmata): The supraspecific taxa. *Annales Sciences Zoologiques, Musee Royal de L'Afrique Central, Tervuren, Belgique* **277**, 3–193.
Ghosh, D. N. (1988). Asymmetry in two rhynchonellid species from the Jurassic formations of Kutch, western India. *Journal of the Geological Society of India* **31**, 476–483.
Glover, J. B. (1996). Larvae of the caddisfly genera *Triaenodes* and *Ylodes* (Trichoptera: Leptoceridae) in North America. *Bulletin of the Ohio Biological Survey*. New Series. **11**(2), vii + 89 pages.
Goss, R. J. (1969). *Principles of Regeneration*. New York: Academic Press.
Govind, C. K. (1989). Asymmetry in lobster claws. *American Scientist* **77**, 468–474.
Govind, C. K., and Pearce, J. (1986). Differential reflex activity determines claw and closer muscle asymmetry in developing lobsters. *Science* **233**, 354–356.
Greer, A. E., and Mills, A. C. (1997). Directional asymmetry in the amphibian pectoral girdle: Additional data and a brief overview. *Journal of Herpetology* **31**, 594–596.
Grüneberg, H. (1935). The causes of asymmetries in animals. *American Naturalist* **69**, 323–343.
Hall, B. K. (1999). *Evolutionary Developmental Biology*. Dordrecht, Netherlands: Kluwer Academic.
Halpern, M. E., Liang, J. O., and Gamse, J. T. (2003). Leaning to the left: Laterality in the zebrafish forebrain. *Trends in Neurosciences* **26**, 308–313.
Hamada, H., Meno, C., Watanabe, D., and Saijoh, Y. (2002). Establishment of vertebrate left-right asymmetry. *Nature Reviews Genetics* **3**, 103–113.
Harper, E. M. (1991). The evolution of the cemented habit in the bivalved molluscs. Ph.D. Thesis. Milton Keynes, England: *The Open University*.
Hashimoto, H., Mizuta, A., Okada, N., Suzuki, T., Tagawa, M., Tabata, K., Yokoyama, Y., Sakaguchi, M., Tanaka, M., and Toyohara, H. (2002). Isolation and characterization of a Japanese flounder clonal line, *reversed*, which exhibits reversal of metamorphic left-right asymmetry. *Mechanisms of Development* **111**, 17–24.
Hashimoto, T. (2002). Molecular genetic analysis of left-right handedness in plants. *Philosophical Transactions of the Royal Society of London Series B* **357**, 799–808.
Hatch, J. J. (1985). Lateral asymmetry of the bill of *Loxops coccineus* (Drepanidae). *Condor* **87**, 546–547.
Heymer, J., Kuehn, M., and Ruther, U. (1997). The expression pattern of nodal and lefty in the mouse mutant Ft suggests a function in the establishment of handedness. *Mechanisms of Development* **66**, 5–11.
Hobert, O., Johnston, R. J., and Chang, S. (2002). Left-right asymmetry in the nervous system: The *Caenorhabditis elegans* model. *Nature Reviews Neuroscience* **3**, 629–640.
Hori, M. (1993). Frequency-dependent natural selection in the handedness of scale-eating cichlid fish. *Science* **260**, 216–219.
Hubbs, C. L., and Hubbs, L. C. (1945). Bilateral asymmetry and bilateral variation in fishes. *Papers from the Michigan Academy of Science, Arts and Letters* **30**, 229–311.

Hudson, A. (2000). Development of symmetry in plants. *Annual Review of Plant Physiology and Plant Molecular Biology* **51**, 349–370.

Ikeno, S. (1923). Erhlichkeitsversuche an einigen Sippen von *Plantago major*. *Japanese Journal of Botany* **1**, 153–212.

Imafuku, M. (1993). Observations on the internal asymmetry of the sternal artery and the cheliped asymmetry in selected decapod crustaceans. *Crustacean Research* **22**, 35–43.

Imai, Y. (1927). The right- and left-handedness of phyllotaxy. *Botanical Magazine* (Tokyo) **41**, 592–596.

Jesson, L. K., and Barrett, S. C. H. (2002a). The genetics of mirror-image flowers. *Proceedings of the Royal Society of London Series B* **269**, 1835–1839.

Jesson, L. K., and Barrett, S. C. H. (2002b). Enantiostyly in *Wachendorfia* (Haemodoraceae): The influence of reproductive systems on the maintenance of the polymorphism. *American Journal of Botany* **89**, 253–262.

Jesson, L. K., and Barrett, S. C. H. (2003). The comparative biology of mirror-image flowers. *International Journal of Plant Sciences* **164**, S237–S249.

Johnson, D. S. (1967). On some commensal decapod crustaceans from Singapore (Palaemonidae and Porcellanidae). *Journal of Zoology* (London) **153**, 499–526.

Jones, D. S., and George, R. W. (1982). Handedness in fiddler crabs as an aid in taxonomic grouping of the genus *Uca* (Decapoda, Ocypodidae). *Crustaceana* **43**, 100–102.

Juste, J., and Ibanez, C. (1993). An asymmetric dental formula in a mammal, the São-Tomé island fruit bat *Myonycteris brachycephala* (Mammalia, Megachiroptera). *Canadian Journal of Zoology* **71**, 221–224.

Kasinov, V. B. (1973). Handedness in Lemnaceae: On the determination of left and right types of development in *Lemna* clones and on its alteration by means of external influences. *Beiträge z. Biol. Pflanzen* **49**, 321–337.

Kihara, H. (1972). Right- and left-handedness in plants: A review. *Seiken Zihô* **23**, 1–37.

Klar, A. J. S. (2002). Fibonacci's flowers. *Nature* **417**, 595.

Knoflach, B. (2000). Copulation and emasculation in *Echinotheridion gibberosum* (Kulczynski, 1899) (Araneae, Theridiidae). In *European Arachnology 2000* (Proceedings of the 19th European Colloquium of Arachnology, Århus, 17–22 July 2000) (S. Toft and N. Scharff, eds.), pp. 139–144. Aarhus, Denmark: Aarhus University Press.

Knoflach, B., and van Harten, A. (2003). Palpal loss, single palp copulation and obligatory mate consumption in *Tidarren cuneolatum* (Tullgren, 1910) (Araneae, Theridiidae). *Journal of Natural History* **34**, 1639–1659.

Koriba, K. (1914). Mechanisch-physiologische Studien über die Drehang der *Spiranthes*. Ähre. *Journal of Colloid Science* **36(3)**, 1–179.

Krebs, L. T., Iwai, N., Nonaka, S., Welsh, I. C., Lan, Y., Jiang, R., Saijoh, Y., O'Brien, T. P., Hamada, H., and Gridley, T. (2003). Notch signaling regulates left-right asymmetry determination by inducing Nodal expression. *Genes and Development* **17**, 1207–1212.

Kriz, J. (2001). Enantiomorphous dimorphism in Silurian and Devonian bivalves; *Maminka* Barrande, 1881 (Lunulacardiidae, Silurian): The oldest known example. *Lethaia* **34**, 309–322.

Kundu, B. C., and Sarma, M. S. (1965). Direction of leaf spiral in *Corchorus capsularis* L. *Transactions of the Bose Research Institute* (Calcutta) **28**, 107–112.

Labadie, L. V., and Palmer, A. R. (1996). Pronounced heterochely in the ghost shrimp, *Neotrypaea californiensis* (Decapoda: Thalassinidea: Callianassidae): Allometry, inferred function and development. *Journal of Zoology* (London) **204**, 659–675.

Landauer, W. (1938). Notes on cross-beak in fowl. *Journal of Genetics* **37**, 51–68.

Larrabee, A. P. (1906). The optic chiasma of teleosts: A study of inheritance. *Proceedings of the American Academy of Arts and Science* **42**, 217–231.

Layton, W. M. J. (1976). Random determination of a developmental process. *Journal of Heredity* **67**, 336–338.

Lilienfeld, F. A. (1959). Dextrality and sinistrality in plants. III. *Medicago tuberculata* Willd. and *M. litoralis* Rohde. *Proceedings of the Japanese Academy of Science* 35, 475–481.

Ludwig, W. (1932). *Das Rechts-Links Problem im Teirreich und beim Menschen.* Berlin, Germany: Springer.

Lugard, W. J. (1931). Quelques observations morphologiques sur le "contonnier Egyptien" au point de vue "Phyllotaxie et Disposition des Petales dans la Carolle." *Bulletin Agriculturale de Congo Belgique* 22, 229–242.

Macdonald, T. A. (2003). Phylogenetic relations among spirorbid subgenera and the evolution of opercular brooding. *Hydrobiologia* 496, 125–143.

Mariappan, P., Balasundaram, C., and Schmitz, B. (2000). Decapod crustacean chelipeds: An overview. *Journal of Biosciences* 25, 301–313.

Markham, J. C. (1985). A new species of *Asymmetrione* (Isopoda: Bopyridae) infesting the hermit crab *Isocheles pilosus* (Holmes) in southern California. *Bulletin Southern California Academy of Sciences* 84, 104–108.

Matsukuma, A. (1996). A new genus and four new species of Chamidae (Mollusca: Bivalvia) from the Indo-West Pacific with reference to transposed shells. *Bulletin du Muséum National d'Histoire Naturelle, Paris, Ser. 4* 18, 23–53.

Mboko, S. K., Kohda, M., and Hori, M. (1998). Asymmetry of mouth-opening of a small herbivorous cichlid fish *Telmatochromis temporalis* in Lake Tanganyika. *Zoological Science* 15, 405–408.

McGhee, G. R., Jr., and McKinney, F. K. (2000). A theoretical morphologic analysis of convergently evolved erect helical colony form in the Bryozoa. *Paleobiology* 26, 556–577.

McKinney, F. K. (1980). Erect spiral growth in some living and fossil bryozoans. *Journal of Paleontology* 54, 597–613.

McKinney, F. K., and Burdick, D. W. (2001). A rare, larval-founded colony of the Bryozoan *Archimedes* from the Carboniferous of Alabama. *Palaeontology* 44, 855–859.

McKinney, F. K., Listokin, M. R. A., and Phifer, C. D. (1986). Flow and polypide distribution in the cheilostome bryozoan *Bugula* and their inference in *Archimedes*. *Lethaia* 19, 81–93.

McManus, I. C. (2002). *Right Hand Left Hand. The Origins of Asymmetry in Brains, Bodies, Atoms and Cultures.* Cambridge, MA: Harvard University Press.

Mellon, de F. Jr. (1981). Nerves and the transformation of claw type in snapping shrimps. *Trends in Neurosciences* 4, 245–248.

Meno, C., Shimono, A., Saijoh, Y., Yashiro, K., Mochida, K., Ohishi, S., Noji, S., Kondoh, H., and Hamada, H. (1998). *lefty-1* is required for left-right determination as a regulator of *lefty-2* and *nodal*. *Cell* 94, 287–297.

Morgan, T. H. (1923). The development of asymmetry in the fiddler crab. *American Naturalist* 57, 269–273.

Mori, M., Biagi, F., and De Ranieri, S. (1994). Morphometric analysis of the size at sexual maturity and handedness in *Nephrops norvegicus* (L.) of the North Tyrrhenian Sea. *Bollettino Dei Musei E Degli Istituti Biologici Dell'universita Di Genova* 60–61, 165–178, illustr.

Morse, J. C. (2001). Phylogeny, classification, and historical biogeography of world species of *Mystacides* (Trichoptera: Leptoceridae), with a new species from Sri Lanka. In *Proceedings of the 10th International Symposium on Trichoptera, Potsdam, Germany* (W. Mew, ed.), pp. 173–186. Keltern, Germany: Goecke & Evers.

Murray, J., and Clarke, B. (1966). The inheritance of polymorphic shell characters in *Partula* (Gastropoda). *Genetics* 54, 1261–1277.

Nakae, M., and Sasaki, K. (2002). A scale-eating triacanthodid, *Macrorhamphosodes uradoi*: Prey fishes and mouth "handedness" (Tetraodontiformes, Triacanthoidei). *Ichthyological Research* 49, 7–14.

Neville, A. C. (1976). *Animal Asymmetry.* London: Edward Arnold.

Newman, W. A. (1989). Juvenile ontogeny and metamorphosis in the most primitive living sessile barnacle, *Neoverruca*, from an abyssal hydrothermal spring. *Bulletin of Marine Science* 45, 467–477.

Newman, W. A., and Hessler, R. R. (1989). A new abyssal hydrothermal verrucomorphan (Cirripedia; Sessilia): The most primitive living sessile barnacle. *Transactions of the San Diego Society for Natural History* **21**, 259–273.

Okada, Y. K. (1933). Remarks on the reversible asymmetry in the opercula of the polychaete *Hydroides*. *Journal of the Marine Biological Association UK* **18**, 655–670.

Otto, E. A., Schermer, B., Obara, T., O'Toole, J. F., Hiller, K. S., Mueller, A. M., Ruf, R. G., Hoefele, J., Beekmann, F., Landau , D., Foreman, J. W., Goodship, J. A., Strachan, T., Kispert, A., Wolf, M. T. T., Gagnadoux, M. M., Nivet, H., Antignac, C., Walz, G., Drummond, I. A., Benzing, T., and Hildebrandt, F. (2003). Mutations in INVS encoding inversin cause nephronophthisis type 2, linking renal cystic disease to the function of primary cilia and left-right axis determination. *Nature Genetics* **4**, 414–420.

Palmer, A. R. (1994). Fluctuating asymmetry analyses: A primer. In *Developmental Instability: Its Origins and Evolutionary Implications* (T. A. Markow, ed.), pp. 335–364. Dordrecht, Netherlands: Kluwer, Dordrecht.

Palmer, A. R. (1996a). From symmetry to asymmetry: Phylogenetic patterns of asymmetry variation in animals and their evolutionary significance. *Proceedings of the National Academy of Sciences USA* **93**, 14279–14286.

Palmer, A. R. (1996b). Waltzing with asymmetry. *Bioscience* **46**, 518–532.

Palmer, A. R. (2004). Symmetry-breaking and the evolution of development. *Science* **306**, 828-833.

Palmer, A. R., and Strobeck, C. (2003). Fluctuating asymmetry analyses revisited. In *Developmental Instability (DI): Causes and Consequences* (M. Polak, ed.), pp. 279–319. Oxford, England: Oxford University Press.

Palmer, A. R., Strobeck, C., and Chippindale, A. K. (1993). Bilateral variation and the evolutionary origin of macroscopic asymmetries. *Genetica* **89**, 201–218.

Parenti, L. R. (1989). A phylogenetic revision of the phallostethid fishes (Atherinomorpha, Phallostethidae). *Proceedings of the California Academy of Sciences* **46**, 243–277.

Pennekamp, P., Karcher, C., Fischer, A., Schweickert, A., Skryabin, B., Horst, J., Blum, M., and Dworniczak, B. (2002). The ion channel polycystin-2 is required for left-right axis determination in mice. *Current Biology* **12**, 938–943.

Perelle, I. B., and Ehrman, L. (1994). An international study of human handedness: The data. *Behavior Genetics* **24**, 217–227.

Policansky, D. (1982). Flatfish and the inheritance of asymmetries. *Behavioral and Brain Sciences* **5**, 262–265.

Poore, G. C. B. (1994). A phylogeny of the families of the Thalassinidea (Crustacea: Decapoda) with keys to the families and genera. *Memoirs of the Museum of Victoria* **54**, 79–120.

Purandare, S. M., Ware, S. M., Kwan, K. M., Gebbia, M., Bassi, M. T., Deng, J. M., Vogel, H., Behringer, R. R., Belmont, J. W., and Casey, B. (2002). A complex syndrome of left-right axis, central nervous system and axial skeleton defects in Zic3 mutant mice. *Development* **129**, 2293–2302.

Purnell, D. J., and Thompson, J. N. J. (1973). Selection for asymmetrical bias in a behavioral character of *Drosophila melanogaster*. *Heredity* **31**, 401–405.

Rama Swamy, N., and Bahadur, B. (1999). Inheritance of seedling handedness in *Triticale* and its parents. *Cereal Research Communications* **27**, 91–97.

Rao, K. L., Bahadur, B., and Satyanarayana, A. (1983). Effect of selection on seedling handedness in pigeon pea. *Current Science* **52**, 608–609.

Rathbun, M. J. (1918). The grapsoid crabs of America. *US National Museum, Smithsonian Institution Bulletin* **97**, 1–461.

Rathbun, M. J. (1930). The cancroid crabs of America of the families Euryalidae, Portunidae, Atelecyclidae, Cancridae and Xanthidae. *US National Museum, Smithsonian Institution Bulletin* **152**, 609.

Raya, A., Kawakami, Y., Rodriguez-Esteban, C., Buscher, D., Koth, C. M., Itoh, T., Morita, M., Raya, R. M., Dubova, I., Bessa, J. G., de la Pompa, J. L., and Belmonte, J. C. I. (2003). Notch activity induces nodal expression and mediates the establishment of left-right asymmetry in vertebrate embryos. *Genes and Development* **17**, 1213–1218.

Rebagliati, M. R., Toyama, R., Haffter, P., and Dawid, I. B. (1998). Cyclops encodes a nodal-related factor involved in midline signaling. *Proceedings of the National Academy of Sciences USA* **95**, 9932–9937.

Richards, O. W. (1927). Sexual selection and allied problems in insects. *Biological Reviews (Cambridge)* **2**, 298–364.

Ringe, F. (1971). Symmetrieverhältnisse bei *Begonia semperflorens 'gracilis'*. Ein Beitrag zum Rechts-Links-Problem und zur geometrischen Symmetriebetrachtung. *Beitraege zur Biologie der Pflanzen* **47**, 453–468.

Roberts, T. R. (1970). Scale-eating American characoid fishes, with special reference to *Probolodus heterostomus*. *Proceedings of the California Academy of Sciences* **38**, 383–390.

Robertson, R. (1993). Snail handedness. The coiling directions of gastropods. *National Geographic Research and Exploration* **9**, 109–119.

Rozhnov, S. V. (1998). The left-right asymmetry in echinoderms. In *Echinoderms: San Francisco.* Proceedings of the Ninth International Echinoderm Conference, San Francisco, California, USA (R. Mooi and M. Telford, eds.), pp. 73–78. Rotterdam, Netherlands: A. A. Balkema.

Savazzi, E. (1984). Adaptive significance of shell torsion in mytilid bivalves. *Paleontology* **27**, 307–314.

Savazzi, E., and Yao, P. (1992). Some morphological adaptations in freshwater bivalves. *Lethaia* **25**, 195–209.

Savilov, A. I. (1961). The distribution of different ecological forms of the by-the-wind sailor, *Velella lata* Ch. and Eys. *Physalia utriculua* (La Martiniere) Esch., in the North Pacific. *Trudy Instituta Okeanologii Akademiya nauk SSSR* **45**, 223–239.

Schlichting, C. D., and Pigliucci, M. (1999). *Phenotypic Evolution: A Reaction Norm Perspective.* Sunderland, MA: Sinauer.

Schochet, J. (1973). Opercular regulation in the polychaete *Hydroides dianthus* (Verrill, 1873); I. Opercular ontogeny, distribution and flux. *Biological Bulletin* **144**, 138–142.

Shannon, L. V., and Chapman, P. (1983). Incidence of *Physalia* on beaches in the south western Cape Province during January 1983. *South African Journal of Science* **79**, 454–458.

Simonson, J. L. (1985). Reversal of handedness, growth, and claw stridulatory patterns in the stone crab *Menippe mercenaria* (Say) (Crustacea: Xanthidae). *Journal of Crustacean Biology* **5**, 281–293.

Skapec, L., and Stys, P. (1980). Asymmetry in the forewing position in Heteroptera. *Acta Entomologica Bohemoslovaca* **77**, 353–374.

Taylor, P. D., and McKinney, F. K. (1996). An *Archimedes*-like cyclostome bryozoan from the Eocene of North Carolina. *Journal of Paleontology* **70**, 218–229.

Timoféeff-Ressovsky, N. W. (1934). Über der Einfluss des genotypischen Milieus und der Aussenbedingungen auf die Realisation des Genotypes. *Nachrichten von der Gesellschaft der Wissenschaften zu Göttingen. Mathematisch-Physikalische Klasse. Fachgruppe* 6 **1**, 53–106.

Tiso, N., Filippi, A., Pauls, S., Bortolussi, M., and Argenton, F. (2002). BMP signalling regulates antero-posterior endoderm patterning in zebrafish. *Mechanisms of Development* **118**, 29–37.

Toar, R. E., Rompas, T. M., and Sudasrip, H. (1979). Non-inheritance of the direction of foliar spiral in coconut. *Experientia* **35**, 1585–1587.

Totten, A. K., and Mackie, G. O. (1956). Dimorphism in the Portuguese Man-of-War. *Nature* **177**, 290.

Tsukui, T., Capdevila, J., Tamura, K., Ruiz-Lozano, P., Rodriguez-Esteban, C., Yonei-Tamura, S., Magallon, J., Chandraratna, R. A. S., Chien, K., Blumberg, B., Evans, R. M., and Belmonte, J. C. I. (1999). Multiple left-right asymmetry defects in Shh(–/–) mutant mice unveil a convergence of the Shh and retinoic acid pathways in the control of Lefty-1. *Proceedings of the National Academy of Sciences USA* **96**, 11376–11381.

Van Valen, L. (1962). A study of fluctuating asymmetry. *Evolution* **16**, 125–142.

Venkateswarlu, T. (1982). Non-inheritance of isomerism in cocoyams. *Proceedings of the Indian Academy of Sciences-Plant Sciences* **91**, 17–23.

Vermeij, G. J. (1973). Biological versatility and earth history. *Proceedings of the National Academy of Sciences USA* **70**, 1936–1938.

Voigt, E. (1987). On new cyclostomate Bryozoa from the Upper Maastrichtian Chalk-Tuff near Maastricht Netherlands. *Palaeontologische Zeitschrift* **61**, 41–56.

Waddington, C. H. (1953). Genetic assimilation of an acquired character. *Evolution* **7**, 118–126.

Welch, C. (2001). Chirality in the natural world: Life through the looking glass. In *Chirality in the Natural World* (J. Lough, ed.), New York: Blackwell, 86–92.

West-Eberhard, M. J. (2003). *Developmental Plasticity and Evolution*. New York: Oxford University Press.

Wiggins, G. B. (1996). *Larvae of the North American Caddisfly Genera (Trichoptera)*. Toronto, Canada: University of Toronto Press.

Yamaguchi, T. (1977). Studies on the handedness of the fiddler crab, *Uca lactea*. *Biological Bulletin* **152**, 424–436.

Yamaguchi, T. (1978). Studies on the decision of handedness in the fiddler crab *Uca lactea*. *Calanus* **6**, 29–51.

Yokoyama, T., Copeland, N. G., Jenkins, N. A., Montgomery, C. A., Elder, F. F. B., and Overbeek, P. A. (1993). Reversal of left-right asymmetry: A situs inversus mutation. *Science* **260**, 679–682.

Yonge, C. M. (1978). On the monomyarian, *Acostaea rivoli* and evolution in the family Etheriidae (Bivalvia: Unionacea). *Journal of Zoology (London)* **184**, 429–448.

Yusa, Y., Yamato, S., and Marumura, M. (2001). Ecology of a parasitic barnacle, *Koleolepas avis*: Relationship to the hosts, distribution, left-right asymmetry and reproduction. *Journal of the Marine Biological Association UK* **81**, 781–788.

Zeleny, C. (1905). Compensatory regulation. *Journal of Experimental Zoology* **2**, 1–102.

CHAPTER 17

Variation in Structure and Its Relationship to Function: Correlation, Explanation, and Extrapolation

ANTHONY P. RUSSELL* AND AARON M. BAUER†

*Department of Biological Sciences, University of Calgary, Calgary, Alberta, Canada
†Department of Biology, Villanova University, Villanova, Pennsylvania, USA

ABSTRACT

Structural variation in organisms has been a topic of long-standing interest in biology and attempts to account for and explain this variation were major contributors to the formulation of Darwin's ideas about evolution. Structural variation is the currency of taxonomy, systematics, functional ecology, and much of evolutionary biology. Whereas it is assumed that changes in structure are related to changes in function, and vice versa, teasing out the intricate relationships of this couplet is a perplexing task.

The relationship of structure to function at the intraspecific level has been investigated chiefly in two distinct, but related ways. *In situ* studies have posed questions about variation as seen in the field at various scales, from the very local to those encompassing large geographic distances. In this overall approach, structural variation has been correlated, in a variety of ways, with environmental variables, both biotic and abiotic. Relying on various theoretical underpinnings of why and how such variation occurs, *in situ* studies have investigated variation in the context of trophic polymorphism within and between populations, clinal variation across broad geographic ranges, and microgeographic variation across smaller ranges.

Ex situ studies have largely examined structural variation in an experimental, laboratory-type context and have focused on performance comparisons between exemplars exhibiting observed structural variations. Such approaches have considered variation in trophic and, more extensively, locomotor performance and attempt to explain them by way of variation in measured performance. Fluctuating asymmetry has recently become of increasing interest in terms of its influence on performance, as have selection experiments that attempt to explore the limits of variation in relation to performance.

There is an increasing tendency to combine *in situ* and *ex situ* components in integrated investigations, but the scale and intensity of such undertakings present new challenges, and also reveal new intricacies and revelations that challenge the extrapolations that tend to be made from single approaches in isolation. Our ability to explain the relationship between structure and function as they relate to variation may be dependent upon the level of detail we expect from the explanatory model.

INTRODUCTION

Variation in the structure of organisms, as perceived by casual or detailed inspection of morphology (including color), has been of long-standing interest in the study of biology and its predecessor, natural history. Indeed, much of the framework for evolutionary biology grew out of the contemplation of structural variation, although the meaning of this variation was not self-evident. Deducing functional correlates and corollaries of such variation has proved a challenging task, and extrapolation from structure to function continues to challenge biologists and their experimental and statistical expertise. These challenges are particularly acute at the intraspecific level, where differences in structure and function are often subtle, continuous within and between populations, and hard to explain, especially in the context of cause and effect, as opposed to correlation.

Essentialism, with its roots in Platonic philosophy and its Aristotelian refinement (Panchen, 1992), dominated the thinking of the Western world well into the nineteenth century (Mayr, 1982). This was particularly problematic for biology as the variable world of phenomena was viewed as the imperfect manifestation of underlying essences, as expressed in form and behavior (Panchen, 1992). Thus variation (in organisms and other entities exhibiting structure) was viewed as being accidental and consequently of no interest to scientific inquiry. Essentialism emphasized discontinuity, constancy, and typical values ("typology"), validating the reality of taxa independently of any explanatory theory (Panchen, 1992) and is directly connected to Linnaeus's insistence on the reality, sharp delimitation, and constancy of species (Mayr, 1982). Linnaeus's view of species was, however, discordant with nature and prompted investigations into the extent and meaning of variation of structure. Naturalists observed and recorded individual variation, and the growing body of evidence cried out for an explanation. Variation thus became an important source of evidence against biological essentialism.

Increasing recognition and contemplation of structural variation eventually manifested itself as a taxonomic problem — what was the extent and pattern of variation within and between species? After Charles Darwin's return to England, following the voyage of the Beagle, he enhanced his understanding and appreciation of anatomy and development in relation to taxonomy by undertaking his monographic study of the Cirripedia (T. H. Huxley, in Darwin, 1892). Charles Darwin noted, "The Cirripedes form a highly varying and difficult group of species to class; and my work was of considerable use to me, when I had to discuss in the *Origin of Species* the principles of natural classification" (Darwin, 1892, p. 41). Charles Darwin carefully noted and recorded the variation of external and internal features in an enormous array of barnacle species (Darwin, 1854), and this comprehensive overview of structural variation within and between species allowed him to recast variation as an important biological phenomenon, and to finally overcome the constraints of essentialist thinking. The transition from variation being a problem for taxonomists to it being of focal significance in evolutionary studies is encapsulated by how Charles Darwin (1859, p. 39) portrayed the manifestation of variation to taxonomists and to those interested in organismic change over time: "... systematists are far from pleased at finding variability in important characters ... there are not many men who will laboriously examine internal and important organs, and compare them in many specimens of the same species."

Charles Darwin advocated that individual differences within species provided the basis for the operation of natural selection. This was echoed by Bateson (1894), who gave primacy to variation in his appraisal of evolution and placed this variation into an environmental context, thus cementing the link between structure, function, and variation. The connection between structural variants and functional differences was not, however, easy to establish in a demonstrable way and left Charles Darwin (1859, p. 137) lamenting, "Not in one case out of a hundred can we pretend to assign any reason why this or that part differs, more or less, from the same part of the parents." Bateson (1894, p. 12) opined that "the only possible test of the utility of a structure must be a quantitative one," but felt that such an approach was likely to be inaccessible to biologists for the foreseeable future.

Almost 110 years later, Lauder (2003) emphasized that intraspecific variation in function remains one of the most important issues in the study of evolution. The predicting of precise aspects of function from morphology (structure), however, depends upon the assumption of a close match between structure and function (Lauder, 1995), an assumption that is frequently violated. Lauder (1995) advocates experimental study of function to measure the functional properties of structures and then careful quantification of the effects of variation. Such assessments subsequently require further evaluation by taking the laboratory (*ex situ*) findings into the field (*in situ*) to examine how studies of performance relate to field-based evaluations of selection (Lauder, 2003). This prognosis revisits the dichotomy explored by Bock and Von Wahlert (1965), in which they advocated the distinction between form and function that could be explored in a laboratory setting, and the biological role that required assessment of the form-function complex within environmental contexts of different scale ("*Umwelt*" and "*Umgebung*") to determine how selection might "see" the particular aspect of form and function in question.

I. BACKGROUND

We herein survey approaches to the understanding of the relationship between variation in structure and its functional consequences at the intraspecific level. In so doing, we outline some of the avenues that have been, and are still being, taken to investigate variation in the absence of functional explanations and then explore how these approaches have lent themselves, either singly or in combination, to correlatory or experimental investigations of the structure-function couplet. Variation itself is accepted as being adaptive (Van Valen, 1965), the fuel that feeds evolutionary change (Stearns, 1989), important for selection (Mather, 1970), and thus critical to the process of speciation (Davis and Gilmartin, 1985). Its magnitude is related to the rate of evolution (Wagner, 1988), and its investigation and documentation have been the subject of research that is assumed to be evolutionarily meaningful simply through the emergence of identifiable patterns, even if these cannot be explained in any functional way.

Until recently, functional morphologists have tended to ignore or downplay structural variation by acknowledging its existence but endeavoring to explain structure and function in the context of some "average" condition. In contrast, evolutionary ecologists have focused on variation in their quest to investigate functional and selective advantages, but have tended to treat structure in a relatively simplistic way, often by reducing it to a series of measurements. Progress in the understanding of structural variation as it relates to functional differences requires a closer integration of morphological form and ecological consequence.

Our consideration of structure and function in the context of variation excludes factors associated with sexual dimorphism. Although morphological variation in such cases is directly implicated in sexual selection, any functional variation that exists is not acted upon by selection to favor the differential success of one sex relative over the other. Our focus is, therefore, on functional attributes that are unlikely to be influenced by sex, although future assessments would do well to investigate the structural and functional implications of the alteration of features that neighbor extravagant sexually dimorphic ornamentation (Emlen, 2001). We also exclude from consideration allometric change through ontogeny (Curren *et al.*, 1993) that brings about functional changes throughout the life history of individuals, such as those documented for the lizard *Basiliscus vittatus* in relation to locomotor proficiency across the surface of water (Laerm, 1974), the shape of the skull of mammals across the functional transition from suckling to adult feeding (Zeldich and Carmichael, 1989), and the changes in the head shield of the amphisbaenian *Monopeltis guentheri* in relation to the effectiveness of burrowing (Gans and Latifi, 1971). We limit ourselves to consideration of variation within single species at developmentally equivalent life-history stages.

Emphasis is placed upon instances in which structural and functional variation have the potential to be of evolutionary relevance, and in which enhanced fitness (Hallgrímsson, 2003) may be expected or demonstrated to result from differences in performance of some kind. Much of the literature in this field is derived from vertebrate examples. As this is the taxonomic group with which we are most familiar, we draw our examples almost exclusively from this clade. We believe, however, that the same general principles apply to other groups of organisms.

II. APPROACHES TO THE STUDY OF STRUCTURAL VARIATION

Variation in structure continues to be a focus of investigation in its own right, and is approached from several directions. Observations resulting from such studies have, in many instances, led to considerations of the functional meaning of the observed variation, or of similar variation observed in other taxa, but in many instances functional inferences have not, and probably cannot, be drawn based upon the available biological information about the taxon in question. We present below some broad categories under which structural variation has been considered, and then document approaches to the functional interpretation of structural variation that show affinity with one or more of these categories in largely *in situ* (Section IV) and *ex situ* (Section V) contexts.

III. VARIATION AS AN OBSERVABLE PHENOMENON

The understanding that structural variation is a general characteristic of biological taxa within and between populations, even in clonal forms (Vasilyeva, 1990), continues to result in documentation of the extent of this variation, even if no interpretation of its functional meaning is forwarded. For example, Smith (1979) examined patterns of morphology in the mouse *Peromyscus californicus* across its range and noted that this matched genetic and historical patterns relating to isolation and subsequent secondary contact. Variation in scalation pattern within species of squamate reptiles is frequently recorded, and sometimes speculative inferences are made about the basis for the observed structural variation. An example of this is Gardner's (1986) examination of the geckos *Phelsuma sundbergi* and *Phelsuma astriata* in the Seychelles and the observation that there is a strong correlation between the number of lamellae under the fourth toe and snout-vent length between islands, but not within islands. It was posited that this may be due to functional reasons related to the effectiveness of adhesion, or may simply be a correlate of developmental constraints. Savage and Donnelly (1988) related differences in chinshield morphology in the snake *Hydromorphus concolor* to potential shifts in diet. Such observations allow functional hypotheses to be erected, but in many cases are difficult to follow up with experimental observations.

In some instances, the variants observed have been correlated with differential survival potential, as is the case for European adders (*Vipera berus*) (Lindell *et al.*, 1993). In this instance, those individuals with the highest number of ventral scales exhibited the highest growth rates and reached the largest sizes. Thus, such variation can be related to selection and fitness, but how the structural features that can be most readily measured and counted are causally related (if at all) to survivability remains unfathomed.

A. Variation and Taxonomic Utility

Structural variation continues to be a major source of characters and character states in taxonomic and systematic investigations (Michaux, 1989; Lauder, 1990). Variation within species is sometimes investigated in the context of functional and ecological relationships. Such is the case with rostral scale and underlying skeletal morphology of the bullsnake, *Pituophis* (Knight, 1986). Variation in structure across the geographic range was related to the comparative ability of eastern and western forms to burrow and excavate mammalian prey. The extent of the observed differences resulted in the question of whether one or more species was involved. Subsequent investigation (Rodriguez-Robles and De Jesus-Escobar, 2000) revealed that two lineages are represented. Accurate taxonomy is thus necessary, as without this, intra- and interspecific differences may be confounded, and causal explanations may be inappropriately directed.

B. Variation Associated with Developmental Plasticity

In salamanders, high incidences of limb skeletal variation have been noted when large samples are examined. Hanken (1983) reported that *Plethodon cinereus* exhibits carpal and tarsal patterns typified by fusions or losses of elements that are unrelatable to genetic causes or functional outcomes (Figure 17-1). Some species of *Thorius* (Hanken, 1982) display similar variation in association with limb reduction, as does the scincid lizard *Hemiergis* (Choquenot and Greer, 1989), and some species of reduced-finned clariid catfishes (Adriaens *et al.*, 2002). Tague (1997) reported a higher level of variability in the first metapodial element of monkeys in which this element is vestigial.

Shubin *et al.* (1995) reported that carpal and tarsal variation in the salamander *Taricha granulosa* was prominent, that not all theoretically possible patterns were expressed, and that the patterns that do occur also occur as derived features in related taxa. This was interpreted as being indicative of limited domains of phenotypic expression that serve as the raw material for selection, even though no functional correlations could be made. Hallgrímsson *et al.* (2002) noted an increase of variation in the pattern of expression of the distal limb skeleton in a variety of taxa. This may be the target of selection, as Hallgrímsson *et al.* (2003) discovered that in prenatal mice skeletal variation decreases with age.

Similar developmental plasticity is encountered in dentitional patterns. Additional premolars are variably present in some felids (Pocock, 1916), a feature that has more recently been related to relative growth patterns between the juvenile and adult dentitions that result in inhibition of tooth germs from becoming established if the space is too restrictive, or permit further development and eruption if it is not (Russell *et al.*, 1995). Although no functional consequences of presence or absence have been advocated, it is likely that the overall growth patterns that result in this variation are adaptive in the context of the dental battery, rather than being interpretable in the context of the particular tooth in question (Russell *et al.*, 1995).

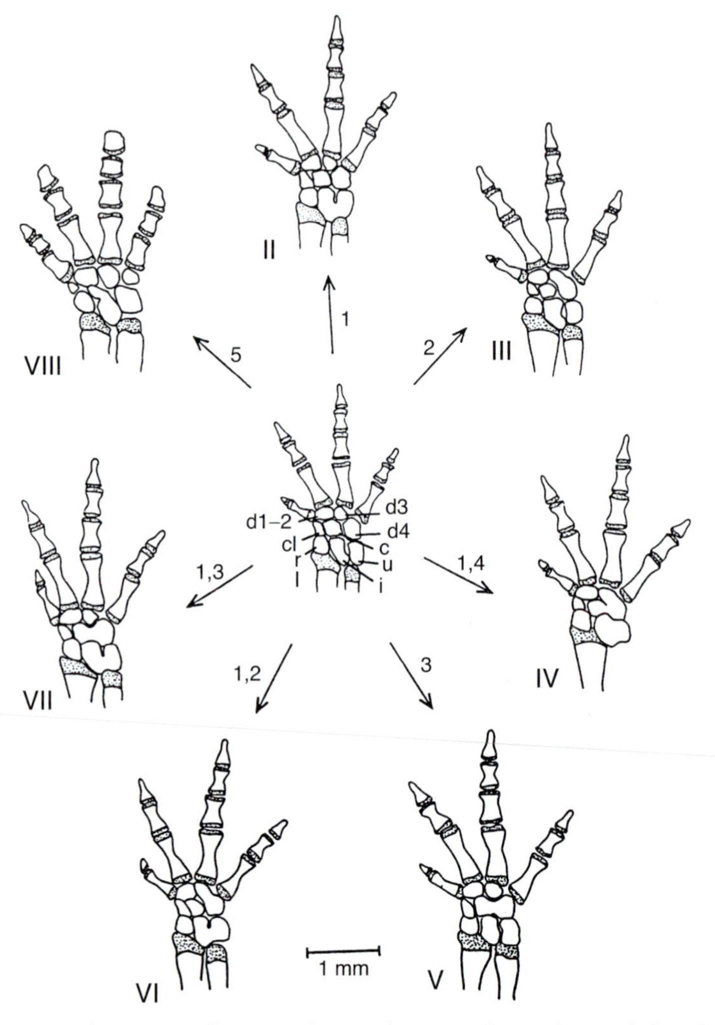

FIGURE 17-1. Alternate carpal patterns discovered in a single population of the salamander *Plethodon cinereus*. Patterns III and IV were encountered on contralateral sides of the same specimen. Patterns are identifiable by different combinations of carpal fusions. Abbreviations: c, centrale; d1-2, distal carpals 1-2; d3, distal carpal 3; d4, distal carpal 4; i, intermedium; r, radiale; u, ulnare. Reprinted with permission from Hanken (1983).

C. Geographically Based Variation

Structural features have been noted to vary in geographically correlated ways, either in a mosaic fashion or across a transect of some kind. For example, the frog *Vanzolinius discodactylus* varies in morphology, color pattern, and calls in a fine-scaled mosaic pattern that matches the pattern of genetic variation (Heyer, 1997). No specific functional correlates of this variation have been discovered, but the fact that this pattern is repeated in other forest-dwelling taxa is suggestive of a common evolutionary cause (Heyer, 1997). Polymorphisms (discontinuous phenotypes) have been noted in several species of the frog genus *Eleutherodactylus* (Lynch, 1966), this parallel polymorphism being referred to as tautomorphism. Again, however, no functional correlates of this pattern of variation have been uncovered.

Geographic variation in structural features is also evident when mainland and island populations are compared (Delany, 1970). The recent colonization by lynx (*Felix lynx*) of Newfoundland has resulted in increased variation in the island population (van Zyll de Jong, 1975), possibly as a result of character release, but without any evident functional correlation. Insular rodents have been found to display high rates of microevolution in morphological characteristics, such as body dimensions, on islands (Pergams and Ashley, 2001), with the effect being particularly marked if the islands are small and/or remote. Common trends were discovered but were not pursued from a functional perspective.

One of the most frequently reported patterns of structural variation is that of clinal variation, manifested by a gradient of continuous (Ridley, 1996) or stepped (Johnson, 1976) variation in a phenotypic or genetic character within a species. For structural features, this is most often documented across a geographic gradient of some sort (such as latitudinal or elevational) and is generally linked to posited adaptive adjustment (Barlow, 1961) to local environmental circumstances. Examination of such clinal variation can encompass quite restricted geographic areas, such as that noted for premolar expression in the badger, *Meles meles* (Szuma, 1994) or can extend over broader expanses, such as the scalation patterns noted for the lizard *Lacerta agilis exigua* (Chirikova *et al.*, 2002) in Kazakhstan or the anguid lizard *Barisia rudicollis* (Zaldivar-Riverón and Nieto-Montes de Oca, 2002) in Mexico. Lynch (1981) noted proportional variation in the salamander *Aniedes flavipunctatus* in northern California associated with latitude, distance inland, and other geographic features and noted that northern forms tended to be more pedomorphic, although no functionally related trends were identified. Altitudinal variation in body and skull proportions were documented for the harsh-furred rat, *Lophuromys flavopunctatus*, in Ethiopia (Bekele and Corti, 1994). Although these variation patterns were associated with phylogeny, no functionally based hypotheses for them were advocated.

Clinal variation is also sometimes combined with polymorphism across a geographic range. For example, the pleuronectiform flatfish, *Platichthys stellatus*, can be right-eyed or left-eyed. The sidedness is inherited and is apparently of no adaptive significance (Policansky, 1981) but is maintained in gradually changing proportions (left-eyed versus right-eyed) across the extent of the range.

IV. *IN SITU* CORRELATIONAL STUDIES OF THE RELATIONSHIP BETWEEN STRUCTURAL VARIATION AND FUNCTIONAL ATTRIBUTES

Various attempts have been made to investigate the relationship between structural variation and functional differentiation within species, and these generally (but not exclusively) fall into two patterns of approach. In the first, the *in situ* approach, data are related directly to environmental factors and thus employ geographic variation as the phenomenon in need of explanation. Such studies tend to be correlational and relate findings about structural variation to broad-scale environmental patterns. They differ somewhat from what we term *ex situ* approaches (Section V), in which performance variables are examined experimentally by endeavoring to control for environmental variables that are thought pertinent to the performance issue at hand. There is a growing trend to unite *in situ* and *ex situ* approaches, but comprehensive examples are still relatively scarce.

A. Trophic Polymorphism and Environmental Fluctuation

Field studies in relatively localized geographic areas have led to the recognition of polymorphisms within or between populations related to the acquisition and processing of dietary resources. The patterns encountered may arise from developmental plasticity (Section III.C, preceding) or from selection in fluctuating environments over a protracted time span.

One example of trophic polymorphism within populations is seen in ambystomatid salamanders. *Ambystoma macrodactylum columbianum* (the Central long-toed salamander) expresses a cannibal morph in the larval phase that is characterized by hypertrophied vomerine teeth and a longer, wider head, when compared with the noncannibal morph (Walls *et al.*, 1993a). Although feeding on conspecifics enhances the morphological differences between the cannibal and noncannibal morphs, initiation of the cannibal morph is mediated through environmental triggers (Walls *et al.*, 1993b).

A similar phenomenon occurs in the tiger salamander (*Ambystoma tigrinum*), where cannibalistic larvae (usually males) have hypertrophied vomerine teeth, large bodies, and wide heads, when compared with noncannibals. The cannibalistic larvae are macrophagous carnivores, whereas the noncannibal morphs are omnivorous planktivores (Lanoo and Bachman, 1984; Lannoo *et al.*, 1989). Cannibalistic individuals metamorphose earlier than typical morphs. Pedersen (1991) investigated the morphological differences between the cannibal and noncannibal morphs and found that the former had longer, recurved teeth, distortion of the underlying vomer, and hypertrophied muscles associated with the feeding apparatus. He postulated that heterochrony between the skull, dentigerous bones, and dentition may be a feature of the cannibal morphs and noted that the morphology develops before cannibalistic behavior is initiated. Pedersen (1991) regarded these morphs as essentially representing functionally different species in a trophic context. The growth of the skull in small cannibals is accelerated (Pedersen, 1993) and is compensatory for the demands of macrophagy.

Aspects of trophic polymorphism are also evident in the larvae of the Southern spadefoot toad (*Spea multiplicata*). Tadpoles of this species occur as a carnivorous and an omnivorous morph, and they occupy different trophic niches (Pfennig, 1992). The carnivorous morph consumes shrimp, and the omnivorous morph detritus. Carnivores arise facultatively through the ingestion of shrimp and the morphological differences—an enlarged orbitohyoideus muscle and a shorter, wider gut—are reversible. The ephemerality of the pond, and thus the number of shrimp, induces the carnivorous morph. Carnivores develop faster but have a smaller size at metamorphosis, fewer fat reserves, and lower survivorship. The closely related Plains spadefoot (*Spea bombifrons*) develops similarly to the Southern spadefoot, but carnivory is further translated into cannibalism (Pfennig *et al.*, 1993). The carnivorous/cannibalistic morph has beak-shaped mouthparts as opposed to flattened, keratinous jaws, hypertrophied jaw depressor muscles, and a morphologically wider and shorter gut and tends to exhibit solitary, rather than gregarious, behavior.

Reversible responses in trophic morphology have also been demonstrated in the cichlid fish *Cichlasoma managuense* (Meyer, 1987) when exposed to different feeding regimens (Figure 17-2). Differences in diet, and possibly in feeding mode, induced phenotypically plastic changes of the feeding apparatus, brought about through retardation of the normal developmental rate. Similar between-population differences in trophic morphology in other cichlid species were documented by Witte *et al.* (1990). Populations feeding habitually on soft foods have delicate pharyngeal jaws and numerous, slender teeth, whereas those feeding on hard food, such as snails, have massive jaws, and heavy, blunt teeth. It is possible that some of the morphological plasticity is triggered by nutritional differences in the diet, as well as by differing mechanical demands of the various foot types (Wimberger, 1993).

In the three-spined stickleback (*Gasterosteus aculeatus*), observed polymorphic variation is heritable, although there is still some plasticity in response to diet (Robinson, 2000). In a study of four populations representing anadromous, stream, lacustrine planktivorous, and benthic feeding groups, Caldecutt and Adams (1998) revealed statistically significant morphological differences between them, as assessed through landmark-based geometric morphometric analysis. The differences in skull form, representing geometric assessment of jaw morphology, were consistent with functional morphological predictions based upon habitat type and dietary data.

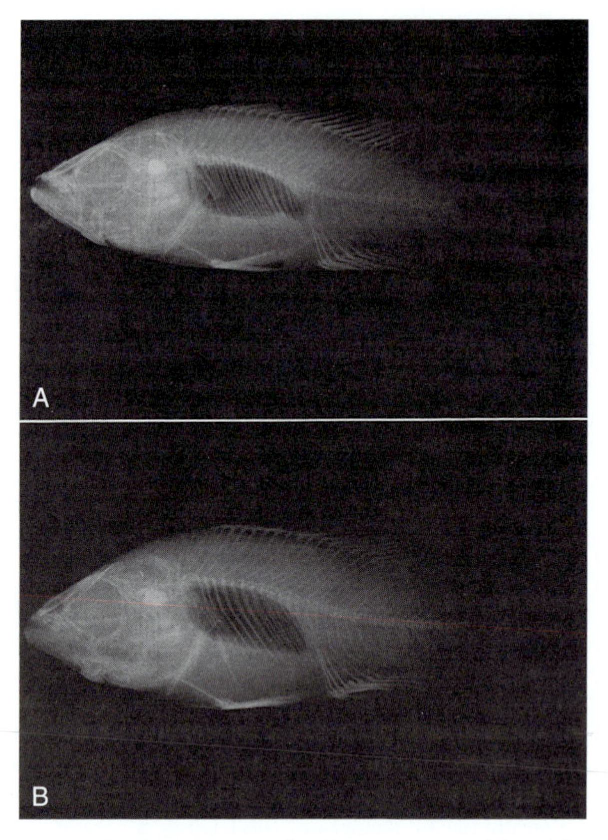

FIGURE 17-2. Radiographs of two specimens of *Cichlasoma manguense* fed on different diets for eight months. The upper illustration (A) exhibits the acutorostral morphology, whereas the lower one (B) depicts the obtusorostral form. Reprinted with permission from Meyer (1987).

More dramatic polymorphism associated with the trophic apparatus has been demonstrated in the teleost fish *Astyanax mexicanus*, which exists as an eyeless cave-dwelling morph and a sighted, surface-active morph (Wilkens, 1988). Eye formation is initiated in the cave-dwelling forms, but subsequently stops, and the eye degenerates (Yamamoto and Jeffery, 2000). Inductive signals from the lens are involved in the surface-dwelling fish's eye formation. Transplantation experiments indicate that the surface fish lens is sufficient to rescue eye development in cave fishes. The differences between surface dwellers and cave dwellers is trophically related and results in a trade-off manifested through modification of developmental pathways. The cave-dwelling forms have larger teeth and more taste buds (Vogel, 2000; Pennisi, 2002), which enhance feeding abilities in the cave environment. Yamamoto and Jeffery (2000) report that the eyed and eyeless forms of this species probably diverged from one another in the last million years and have been able to exploit food resources in environments requiring different sensory modalities. Eye loss need not be explained simply as a result of neutral mutation or drift, but instead seems to be related to a relaxation of stabilizing selection and the favoring of smaller eyes, larger teeth, and more taste buds in the cave-dwelling forms. The result is polymorphism and a stepped cline.

Selection in the field in the context of trophic polymorphism has also been the subject of intense study of Darwin's finches on the Galapagos islands. Inherent variation in heritable characteristics, such as beak morphology in *Geospiza*, permit selection to act in response to rapid changes in food availability associated with climatic extremes, such as droughts (Grant, 1991). The intensity of selection in these situations is enhanced by small population sizes. In certain situations, where environmental changes are dramatic, this inherent variability can lead to more firmly established and divergent patterns of form. For *G. conirostris* on Isla Genovesa, genealogies of family groups were established and heritability of traits studied (Grant and Grant, 1979, 1989). Feeding patterns of as many birds as possible throughout their life span were quantified and the relationship of beak size to survival and reproduction established. Size and proportions of the beak were found to be highly heritable and directly correlated with the capacity to feed on particular foods; slender bills were more effective at penetrating soft fruit and feeding on nectar, whereas the most massive bills were associated with cracking open cactus seeds and stripping bark from trunks to search for insects. In periods of stable climate, adult survivorship was generally unrelated to phenotype, but in extreme drought conditions and associated vegetational changes, there was strong selection on bill size and shape. A clear functional relationship between bill type and food manipulation was demonstrated. Had the drought conditions persisted, eventual fixation of the alleles producing the largest possible sized beaks may have resulted. Variation in structural features, such as bill size, and the functional associations thereof, thus appear to permit survival in fluctuating environments.

B. Clinal Variation

Across broad latitudinal or altitudinal scales, structural differences in size and correlated variables have long been associated with a gradient of thermal requirements. For endotherms, Bergmann's rule (endothermic vertebrates tend to be represented by larger forms in the colder parts of their range than in the warmer parts; Lincoln *et al.*, 1982) and the associated Allen's rule (the extremities of endotherms tend to be relatively shorter in colder climates; Lincoln *et al.*, 1982) predict that bird and mammal surface areas will be decreased in colder climates, and these rules are reasonably well established (James, 1970; Ashton *et al.*, 2000). Such intraspecific trends can thus be related to the functional demands of an endothermic physiology and the energetic costs of this.

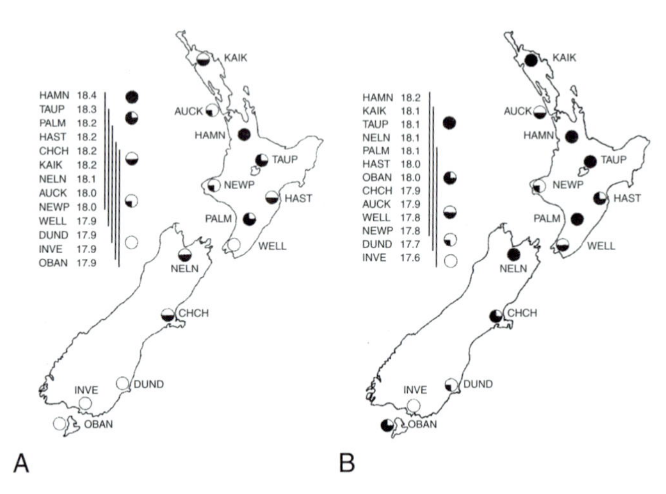

FIGURE 17-3. Sum of squares simultaneous test procedure results for humerus length of (A) male and (B) female New Zealand house sparrows. Vertical lines define maximally nonsignificant subsets of locality means. Variously shaded circles indicate trends in character variation (ranging from smallest, denoted by open circles, to largest, represented by solid circles). Reprinted with permission from Baker (1980).

This relationship has been examined in the context of recent colonizations of previously unoccupied areas by widespread and adaptable species (Huey *et al.*, 2000) to test its potential veracity. Johnston (1969) and Johnston and Selander (1971, 1973) examined large numbers of specimens of the house sparrow (*Passer domesticus*) in Europe and from across its introduced range in North America. Since its introduction into North America in the nineteenth century, there have been relatively few generations of sparrows, but size variables follow the patterns predicted by Bergmann's and Allen's rules, and variation with North American populations is comparable with that expressed in the ancestral European populations. For house sparrows introduced into New Zealand, interlocality variation in size and shape is clinal (Figure 17-3) and related to climatic patterns (Baker, 1980), but less variability is present than in the North American populations, because of the small size of the founding population in New Zealand. Similarly, Johnston (1994) found that body size varies with latitude in feral pigeons (*Columba livia*), both in Europe and in introduced populations in North America.

Marchand (1987) considered the size relationships predicted by Bergmann's and Allen's rules and noted that they tend to hold for some large mammals and nonmigratory birds, but noted that for small mammals, especially those active in the subnivean environment, there is no evidence supporting an increase in body size with increasing latitude (also noted by Geist, 1987). He also indicated that, for some taxa, increase in body size with increase in latitude is more clearly correlated with release from competition than it is with climate. In western Canada, the marten (*Martes americana*) exhibits an increase in size only north of about 62° N, which corresponds with the range limit of the larger fisher (*Martes pennanti*). In Labrador, however, where the fisher drops out at 52° N, the marten shows a corresponding increase in size immediately. Within the genus *Martes*, clinal variations in body size of *M. martes* and *M. foina* do not follow Bergmann's rule, but may instead be related to characteristics of prey size (Reig, 1992). Functional correlates predicted by Bergmann's and Allen's rules may thus be compromised by other environmental variables.

James (1970) proposed that for birds the combination of temperature and humidity may establish a more predictive gradient for the interpretation of size variation, heat load being the functional variable tracking the latitudinal gradient. Niles (1973) found this to be the case for larks (*Eremophila alpestris*), with the critical factor being the reduction of nonevaporative heat loss in dry areas.

For migratory birds, different potential functional associations have been found. For white-throated swifts, Behle (1973) demonstrated a trend toward larger size in the north and related this to the migratory behavior of these northern populations versus the smaller, nonmigratory southern populations. Telleria and Carbonell (1999) investigated Iberian blackcaps (*Sylvia atricapilla*) and found morphometric variation in five populations across a latitudinal gradient. The northern and central populations had more pointed wings, were smaller, and had longer tarsi than the southern groups, confounding the predictions of Bergmann's and Allen's rules. They explained these discrepancies by relating the differences to breeding and migratory habits, the southern populations remaining sedentary throughout the winter, whereas the northern and central ones migrate.

Bergmann's and Allen's rules have also been considered for ectotherms (Lindsey, 1966; Van Voorhies, 1996; Atkinson and Sibly, 1997; Ashton, 2002; Ashton and Feldman, 2003). Although heat conservation is clearly inapplicable as an adaptive explanation for such trends in ectotherms, other factors, such as the relationship between developmental temperature and size (Atkinson, 1994), fasting endurance (Ashton, 2002), or hydric considerations (Nevo, 1973) might be at play. Ashton (2002) demonstrated that amphibians, as assessed on limited data available for 34 species, generally follow a latitudinal increase in body size, a trend that is stronger in salamanders than in frogs. The trend is also less pronounced if environmental temperature, rather than latitude, is considered. Turtles also follow the trends predicted by Bergmann's rule, but squamates show a reversed trend (Ashton and Feldman, 2003), perhaps in association with the benefits of rapid head gain in colder areas.

In some instances, questions relating to patterns of clinal variation have been approached functionally in the context of diet. Lindsay (1986) examined mensural variation in the skull of Douglas's squirrels (*Tamiasciurus douglasi*) and revealed a geographic pattern that corresponded to differences in conifer cone size, toughness, and caloric content. The functional aspects that varied were related to the ability of squirrels to use caches of cones stored for the winter. Similar variation in cranial features of *Dipodomys microps* was found in relation to the mechanical properties of diet (Csuti, 1979). In contrast to this, Lindsay (1987) discovered that the Hudson's squirrel (*Tamiasciurus hudsonicus*), a close relative of *T. douglasi*, does not exhibit size variation with geography and uses cones of similar size across its range. Variation in this species displayed characteristics consistent with Allen's rule, but showed no other functionally specific environmental adaptations. Comparisons such as these might reveal aspects of character displacement (Gallagher *et al.*, 1986) associated with variation and may be related to competition between species.

C. Microgeographic Variation

In recent years, considerable effort has been directed toward studying variation in form across ranges smaller than those normally considered for the investigation of such phenomena as Bergmann's and Allen's rules. Such studies have endeavored to correlate structural variation with environmental variables (King, 1997) and to sample intensively across the area of study. Such studies tend to lack explicit documentation of the relationship between structure and function, but they typically provide a "real world" context into which the observed variation may be placed, and at least heuristically offer a plausible evolutionary context within which variation may be acted upon by selection. Although similar in some respects to some of the trophic polymorphism studies cited in the preceding text, the investigations of microgeographic variation tend to examine large numbers of covarying characters rather than a select few morphological variables.

Over a fairly broad range, Thorpe (1984) found that clinal variation in the grass snake, *Natrix natrix helvetica*, reflected latitudinal variation, but in this case not only with current climatic regimes, but also with the entire Pleistocene history of Europe.

More typical microgeographic studies are those conducted by Malhotra and Thorpe (1991a,b) on the lizard *Anolis oculatus* on the West Indian island of Dominica. They examined 45 characters across 33 localities, from 15 to 750 meters above sea level, to examine the ecogenetic origin of the observed variation. They were able to demonstrate a complex pattern of morphological variation associated with ecogeographic features and implied causal relationships between these. Increase in scale number was associated with less humid conditions, a trend also seen in other lizards. The functional relationship between these two factors has not been explained mechanistically, but has been demonstrated experimentally to have differential survival value in the appropriate environment (Malhotra and Thorpe, 1991b). The inference is that morphological variation is maintained by ecological adaptation, even though gene flow exists. Mitochondrial DNA variation parallels these patterns of morphological variation and correlates significantly with moisture gradients in *Anolis oculatus* on Dominica, indicating congruence between morphological and molecular patterns of variation (Malhotra and Thorpe, 1994).

In a study of the snake *Trimeresurus stejnegeri*, Castellano *et al.* (1994) revealed a causal relationship between morphological variation and a diversity of ecogeographic and climatic factors in Taiwan, using canonical analysis combined with Mantel tests. Similar adaptations to current habitat types (rather than historical factors), have been identified in the lizards *Gallotia galloti, Gallotia stehlini,* and *Tarentola delalandii* on Tenerife (Thorpe and Brown, 1989; Thorpe, 1991; Thorpe and Baez, 1993). Variation is more likely to be correlated between characters if this relates to phylogenetic patterns of relationship than to ecogeographic factors, which may influence different structures independently (Figure 17-4) (Thorpe and Baez, 1993).

Such studies of microgeographic variation on islands relate to the general study of insular patterns of geographic variation (Section III.D, preceding) but are usually not conducted in the context of employing comparisons with mainland forms (if such exist). Instead, these studies endeavor to understand the fine-scale basis of multivariate variation as it relates to fine-scale terrain and climatic conditions. In most cases, such studies reveal correlations between structural features and ecogeographic (and sometimes phylogenetic) patterns, but questions about why a particular structural variant is particularly suited to a particular set of environmental circumstances are usually not posed, although they open up avenues for future investigation.

Microgeographic variation associated with congeneric competition has been demonstrated for dewlap color in *Anolis*. Dewlap color and pattern are important social signals in *Anolis* lizards, and variability in color can be greater, or at least different, when not constrained by the presence of congeners (Case, 1990). Other color variants may also vary on a microgeographic scale. Background color matching is generally assumed to provide a protective function and may be related in a patchy way to the distribution of particular substrate types, such as darker volcanic rocks versus lighter-colored rocks. Such patterns are evident in the squamate taxa *Uta* and *Pituophis*. In the case of mimetic snakes, protective function is related to the ability of nonvenomous or mildly venomous mimics to color match various models. For coral snake mimicry, Greene and McDiarmid (1981) found that within the single species *Lampropeltis triangulum*, geographic matches of several coral snake species occur across the range of this widespread mimic. The same holds true for several other mimetic species, such as *Erythrolamprus guentheri,* which mimics several *Micrurus* species, and *Pliocercus elapoides,* which matches several elapids in color pattern and general body habitus.

V. *EX SITU* STUDIES OF THE RELATIONSHIP BETWEEN STRUCTURAL VARIATION AND PERFORMANCE

The largely *in situ* studies outlined in the previous section address issues of structural variation with respect to trends across the geographic ranges of species and endeavor to elucidate functional correlates of this variation by placing them into the context of environmental variables. In general, however, performance attributes

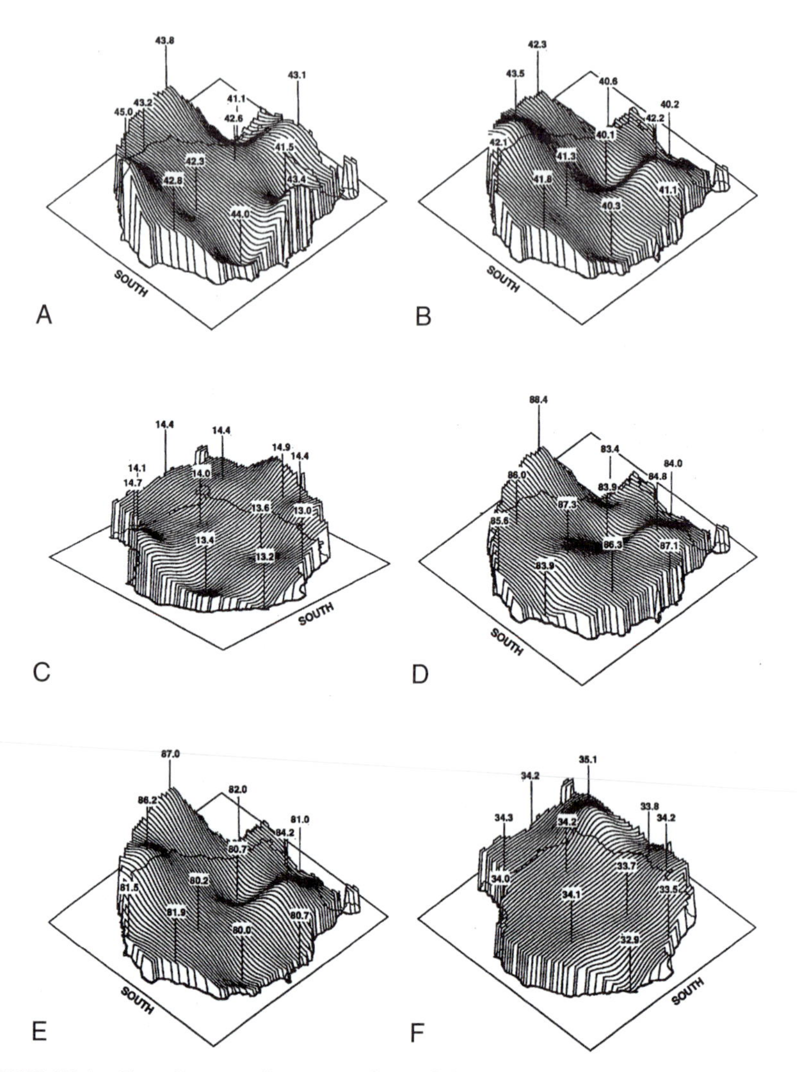

FIGURE 17-4. Three-dimensional isometric plots with locality sample means for various scalation variables of the lizard *Gallotia stehlini* superimposed on an outline map of the island of Gran Canaria. A, mean number of gular scales in males; B, mean number of gular scales in females; C, mean number of collar scales in both sexes; D, mean number of dorsal scales around the trunk in males; E, mean number of dorsal scales around the trunk in females; F, mean number of ventral scales along the trunk in both sexes. Reprinted with permission from Thorpe and Baez (1993).

of the structural variants are not directly measured, but are evaluated in terms of association with presumed functional challenges. In contrast, structural variation within species has also been investigated by taking animals out of their natural environment and measuring performance predicted to be related to particular structural variables (Irschick, 2002) and survival potential. This *ex situ* approach endeavors to investigate the variation under more controlled experimental situations and thus interprets performance differences in the context of environmental differences across the range. It is evident that there is no sharp distinction between *in situ* and *ex situ* investigations, and some approaches endeavor to combine both. Our categorization is thus based upon major emphasis rather than absolute categorization. Both approaches assume that performance is the link between phenotype and relative ecological success (Koehl, 1996), but the *ex situ* approach attempts, more directly, to measure the performance part of the relationship (Irschick, 2002). Although in some instances small changes can lead to major functional shifts, it is also evident that in other cases larger changes have not been able to be correlated with any functional implications. Performance-based, *ex situ* studies can directly address whether or not structural changes actually do translate into their hypothesized functional differences.

For *in situ* studies, considerable emphasis has been placed upon trophic modifications (Section IV.A, preceding), largely because the potential availability of food resources can be assessed across broad or narrow geographic ranges. In contrast, much of the performance-based, *ex situ* data on variation has focused on locomotor attributes, probably largely because data are relatively uncomplicated to collect, and body measurements associated with performance data can be collected without sacrificing the experimental subjects.

A. Variation in Trophic Performance

In their examination of the trophic morphology of the snakes *Vipera berus* and *Thamnophis sirtalis,* Forsman and Shine (1997) investigated the relationship between head and body size in relation to prey handling. Snakes must ingest prey by use of the head alone, and other body regions cannot directly assist in this. Prey cannot be reduced in size and so must be consumed whole. Thus, snakes are gape limited in terms of the size of prey that can be consumed. Forsman and Shine (1997) found that the allometric relationships between head and body size in these species are adaptive and are related, across the range, to prey availability and handling ability, but are not explained by geographic proximity of populations.

Dalrymple (1977) described extreme variation of the skull of the turtle *Apalone ferox* and uncovered a complex of interacting factors. Some of the morphological variation was found to be induced by changes in diet through growth (and was thus plastic), and some was attributed to allometry. He did not find a high degree of correlation with specific food items and skull features, thus making a linkage between skull form and performance difficult to establish.

B. Locomotor Performance

As outlined in the preceding text (Section IV.B), considerable effort has been devoted to the study of clinal variation in birds, and much of this has been directed toward house sparrows, because of their broad geographic range and repeated colonization of extensive ranges in other continents. The adaptive significance of size variation has been extensively debated but rarely examined in terms of actual performance. Blem (1975), however, noted that wing length dimensions only are usually included in such studies of clinal variation and that consideration of the relationship between body weight, wing length, and wing area, which will collectively affect wing loading, would be more informative. He investigated these factors by exercising sparrows in a flyway, weighting the bodies of individuals from various regions, and assessing their flight capabilities. The conclusions reached were that wing loading in sparrows at the far northerly extent of their range can be extreme (Blem, 1975) and is associated with the larger body size of such individuals (as predicted by Bergmann's rule). This is exacerbated by such individuals requiring more fat and water to deal with the long, cold winter nights. Although selection may favor increases in wing length and flight muscle size, Blem (1975) surmised that individuals at the northernmost extent of their range can only persist there by associating with food sources made available by human activities (such as agriculture). Thus at these latitudes physiological and locomotor demands are critical, and appropriate flight performance can only be maintained in the presence of easily available dietary resources that diminish the effort that must go into foraging.

Squamate reptiles have been the subject of many studies of performance parameters. Garland (1985) examined size, shape, and speed relationships in the Australian agamid *Ctenophorus nuchalis*. Maximum running speed was recorded in a race track apparatus and examined in the context of body mass, snout-vent length, tail length, forelimb and hindlimb lengths, limb spans, and thigh muscle mass across the entire posthatching ontogenetic range of size. Allometric patterns of body proportions were revealed, with the limbs becoming relatively shorter, whereas tail length and thigh muscle masses became relatively larger. Repeatable differences in performance between individuals was found, with 59% of maximum speed being related to body mass. Variation in maximum speed was high and was found to be dissociated from variation in shape. Garland (1985) advocated that maximal sprint speed is evolutionarily important in terms of fitness and that it is acted upon directly by selection, whereas selection acts indirectly on morphology (structure) via sprint speed, only if morphology correlates with features related to fitness. This was an early indication that limb dimensions, in themselves, may not be a good predictor of maximal sprint speed but that a constellation of factors may be related to performance variables and therefore selection.

Arnold and Bennett (1988), in a similar fashion, examined potential morphological correlates of locomotor performance (burst speed, and midrange speed, and antipredator behavior) in the snake *Thamnophis radix*. Performance was found to be mass dependent. For neonates, 24% of locomotor performance was accounted for by morphological variation, including a weak, but significant, correlation between vertebral number and locomotor performance. Many effects were shown to be borderline significant.

The potential role of genetic variability in locomotor performance attributes was investigated relatively early, and Van Berkum and Tsuji (1987) examined interfamiliar difference in sprint speeds of the lizard *Sceloporus occidentalis*. Significant size differences were found between families, but the speeds between families remained different when corrected for size, and 27% of variation in speed was explained by family relations. This study again points to performance being related to some degree to mensurable differences, but it is cautionary in that evidence was provided that heritable aspects of performance cannot be explained simply by morphometric differences. The same species was examined by Garland *et al.* (1990), who investigated the relationships between endurance, maximal sprint speed, and dominance in males. The most dominant males had the fastest sprint speed in 14 out of 20 cases, but dominant males exhibited no differences in locomotor stamina over submissive ones. This study spawned further projects that morphometrically examined the relationship between performance and dominance (see the following text).

Jumping and running performance of the agamid lizard *Leiolepis belliana* was investigated by Losos *et al.* (1989), who found that limb length, when corrected for size, was not correlated with locomotor performance. The role of experimental error in measuring performance capabilities was raised. It is now recognized that performance measures may not be as reliable or as representative as had initially been hoped (Losos *et al.*, 2002), partly because of variance in actual performance output of individuals (Djawdan and Garland, 1988), and partly because of too few trials being performed (in some instances) that cast doubt about maximal performance values being reported.

Plasticity in the face of environmental variation was investigated by Van Buskirk and Relyea (1998) for tadpoles of the frog *Rana sylvatica*. Tadpoles were raised in the presence and absence of *Anax* predators to test the hypothesis that plasticity is an adaptation to variable predator environments. Changes in the morphological expression of the tail were noted in those tadpoles exposed to the predator, these changes enhancing escape performance. This study was followed up by an investigation of free-living tadpoles (Van Buskirk, 2002) in the presence and absence of predators. Predation was again found to induce behavioral and morphological changes that enhanced escape, the former more rapidly than the latter. Thus changes in behavior may be triggers that induce morphological variation via mechanical feedback.

Anolis lineatopus was used by Macrini and Irschick (1998) to examine sprinting ability in a variety of locomotor arenas (rods of various diameters). It was predicted that longer-legged individuals should be able to sprint faster on broad rods and should exhibit a decline in speed on narrower perches. No significant differences, however, were found across individuals of different limb dimensions within size classes.

In an attempt to investigate locomotor characteristics with selection, Losos *et al.* (1997) documented the time course of the introduction of *Anolis sagrei* onto a series of small, lizard-free islands in the Bahamas. On the smallest islands, the lizards disappeared rapidly, but on larger islands, breeding populations became established. Over a period of 10 to 14 years, differentiation was noted, with divergence in limb length measurements being greatest in situations in which the vegetation on the colonized island was structurally the most different from that of the vegetation associated with the source population. The suggestion was that limb length within this species may have changed adaptively, but it was also recognized that plasticity (see Section III.C, preceding) could have played a role. Subsequent study (Losos *et al.*, 2000) demonstrated that, indeed, plasticity was the most likely explanation and that *A. sagrei* responds to different substrate types and develops different expressions of locomotor features commonly examined morphometrically in performance studies (Figure 17-5). It is evident, however, that such plasticity may be adaptive in itself and may play a role in the evolution and radiation of lineages.

Colonization of islands relates to variation in general (see Section III.D, preceding) and has been studied in the context of its relationship to locomotor performance. One example of this is the introduction experiments outlined in the preceding text (Losos *et al.*, 1997). Another was conducted by Van Damme *et al.* (1998), who examined two subspecies of the lacertid lizard *Podarcis hispanica*. In racetrack trials, it was found that mainland forms were slower than island forms and that this was because of gait differences (relating to stride length for mainland and stride frequency for island forms). The differences were suggested to be the result of lower predation pressures on the island populations, but morphometry revealed no limb differences between the two subspecies, other than that the mainland lizards had relatively larger hind feet. Thus examples of performance differential exist that are not relatable to any structural causes or correlations.

Sinervo (1990) and Sinervo and Huey (1990) attempted to investigate the effects of body size versus adaptation to local circumstance, as they relate to locomotor performance, by manipulating embryos of *Sceloporus occidentalis* (Figure 17-6). Southern populations (California) of this species have long limbs, high burst speed, and good stamina relative to northern populations (Oregon, Washington).

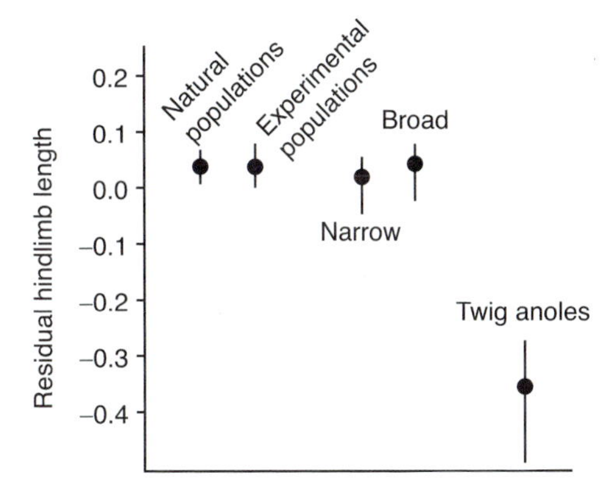

FIGURE 17-5. Plots of relative hindlimb length for various "populations" of the lizard genus *Anolis*. Natural populations of *Anolis sagrei* from the Bahamas; experimental populations of *Anolis sagrei* introduced onto small islands near Staniel Cay, Bahamas; narrow and broad populations of hatchlings raised in terraria with only broad or narrow perch surfaces available; twig anoles—Caribbean *Anolis* species specialized for using very narrow surfaces. Reprinted with permission from Losos *et al.* (2000).

It was hypothesized that the performance differences were the result of size or alternatively that they were related to other physiological or morphological differences. Southern individuals of the same size as northern ones were produced by removing yolk from eggs, resulting in smaller hatchlings. Individuals from both regions (southern and northern) were raised and measured at 3 weeks of age for burst speed (relative locomotor performance was found to be highly repeatable at given temperatures for this species—Van Berkum *et al.*, 1989) and later for endurance. Individuals were measured at the same age and compared at the same snout-vent lengths.

Some interpopulational differences in stamina persisted, but burst speed differences disappeared, although the miniaturized southern hatchlings still had relatively longer legs. These manipulations indicated that interpopulation differences in this species involve additional evolved factors and cannot be accounted for simply by differences in size. Within-population differences may be related to differences in performance, but between-population differences are more complex and appear to result from a constellation of adaptations to local circumstance (Sinervo and Losos, 1991). Growth rate has genetic, behavioral, and physiological components, and resultant size has implications for performance, such as sprint speed (Sincervo and Adolph, 1989), but within species these interactions might be population specific.

Locomotor performance in lizards, and its variation, is now recognized to be controlled by a variety of factors, and recent studies have attempted to factor out the relative importance of these. Development is known to play a role, and

FIGURE 17-6. A graded size series, with a two-fold difference in mass from smallest to largest, of hatchlings from a single clutch of the lizard *Sceloporus occidentalis* manipulated by removing various amounts of yolk from some eggs. Reprinted with permission from Sinervo, B., and Huey, R.B. (1990). Allometric engineering: an experimental test of the causes of interpopulational differences in performance. Science 248, 1106–1109. Copyright (1990) AAAS.

incubation temperature and humidity have been investigated. Vanhooydonck *et al.* (2001) reported variation in sprint speed, climbing, and clambering performance in the lacertid *Gallotia galloti* and noted that sprint speed was highest for those individuals raised at the lowest incubation temperatures. Climbing ability was affected differently by size and limb length in different populations and was found to be genetically related.

Qualls and Shine (1998) also demonstrated that incubation temperature influences phenotype and results in variation. They studied hatchlings of the lizard *Lampropholis guichenoti* and found that interlimb length, body shape, tail length, and locomotor performance were related to thermal regimen. Flatt *et al.* (2001) found similar effects in the lizard *Bassiana duperreyi*, with temperature, but not hydric features, being associated with phenotypic and performance differences.

Warner and Andrews (2002) examined phenotype and survival in hatchling *Sceloporus undulatus* and investigated the effects of incubation moisture, maternal yolk investment, and clutch (genotype) on performance. Eggs were raised under two moisture regimens and with the yolk left unmanipulated or with some removed. After hatching, a number of size, shape, and performance parameters were measured and then individuals were released into a test plot to determine survival under natural conditions. All phenotypes were affected by clutch, and clutches with larger individuals had higher survival potential. Slower-growing, faster-running individuals survived better than rapidly growing, slower-running individuals. Clutch was found to be of overriding importance and thus contributes to variation in structure and function. This study is one of the few that successfully combines laboratory experimental and field approaches in attempting to account for the causes and outcomes of variation.

Studies of locomotor performance and its contribution to selection are not exclusive to investigations of squamates. Schulte-Hostedde and Millar (2002) examined the effects of mass and structural size on running speed in chipmunks (*Tamias amoenus*), a species with female-biased sexual dimorphism. It was found that structurally large males had the fastest running speeds, but that smaller males are aggressively dominant over larger ones. This was found to lead to a trade-off between running speed (mating success) and dominance. Structural size (which is perhaps related to stride length and is a measure of linear dimensions), but not body mass, explained a significant amount of the variation, and speed was found to be independent of mass.

C. Fluctuating Asymmetry and Variation in Performance

Fluctuating asymmetry is manifested in many anatomical structures in vertebrate development (Hallgrímsson *et al.*, 2003) and has received some attention in terms of its potential impact on function. In birds, variation brought about by fluctuating asymmetry in wing length negatively affects the effectiveness of flight. Thomas (1993) examined asymmetry in the wings and tails of birds in relation to aerodynamic theory and found that wing asymmetry is much more costly than tail asymmetry in relation to flight efficiency and turning effectiveness. Swaddle (1997) measured speed and maneuverability in European starlings (*Sturnus vulgaris*) and reported that small differences in performance are detectable in concordance with patterns of fluctuating asymmetry.

Martin and Lopez (2001) discovered that fluctuating asymmetry of the femur is associated with a reduction in escape speed in the lacertid lizard *Psammodromus algirus*, but that no such effects are detectable for crural asymmetry. They advocated that larger asymmetries could be tolerated in less functionally important structures, a point also made by Hallgrímsson (2003) considering a report of asymmetries reported for forelimbs and hindlimbs of bats (Gummer and Brigham, 1995), and intimated that the feet were the most important unit in the lizard hindlimb because, in *Psammodromus*, they displayed the smallest degree of fluctuating asymmetry. This is somewhat circular reasoning and fails to take into account the morphological and kinematic complexities that characterize the lizard hindlimb (Russell and Bels, 2001).

In an investigation of *Lacerta monticola*, Lopez and Martin (2002) indicated that trade-offs between reproductive success and survival are important in determining structural attributes. They examined maximal sprint speed (this being equated with the ability to escape predators) and investment in dominance traits (as manifested by head dimensions) in males. They staged agonistic encounters between pairs of males and noted that dominance is strongly correlated with head depth, whereas speed is negatively correlated with head depth and limb asymmetry. Males with the greatest head depth also had the most asymmetrical femora. They interpreted this pattern of structural variation as being indicative of a situation in which access to females comes at the cost of diminished escape capabilities and thus cautioned that a simple measure of locomotor performance may not be sufficient to enable inference of whole-animal performance. Lopez and Martin (2002) could not fully explain the revealed relationships but suggested that the disequilibrium created by fluctuating asymmetry of the femora necessitated those animals taking slower steps to compensate. Such suggestions may be able to be assessed by undertaking detailed kinematic investigations.

D. Selection Experiments and the Investigation of the Limits of Variability

Within species, it has been demonstrated that performance can vary between individuals and, on average, between populations occupying different environmental circumstances. In laboratory situations, Garland (2003) and his coworkers have begun to investigate the ways that selective breeding may influence structure and performance. The experimental protocol involves artificial selection with outbred house mice, with the trait of voluntary wheel running being the focus of attention, a heritable response measurable over a tractable number of generations. Eight breeding lines were established of 10 pairs of mice each. Four of these lines were chosen to experience selection, and in these the highest-performing male and female from each generation were paired to produce the next generation. In the controls, the parents were randomly selected. After 16 generations, differences had stabilized (probably because of the small population size, intense selection, and lack of new genetic variance). The major contributing factor to differentiation was the average speed of running, not the duration of running (Garland, 2003). No changes in maximal aerobic or resting metabolic rates were noted, but the selected lines had more symmetrical hindlimb bone lengths, physiological differences in some hindlimb muscles, smaller body mass, and less body fat. Some of the variation may be reflective of genetic adaptations, and certain muscle changes, in terms of size and oxidative capacity, may be adaptive for high-speed running.

In a similar approach, Reznick *et al.* (1997) examined the effects of predation rates on guppies in a laboratory simulation of natural selection. Under high predation rate regimens, the guppies matured at a smaller size and more rapidly, produced larger litters, and produced more litters per year. Structurally the selected lines rapidly developed color changes to become either more cryptic or, in the case of males, more showy.

E. Other Measures of Structural and Functional Variation

Most performance-based studies focus on locomotor attributes, and trophic modifications and geographic variation of easily recorded traits contribute disproportionately to correlational, *in situ* studies of structural variation. Emphasis is thus on musculoskeletal and integumentary characteristics, and other anatomical systems are rarely assessed in this regard, possibly because data are not accessible without special means of specimen preparation.

One system that has received a little attention in this regard, however, is the circulatory system. Hillen (1987) investigated patterns of variation in the Circle of Willis in humans. He took 19 measurements from 100 circles and examined the relationships between vessel sizes. An inverse relationship was found in vessels with at least partially identical irrigation areas. This led Hillen (1987) to a hemodynamic hypothesis founded on the prediction that the relations of individual blood vessel diameters reflect flow patterns, and the characteristics of the area being served determines total diameter. Within a particular area, the apportioning of diameter across specific vessels is variable, but the total diameter remains essentially constant. Thus variation is limited by flow dynamics, but vessel pattern may vary within an area, with apparently no functional differentiation resulting.

In a study of variation in the aortic arches of rabbits, Angell-James (1974) found a high degree of variation. Twenty-two variants were found in the 31 rabbits studied. The arch can give rise from two to up to six major vessels, apparently without functional consequence. Again, hemodynamics of flow probably determine the viability of alternate patterns.

VI. CONCLUDING REMARKS

This review of variation of structure and function has by no means been exhaustive. We have, however, attempted to give some perspective of how this topic has been and continues to be approached. Despite extensive study of variability in structure and the relationship between structure and function, surprisingly little is known (Hallgrímsson, 2003). Indeed, as investigative techniques become more sophisticated, the number of variables to be considered increases, and the ability to draw definitive conclusions appears to diminish.

Some of this inconclusiveness may stem from the way in which data are collected. Measurements are often not comparable across studies, even in situations in which the same species is employed, making cross-comparison difficult. Little consideration is given to measurement error, which could blur the differences being reported and the functional conclusions based thereon. Performance data, too, pose a problem in terms of comparing output between individuals and knowing when maximal performance has been delivered (Losos *et al.*, 2002).

Functional correlations and causations are usually implicit rather than explicit, as might be expected with complex and integrated biological systems (Bock and von Wahlert, 1965; Zweers, 1979). Correlations sought are very often found but in many cases are general rather than specific, and causal connections are difficult to envisage.

Despite these drawbacks, the study of the relationship between structure and function as a component of variation is of paramount importance to our understanding of the operation of selection. In many instances, we invoke structural variation associated with some functionally variable task as the couplet that determines the relative success with which biological roles are achieved (Bock and von Wahlert, 1965), and ultimately, therefore, with differential reproductive success.

Part of our success in understanding the relationship between structure and function as it relates to variation within species will depend upon the level of explanation we expect from such studies and our abilities to integrate *in situ* and *ex situ* investigations (Warner and Andrews, 2002). An example of this is encapsulated by Lauder (1995) in his study of centrarchid pumpkinseed fishes. These fishes eat snails, and in southern Michigan there is interlake variation in prey abundance. Lauder (1995) reported a population level examination of functional morphological variation by examining populations from lakes with sparse and abundant populations of snails. The pumpkinseed fish in these lakes exhibit distinctive trophic polymorphisms—individuals from lakes with abundant snails have hypertrophied pharyngeal muscles compared with those from lakes with sparse snail populations.

Lauder (1995) examined the mechanism of snail crushing to investigate the correlation between structure and function in recently trophically divergent populations (the trophic resource base differences had only been established for 20 years). The mechanics of snail capture, crushing, segregation of shell from soft parts, swallowing, and shell fragment ejection were examined in detail by careful anatomical observation and kinematic studies. Pharyngeal morphology was then compared for ten muscles and five skeletal elements important in food processing, and analysis of covariance was employed to assess interpopulational differences in muscle and bone mass. Significant differences were found in three of the five bone masses, and six of ten muscle masses, all of the values being significantly greater for fish from populations exposed to high levels of availability of snails.

Electromyographic investigation of the activity patterns of the muscles involved in snail crushing were then conducted, with 16 timing and intensity variables measured. The results were analyzed within a nested analysis of variance (ANOVA) to test for lake effect. Despite all of the morphological predictors, the outcome was that there was little relationship between structure and function of the jaw apparatus. Muscles that showed the greatest differences in mass between lakes showed some of the lowest values of functional differences, and vice versa.

The conclusions drawn from this (Lauder, 1995) were that whereas relative size of structures is often used as an indicator of functional, performance, or behavioral difference, it is not axiomatic from these that one can deduce how such features are actually used during behaviors. Quantifying behavior may thus be of paramount importance in performance studies (Irschick and Garland, 2001; Irschick, 2002). General level predictions allow accurate inferences about differences between populations, in this case in the context of diet and ecological situation, but the details are more confusing. Snail-eating populations have hypertrophied bones and muscles compared with populations exposed to fewer snails, but these correlations do not necessarily extrapolate to expected functional details.

Fine-grained subtleties of differences within species thus continue to present many challenges in our attempts to understand variation as it relates to structure and function. It is evident that structures are employed in multiple functions (Bock and von Wahlert, 1965) and that such integration and multiplicity must be taken into account when trying to address structure–function questions (Irschick and Garland, 2001). Nonmusculoskeletal features are difficult to quantify and are rarely examined, but these are also integral parts of systems being examined in the context of form–function relationships. Organisms are integrated functional wholes, and atomization without subsequent reassembly may impose limitations on how much we can understand about patterns of variation that are observed.

REFERENCES

Adriaens, D., Devaere, S., Teugels, G. G., Dekegel, B., and Verraes, W. (2002). Intraspecific variation in limblessness in vertebrates: A unique example of microevolution. *Biological Journal of the Linnean Society* **75**, 367–377.

Angell-James, J. E. (1974). Variations in the vasculature of the aortic arch and its major branches in the rabbit. *Acta Anatomica* **87**, 283–300.

Arnold, S. J., and Bennett, A. F. (1988). Behavioural variation in natural populations. V. Morphological correlates of locomotion in the garter snake (*Thamnophis radix*). *Biological Journal of the Linnean Society* **34**, 175–190.

Ashton, K. G. (2002). Do amphibians follow Bergmann's rule? *Canadian Journal of Zoology* **80**, 708–716.

Ashton, K. G., and Feldman, C. R. (2003). Bergmann's rule in nonavian reptiles: Turtles follow it, lizards and snakes reverse it. *Evolution* **57**, 1151–1163.

Ashton, K. G., Tracy, M. C., and de Queiroz, A. (2000). Is Bergmann's rule valid for mammals? *American Naturalist* **156**, 390–415.

Atkinson, D. (1994). Temperature and organism size: A biological law for ectotherms? *Advances in Ecology Research* **25**, 1–58.

Atkinson, D., and Sibly, R. M. (1997). Why are organisms usually bigger in colder environments? Making sense of a life history puzzle. *Trends in Ecology and Evolution* **12**, 235–239.

Baker, A. J. (1980). Morphometric differentiation in New Zealand populations of the house sparrow (*Passer domesticus*). *Evolution* **34**, 638–653.

Barlow, G. W. (1961). Causes and significance of morphological variation in fishes. *Systematic Zoology* **10**, 105–117.

Bateson, W. (1894). *Materials for the Study of Variation Treated with Especial Regard to Discontinuity in the Origin of Species*. London: Macmillan.

Behle, W. H. (1973). Clinal variation in white-throated swifts from Utah and the Rocky Mountain region. *Auk* **90**, 299–306.

Bekele, A., and Corti, M. (1994). Multivariate morphometrics of the Ethiopian populations of harsh-furred rat (*Lophuromys*: Mammalia, Rodentia). *Journal of Zoology London* **232**, 675–689.

Blem, C. R. (1975). Geographic variation in wing-loading of the house sparrow. *Wilson Bulletin* **87**, 543–549.

Bock, W. J., and von Wahlert, G. (1965). Adaptation and the form-function complex. *Evolution* **19**, 269–299.

Caldecutt, W. J., and Adams, D. C. (1998). Morphometrics of trophic osteology in the threespine stickleback, *Gasterosteus aculeatus*. *Copeia* **1998**, 827–838.

Case, S. M. 1990. Dewlap and other variation in the lizards *Anolis distichus* and *A. brevirostris* (Reptilia: Iguanidae). *Biological Journal of the Linnean Society* **40**, 373–393.

Castellano, S., Malhotra, A., and Thorpe, R. S. (1994). Within-island geographic variation of the dangerous Taiwanese snake, *Trimeresurus stejnegeri*, in relation to ecology. *Biological Journal of the Linnean Society* **52**, 365–375.

Chirikova, M. A., Dubjansky, V. M., and Dujsebayeva, T. N. (2002). Morphological variation of the eastern sand lizard, *Lacerta agilis exigua* Eichwald, 1831 (Squamata: Lacertidae) in Kazakhstan. *Russian Journal of Herpetology* **9**, 1–8.

Choquenot, D., and Greer, A. E. (1989). Intrapopulational and interspecific variation in digital limb bones and presacral vertebrae of the genus *Hemiergis* (Lacertilia, Scincidae). *Journal of Herpetology* **23**, 274–281.

Csuti, B. A. (1979). Patterns of adaptation and variation in the Great Basin kangaroo rat (*Dipodomys microps*). *University of California Publications in Zoology* **3**, Paper Vol. 3.

Curren, K., Bose, N., and Lien, J. (1993). Morphological variation in the harbour porpoise (*Phocoena phocoena*). *Canadian Journal of Zoology* **71**, 1067–1070.

Dalrymple, G. H. (1977). Intraspecific variation in the cranial feeding mechanism of turtles of the genus *Trionyx* (Reptilia, Testudines, Trionychidae). *Journal of Herpetology* **11**, 255–285.

Darwin, C. (1854). *A Monograph on the Sub-class Cirripedia, with Figures of all the Species*. Vol. 2. The Balalnidae (or Sessile Cirripedes); the Verrucidae, etc., London: The Ray Society.

Darwin, C. (1859). *The Origin of Species by Means of Natural Selection, or the Preservation of Favoured Races in the Struggle for Life*. London: John Murray.

Darwin, F. (ed.). (1892). *Charles Darwin: His Life Told in an Autobiographical Chapter and in a Selected Series of His Published Letters*. New York: D. Appleton.

Davis, J. I., and Gilmartin, A. J. (1985). Morphological variation and speciation. *Systematic Botany* **10**, 417–425.

Delany, M. J. (1970). Variation and ecology of island populations of the long-tailed field mouse (*Apodemus sylvaticus* (L.)). *Symposium of the Zoological Society of London* **26**, 283–295.

Djawdan, M., and Garland, T., Jr. (1988). Maximal running speeds of bipedal and quadrupedal rodents. *Journal of Mammalogy* **69**, 765–772.

Emlen, D. J. (2001). Costs and the diversification of exaggerated animal structures. *Science* **291**, 1534–1536.

Flatt, T., Shine, R., Borges-Landaez, P. A., and Downes, S. J. (2001). Phenotypic variation in an oviparous montane lizard (*Bassiana duperreyi*): The effects of thermal and hydric incubation environments. *Biological Journal of the Linnean Society* **74**, 339–350.

Forsman, A., and Shine, R. (1997). Rejection of non-adaptive hypotheses for intraspecific variation in trophic morphology in gape-limited predators. *Biological Journal of the Linnean Society* **62**, 209–223.

Gallagher, D. S., Jr., Dixon, J. R., and Schmidly, D. J. (1986). Geographic variation in the *Kentropyx calcarata* species group (Sauria: Teiidae): A possible example of morphological character displacement. *Journal of Herpetology* **20**, 179–189.

Gans, C., and Latifi, M. (1971). Redescription and geographical variation of *Monopeltis guentheri* Boulenger (Amphisbaenia, Reptilia). *American Museum Novitates* **2464**, 1–21.

Gardner, A. S. (1986). Morphological evolution in the day gecko *Phelsuma sundbergi* in the Seychelles: A multivariate study. *Biological Journal of the Linnean Society* **29**, 223–244.

Garland, T., Jr. (1985). Ontogenetic and individual variation in size, shape and speed in the Australian agamid lizard *Amphibolurus nuchalis*. *Journal of Zoology, London* **207**, 425–439.

Garland, T. J., Jr. (2003). Selection experiments: An under-utilized tool in biomechanics and organismal biology. In *Vertebrate Mechanics and Evolution* (V. Bels, J.-P. Gasc, and A. Casinos, eds.), pp. 23–56. Oxford, England: BIOS Scientific Publishers.

Garland, T., Jr., Hankins, E., and Huey, R. B. (1990). Locomotor capacity and social dominance in male lizards. *Functional Ecology* **4**, 243–250.

Geist, V. (1987). Bergmann's rule is invalid. *Canadian Journal of Zoology* **65**, 1035–1038.

Grant, B. R., and Grant, P. R. (1979). Darwin's finches: Population variation and sympatric speciation. *Proceedings of the National Academy of Science USA* **76**, 2359–2363.

Grant, B. R., and Grant, P. R. (1989). *Evolutionary Dynamics of a Natural Population*. Chicago: University of Chicago Press.

Grant, P. R. (1991). Natural selection and Darwin's finches. *Scientific American* **265**, 82–87.

Greene, H. W., and McDiarmid, R. W. (1981). Coral snake mimicry: Does it occur? *Science* **213**, 1207–1212.

Gummer, D. L., and Brigham, R. M. (1995). Does fluctuating asymmetry reflect the importance of traits in little brown bats (*Myotis lucifugus*)? *Canadian Journal of Zoology* **73**, 990–992.

Hallgrímsson, B. (2003). Variation. In *Key Words and Concepts in Evolutionary Developmental Biology* (B. K. Hall and W. M. Olson, eds.), pp. 368–377. Cambridge, MA: Harvard University Press.

Hallgrímsson, B., Willmore, K., and Hall, B. K. (2002). Canalization, developmental stability, and morphological integration in primate limbs. *Yearbook Physical Anthropology* **45**, 131–158.

Hallgrímsson, B., Miyake, T., Willmore, K., and Hall, B. K. (2003). Embryological origins of developmental stability: Size, shape and fluctuating asymmetry in prenatal random bred mice. *Journal of Experimental Zoology (Molecular and Developmental Evolution)* **296B**, 40–57.

Hanken, J. (1982). Appendicular skeletal morphology in minute salamanders, genus *Thorius* (Amphibia: Plethodontidae): Growth regulation, adult size determination, and natural variation. *Journal of Morphology* **174**, 57–77.

Hanken, J. (1983). High incidence of limb skeletal variants in a peripheral population of the red-backed salamander, *Plethodon cinereus* (Amphibia: Plethodontidae), from Nova Scotia. *Canadian Journal of Zoology* **61**, 1925–1931.

Heyer, W. R. (1997). Geographic variation in the frog genus *Vanzolinius* (Anura: Leptodactylidae). *Proceedings of the Biological Society of Washington* **110**, 338–365.

Hillen, B. (1987). The variability of the *Circulus arteriosus* (Willisii): Order or anarchy? *Acta Anatomica* **129**, 74–80.

Huey, R. B., Gilchrist, G. W., Carlson, M. L., Berrigan, D., and Serra, L. (2000). Rapid evolution of a geographic cline in size in an introduced fly. *Science* **287**, 308–309.

Irschick, D. J. (2002). Evolutionary approaches for studying functional morphology: Examples from studies of performance capacity. *Integrated Comparative Biology* **42**, 278–290.

Irschick, D. J., and Garland, T., Jr. (2001). Integrating function and ecology in studies of adaptation: Investigations of locomotor capacity as a model system. *Annual Review of Ecological Systematics* **32**, 367–396.

James, F. C. (1970). Geographic size variation in birds and its relationship to climate. *Ecology* **51**, 365–390.

Johnson, C. (1976). *Introduction to Natural Selection.* Baltimore: University Park Press.

Johnston, R. F. (1969). Character variation and adaptation in European sparrows. *Systematic Zoology* **18**, 206–231.

Johnston, R. F. (1994). Geographic variation in size of feral pigeons. *Auk* **111**, 398–404.

Johnston, R. F., and Selander, R. K. (1971). Evolution in the house sparrow. II. Adaptive differentiation in North American populations. *Evolution* **25**, 1–28.

Johnston, R. F., and Selander, R. K. (1973). Evolution in the house sparrow. III. Variation in size and sexual dimorphism in Europe and North and South America. *American Naturist* **107**, 373–390.

King, R. B. (1997). Variation in brown snake (*Storeria dekayi*) morphology and scalation: Sex, family, and microgeographic differences. *Journal of Herpetology* **31**, 335–346.

Koehl, M. A. R. (1996). When does morphology matter? *Annual Review of Ecological Systems* 501–542.

Knight, J. L. (1986). Variation in snout morphology in the North American snake *Pituophis melanoleucus* (Serpentes: Colubridae). *Journal of Herpetology* **20**, 77–79.

Laerm, J. (1974). A functional analysis of morphological variation and differential niche utilization in basilisk lizards. *Ecology* **55**, 404–411.

Lannoo, M. J., and Bachmann, M. D. (1984). Aspects of cannibalistic morphs in a population of *Ambystoma t. tigrinum* larvae. *American Midland National* **112**, 103–109.

Lannoo, M. J., Lowcock, L., and Bogart, J. P. (1989). Sibling cannibalism in noncannibal morph *Ambystoma tigrinum* larvae and its correlation with high growth rates and early metamorphosis. *Canadian Journal of Zoology* **67**, 1911–1914.

Lauder, G. V. (1990). Functional morphology and systematics: Studying functional patterns in an historical context. *Annual Review of Ecology* **21**, 317–340.

Lauder, G. V. (1995). On the inference of function from structure. In *Morphology in Vertebrate Paleontology* (J. Thomason, ed.), pp. 1–18. Cambridge, England: Cambridge University Press.

Lauder, G. V. (2003). The intellectual challenge of biomechanics and evolution. In *Vertebrate Biomechanics and Evolution* (V. Bels, J.-P. Gasc, and A. Casinos, eds.), pp. 319–325. Oxford, England: BIOS Scientific.

Lincoln, R. J., Boxshall, G. A., and Clark, P. F. (1982). *A Dictionary of Ecology, Evolution and Systematics.* Cambridge, England: Cambridge University.

Lindell, L. E., Forsman, A., and Merilä, J. (1993). Variation in number of ventral scales in snakes: Effects on body size, growth rate and survival in the adder, *Vipera berus*. *Journal of Zoology, London* **230**, 101–115.

Lindsay, S. L. (1986). Geographic size variation in *Tamiasciurus douglasii*: Significance in relation to conifer cone morphology. *Journal of Mammalogy* **67**, 317–325.

Lindsay, S. L. (1987). Geographic size and non-size variation in Rocky Mountain *Tamiasciurus hudsonicus*: Significance in relation to Allen's rule and vicariant biogeography. *Journal of Mammalogy* **68**, 39–48.

Lindsey, C. C. (1966). Body sizes of poikilotherm vertebrates at different latitudes. *Evolution* **20**, 456–465.

Lopez, P., and Martin, J. (2002). Locomotor capacity and dominance in male lizards *Lacerta monticola*: A trade-off between survival and reproductive success. *Biological Journal of the Linnean Society* **77**, 201–209.

Losos, J. B., Creer, D. A., and Schulte II, J. A. (2002). Cautionary comments on the measurement of maximum locomotor capabilities. *Journal of Zoology, London* **258**, 57–61.

Losos, J. B., Papenfuss, T. J., and Macey, J. R. (1989). Correlates of sprinting, jumping and parachuting performance in the butterfly lizard, *Leiolepis belliani* [sic]. *Journal of Zoology. London* **217**, 559–568.

Losos, J. B., Warheit, K. I., and Schoener, T. W. (1997). Adaptive differentiation following experimental island colonization in *Anolis lizards*. *Nature* **387**, 70–73.

Losos, J. B., Creer, D. A., Glossip, D., Goellner, R., Hampton, A., Roberts, G., Haskell, N., Taylor, R., and Ettling, J. (2000). Evolutionary implications of phenotypic plasticity in the hindlimb of the lizard *Anolis sagrei*. *Evolution* **54**, 301–305.

Lynch, J. D. (1966). Multiple morphotypy and parallel polymorphism in some neotropical frogs. *Systematic Zoology* **15**, 18–23.

Lynch, J. F. (1981). Patterns of ontogenetic and geographic variation in the black salamander, *Aneides flavipunctatus* (Caudata: Plethodontidae). *Smithsonian Contributions to Zoology* **324**.

Macrini, T. E., and Irschick, D. J. (1998). An intraspecific analysis of trade-offs in sprinting performance in a West Indian lizard species (*Anolis lineatopus*). *Biological Journal of the Linnean Society* **63**, 579–591.

Malhotra, A., and Thorpe, R. S. (1991a). Microgeographic variation in *Anolis oculatus* on the island of Dominica, West Indies. *Journal of Evolutionary Biology* **4**, 321–335.

Malhotra, A., and Thorpe, R. S. (1991b). Experimental detection of rapid evolutionary response in natural lizard populations. *Nature* **353**, 347–348.

Malhotra, A., and Thorpe, R. S. (1994). Parallels between island lizards suggests selection on mitochondrial DNA and morphology. *Proceedings of the Royal Society of London* **B257**, 37–42.

Martin, J., and Lopez, P. (2001). Hindlimb asymmetry reduces escape performance in the lizard *Psammodromus algirus*. *Physiological and Biochemical Zoology* **74**, 619–624.

Marchand, P. J. (1987). *Life in the Cold: An Introduction Winter Ecology.* Hanover, NH: University Press of New England.

Mather, K. (1970). The nature and significance of variation in wild populations. *Symposium of the Zoological Society London* **26**, 27–39.

Mayr, E. (1982). *The Growth of Biological Thought: Diversity, Evolution, and Inheritance.* Cambridge, MA: The Belknap Press of Harvard University Press.

Meyer, A. (1987). Phenotypic plasticity and heterochrony in *Chiclasoma managuense* (Pisces, Cichlidae) and their implications for speciation in cichlid fishes. *Evolution* **41**, 1357–1369.

Michaux, B. (1989). Morphological variation of species through time. *Biological Journal of Linnean Society* **38**, 239–255.

Nevo, E. (1973). Adaptive variation in size of cricket frogs. *Ecology* **54**, 1271–1281.

Niles, D. M. (1973). Adaptive variation in body size and skeletal proportions of horned larks of the southwestern United States. *Evolution* **27**, 405–426.

Panchen, A. L. (1992). *Classification, Evolution and the Nature of Biology.* Cambridge, MA: Cambridge University Press.

Pedersen, S. C. (1991). Dental morphology of the cannibal morph in the tiger salamander, *Ambystoma tigrinum*. *Amphibia-Reptilia* **12**, 1–14.

Pedersen, S. C. (1993). Skull growth in cannibalistic tiger salamanders, *Ambystoma tigrinum*. *Southwestern Naturalist* **38**, 316–324.

Pennisi, E. (2002). Evo-devo enthusiasts get down to details. *Science* **298**, 953–955.

Pergams, O. R. W., and Ashley, M. V. (2001). Microevolution in island rodents. *Genetica* **112–113**, 245–256.

Pfennig, D. W. (1992). Polyphenism in spadefoot toad tadpoles as a locally adjusted evolutionarily stable strategy. *Evolution* **46**, 1408–1420.

Pfennig, D. W., Reeve, H. K., and Sherman, P. W. (1993). Kin recognition and cannibalism in spadefoot toad tadpoles. *Animal Behavior* **46**, 87–94.

Pocock, R. I. (1916). Some dental and cranial variations in the Scotch wild cat (*Felis sylvestris*). *Annals and Magazine of Natural History* **8**, 272–277.

Policansky, D. (1982). The asymmetry of flounders. *Scientific American* **246**, 116–122.

Qualls, F. J., and Shine, R. (1998). Geographic variation in lizard phenotypes: Importance of the incubation environment. *Biological Journal of the Linnean Society* **64**, 477–491.

Reig, S. (1992). Geographic variation in pine marten (*Martes Martes*) and beech marten (*M. foina*) in Europe. *Journal of Mammalogy* **73**, 744–769.

Reznick, D. N., Shaw, F. H., Rodd, F. H., and Shaw, R. G. (1997). Evaluation of the rate of evolution in natural populations of guppies (*Poecilia reticulata*). *Science* **275**, 1934–1937.

Ridley, M. (1996). *Evolution* 2nd ed. Cambridge, MA: Blackwell Science.

Robinson, B. W. (2000). Trade-offs in habitat-specific foraging efficiency and the nascent adaptive divergence of sticklebacks in lakes. *Behavior* **137**, 865–888.

Rodríguez-Robles, J. A., and De Jesús-Escobar, J. M. (2000). Molecular systematics of New World gopher, bull, and pinesnakes (*Pituophis*: Colubridae): A transcontinental species complex. *Molecular Phylogenetics and Evolution* **14**, 35–50.

Russell, A. P., and Bels, V. (2001). Biomechanics and kinematics of limb-based locomotion in lizards: Review, synthesis and prospectus. *Comparative Biochemistry and Physiology A: Physiology* **131**, 89–112.

Russell, A. P., Bryant, H. N., Powell, G. L., and Laroiya, R. (1995). Scaling relationships within the maxillary tooth row of the Felidae, and the absence of the second upper premolar in *Lynx*. *Journal of Zoology, London* **236**, 161–182.

Savage, J. M., and Donnelly, M. A. (1988). Variation and systematics in the colubrid snakes of the genus *Hydromorphus*. *Amphibia-Reptilia* **9**, 289–300.

Schulte-Hostedde, A. I., and Millar, J. S. (2002). Effects of body size and mass on running speed of male yellow-pine chipmunks (*Tamias amoenus*). *Canadian Journal of Zoology* **80**, 1584–1587.

Shubin, N., Wake, D. B., and Crawford, A. J. (1995). Morphological variation in the limbs of *Taricha granulose* (Caudata: Salamandridae): Evolutionary and phylogenetic implications. *Evolution* **49**, 874–884.

Sinervo, B. (1990). The evolution of maternal investment in lizards: An experimental and comparative analysis of egg size and its effects on offspring performance. *Evolution* **44**, 279–294.

Sinervo, B., and Adolph, S. C. (1989). Thermal sensitivity of growth rate in hatchling *Sceloporus* lizards: Environmental, behavioral and genetic aspects. *Oecologia* **78**, 411–419.

Sinervo, B., and Huey, R. B. (1990). Allometric engineering: An experimental test of the causes of interpopulational differences in performance. *Science* **248**, 1106–1109.

Sinervo, B., and Losos, J. B. (1991). Walking the tight rope: Arboreal sprint performance among *Sceloporus occidentalis* lizard populations. *Ecology* **72**, 1225–1233.

Smith, M. F. (1979). Geographic variation in genic and morphological characters in *Peromyscus californicus*. *Journal of Mammalogy* **60**, 705–722.

Stearns, S. C. (1989). The evolutionary significance of phenotypic plasticity. *Bioscience* **39**, 436–445.

Swaddle, J. (1997). Within-individual changes in developmental stability affect flight performance. *Behavioral Ecology* **8**, 601–604.

Szuma, E. (1994). Quasi-continuous variation of the first premolars in the Polish population of the badger *Meles meles*. *Acta Theriologica* **39**, 201–208.

Tague, R. G. (1997). Variability of a vestigial structure: First metacarpal in *Colobus guereza* and *Ateles geoffroyi*. *Evolution* **51**, 595–605.

Telleria, J. L., and Carbonell, R. (1999). Morphometric variation of five Iberian blackcap *Sylvia atricapilla* populations. *Journal of Avian Biology* **30**, 63–71.

Thomas, A. L. R. (1993). The aerodynamic costs of asymmetry in the wings and tail of birds: Asymmetric birds can't fly round tight corners. *Proceedings of the Royal Society of London* **B254**, 181–189.

Thorpe, R. S. (1984). Geographic variation in the western grass snake (*Natrix natrix helvetica*) in relation to hypothesized phylogeny and conventional subspecies. *Journal of Zoology, London* **203**, 345–355.

Thorpe, R. S. (1991). Clines and cause: Microgeographic variation in the Tenerife gecko (*Tarentola delalandii*). *Systematic Zoology* **40**, 172–187.

Thorpe, R. S., and Baez, M. (1993). Geographic variation in scalation of the lizard *Gallotia stehlini* within the island of Gran Canaria. *Biological Journal of the Linnean Society* **48**, 75–87.

Thorpe, R. S., and Brown, R. P. (1989). Microgeographic variation in the colour pattern of the lizard *Gallotia galloti* within the island of Tenerife: Distribution, pattern and hypothesis testing. *Biological Journal of the Linnean Society* **38**, 303–322.

van Berkum, F. H., and Tsuji, J. S. (1987). Inter-familiar differences in sprint speed of hatchling *Sceloporus occidentalis* (Reptilia: Iguanidae). *Journal of Zoology, London* **212**, 511–519.

van Berkum, F. H., Huey, R. B., Tsuji, J. S., and Garland, T., Jr. (1989). Repeatability of individual differences in locomotor performance and body size during early ontogeny of the lizard *Sceloporus occidentalis* (Baird & Girard). *Functional Ecology* **3**, 97–105.

Van Buskirk, J. (2002). Phenotypic lability and the evolution of predator-induced plasticity in tadpoles. *Evolution* **56**, 361–370.

Van Buskirk, J., and Relyea, R. (1998). Selection for phenotypic plasticity in *Rana sylvatica* tadpoles. *Biological Journal of the Linnean Society* **65**, 301–328.

Van Damme, R., Aerts, P., and Vanhooydonck, B. (1998). Variation in morphology, gait characteristics and speed of locomotion in two populations of lizards. *Biological Journal of the Linnean Society* **63**, 409–427.

Van Valen, L. (1965). Morphological variation and width of ecological niche. *American Naturalist* **99**, 377–390.

Van Voorhies, W. A. (1996). Bergmann size clines: A simple explanation for their occurrence in ectotherms. *Evolution* **50**, 1259–1264.

van Zyll de Jong, C. G. (1975). Differentiation of the Canada lynx, *Felis (Lynx) Canadensis subsolana*, in Newfoundland. *Canadian Journal of Zoology* **53**, 699–705.

Vanhooydonck, B., Van Damme, R., Van Dooren, T. J. M., and Bauwens, D. (2001). Proximate causes of intraspecific variation in locomotor performance in the lizard *Gallotia galloti*. *Physiological and Biochemical Zoology* **74**, 937–945.

Vasilyeva, E. D. (1990). Morphological variability of vertebrate clonal species: Polyploid spined loaches (*Cobitis*) and crucian carp (*Carassius auratus gibelio*) [in Russian]. *Akademii Nauk USSR Zhurnal Obshchei Biologii* **51**, 775–782.

Vogel, G. (2000). A mile-high view of development. *Science* **288**, 2119–2120.

Wagner, G. P. (1988). The influence of variation and of developmental constraints on the rate of multivariate phenotypic evolution. *Journal of Evolutionary Biology* **1**, 45–66.

Walls, S. C., Belanger, S. S., and Blaustein, A. R. (1993a). Morphological variation in larval salamander: Dietary induction of plasticity in head shape. *Oecologia* **96**, 162–168.

Walls, S. C., Beatty, J. L., Tissot, B. N., Hokit, D. G., and Blaustein, A. R. (1993b). Morphological variation and cannibalism in a larval salamander (*Ambystoma macrodactylum columbianum*). *Canadian Journal of Zoology* **71**, 1543–1551.

Warner, D. A., and Andrews, R. M. (2002). Laboratory and field experiments identify sources of variation in phenotypes and survival of hatchling lizards. *Biological Journal of the Linnean Society* **76**, 105–124.

Wilkens, H. (1988). Evolution and genetics of epigean and cave *Astyanax fasciatus* (Characidae, Pisces): Support of the neutral mutation theory. *Evolutionary Biology* **23**, 271–367.

Wimberger, P. H. (1993). Effects of vitamin C deficiency on body shape and skull osteology in *Geophagus brasiliensis*: Implications for interpretations of morphological plasticity. *Copeia* **1993**, 343–351.

Witte, F., Barel, C. D. N., and Hoogerhoud, R. J. C. (1990). Phenotypic plasticity of anatomical structures and its ecomorphological significance. *Netherlands Journal of Zoology* **40**, 278–298.

Yamamoto, Y., and Jeffery, W. R. (2000). Central role for the lens in cave fish eye degeneration. *Science* **289**, 631–633.

Zaldívar-Riverón, A., and Nieto-Montes de Oca, A. (2002). Variation in the rare lizard *Barisia rudicollis* (Wiegmann) (Anguidae) with description of a new species from central Mexico. *Herpetologica* **58**, 313–326.

Zeldich, M. L. and Carmichael, A. C. (1989). Growth and intensity of integration through postnatal growth in the skull of *Sigmodon fulviventer*. *Journal of Mammalogy* **70**, 477–484.

Zweers, G. A. (1979). Explanation of structure by optimization and systemization. *Netherlands Journal of Zoology* **29**, 418–440.

CHAPTER 18

A Universal Generative Tendency toward Increased Organismal Complexity

DANIEL W. MCSHEA
Biology Department, Duke University, Durham, North Carolina, USA

INTRODUCTION

Consider a simple organism consisting of a number of repeated identical parts. The organism could be a multicellular alga, and the parts could be identical cells, for example. Or it could be a plant or animal, and the parts could be identical leaves or segments. Now suppose that a random heritable variation occurs in the next generation, one that modifies the morphology of one or more of the parts in the offspring. The result is that the degree of differentiation among the offspring's parts, or its "internal variance," will be greater than the parent's. (Here I use the term variance not in its formal sense, to mean a sum of squared deviations, but to refer more generally to something like "amount of variation" or "degree of differentiation.")

Then suppose that more heritable random variations occur as time passes, and the lineage is extended. Some variations may cause all parts to vary in precisely the same way, producing no change in internal variance. And a few may by chance reduce the internal variance by making the parts more similar to each other. But these will be far outnumbered by variations that make them more different. There are simply many more ways to increase internal variance than to decrease it, more ways for a set of parts to diverge from each other than to converge, and therefore random changes will produce mostly increases. The principle is simple and quite general. A uniformly painted picket fence increases its internal variance as the paint on each of the posts acquires a different pattern of wear. This is true not just initially, when all parts are identical or nearly so, but also later after considerable differentiation has occurred. Indeed, so long as some degree of order is present and introduced variation is random, internal variance should increase indefinitely. Further, the argument does not depend on starting with repeated or even similar parts.

The suggestion so far is that the accumulation of random variations should produce a tendency, a vector, toward increasing internal variance. What about selection? What will be its effect? There are three possibilities. One is that selection will act neutrally with respect to internal variance. Only a small subset of variations arising will be favored by selection, of course. If selection is neutral, then variations producing greater internal variance will be represented in the advantageous subset in proportion to the frequency with which they occur. To put it another way, all variations, whether they produce increases or decreases in internal variance, have the same (low) probability of meeting the functional and ecological demands on the organism. If so, if the variations favored are an unbiased sample of the variations occurring, then most will be increases in internal variance. A second possibility is that internal-variance increases will have a greater probability of being favorable, and in that case, selection will just reinforce the internal-variance vector.

The third possibility is that decreases will have a greater probability of being favorable, so that selection acts against the vector, perhaps overpowering it, with the result that no actual bias toward increase is manifest in real lineages. Gravity imparts to helium-filled balloons a tendency to fall, but they do not fall, because another force intervenes. One might imagine that the case for selection canceling the internal variance vector is overwhelming, because most increases in internal variance will be deleterious, and we know that selection has favored mechanisms in organisms to buffer or canalize development. If so, most of the variations that would produce increases in internal variance will never be expressed. However, decreases are also likely to be deleterious. What buffering mechanisms oppose is *departure* from a functional norm, not greater internal variance per se, and therefore they should oppose the expression of both increases and decreases equally. Therefore, the variations that are actually expressed, the unbuffered fraction, should reveal the generative tendency, that is, they should mostly have higher internal variance.

Other, related arguments are discussed later. The conclusion will be that, unless overwhelmed by selection, a general or pervasive tendency for internal variance to increase should exist in evolution, across all lineages, and over the entire history of life.

Further, I will argue that internal variance can also be understood as a type of complexity. Here I use the term "complexity" in a restricted sense, justified at length later, to refer to systems with many different part types, or with parts that are well differentiated from each other. Complexity in this sense is a purely structural property, independent of function. And therefore, one can replace internal variance with complexity in the preceding argument, and the claim here becomes that there is a universal tendency for complexity to increase.

If such a complexity vector exists, it demands what for some may be a dizzying reversal of intuition. Explaining the enormous structural complexity of modern organisms has been considered one of the central problems of evolutionary biology. Equally perplexing has been the apparent trend in the complexity of animals over their history, roughly the past 600 million years. To explain these appearances, various mechanisms have been invoked. For example, the suggestion has been made that natural selection generally favors increased differentiation on account of the advantages of internal division of labor. However, the internal-variance principle suggests that increasing complexity in a purely structural sense, independent of function, is an expected or generic feature of evolution, that it is not a puzzle to be explained but that it follows from the simple accumulation of variations.

Importantly, the suggestion is not that the internal-variance principle replaces natural selection as the explanation for complex structures such as vertebrate eyes. Rather, it is that the variational possibilities, the raw materials so to speak, supplied to selection will be increasingly complex, and an eye is the result of the differential survival of the subset of these ever more complex options that are functional. In effect, the claim is that in the evolution of an eye most of the less-functional and nonfunctional alternatives rejected by selection *were also complex*.

In recent decades, some have raised doubts that any trend in complexity has occurred, doubting for example that modern animals are more complex on average than ancient ones (Simpson, 1967; McShea, 1996). Interestingly, the internal-variance principle disturbs this view as well, raising the question of why complexity has been so resolutely opposed by selection, or to put it another way, what's wrong with complexity?

Notice that the internal-variance principle is different from the notion that a trend in complexity could result from what has been called "diffusion" in the presence of a lower bound. According to the diffusion notion, the first organisms were as simple as possible, and later-arising lineages had "nowhere to go but up" (Maynard Smith, 1970). Then, as diversity increased, complexity in descendant lineages both increased and decreased, but the spread or diffusion of lineages was asymmetrical on account of the lower bound. Thus the mean increased (McShea, 1996; Gould, 1996). In other words, the diffusion mechanism suggests that the trend is "passive" (*sensu* McShea, 1994). The internal-variance principle does invoke boundaries in the sense that variances cannot be negative, i.e., they are bounded at zero. Also, the principle invokes diffusion in the sense that the morphologies of parts within an organism are understood to diffuse in morphospace. But at the lineage level, the principle predicts a pervasive increasing tendency. It predicts that in all lineages, from the simplest to the most complex, parts should diffuse, and variances should be bounded, so that internal variances will increase in all or most of them. In other words, the principle predicts that the trend should be "driven" (*sensu* McShea, 1994).

The idea behind the internal-variance principle is not novel in biology. It was central in Herbert Spencer's metaphysic (1900, 1904), an instance of a higher principle he called "the instability of the homogeneous" (see discussion in McShea, 1991). (Indeed, he invoked it to explain what he took to be a trend in complexity of internal structure of organisms.) A related idea is implicit in the notion of duplication and differentiation of parts, which has been recognized in biology at all levels, from gene to ecosystem. The principle also lies just below the surface in the modern treatments of morphological evolution as a diffusive or Markov process (e.g., Raup, 1977). And it is present in at least one of the recent schools of evolutionary thought based on the second law of thermodynamics (Wicken, 1987; Brooks and Wiley, 1988). Finally, the principle seems to fall into the category of analytical truths we call truisms, truths so obvious as not to require demonstration. For example, the principle explains, in an almost trivial way, the increase over time in the surface complexity of the moon as result of the accumulation of variations from meteorite impacts.

Still the principle's application to the complexity of organisms has not been widely appreciated. Spencer's contribution has been forgotten, for the most part, and the thermodynamic version of the principle has not received much attention in biology partly because, I believe, it has been expressed mainly in the language of physics. Also, the thermodynamic school applied the principle mainly to ecology, with the consequence that its major prediction, increasing diversity, attracted little attention. Diversity increase over the history of life is considered already well explained in conventional Darwinian terms. In contrast, for organismal complexity, the Darwinian view is more uncertain in its predictions. (Selection for increased specialization might favor complexity, but generalized adaptation might require simplicity of structure.) My purpose here is partly to revive and further explore Spencer's principle. It is also partly to translate the argument of the thermodynamic school into more conventional biological terms and to show more clearly how it applies to individual organisms.

I. INTERNAL VARIANCE AS COMPLEXITY

Characterizing internal variance as complexity needs justification, because in colloquial usage, complexity connotes so much more. A complex organism is ordinarily understood to be not just more internally varied, or more differentiated, but more capable as well. We think of the human brain as complex not simply because it has many cell types, but because of its impressive functional capabilities, because of what it can do. Thus, as conventionally understood, complexity depends on both structure and function.

However, in biology, a narrower view has been adopted (McShea, 1996; Bell and Mooers, 1997; Carroll, 2001), one that deliberately divorces complexity from any notion of functionality, making it a property purely of structure. In this view, complexity refers to number of part types, or degree of differentiation among parts. Thus a watch is functional and complex, because it has many part types, most very different from most others. And a smashed watch is nonfunctional but still complex, perhaps even more so than the original watch. Conversely, a chunk of raw iron ore is simple and nonfunctional (ordinarily, uses can be imagined). A hammer fashioned from iron ore is also simple and functional (for smashing watches, for example). Likewise for organisms: The complex ones are those with many types of parts, well differentiated, while the simple ones are the opposite. And at least in principle, as a matter of logic, there is no reason that a simple organism cannot be just as functional, just as capable (or even more so), as a complex one.

Divorcing structural complexity and function has two main virtues. First, part of the motivation for investigating complexity is to investigate its relationship with function. Are more complex organisms also more functional, i.e., more fit? To answer questions like that, we need to be able to assess them independently, to plot complexity and functionality on separate axes so to speak. Second, by defining complexity in terms of structure alone, we can actually measure it sometimes in organisms. Cisne (1974) measured complexity in aquatic arthropods as a function of number limb-pair types. Bonner (1988) and Bell and Mooers (1997) measured the complexity of multicellular organisms as the number of cell types. I investigated vertebral-column complexity using a measure based on degree of differentiation among the vertebral elements within a column (McShea, 1993). Saunders *et al.* (1999) devised a measure of complexity in ammonoids, ancient relatives of the modern chambered nautilus, based on the number of curvilinear components in the septal sutures marking the boundaries between successive body chambers. And Sidor (2001) measured skull complexity as the number of types of bony skull elements.

The list of metrics reveals two definitional themes, complexity as an increasing function of number of different types of parts, where types are distinct, and complexity as degree of differentiation among parts, where types intergrade. The two themes correspond to the canonical separation in biology between discrete and continuous variation, a separation that I will maintain in this discussion. Thus, here, an organism with three cell types has greater internal variance and is more complex than one with two. And an organism with two highly differentiated cell types is more complex than one with two nearly identical types.

Calling a smashed watch complex might sound odd for another reason, besides its lack of function. Smashing a watch can produce any of a vast number of possible specific results, specific configurations of part types (i.e., fragments). And because we are ordinarily indifferent to the details of each result, we think of them as equivalent. That is, we tend to think of this vast number of equivalent alternative smashings as an ensemble, rather than as individual cases. And "complex" sounds like the wrong word for such an ensemble. Considered together, indifferent to the differences among them, they seem "disordered," "random," or "entropic," but not complex. So it might seem that what the internal-variance principle predicts is increasing disorder or entropy, not complexity.

However, consider instead the result of a single smashing event, producing a single assortment of part types. And suppose that we have decided to turn the results of this event into a work of art. Now every detail of its composition could be a matter of great interest. And the description of each part type would be quite a long one. Now the word complex seems more appropriate. More generally, for systems with high internal variance, the common intuition is that ensembles are disordered, but specific instances are complex. So in biology the internal-variance principle does predict disorder at the level of the ensemble, across all organisms everywhere over all time, and this we see in the tremendous diversity of organismal forms. However, what it predicts in each specific instance, each specific organism (or species), is complexity.

On this understanding, describing the increase in internal variance in evolution as a consequence of the second law of thermodynamics would be misleading (at least, for the statistical–mechanical interpretation of the second law). The second law predicts an increase in entropy through time, but entropy is a property of an ensemble. More precisely, entropy refers to the number of microstates corresponding to a given macrostate. In contrast, the internal-variance principle, like Spencer's principle of the instability of the homogeneous, predicts an increase in the variance among the parts in a single entity, an organism. We cannot call it entropic, because there is no ensemble, no microstate–macrostate relationship. Rather, for the organism with many differentiated parts, we say that it is complex. And if it has survived the selective filter, we say it is complex and functional as well.

Complexity has other aspects besides number of part types. For example, there is complexity of spatial arrangement of parts, a kind of second-order complexity (where number of part types is first order), and number of types of connections among parts (McShea, 1996). I do not consider either of these aspects here, but I expect that a parallel argument could be developed for them. Complexity is also used in a hierarchical sense to refer to the degree of nestedness of a system, or in other words, the number of levels of parts within wholes that it contains (Salthe, 1985; Pettersson, 1996; McShea, 2001). However, the internal-variance principle would seem to apply only to complexity in its nonhierarchical sense, to differentiation among parts at roughly the same hierarchical level, such as cells within a multicellular organism, or to what I have elsewhere called nonhierarchical object complexity (McShea, 1996).

My argument makes two major claims. The first is that a vector of increasing internal variance exists in evolution (although I make no claim about the magnitude or direction of the selective vector nor therefore about the resultant). The second is that internal variance is a type or aspect of complexity. Importantly, the two claims are independent of each other. Thus even those who find the second unconvincing can still accept the first.

II. THREE SIMPLE MODELS

Three simple models will be introduced to illustrate the internal-variance principle and also to reveal its robustness. In each successive model, the variations introduced are more finely tuned in such a way as to negate or overcome the internal-variance principle. But in each case, the principle survives, that is, internal variance increases, on average.

A. Model 1

Five parts of an organism are plotted as points on an axis corresponding to some dimension, say, their length. Initially all five parts are identical, with their lengths set at the same value (arbitrarily, 10 mm). In the first generation, random heritable variation is added to each part, so that its length increases or decreases by the same factor, divided by 0.9 for increases and multiplied by 0.9 for decreases, with the direction of change chosen at random (probability of increase = probability of decrease = 0.5). Each part is treated independently. And in each subsequent generation, the process is repeated, with new values for each of the five parts calculated based on their values in the previous generation, so that change is cumulative.

Internal variance is measured as the standard deviation among the logged lengths of the parts. (Thus the model captures internal variance in its continuous sense only.) Figure 18-1 shows the distribution of logged part-lengths in a single run of the model. In the initial distribution in Figure 18-1A, the five data points overlap at the same coordinate on the log-length axis. (The log of 10 is 1.0.) All have the same length so the standard deviation is zero. The distributions of standard deviations after 1, 2, 5, 10, and 20 generations are shown in Figures 18-1B–F. Figure 18-2A shows its trajectory over the entire run, and Figure 18-2B shows the mean standard deviation over 1000 runs, with error bars (i.e., the standard deviation of the standard deviation). Analytically, it can be shown that mean rises as the square root of the number of generations.

Notice that the model shows the internal variance rising even when parts have become highly differentiated. That is, even when complexity is high, the expectation is further increase. The curve in Figure 18-2B increases without limit.

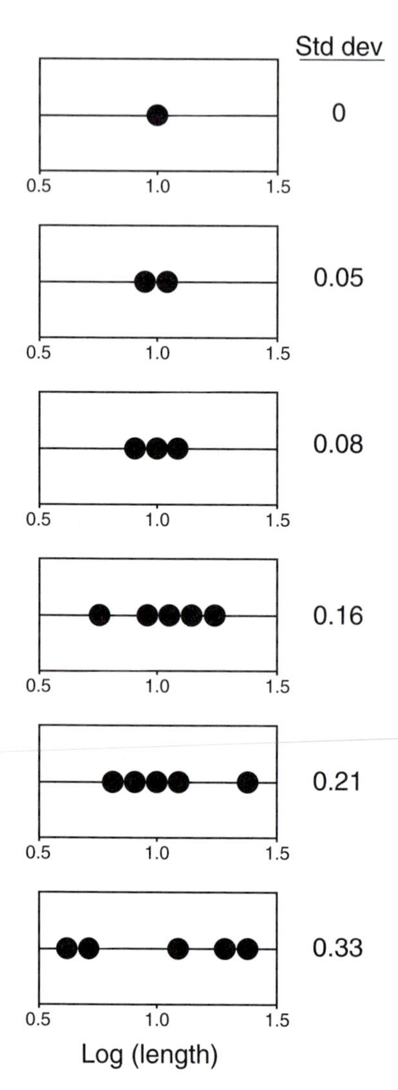

FIGURE 18-1. Model 1. Each dot represents one part, and there are initially five identical parts. In each time interval (each generation), a variation introduced to each part changes its length by a factor of 0.9 (length is multiplied by 0.9 for decreases and divided by 0.9 for increases), with increases and decreases occurring equally frequently. Internal variance is measured as the standard deviation among log lengths. The numbers show the increase in internal variance at time zero (top box), and then after 1, 2, 5, 10, and 20 time intervals.

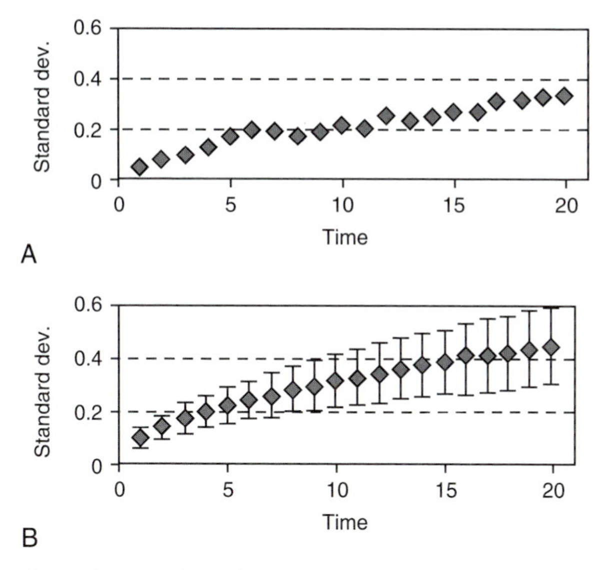

FIGURE 18-2. The top figure, A, shows the trajectory of the standard deviation for a single run of model 1 over 20 time intervals. The bottom figure shows the trajectory averaged over 1000 runs. Error bars show one standard deviation of the internal variance (i.e., the standard deviation of the standard deviation).

B. Model 2

In the first model, parts vary independently, which may often be realistic. We have some reason in theory to think that parts with independent function, for example, will tend to be developmentally independent (Wagner and Altenberg, 1996) and therefore to vary independently. However, in some cases the assumption is clearly wrong, perhaps especially where parts develop as part of a single unit or collaborate in the performance of a single function. Obvious cases include homologous series such as limb series in arthropods, or tooth rows and vertebral columns in vertebrates, where graded change often occurs in regions of the series rather than in single elements independently. An example is the gradual increase in size of vertebral bodies through the lumbar region of many mammals.

The evolutionary origin of such patterns is unknown, but it is easy to imagine they are produced by the introduction in evolution of variations with regional, graded effects. Does the introduction of such variations also tend to produce increases in internal variance? Figure 18-3A shows the pattern of change in a single dimension along the vertebral column of a cheetah, from the axis to the last lumbar vertebra. The dimension is the altitude of the neural process, an arch of bone lying

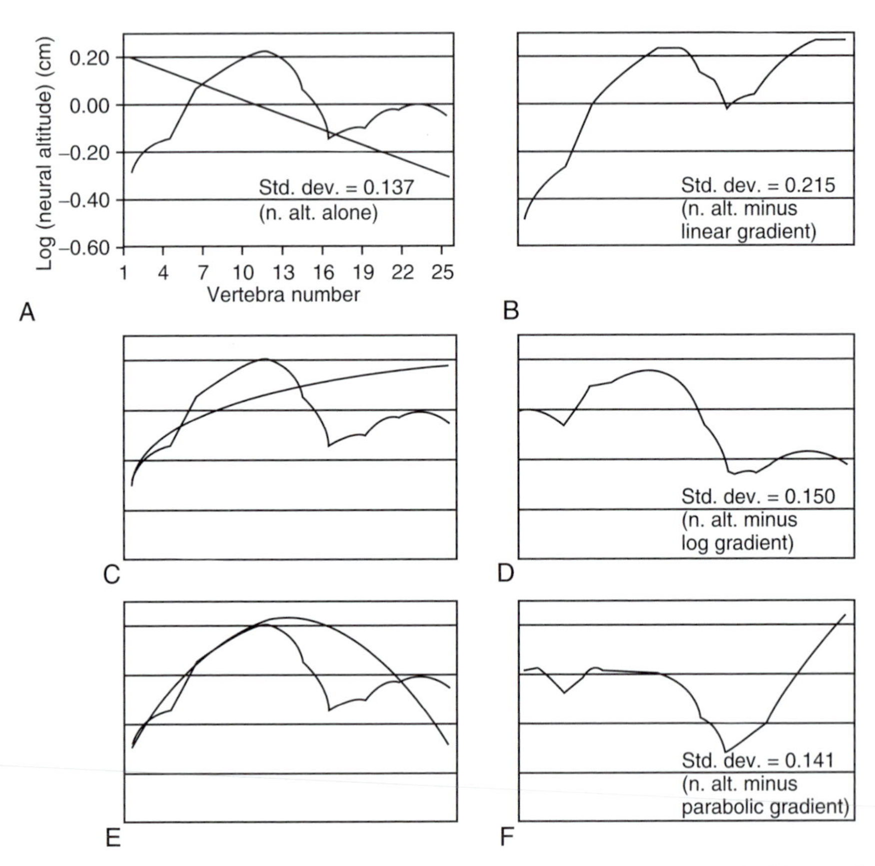

FIGURE 18-3. Model 2. The irregular curves in A, C, and E show the log of the length of the neural process (log neural altitude) for the vertebral elements in the vertebral column of a cheetah specimen. The smooth curves in these figures show the variation introduced: in A, a log-linear gradient; in C, a log gradient; in E, a parabolic gradient. The figures to the right of each of these, B, D, and F, show the resultant in each case when the introduced variation is subtracted. In all three cases, the standard deviation increased.

above the vertebral body and enclosing the nerve cord. The figure shows log neural altitude, size corrected by dividing by the mean. The curve increases over the neck and anterior thoracic regions (numbered 1–11), declines in the posterior thoracic and anterior thoracic and lumbar regions (12–16), and rises again slightly in the posterior lumbars (17–25). The standard deviation of logged neural altitudes is 0.137.

Figure 18-3A also shows a hypothetical linear gradient, such as might be introduced as a random variation, analogous to the random changes introduced in each element in the first model. However, here a single variation affects the whole series, so that the elements change in a coordinated way. Figure 18-3B shows the effect of subtracting this linear gradient from the original data series. (If one prefers to think of variations as added, rather than subtracted, the gradient could be inverted.) The result is an increase in the standard deviation from 0.137 to 0.215. In Figure 18-3C, the introduced variation is a log gradient (actually a function of the log of the vertebra number). Figure 18-3D shows the effect of subtracting this gradient, again an increase in standard deviation, this time from 0.137 to 0.150. In Figure 18-3E, a parabolic gradient is introduced, and Figure 18-3F shows the effect, an increase to 0.141.

Notice that all of the introduced gradients in Figure 18-3 were scaled and oriented so that they matched the actual trajectory fairly closely, increasing the likelihood that they would produce decreases in internal variance. In other words, these functions were somewhat engineered to reduce complexity, but complexity increased nonetheless. Presumably, truly randomly chosen gradients would be even more likely to produce increases.

C. Model 3

Consider finally a gradient that is even more deliberately chosen for its likelihood of reducing internal variance, a sine function. Table 18-1 shows the effect of adding a sine function from the logged cheetah data. The top part of the table shows the change in standard deviation, relative to the original standard deviation of the cheetah data (0.137), which results from adding sine curves of various wavelengths. Wavelength values range from 0.25 to 2.0, where 1.0 represents a sine curve that cycles exactly once over the length of the vertebral series, i.e., over 25 units. The table also shows results for alternative phase shifts of the wave, in three-vertebra increments. The amplitude of the wave was fixed at a value that gave it roughly the same range in the vertical dimension as the original data. As the table shows, adding the sine function does produce some decreases (12), but most of the results are increases (60).

The bottom part of Table 18-1 shows the effect of varying the amplitude of the wave, this time with the wavelength fixed at 0.75. Again some decreases occurred (24), but again most of the entries are increases (48). Figure 18-4 shows a single and somewhat extreme case in which a sine curve produced a decrease. With this combination of parameters (wavelength = 0.75, phase shift = 3, amplitude = 0.15), the peaks of the sine curve correspond roughly to the troughs in the original data and vice versa (Figure 18-4A), producing a new curve (Figure 18-4B) with a standard deviation of 0.075, well below that of the original data, 0.137.

It is worth restating that the choices of parameters and functions in models 2 and 3 (and especially in the case in Figure 18-4) were rigged against the internal-variance

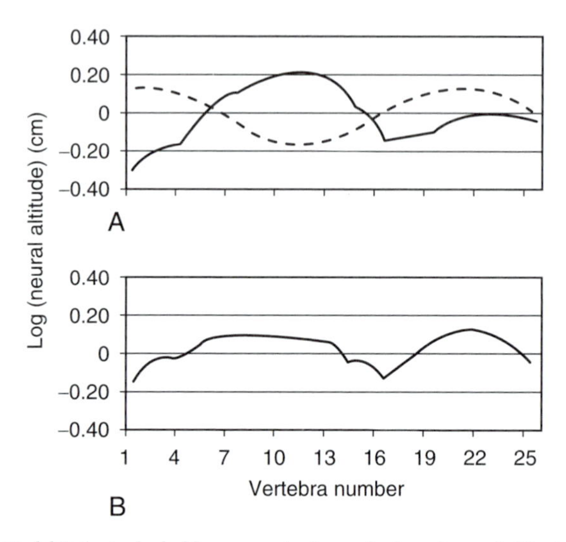

FIGURE 18-4. Model 3. In A, the bold curve again shows the log of neural altitude for the cheetah, and the thin curve shows an introduced variation, a sine curve. In this case, the figure shows a rare case in which the frequency and phase alignment of the introduced variation offset the neural altitude curve very well, so that adding the sine curve (equivalent to subtracting its inverse) had a smoothing effect, and the standard deviation decreased (shown in B, see text).

principle. This was done for the three monotonic functions in model 2 by restricting their ranges to roughly that of the original data. The set of real possible functions includes a vast number with much greater ranges, all of which would likely produce increases in the standard deviation. Also the sine-curve functions were chosen because the original data seemed to follow something like a sine curve, and it seemed likely that such functions would be likely to produce decreases. (Indeed, for the simulations in the bottom of Table 18-1, the wavelength parameter was set at a value that seemed likely to produce peak-to-trough matching and therefore decreases.) However, despite deliberate engineering of these functions to produce decreases, increases predominated.

A comment might be needed on the use of standard deviation as a measure of internal variance, because it seems to produce counterintuitive results in some cases. In particular, for a given total range of variation, standard deviation is maximized when half of the parts lie at one extreme and half at the other. Intuitively, such a distribution might not seem to correspond to maximal internal variance, or maximal complexity, because it consists of two groups that, although very different from each other as groups, consist of identical parts. Rather, intuitively, internal variance of the organism would be higher if the parts were more evenly distributed. One solution would be to adopt a different metric, one that captures complexity

TABLE 18-1. **The Effect on Internal Variance of Adding a Sine Curve to Cheetah Vertebral-Column Neural Attitude**[a]

	Wavelength									Amplitude							
Phase-shift	0.25	0.5	0.75	1	1.25	1.5	1.75	2	Phase-shift	0.05	0.10	0.15	0.20	0.25	0.30	0.35	0.40
0	+	+	+	+	+	+	+	+	0	–	–	+	+	+	+	+	+
3	+	+	–	+	+	+	+	+	3	–	–	–	–	–	–	+	+
6	+	+	–	+	+	+	+	+	6	–	–	–	–	–	+	+	+
9	+	+	+	–	+	+	+	+	9	+	+	+	+	+	+	+	+
12	+	+	+	+	–	+	+	+	12	+	+	+	+	+	+	+	+
15	+	+	+	+	+	–	+	+	15	+	+	+	+	+	+	+	+
18	+	+	+	+	+	–	–	+	18	+	+	+	+	+	+	+	+
21	+	+	–	+	+	+	–	–	21	–	–	–	–	–	+	+	+
24	+	+	–	+	+	+	+	–	24	–	–	–	–	–	–	+	+

[a]Wavelength refers to the fraction of column length over which the sine curve cycles, phase shift refers to the horizontal position of the curve (in number of vertebra from an arbitrary starting point) with respect to the cheetah data, and amplitude is a vertical scaling coefficient. In the body of the table, "+" means that adding the sine curve to the cheetah data produced an increase in the standard deviation, and "–" means that it produced a decrease.

in its discrete sense, as number of different types. To do this, we could divide the range of variation into bins, so that parts achieving sufficient difference from each other would be designated as different types, and then internal variance would be simply the number of different types. My expectation is that results for the three models would be qualitatively the same.

III. THE EFFECT OF INCREASED DIMENSIONALITY

The models can be extended easily to show that with more dimensions, internal variance increases even faster. That is, if the parts are plotted in a multidimensional space, with axes defined by length, width, color, permeability, and other properties of the parts, the rate of expansion of the points or the rate at which they differentiate from each other will be greater than in one dimension. How much greater will depend on the number of dimensions and how we measure internal variance. If we measure it as a sum of univariate standard deviations, the rate will grow linearly with number of dimensions, all else being equal. Other metrics—such as the standard deviation of distances from the multivariate centroid—would yield a different relationship, but in all cases rate of expansion would increase with dimensionality. In other words, the more detailed our characterization of the parts, the higher the dimensionality, the more rapidly we can expect them to differentiate from each other, and the greater the magnitude of the vector tending to increase complexity.

IV. APPARENT DIFFICULTIES

I can foresee two major objections to the internal-variance principle. First, the principle might seem to claim that generative processes producing complexity must be stronger than selection. And if so, the implication would seem to be that the complex organisms we see around us are maladaptive or at least suboptimal in some way. Although this could certainly be the case, it does not follow from the principle. The principle argues instead that the range of options presented to selection will contain mostly organisms that are more complex than their parents or ancestors, and therefore unless there is some selective preference for simple over complex, those passing the selective filter will more often be complex.

Second, it is widely acknowledged in the evolutionary literature that losses of structures are easier to achieve developmentally than gains (cf. Saunders and Ho, 1976, who actually make the opposite argument). One conventional line of thinking is that most mutations affecting a given developmental pathway will be more likely to disrupt it, to cause its failure, than to enhance or extend it, and therefore losses or reductions of the terminal structure produced by that pathway are more likely than elaborations of it or additions to it. A classic example raised in this context is the reduction of eyes and loss of sight in certain cave fish. I think this logic is partly faulty and that the fault lies in a confusion between loss of function and loss of structure. It does seem clear that loss of function must be more likely than gain. This asymmetry is the basis of Darwin's intuition that most variations will be deleterious.

However, it is not obvious that the loss of function means loss of structure, i.e., loss or reduction of parts. It seems to me equally likely that random modification of a developmental pathway would lead to elaboration of a terminal structure, that is, to an increase in its complexity. It is true that eliminating a generative instruction might lead to a failure of an inductive tissue contact, elimination of a differentiation event, or a failure of the production of a fold. But given the dynamism of development, would it not be equally likely to lead to an absence of an inductive barrier, to an extraneous differentiation, or to a smoothing-mechanism failure? The earliest variations arising in cave-fish eyes (when they first enter caves) would certainly reduce their functionality, but would those variations also render their eyes structurally simpler? Are not developmental aberrations, "monstrosities," often more complex structurally? I think these are currently open questions. (Of course, in the long run, perhaps cave-fish eye complexity will decline. Selection for economy in development and for reduced probability of infection, for example, might be expected to reduce the size of unused eyes, and loss of structure might tend to accompany size reduction (Bonner, 1988). However, this is a new argument, one that invokes selection, which the earlier objection does not. Later I address the possibility that selection is not complexity neutral.)

V. IS THERE AN UPWARD BIAS IN REAL LINEAGES?

Before addressing this question, I must distinguish an upward tendency or vector, such as predicted by the internal-variance principle, from the actual upward movement of real lineages, or what I call a bias. An internal-variance vector might exist, but no actual bias would be produced if selection opposes and overwhelms it.

Here is what we know about the existence of an upward bias in nonhierarchical complexity in metazoans (see also McShea, 1996, and references therein). A trend in mean complexity of septal sutures occurred in Paleozoic ammonoids, and it was produced by a bias, i.e., suture complexity increased significantly more often than it decreased among lineages (Saunders *et al.*, 1999). On the other hand, in the vertebral column, a trend in mean complexity probably occurred without any bias. Among mammalian lineages, at least, increases and decreases in complexity occurred about equally often, suggesting that the trend was probably diffusive (McShea, 1993; for other evidence, see McShea, 1994). That is, the fish condition constituted a kind of boundary, and there was nowhere for the mean to go but up.

Other cases are uncertain. The maximum and mean number of cell types in metazoan individuals rose over the Phanerozoic, a trend that Valentine *et al.* (1994)

interpreted as diffusive but only tentatively, based on the shape the trend trajectory. In other cases, such as the Phanerozoic trend in mean and maximum complexity of limb series in aquatic arthropods (Cisne, 1974), the evidence is equally consistent with the presence or absence of an upward bias. In sum, it is impossible to generalize at this point. We cannot say whether, across the history of metazoans, complexity change among lineages is biased upward.

VI. IF SO, THE PRINCIPLE IS SUPPORTED

However, if such a bias could be documented, it would support the principle, and the principle would in turn explain the bias. Importantly, however, the principle would not explain any particular case of increase, only the on-average bias. Each instance of complexity increase would require its own unique explanation. Notice too that a bias does not rule out selection. Selection might reinforce the internal-variance vector, on average, perhaps for the standard reason that greater complexity means greater division of labor. To detect this, we would have to find some way to estimate the magnitude of the internal-variance vector.

VII. IF NOT, WHY NOT?

The vector could be opposed by a selective vector of equal magnitude. How might selection do this? One possibility is that variation might be limited by selection against parts that are phenotypically extreme in the dimension of interest, that is, that selection might impose boundaries or limits. A wing that is too large or too small to function properly will be selected against. The implication would be that model one needs to be modified, because in real lineages the spread of points is bounded on both sides. The result would be that internal variance would stop increasing when points become maximally dispersed between the two limits. Notice that, for this mechanism to limit internal variance in real organisms, limits must exist and maximal dispersion must be reached in *all* dimensions. If dispersion can continue in even one dimension, then internal variance should increase. The suggestion that selection imposes limits in all dimensions is plausible (perhaps inescapable). But the idea that dispersion between limits is maximal in *all* dimensions is less plausible. It would have to be the case that in organisms the distribution of parts in every dimension fills the entire functionally permissible range.

Another possibility is that selection might favor losses of specialized parts just often enough to offset the increase in internal variance occurring among the remainder. For example, losses of parts might occur as a side effect of large reductions in body size, perhaps such as those associated with parasitism. For this to work to limit internal variance, losses would have to outnumber gains, i.e., the gains associated with body size increases (Bonner, 1988). Importantly, notice too that, for part losses to reduce internal variance they must occur preferentially among the more specialized parts. That is, lost parts must be drawn from the phenotypic extremes. Random removal of parts would not tend to decrease the standard deviation, on average.

Finally, it could be that ecology is limiting, that ecological opportunities for origination are about the same for simple and complex species. In other words, the generative bias that produces more increases in internal variance is negated by a selective filter that allows passage to an equal number of both, on average. (For this to work, the filter must allow increases and decreases in equal numbers, not equal proportions, which would be equivalent to selective neutrality.) It is easy to imagine many reasons why reductions in complexity might be generally advantageous. Obvious instances include the reductions associated with the evolution of parasitism or the evolution of the so-called interstitial fauna or meiofauna. However, the range of opportunities for reduction is really much broader. For example, consider the evolution of the mammalian vertebral column in the transitions from terrestrial to aquatic living. Terrestrial ancestors typically had complex, well-differentiated columns suitable for quadrupedal locomotion on land (e.g., with special localized modifications of the column for attachment of the hindlegs and for support of the neck), but their aquatic descendants tended to have simpler, more uniform, fishlike columns, suitable for undulatory propulsion in the water (McShea, 1991). Selection favored simplicity, not for any reason having to do with simplicity itself, or even with body size, but on account of an accidental association between simplicity and a particular functional mode. Such associations must be quite common.

The suggestion here is that, if no bias in complexity exists in the history of life, it could be that ecological conditions requiring decrease are just as common as those requiring increase. Would this contradict the internal-variance principle? The answer is no, because it could still be the case that most variations arising are increases. That is, decreases in internal variance might arise only rarely, but selection seizes on them disproportionately often, as required to meet functional demands.

A similar line of reasoning leads to the suggestion that extremely complex organisms might be more "delicate" in a developmental, functional, or ecological sense. This would be consistent with a long-standing but undemonstrated notion in paleobiology that structurally complex species are more prone to extinction than simple ones. Schopf *et al.* (1975) found just such a correlation between structural complexity and extinction rate. Interestingly, however, they interpreted it as an artifact, arguing that anagenetic change is more easily observed in complex morphologies, and therefore turnover rates for complex species only appear to be higher.

VIII. TESTING THE PRINCIPLE

The internal-variance principle predicts the existence of a vector tending to increase the complexity of any arbitrarily chosen set of organismal parts sharing some set of common dimensions. Unfortunately testing is not straightforward, because—as discussed—whether this vector manifests itself as a bias among lineages depends on the direction and magnitude of selection, which is unknown. However, some progress might be made by investigating changes in internal variance in special cases where the effect of selection should be minimal. For example, we might measure a number of dimensions in adult teeth in a tooth row and compare the standard deviations of this measure in a large sample of parent–offspring pairs. Because these teeth do not emerge and function until adulthood, they are arguably less subject than other structures to selection during development. Also, comparable measurements could be obtained from parents and offspring at the same life stage, ideally when the teeth are just erupting. The internal-variance principle predicts that the standard deviation for the offspring will be higher, on average. Of course, the test is imperfect in that selection is not eliminated entirely: Teeth do have a role in development and presumably are subject to selection during that time.

IX. A REVERSAL OF INTUITION

For many evolutionists, the apparent rise of adaptive complexity over the history of life is a source of wonder and inspiration. What is wonderful is that it happened, that organisms so exquisitely complex in their structure and apparently so magnificently engineered for survival and reproduction could have evolved. What is inspiring is the promise held out by evolutionary thought since Darwin that we might understand how and why it happened. However, adaptive complexity is also a problem to be explained, to skeptical creationists and students, of course, but also at least occasionally to ourselves. The power of natural selection and the enormity of geological time may be demonstrably sufficient, but the intuition sometimes stubbornly resists. Is selection really sufficient to explain the highly differentiated, specialized, and functional cell and tissue structure of, say, a modern arthropod?

The suggestion here is that it does not need to, that in posing the question this way we have unnecessarily burdened selection with one too many missions. To see this, we need make a clean conceptual separation between complexity and function. Having done so, it emerges that, although selection does need to solve the problem of adaptation, it does not need to solve the problem of complexity. Complexity is internal, variance and internal variance increases spontaneously in evolution, with the simple accumulation of variations. If arthropods increased in complexity over time, perhaps it was not because complexity was especially advantageous, but because most of the variational options available to selection were increasingly complex. And from such a range of increasingly complex variational options, selection found the few among them that were functional, that worked. In conclusion we might say—turning the common intuition halfway inside out—that selection produced adaptation but was stuck with complexity.

REFERENCES

Bell, G., and Mooers, A. O. (1997). Size and complexity among multicellular organisms. *Biological Journal of the Linnean Society* **60**, 345–363.

Bonner, J. T. (1988). *The Evolution of Complexity.* Princeton, NJ: Princeton University.

Brooks, D. R., and Wiley, E. O. (1988). *Evolution as Entropy: Toward a Unified Theory of Biology.* Chicago: University of Chicago Press.

Carroll, S. B. (2001). Chance and necessity: The evolution of morphological complexity and diversity. *Nature* **409**, 1102–1109.

Cisne, J. L. (1974). Evolution of the world fauna of aquatic free-living arthropods. *Evolution* **28**, 337–366.

Gould, S. J. (1996). *Full House: The Spread of Excellence from Plato to Darwin.* New York: Harmony.

Maynard Smith, J. (1970). Time in the evolutionary process. *Studium Generale* **23**, 266–272.

McShea, D. W. (1991). Complexity and evolution: What everybody knows. *Biology and Philosophy* **6**, 303–324.

McShea, D. W. (1993). Evolutionary change in the morphological complexity of the mammalian vertebral column. *Evolution* **47**, 730–740.

McShea, D. W. (1994). Mechanisms of large-scale trends. *Evolution* **48**, 1747–1763.

McShea, D. W. (1996). Metazoan complexity and evolution: Is there a trend? *Evolution* **50**, 477–492.

McShea, D. W. (2001). The hierarchical structure of organisms: A scale and documentation of a trend in the maximum. *Paleobiology* **27**, 405–423.

Pettersson, M. (1996). *Complexity and Evolution.* Cambridge, England: Cambridge University Press.

Raup, D. M. (1977). Stochastic models in evolutionary palaeontology. In *Patterns of Evolution as Illustrated by the Fossil Record* (Hallam, A., ed), pp. 59–78. Amsterdam: Elsevier.

Salthe, S. N. (1985). *Evolving Hierarchical Systems.* New York: Columbia University.

Saunders, P. T., and Ho, M. W. (1976). On the increase in complexity in evolution. *Journal of Theoretical Biology* **63**, 375–384.

Saunders, W. B., Work, D. M., and Nikolaeva, S. V. (1999). Evolution of complexity in Paleozoic ammonoid sutures. *Science* **286**, 760–763.

Schopf, T. J. M., Raup, D. M., Gould, S. J., and Simberloff, D. S. (1975). Genomic versus morphologic rates of evolution: Influence of morphologic complexity. *Paleobiology* **1**, 63–70.

Sidor, C. A. (2001). Simplification as a trend in synapsid cranial evolution. *Evolution* **55**, 1419–1442.

Simpson, G. G. (1967). *The Meaning of Evolution.* New Haven, CT: Yale University Press.

Spencer, H. (1900). *The Principles of Biology.* New York: Appleton.

Spencer, H. (1904). *First Principles.* New York: Hill.

Valentine, J. W., Collins, A. G., and Meyer, C. P. (1994). Morphological complexity increase in metazoans. *Paleobiology* **20**, 131–142.

Wagner, G. P., and Altenberg, L. (1996). Complex adaptations and evolution of evolvability. *Evolution* **50**, 967–976.

Wicken, J. S. (1987). *Evolution, Thermodynamics, and Information.* New York: Oxford University Press.

CHAPTER 19

Variation and Versatility in Macroevolution

V. Louise Roth
Biology Department, Duke University, Durham, North Carolina, USA

I. PRINCIPLES

A. To Vary Is Easy

Evolution is manifest as both themes and variations, with components of both continuity and change: Natural selection acting upon *heritable variation* may produce evolution, which is described as *descent* with *modification*. Thus processes that on the one hand insure faithful replication, which is the source of continuity, inheritance, and what we infer to be homology, are on the other compromised by imprecision and errors in the copying process, which introduce variability. Throughout the history of life, the dynamic tension between these antagonistic processes—maintenance of fidelity and the introduction of novelty—has been creative, because evolution requires both.

The ability to self-replicate is a defining feature of living things. The processes of copying take place at multiple levels in the hierarchy of biological organization, including nucleic acids, cells, organisms, populations, and lineages (Roth, 2001); but variation, as the source of diversity, is what permitted multiple levels to arise in the first instance.

Between the two kinds of processes, one set assuring fidelity of replication and another producing change, it is the maintenance of faithful replication that is the more challenging to explain and is mechanistically more complex: There is but a single way for a copy to be correct, but many ways for it to be flawed. When the rate at which deleterious errors accumulate exceeds the rate of faithful copying, the result is death, or at the population level, extinction. As a consequence, the size of the simplest replicating systems is limited by an "error threshold" (Eigen, 1971), but as Maynard Smith and Szathmáry (1995) have shown, replicating systems of greater size may be accommodated if replicators are interdependent and compartmentalized.

At any level of organization, from molecule to organism, the process of replication is contingent on interaction, under precise temporal and spatial conditions, of multiple entities—be they, in the case of a nucleic acid, polymerases and a supply of nucleotides interacting with the template of the original; or in the case of cells and organisms, the full array of still-to-be-fully explicated processes required for metazoan development, such as cell-cell signaling and the temporal and spatial regulation of gene expression, cell cycle, cell shape, and cell movement, as well as macroscopic manifestations of these processes, such as morphogenetic movements and induction. By contrast, a disruption can be simple. The existence of variation, from one perspective, is therefore not surprising. In a macroevolutionary context, perhaps the answer to the question G. E. Hutchinson (1959) posed in an ecological one, "why are there so many kinds of animals?" is "why not?"

B. Evolvability and Versatility

What begs explanation, of course, is not the ability of living systems to produce variation, but rather their ability to produce variation that is consistent with persistence—the survival and replication of organisms and lineages. "The genome's ability to produce adaptive variants when acted upon by the genetic system" (Wagner and Altenberg, 1996), "an organism's capacity to generate heritable phenotypic variation" [stated also as "the capacity of a lineage to generate heritable, selectable phenotypic variation"] (Kirschner and Gerhart, 1998), or the "intrinsic capacity of an organism for evolutionary change" (Yang, 2001) have been termed "evolvability."

The emphasis in these definitions is on the kind and quantity of variation produced, an emphasis that reflects a shift initiated a quarter century ago toward consideration of "developmental constraints" (Gould and Lewontin, 1979; Maynard Smith *et al.*, 1985). As the study of adaptation, the fit between phenotypes and their environment, became tempered with recognition that the phenotypes available for selection to act upon are limited (Rose and Lauder, 1996), a growing appreciation for the origin of variation served as a counterpoise to the emphasis that in previous decades had seemed focused exclusively on its fate. Evolution is a product of the processes through which variation is produced, as well as of those through which it is selectively preserved.

For example, Nijhout *et al.* (2003) distinguished two components of evolvability: (1) the availability of genetic variation and (2) the ability of a developmental-genetic mechanism to generate potentially adaptive phenotypic variation. Evolution requires genetic variability of a sort that results in phenotypic variability. Implicit in this and other formulations of evolvability, but not commonly examined in that context, a third level, beyond the ability to *generate* phenotypic variation, must be distinguished: the potential of such variability to be *adaptive*. If it ultimately is to have macroevolutionary consequences, evolvability must be both (1) "a function of the way genetic variation interacts with the molecular and developmental mechanisms that produce the phenotype" (Nijhout *et al.*, 2003, p. 292) and (2) a function of how well those phenotypes function in their environments.

The example analyzed by Nijhout *et al.* was a mathematical model of the mitogen-activated protein kinase (MAPK) signal transduction cascade. The MAPK system is a highly conserved pathway that mediates transmission of signals from the surface of the cell to the cytoplasm and nucleus and is involved in regulating a wide range of cellular and developmental processes. In the simulation model Nijhout *et al.* (2003) developed, parameters were varied within a set of differential equations representing synthesis, activation, and deactivation of cell-surface receptors; a cascade of phosphorylation/dephosphorylation reactions; and negative feedback of the cascade from its final product. Decades ago, much in the way these authors examined the phenotypic consequences of varying the rate parameters of their molecular model, Vermeij (1974) had commented upon the mathematical representation of coiled shells (Raup, 1966). He observed that the potential "versatility" of form of a higher taxon or body plan "depends on the number and range of independent morphogenetic parameters." (It also depends importantly on the nature of morphogenetic interactions—see the following text.) The basic form produced in shells by mollusks or brachiopods—accretionary growth at an aperture of expanding diameter that traces through space any one of a family of curves—is geometrically versatile, as the curves vary from the linear trajectory that produces a cone, to a spiral producing a flat coil, to a helix that produces a spired coil. They are also mechanically versatile, as the resulting shapes vary in strength of construction, gravitational stability, intimacy and area of contact permitted between the aperture and the substrate, streamlining, etc. (Vermeij, 1974). All of these traits are potentially important for survival, and all are important in different ways for animals of different habits (predation, suspension feeding, etc.)

living in different environments (buried in soft substrates, clinging to hard ones, exposed to strong wave action, etc.).*

If we consider the concept of evolvability to have three layers—as identified in the preceding text, dependence on (1) genetic variability that produces (2) phenotypic variation that is preserved by (3) natural selection—the examples of Nijhout *et al.* and Vermeij cover different ground and meet in the middle. The beauty of the MAPK example is that (1) the parameters in this model of physiological processes at the cellular level correspond in a plausible way to traits affected by variation or allelic substitution in individual genes. (2) Some combinations of parameter values yield (phenotypic) levels of the terminal kinase in the cascade that are relatively insensitive to variation in the individual parameters (which represent rate constants in the pathway) and thereby exhibit robustness; yet changes in other regions of parameter space show more phenotypic effects and therefore greater potential for evolvability. Differences in the pathway have been observed between species, suggesting that the cascade has evolved (Riley *et al.*, 2003). However, variation in the MAPK pathway has not been linked to (3) large-scale macroevolutionary change or functional differences between lineages that are phylogenetically deeply divergent.

The beauty of Vermeij's shell example, at the other end, is the clear association between (1) major differences in phenotype and (2) their functional consequences. However, although phenotypic differences resulting from evolutionary divergence on this scale are doubtless associated with (3) changes in genotype, the link between developmental mechanisms at the genetic level and shape variation in mollusks or brachiopods has not been well characterized. Previously proposed models of shell coiling (Thompson, 1942; Raup, 1966), though geometrically elegant, were not good models of the growth process that produces them (Ackerly, 1990).

Why are studies of evolvability that explicitly link all three of these levels uncommon? The goal of modeling morphogenesis in a way that mirrors mechanism, rather than phenomenology—clarifying the "genotype-phenotype map" (Wagner and Altenberg, 1996)—is a longstanding challenge. Even the metaphor of a map, which brings to mind a mathematical function (Lipschutz and Lipson, 1997) pairing each or even several genotypes with a single phenotype, is problematic because the "mapping" is actually many-to-many; although similar

*"Versatility" has several definitions and has been used in varying ways. Linking it to the number and range of morphogenetic parameters (Vermeij, 1973, 1974) follows the sense of "versatile" as "1: changing or fluctuating readily: variable" (Webster, 1976). Vermeij (1970) also discussed "adaptive versatility," but used "adaptive" in the physiological sense of plastic, or readily adjusting during the ontogeny of an individual. Of the three levels or layers of evolvability I identify here, Vermeij's usage of "versatility" makes reference most directly to the lower two—showing genetic or developmental variation that contributes to phenotypic variation. However "versatile" can also mean "4: having many uses or applications" (Webster, 1976), as when Liem (1984) contrasted the high "functional versatility" of teleost jaws with lower versatility within terrestrial vertebrate species, which is reflected in their narrower trophic specializations.

phenotypes may be manifestations of a variety of different genotypes, a single genotype may also underlie a variety of different phenotypes (Müller, 1990; Schlichting and Pigliucci, 1998; West-Eberhart, 2003). It is work enough to clarify the effects of the activity of one locus on another. It is a larger job to combine these into a network of (often nonlinear) interactions extending to an entire genome. Add environmental variables to the analysis and the input of a variable environment, and any goal of "computing the organism" (Lewontin, 2000) becomes still more remote. Extending the problem of the genotype-phenotype "map" to one of genotype-phenotype-adaptive landscape is a complex task of even higher order (Kauffman, 1993; Gavrilets, 1997), particularly where the study of variation is extended into the realm of macroevolution.

The question I would like to address in this chapter is whether any general statement can be made about the types of characters or developmental architecture that in a macroevolutionary context are evolutionarily versatile. But that task is too big to be accomplished definitively here. Important ontological aspects of the problem remain amorphous and need to be more clearly delineated: What are the relevant entities and categories? By what criteria should characters be defined or grouped, and what is the appropriate language (mathematical or otherwise) for representing their relationships (Stadler *et al.*, 2001; Wagner and Stadler, 2003)? How should ecological and functional attributes be arranged into an adaptive landscape, and how can such a landscape be described or measured?

What ultimately will be needed is a conceptual and methodological framework as well as empirical case studies. Elements of the problem have been broached successfully by many workers in various systems. For example, macroevolutionary variation in lepidopteran wing patterns has been analyzed in terms of the distribution and patterning of pigments (Nijhout, 1991). It has been dissected to the level of signaling pathways (Keys *et al.*, 1999) and examined morphometrically for covariance structure in a quantitative genetic framework (Paulsen, 1996). Approaching the question from the other direction in other animals, many productive programs of research have yielded insights into the macroevolutionary diversification of functional complexes such as those used in feeding and locomotion in vertebrates (e.g., Wake and Roth, 1988), and they have done so by characterizing the types of changes in structural relationships and form that have functional consequences. In these studies concepts such as developmental, morphological, or functional complexity, redundancy, diversity, and integration have been made operational and compared within particular clades and between sister taxa, allowing tests to proceed on a case by case basis (e.g., Lauder, 1981; Schaefer and Lauder, 1996; Alfaro *et al.*, 2004).

However, generalizing among examples requires a level of abstraction that makes formulating hypotheses difficult—"it may be possible to describe all squirrel species by a combination of a character states of the set of 'squirrel characters,' but there is no such set of characters which would describe the phenotypic disparity of all metazoans" (Wagner and Stadler, 2003, p. 509)—so I set here a more modest goal. In an attempt to give shape to the discussion, I will focus on the third "layer" of evolvability identified in the preceding text—phenotypic variation that has functional consequences—and treat two examples.

One example is a character complex. Mammalian teeth are a system of evolving characters that exhibits phenotypic differences that are functionally and ecologically meaningful and also for which, at multiple levels of organization and for each of the three layers of evolvability identified in the preceding text, an understanding of underlying variational properties during development is growing. The other example, the Sciuridae, is a taxonomic group and clade whose diversification and current diversity appear to have been enhanced not by variation in a single trait, but rather by a combination of phenotypic attributes whose variability has been expressed on different macroevolutionary time scales.

II. EXAMPLES

A. Elephantid Teeth

Mammalian dentitions are noteworthy in their versatility: Teeth of different taxa (often to the level of species) are morphologically distinctive, their morphological differences have well-understood functional consequences, and morphological variants are the product of fairly simple changes in the sizes, numbers, and arrangement of a modular unit, the cusp. Modularity and reiteration are organizational features of teeth at multiple structural levels. Like many organs in the vertebrate body (kidney, limb bud, feathers, and other cutaneous structures), teeth are products of epithelial–mesenchymal interactions during development (Gilbert, 1997). Within a tooth bud, reiteration of the enamel knot—a center of nonproliferating epithelial cells and signaling—creates an array of cusps in the growing epithelium (Jernvall, 1995). The signaling molecules involved are also used in many other developing tissues, and they combine in their interaction to form modules that are used reiteratively at different stages of development of the tooth, from its initiation to the formation of cusps (Jernvall and Thesleff, 2000). The tooth itself is a reiterated unit, repeated along the tooth row and varying in morphology according to serial position. Phylogenetically, teeth may have originated as variations on a theme reiterated over the entire oral, pharyngeal, or body surfaces as denticles or scales of ancestral fishes (Smith, 2003). Thus modularity, duplication, divergence, and ultimately individuation of morphologically distinct units are repeating themes in the development and evolution of dental diversity in mammals.

Salazar-Ciudad and Jernvall (2004; see also Salazar-Ciudad *et al.*, 2003) have proposed that complex and widely varying patterns of morphology can be produced without a corresponding increase in the number of developmental instructions ("genetic information" or "molecular variation") if the developmental mechanisms are "morphodynamic" rather than "morphostatic." In morphostatic mechanisms, morphogenesis and growth of a structure occur subsequent to inductive interactions that establish a prepattern. In such systems, the reiteration of a structure requires elements to be added. By contrast, in morphodynamic mechanisms like those that shape the enamel surface of a developing tooth, induction, growth, and morphogenesis occur concurrently and therefore interact. In teeth, cusps arise through differential growth of the enamel epithelium. Cells of the enamel knot, located at the tip of a developing cusp, do not themselves proliferate, but serve as a signaling center that stimulates growth around them; additional, secondary enamel knots form in areas beyond the range of inhibition. Timing and spatial relationships are key. Because growth alters the context in which induction occurs and induction can affect growth, new features (e.g., additional cusps) can arise relatively simply through quantitative shifts in the values of developmental parameters characterizing the reciprocal interactions of morphology and gene expression (Salazar-Ciudad and Jernvall, 2002). For developing teeth, many of the relevant signaling molecules are known (Jernvall and Thesleff, 2000), the cell- and tissue-level interactions have been modeled effectively (Jernvall, 2000; Salazar-Ciudad and Jernvall, 2002), and functions of different morphologies are well studied (e.g., Hillson, 1986; Janis and Fortelius, 1988).

The versatility of the cusp-generating mechanism in teeth is evident from the evolution of teeth of disparate shape and function, ranging from the single cusp of the canine that a shrew uses to puncture insect cuticle to the enormous grinding apparati of megaherbivores (Jernvall, 1995). The world's largest teeth are produced by members of the elephant and mammoth family, Elephantidae. In the cheek teeth of elephantids, the reiterated unit of the cusp is arrayed transversely across the tooth and fused into lamellae that themselves are reiterated along the tooth's longitudinal axis (Home, 1799; Rose, 1894; Bolk, 1919). A single tooth of an elephant can exceed 8 kg in mass and be voluminous enough to occupy an entire quadrant of the jaw (Roth and Shoshani, 1988). The structure of each tooth appears to have been produced by an enamel epithelium thrown into tall, broad, vertical invaginations, which form the lamellae. Dentin fills each lamella, surrounding and obliterating the part of the pulp cavity that extends into each from the base of the tooth (Figure 19-1). Surrounding the outer surface of the tooth, binding it together, and filling the valleys between lamellae is cementum. Thus the tooth of an elephant is constructed of the same structural (cusps) and material (enamel, dentin, cementum) elements in the same topological relationship as any other mammalian tooth, but their quantity and arrangement produce a durable array of transverse shearing blades on the chewing surface throughout the functional life of the tooth. As they are sectioned horizontally with wear in the occlusal plane, lamellae

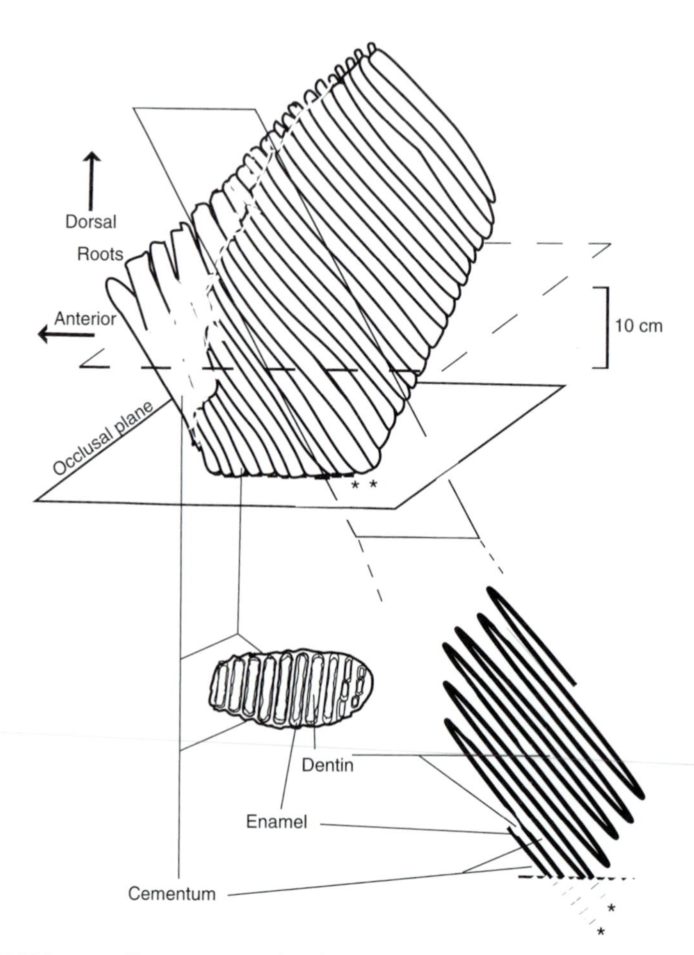

FIGURE 19-1. Partially worn upper molar of mammoth, in lateral view (above), with occlusal view of wear surface shown below. At right is a schematic illustration of the enamel of six adjacent lamellae as it would appear in a vertical longitudinal section of the tooth. Lamellae are broad elongate pouches of enamel surrounded by cementum and filled with dentin; regions filled with these tissues are labeled. Tooth wear from mastication sections the tooth in an approximately horizontal plane. As wear continues, the occlusal plane advances toward the roots (dashed plane), the tooth continues to erupt and move forward, and lamellae are lost from the front of the tooth. Asterisks indicate the tips of two lamellae removed by wear. The tooth as shown here has been removed from the bone of the jaw, which otherwise surrounds it to within 1 or 2 cm of the occlusal surface. Both the tooth and the jaw that holds it curve in three dimensions.

are exposed as transverse, longitudinally compressed loops of enamel that form horizontal blades that alternate with softer dentin or cementum—a rugged surface effective for shearing the animals' diet of vegetation. The large height and volume of the tooth render it durable despite the wear produced by large quantities of abrasive grit ingested with the food.

The teeth of the set that constitutes a tooth row also maintain the same spatial relationships to one another as in any other mammals. Six cheek teeth corresponding to adult second, third, and fourth deciduous premolars (dp) and molars (M) #1 through #3 are positioned in series from anterior to posterior in each jaw quadrant, except that the enormous size of each tooth precludes their erupting, or even forming, in the jaw together. Instead, the anterior-to-posterior gradient in development of adjacent teeth that is characteristic of most mammalian neonates is exaggerated, and the timing of eruption (in sequence from dp2 through M3) is retarded to the point that it spans most of the animal's life. Attrition through wear is also manifest in an anterior-to-posterior gradient: At the front of the jaw, as dental roots are resorbed; as worn lamellae erode, fragment, and are shed; and as wear removes material from the occlusal surface of an erupted tooth, the remainder of the tooth moves forward, new dental material erupts into place, and new teeth move in from behind. The gradual horizontal motion of teeth through the jaw in elephants and the replacement of each cheek tooth from behind by an adjacent, more posterior one is referred to as progression. In elephants, dental development continues throughout many decades of the life of an individual, and a single tooth may at its anterior end be fully formed, in use, and partially worn when its more posterior lamellae are still enclosed within the bone of the jaw and not yet fully formed or mineralized.

Delay in the mineralization of the posterior lamellae enhances the morphological plasticity of the structure before it hardens, allowing the tooth to mold to the shape of the growing jaw in which it is erupting and to be subjected to the forces of mastication before it has fully formed. It is difficult to imagine a morphostatic mechanism (in which a prepattern for the dentition or for even an entire single tooth is established in a separate stage preceding morphogenesis and growth) that could produce this pattern of dental development and progression. Whereas growth of most mammals is complete within a year or so and morphogenesis of teeth is confined to a brief period very early in ontogeny, growth and dental ontogeny in elephants continue for decades. (Even in "ever-growing" teeth like rodents' incisors, elephants' tusks, or the molars of some ungulates, the trajectory of growth and eruption is much simpler than that for elephants' cheek teeth; Hillson, 1986.) Throughout the prolonged process, the morphodynamic character of dental development permits coordination of the formation, eruption, and progression of a succession of enormous, hardened, and structurally complex objects with one another and with the changing size and shape of a growing jaw.

One consequence is that the size and shape of elephant teeth vary intraspecifically to a greater degree than do the teeth of other mammals (Roth, 1992). Another is that coordination is not always perfect, and abscesses or other impediments to progression, malocclusion, and nutritional stress can produce teeth that are mechanically distorted and morphologically malformed (Roth, 1989).

Perhaps more striking than occasional anomalies in the teeth, however, is their versatility. The body mass of the largest elephantids approaches 10 metric tons; both neonates of large species and adult dwarfed forms that evolved rapidly in the Pleistocene on various islands free of mammalian predators could weigh 100 kg or less (Roth, 1990). Elephantid teeth have been versatile in their ability to accommodate, in both development and function, the increases of body mass in ontogeny and rapid phylogenetic decreases, which each span two orders of magnitude.

B. Disparity and Versatility in Sciuridae

By any intuitive measure of evolutionary success, squirrels have succeeded: The family has endured through to the present from over 30 million years ago (Emry and Thorington, 1984); it has diversified into close to 300 extant species (Hoffmann *et al.*, 1993); it has spread to occupy all continents except Australia and Antarctica; and some species are abundant (e.g., Manski *et al.*, 1981). The most ecologically salient axes of morphological variation for sciurids, as for any animal taxon, are indices of diet, locomotor habits, and body size (Hafner, 1984, Van Valkenburgh, 1994). In squirrels, as I will describe in the following text, each of these axes shows a different pattern and has varied on a different temporal scale: Major locomotor innovations have occurred infrequently and tend to characterize major clades. Body size has been evolutionarily labile, shifting frequently among closely related species and genera, and versatility of diet is a feature of individual squirrels.

Within the Sciuridae there are three broad categories of locomotor types: tree squirrels, which tend to live in forests and nest in trees or fallen logs; ground squirrels, which inhabit more open habitats and nest in burrows; and the inaccurately named "flying" squirrels, which unlike the other two forms typically are nocturnal and capable of gliding between trees using a patagium stretching from wrist to ankle as an airfoil. Variation exists within and to some extent blurs differences between these categories: Flying squirrels are otherwise arboreal. Among arboreal squirrels, foraging may be stratified, with some species spending time high in the canopy and others foraging mainly on the ground (Emmons, 1980). Although they are fossorial, chipmunks have been viewed as intermediate between tree and ground squirrels (Black, 1963) morphologically and because they inhabit woodlands and have modest climbing abilities. However, distinctions among the three main types can be cleanly drawn, are widely recognized, and correlate with anatomical differences. Flying squirrels bear a cartilaginous support for the gliding membrane at the wrist (Thorington, 1983), and limb lengths, digital proportions, and articular orientations distinguish tree from ground squirrels (Emry and Thorington, 1982).

Evolutionary transitions between the locomotor categories have been few: The anatomy of the earliest known fossil squirrel suggests arboreality (Emry and Thorington, 1982), and all modern genera of flying squirrels arose from a single common ancestor not shared with other forms (Thorington, 1983; Mercer and Roth, 2003). Fossoriality arose within just one of five major lineages of squirrels (clade IV, Figure 19-2). Within this clade the relationships among genera are not fully resolved, but they are consistent with the possibility either of several separate origins of fossoriality within the clade or of a single origin at the base of the clade followed by a single resumption of arboreality by the common ancestor of African tree squirrels. (Similar conclusions were drawn by Steppan *et al.*, 2004, using a partial sampling of genera.)

By contrast, body size among genera and species of squirrels is labile. The contrast is evident in the patterns of shading for the columns representing locomotion and size in Figure 19-2: Whereas locomotor categories characterize large clades and are represented opposite the tree by long strips of a single shade, the column representing body size is fragmented into a mosaic of variation despite the fact that the boundaries of size categories were originally chosen to minimize the number of borderline taxa. The five genera of pygmy squirrels (which are those with head + body lengths less than 11 cm) are distributed among four of the five major lineages. Likewise, the six giant genera (whose head + body lengths exceed 31 cm) are segregated among three major lineages, including several distinct subclades of clade V. A parsimony mapping (Maddison and Maddison, 2002) of size categories onto the tree (pruned to one branch per genus, rooting the tree with the earliest fossil squirrel, rather than *Aplodontia*, and treating size as an ordered character) yields 24 steps out of a possible range of 3–26, a consistency index of 0.13, and a retention index of only 0.09 (c.f. Archie, 1996). I obtained the same number of steps (24) as an average for the distribution produced by repeatedly remapping size categories onto the tree after reshuffling the (taxon × size) matrix at random (Maddison and Maddison, 2004). Differences among species within genera are not treated here, but body-size variability is also evident at this level where it is responsible for the confused taxonomy of polyphyletic "*Microsciurus*" (J. M. Mercer and V. L. Roth, in preparation).

Although the homoplasy for body size is considerable, there is no indication of size having evolved particularly rapidly. Using divergence dates estimated from a well-calibrated molecular clock (Mercer and Roth, 2003), I estimated rates of evolution in darwins for pairs of terminal taxa whose members fell in different size categories. Values were at most 0.22 darwins (and that only if all change was assumed to occur on one branch of a pair and size on the other was assumed to be static), which corresponds to a change of 25% per million years, well within the range of values typically reported for time intervals of durations of a few million years (Gingerich, 1983). But whatever its rate of change, body size has its consequences.

Aplodontia
[Outgroup]
Funambulus
Funambulus
III
Exilisciurus
Exilisciurus
Rubrisciurus
Prosciurillus
Hyosciurus
Tamiops
97
Dremomys
94
Nannosciurus
48
Sundasciurus
96
Rhinosciurus
Menetes
85
Lariscus
95
Glyphotes
Callosciurus
Funisciurus
Paraxerus
91
Epixerus
54
Protoxerus
Heliosciurus
Myosciurus
56
Spermophilus
Marmota
99
Cynomys
61
Ammospermophilus
Tamias
Tamias
IV
Sciurotamias
Spermophilopsis
Xerus
Atlantoxerus
Aeretes
Trogopterus
Belomys
99
Pteromyscus
99
Aeromys
Eupetaurus
73
Pteromys
Petaurista
Eoglaucomys
Glaucomys
84
Hylopetes
Petaurillus
V
Petinomys
Iomys
Microsciurus
Syntheosciurus
97
Rheithrosciurus
87
Sciurus
Tamiasciurus
II
Ratufa
Ratufa
I
Sciurillus
Sciurillus
0.05 changes
Geography
South America
North America
Eurasia
Africa
Locomotion
Burrowing
Arboreal
Gliding
Size
Midpoint of range for head+body length
<11cm
11–18cm
18–13cm
>31cm

FIGURE 19-2. Phylogenetic relationships of modern genera of the Sciuridae based on Bayesian analysis of nuclear (IRBP) and mitochondrial (16s, 12s) DNA sequences (modified from Mercer and Roth, 2003). Posterior probabilities at all nodes are 100% unless indicated. (Where possible, to mitigate artifacts of long-branch attraction for genera with relatively long, basally emerging branches, two species were included in the analysis.) Basal radiation produced two monotypic lineages (I, II) and three major clades (III–V). In vertical strips to the right of the tree, two traits—locomotor mode and body size—are coded as shades of gray in blocks opposite each genus; geographic distribution is shown in the column between them. Body-size categories were defined to produce approximately even spacing on a logarithmic scale, and boundaries were set to minimize the number of borderline taxa. Note that *Spermophilus*, *Tamias*, and *Marmota* occur both in North America and Eurasia. The existence of *Microsciurus* in both North and South America and of *Sciurus* in North and South America and Eurasia is an artifact of polyphyly and paraphyly in these genera (Mercer and Roth, in preparation). Parsimony mapping of traits onto this topology yields the following total numbers of steps for each: locomotor category, 3 transitions (where $p = 0.001$ for 15 or fewer transitions in a distribution produced by remapping of random permutations of this data matrix; Maddison and Maddison, 2004); geography, 10 transitions among continental landmasses (plus additional transitions for species within each of the genera *Marmota*, *Spermophilus*, *Tamias*, and *Sciurus* + *Microsciurus*; $p = 0.05$ for 15 or fewer transitions and $p = 0.01$ for 14 or fewer); size, 24 category steps (see text). (Reprinted figure with permission from Mercer J., and Roth, V.L. (2003). The effects of Cenozoic Global Change on Squirrel Phylogeny, *Science* 299:1568–1572. Copyright (2003) AAAS.)

Body masses of squirrels span nearly three orders of magnitude (from 16.5 g *Myosciurus* to 7.5 kg *Marmota;* Nowack, 1991), and this should have implications for the many physiological, mechanical, and life-history variables that scale with size (Peters, 1983; Calder, 1984; Schmidt-Nielsen, 1984), as well as for locomotion and diet. For instance, small animals can traverse smaller branches, but because a muscle's force scales with its cross-sectional area, an animal with small jaws generally cannot apply as much force with its teeth as a larger individual.

Versatility of diet is a feature of individual squirrels. The masticatory apparatus of a squirrel can process foods that are soft (such as fungi, buds, small vertebrates and other animals) or hard (such as nuts, shelled fruits, and seeds). With their distinctive arrangement of jaw muscles that allows especially forceful application of the incisors in gnawing (Druzinsky, 1989), squirrels have access to food items such as nuts, seeds, insects, and exudates that are protected by tough hulls or hidden beneath layers of bark. As a consequence, many species are dietary generalists (Nowack, 1991). Some lineages of squirrels have produced dietary specialists. Long-nosed *Rhinosciurus* appears to be especially well adapted for apprehending insects (Nowack, 1991); the physiology and dentition of the giant woolly flying squirrel, *Eupetaurus,* permit it to subsist on pine needles (Zahler and Kahn, 2003). While the large size and robust jaws of *Rheithrosciurus* allow it to process nuts that are large and tough, the small size of the pygmy tree squirrels *Sciurillus, Exilisciurus, Nannosciurus,* and *Myosciurus* may limit their ability to deal with mechanically challenging foods and relegate them to diets of gleaning under bark (Emmons, 1980, and personal communication; personal observations). But more commonly, even where multiple species of squirrels coexist in a region or community, they may consume the same foods: Although sympatric squirrels may select them in different proportions, the differences tend to be quantitative rather than categorical (Emmons, 1980; Ball and Roth, 1995). Detailed comparisons of mandibular and masticatory muscle morphology have been made among species of squirrels (Bryant, 1945; Ball and Roth, 1995; Thorington and Darrow, 1999; Cardini, 2003), but not much of this variation has been associated with dietary differences. The jaws of squirrels are a tool whose versatility for single individuals (particularly for dietary generalists) is more impressive than are the differences in utilization among most species.

Thus a taxon composed of ecological generalists can show versatility of function even if it is morphologically conservative. In addition, versatility in the sense closer to that used by Vermeij (1974)—showing functionally and ecologically relevant *variation*—is most evident in the Sciuridae not in single traits, but when traits that manifest variability on different phylogenetic and temporal scales are combined. The combinatorial possibilities produced by two traits that differ in their patterns of variation are greater than those permitted by any single trait (or by two traits that covary) (Figure 19-2).

Importantly, this variation does not arise just in morphospace: It is also structured geographically. In multiple instances, the initiation of events of diversification in clades of squirrels have coincided in time and place with geological events that permitted the arrival of a lineage of squirrels on a new land mass (Mercer and Roth, 2003). Yet another pattern of variation (intermediate in variability between locomotion and size categories in Figure 19-2) is evident when geographic distribution (categorized by continent) and organismal traits are overlain (Figure 19-2). Geography provides the context in which versatility is repeatedly expressed and is a key factor in the mechanisms of diversification that produce it.

III. OVERVIEW AND CONCLUSION

Studies of macroevolutionary change are uniquely focused on events and processes that require time, including events that occur infrequently (or just once), or processes that are long in duration. With respect to phenotypic variation, macroevolution is typically the domain of large differences (whose study in aggregate becomes the study of disparity; e.g., Foote, 1997; Ciampaglio, 2002), of the origin of novelties (features that do not correspond to characters or structures present in an ancestor; Müller and Wagner, 1991), or of differences (of whatever magnitude or character) that are associated with taxonomic diversification.

For variation to be observed, it must not only have been produced but also permitted to persist. The expression and persistence of macroevolutionary variation in a trait is a manifestation of that trait's versatility. My goal here has been

(a) to draw attention to the "third component" of evolvability, which has to do with macroevolutionary production of variation that is ecologically and functionally relevant, and (b) to consider how two organismal systems on which I have worked might illustrate such versatility.

The examples I treated include in one case a structure—teeth in elephantids—whose morphology is extremely divergent from that in most other mammals and very different from their primitive state. Yet by virtue of their reiterated modular structure and dynamic mode of development, the enormous teeth of elephants have readily adjusted to developing and functioning within jaws of animals whose body sizes have undergone radical change. Does this observation allow a more general statement about attributes that confer macroevolutionary versatility?

The hypothesis that morphodynamic development confers versatility is suggested by simulations (Salazar-Ciudad and Jernvall, 2004) and remains to be rigorously tested. It must first be established for actual examples whether the quality of being morphostatic or morphodynamic is a difference of type, degree, or phases of development, and the distinction must then be applied to systems other than teeth. However, there are indications that versatility is conferred by reiteration and modularity. The tendency—now known as Williston's Rule—for duplicated structures such as limbs or segments of arthropods or leaflike appendages in plants to diverge morphologically, specialize, and reduce in number was recognized a century ago (Minkoff, 1983); the contribution of modular development to the evolvability of many lineages is now widely discussed (e.g., Raff, 1996; von Dassow and Munro, 1999; Yang, 2001); and the role of divergence of duplicated genes in genome evolution is regarded as fundamental (Ohno, 1970; Lynch and Force, 2000).

It is both revealing and significant that the terminology applied to discussions of the evolution of molecular and cellular interactions during development (levels 1 and 2), on the one hand, and the organization and functional relationships of morphological structures (levels 2 and 3), on the other, is similar: Both realms make use of the concepts of modularity, duplication, decoupling, and integration (c.f. Lauder *et al.*, 1984; Gerhart and Kirschner, 1997). Modular sets of phenotypic traits may be delineated by functional, developmental, or genetic correlations. Traits delineated in these ways sometimes do and sometimes do not coincide (e.g., Olson and Miller, 1958; Zelditch *et al.*, 1990; Cheverud, 1996). However, for macroevolutionary explanation, all levels are relevant.

Versatility may be attained by other routes (c.f. Raup, 1966). The other example explored here, the Sciuridae, is a clade of morphologically conservative mammals (Emry and Thorington, 1984) that, like the Elephantidae, have proliferated across several orders of magnitude in body mass but have persisted for longer (36 my vs. 4–5 my). Though their variability in some dimensions of morphospace is comparatively modest, sciurids exhibit versatility in the wide geographic deployment of a *combination* of functionally and ecologically important attributes that vary on different temporal scales. It remains to be seen whether these contrasts in the patterns of variation among the major axes of functional morphospace are a feature common to other clades of rodents, mammals, or organisms and whether their presence is generally associated with comparatively high levels of diversification, persistence, or ecological disparity. In the interim, in considering variation and versatility in macroevolutionary context, it is worth remembering Van Valen's (1976, p. 180) appealingly terse and compelling assertion, "evolution is the control of development by ecology,"—modulating it with recognition that influences among the processes of evolution, ecology, and development are not only pervasive but also mutual.

ACKNOWLEDGMENTS

Thanks to Dan McShea, John Mercer, and Fred Nijhout for comments on the manuscript and discussion; to Ted Garland, Daniel Henk, and Wendy Hodges for initiation with Mesquite; and to NSF (DEB 791482 and 9726855) for support of research.

REFERENCES

Ackerly, S. C. (1990). Using growth functions to identify homologous landmarks on mollusc shells. *Special Publications of the University of Michigan Museum of Zoology* **2**, 339–344.

Alfaro, M. E., Bolnick, D. I., and Wainwright, P. C. (2004). Evolutionary dynamics of complex biomechanical systems: An example using the four-bar mechanism. *Evolution* **58**, 495–503.

Archie, J. W. (1996). Measures of homoplasy. In *The Character Concept in Evolutionary Biology* (M. J. Sanderson and L. Hufford, eds.), pp. 153–188. San Diego, CA: Academic Press.

Ball, S. S., and Roth, V. L. (1995). Jaw muscles of New World squirrels. *Journal of Morphology* **224**, 265–291.

Black, C. G. (1963). A review of the North American tertiary Sciuridae. *Bulletin of the Museum of Comparative Zoology (Harvard)* **130**, 100–248.

Bolk, L. (1919). Over de ontwikkeling van het gebit van *Elephas africanus. Amsterdam Verhandelingen der Koninklijke Akademie van Wetenschappen Afdeeling Natuurkunde* **27**, 1056–1070.

Bryant, M. D. (1945). Phylogeny of Nearctic Sciuridae. *American Midland Naturalist* **33**, 257–390.

Calder, W. A. (1984). *Size, Function, and Life History.* Cambridge, MA: Harvard University Press.

Cardini, A. (2003). The geometry of the marmot (Rodentia: Sciuridae) mandible: Phylogeny and patterns of morphological evolution. *Systematic Biology* **52**, 186–205.

Cheverud, J. M. (1996). Developmental integration and the evolution of pleiotropy. *American Zoology* **36**, 44–50.

Ciampaglio, C. N. (2002). Determining the role that ecological and developmental constraints play in controlling disparity: Examples from the crinoid and blastozoan fossil record. *Evolution and Development* **4**, 170–188.

Druzinsky, R. E. (1989). Incisal biting in (*Aplodontia rufa*) and (*Marmota monax*). Ph.D. dissertation, University of Illinois, Chicago. Ann Arbor: University Microfilms International.

Eigen, M. (1971). Self-organization of matter and the evolution of biological macromolecules. *Die Naturwissenschaften* **58**, 465–523.

Emmons, L. H. (1980). Ecology and resource partitioning among nine species of African rain forest squirrels. *Ecological Monographs* **50**, 31–54.

Emry, R. J., and Thorington, R. W. (1982). Descriptive and comparative osteology of the oldest fossil squirrel, *Protosciurus* (Rodentia: Sciuridae). *Smithsonian Contributions to Paleobiology* 47, 1–35.

Emry, R. J., and Thorington, R. W. (1984). The tree squirrel *Sciurus* as a living fossil. In *Living Fossils* (N. Eldredge and S. Stanley, eds.), pp. 23–31. New York: Springer-Verlag.

Foote, M. (1997). The evolution of morphological diversity. *Annual Review of Ecological Systems* **28**, 129–152.

Gavrilets, S. (1997). Evolution and speciation on holey adaptive landscapes. *Trends in Ecology and Evolution* **12**, 307–312.

Gerhart, J., and Kirschner, M. (1997). *Cells, Embryos, and Evolution: Toward a Cellular and Developmental Understanding of Phenotypic Variation and Evolutionary Adaptability.* Malden, MA: Blackwell Science.

Gilbert, S. F. (1997). *Developmental Biology.* 5th ed. Sunderland, MA: Sinauer.

Gingerich, P. D. (1983). Rates of evolution: Effects of time and temporal scaling. *Science* **222**, 159–161.

Gould, S. J., and Lewontin, R. C. (1979). The spandrels of San Marco and the Panglossian paradigm: A critique of the adaptationist programme. *Proceedings of the Royal Society of London Series B, Biological Science* **205**, 581–598.

Hafner, D. J. (1984). Evolutionary relationships of the Sciuridae. In *The Biology of Ground-Dwelling Squirrels*. (J. A. Murie and G. R. Michener, eds.), pp. 3–23. Lincoln: University of Nebraska Press.

Hillson, S. (1986). *Teeth.* Cambridge, England: Cambridge University Press.

Hoffmann, R. S., Anderson, C. G., Thorington, R. W., and Heaney, L. R. (1993). Family Sciuridae. In *Mammal Species of the World.* 2nd ed. (D. E. Wilson and D. M. Reeder, eds.), pp. 419–465. Washington, DC: Smithsonian Institution Press.

Home, E. (1799). On the structure of the teeth of gramnivorous quadrupeds, particularly those of the elephant and *Sus aethiopicus*. *Philosophical Transactions of the Royal Society of London* **89**, 519–527.

Hutchinson, G. E. (1959). Homage to Santa Rosalia or why are there so many kinds of animals? *American Naturalist* **93**, 145–159.

Janis, C. M., and Fortelius, M. (1988). On the means whereby mammals achieve increased functional durability of their dentitions, with special reference to limiting factors. *Biological Reviews* **63**, 197–230.

Jernvall, J. (1995). Mammalian molar cusp patterns: Developmental mechanisms of diversity. *Acta Zoologica Fennica* **198**, 1–61.

Jernvall, J. (2000). Linking development with generation of novelty in mammalian teeth. *Proceedings of the National Academy of Sciences USA.* **97**, 2641–2645.

Jernvall, J., and Thesleff, I. (2000) Reiterative signaling and patterning during mammalian tooth morphogenesis. *Mechanisms of Development* **92**, 19–29.

Kauffman, S. A. (1993). *The Origins of Order: Self-Organization and Selection in Evolution.* New York: Oxford University Press.

Keys, D. N., Lewis, D. L., Selegue, J. E., Pearson, B. J., Goodrich, L. V., Johnson, R. L., Gates, J., Scott, M. P., and Carroll, S. B. (1999). Recruitment of a hedgehog regulatory circuit in butterfly eyespot evolution. *Science* **283**, 532–534.

Kirschner, M., and Gerhart, J. (1998). Evolvability. *Proceedings of the National Academy of Sciences USA* **95**, 8420–8427.

Lauder, G. V. (1981). Form and function: Structural analysis in evolutionary morphology. *Paleobiology* 7, 430–442.

Lauder, G. V., Crompton, A. W., Gans, C., Hanken, J., Liem, K. F., Maier, W. O., Meyer, A., Presley, R., Rieppel, O. C., Roth, G., Schluter, D., and Zweers, G. A. (1984). Group report: How are feeding systems integrated and how have evolutionary innovations been introduced? In *Complex Organismal Functions: Integration and Evolution in Vertebrates: Report of the Dahlem Workshop on Complex Organismal Functions—Integration and Evolution in Vertebrates* (D. B. Wake and G. Roth, eds.), pp. 97–115. Chichester, England: Wiley.

Lewontin, R. S. (2000). Computing the organism. *Natural History* **109**(3), 94.

Liem, K. F. (1984). Functional versatility, speciation, and niche overlap: Are fishes different? In *Trophic Interactions within Aquatic Ecosystems* (D. G. Meyers and J. R. Strickler, eds.), pp. 269–305. Boulder, CO: Westview Press.

Lipschutz, S., and Lipson, S. (1997). *Schaum's Outline of Theory and Problems of Discrete Mathematics.* 2nd ed. New York: McGraw-Hill.

Lynch, M., and Force, A. (2000) The probability of duplicate gene preservation by subfunctionalization. *Genetics* **154**, 459–473.

Maddison, W. P., and Maddison, D. R. (2002). *MacClade 4.* Version 4.05. Sinauer, Sunderland, MA.

Maddison, W. P., and Maddison, D. R. (2004). *Mesquite: A Modular System for Evolutionary Analysis.* Version 1.03 (http://mesquiteproject.org).

Manski, D. A., Van Druff, L. W., and Flyger, V. (1981). Activities of gray squirrels and people in a downtown Washington, D.C. park: Management implications. *Transactions of the North American Wildlife and Natural Resources Conference* **46**, 439–454.

Maynard Smith, J., and Szathmáry, E. (1995). *The Major Transitions in Evolution.* Oxford, England: W. H. Freeman.

Maynard Smith, J., Kauffman, S., Alberch, P., Campbell, J., Goodwin, B., Lande, R., Raup, D., and Wolpert, L. (1985). Developmental constraints and evolution. *Quarterly Review of Biology* **60**, 265–287.

Mercer, J. M., and Roth, V. L. (2003). The effects of Cenozoic global change on squirrel phylogeny. *Science* **299**, 1568–1572.

Minkoff, E. C. (1983). *Evolutionary Biology.* Reading, MA: Addison-Wesley.

Müller, G. B. (1990). Developmental mechanisms at the origin of morphological novelty: A side-effect hypothesis. In *Evolutionary Innovations* (M. H. Nitecki, ed.), pp. 99–130. Chicago: University of Chicago Press.

Müller, G. B., and Wagner, G. P. (1991). Novelty in evolution: Restructuring the concept. *Annual Review of Ecology and Systematics* **22**, 229–256.

Nijhout, H. F. (1991). *The Development and Evolution of Butterfly Wing Patterns.* Washington, DC: Smithsonian Institution Press.

Nijhout, H. F., Berg, A. M., and Gibson, W. T. (2003). A mechanistic study of evolvability using the mitogen-activated protein kinase cascade. *Evolution and Development* **5**, 281–294.

Nowack, R. M. (1991). *Walker's Mammals of the World.* 5th ed. Baltimore: Johns Hopkins University Press.

Ohno, S. (1970). *Evolution by Gene Duplication.* New York: Springer-Verlag.

Olson, E. C., and Miller, R. L. (1958). *Morphological Integration.* Chicago: University of Chicago Press.

Paulsen, S. M. (1996). Quantitative genetics of the wing color pattern in the buckeye butterfly (*Precis coenia* and *Precis evarete*): Evidence against the constancy of G. *Evolution* **50**, 1585–1597.

Peters, R. H. (1983). *The Ecological Implications of Body Size.* Cambridge, England: Cambridge University Press.

Raff, R. (1996). *The Shape of Life: Genes, Development, and the Evolution of Animal Form.* Chicago: University of Chicago Press.

Raup, D. M. (1966). Geometric analysis of shell coiling: General problems. *Journal of Paleontology* **40**, 1178–1190.

Riley, R. M., Jin, W., and Gibson, G. (2003). Contrasting selection pressures on components of the Ras-mediated signal transduction pathway in *Drosophila. Molecular Ecology* **12**, 1315–1323.

Rose, C. (1894). Über den Zahnbau und Zahnwechsel von *Elephas indicus. Morphologische Arbeiten* **3**, 173–194.

Rose, M. R., and Lauder, G. V. (1996). *Adaptation.* San Diego, CA: Academic Press, San Diego.

Roth, V. L., and Shoshani, J. (1988). Dental identification and age determination in *Elephas maximus. Journal of Zoology* **214**, 567–588.

Roth, V. L. (1989). Fabricational noise in elephant dentitions. *Paleobiology* **15**, 165–179.
Roth, V. L. (1990). Insular dwarf elephants: A case study in body mass estimation and ecological inference. In *Body Size in Mammalian Paleobiology: Estimation and Biological Implications* (J. Damuth and B. J. MacFadden, eds.), pp. 151–180. Cambridge, England: Cambridge University Press.
Roth, V. L. (1992). Quantitative variation in elephant dentitions: Implications for the delimitation of fossil species. *Paleobiology* **18**, 184–202.
Roth, V. L. (2001). Character replication. In *The Character Concept in Evolutionary Biology* (G. P. Wagner, ed.), pp. 81–107. San Diego, CA: Academic Press.
Salazar-Ciudad, I., and Jernvall, J. (2002). A gene network model accounting for development and evolution of mammalian teeth. *Proceedings of the National Academy of Sciences USA* **99**, 8116–8120.
Salazar-Ciudad, I., and Jernvall, J. (2004). How different types of pattern formation mechanisms affect the evolution of form and development. *Evolution and Development* **6**, 6–16.
Salazar-Ciudad, I., Jernvall, J., and Newman, S. A. (2003). Mechanisms of pattern formation in development and evolution. *Development* **130**, 2027–2037.
Schaefer, S. A., and Lauder, G. V. (1996). Testing historical hypotheses of morphological change: Biomechanical decoupling in loricarioid catfishes. *Evolution* **50**, 1661–1675.
Schlichting, C. D., and Pigliucci, M. (1998). *Phenotypic Evolution: A Reaction Norm Perspective.* Sunderland, MA: Sinauer, Sunderland.
Schmidt-Nielsen, K. (1984). *Scaling: Why Is Animal Size So Important?* Cambridge, England: Cambridge University Press.
Smith, M. M. (2003). Vertebrate dentitions at the origin of jaws: When and how pattern evolved. *Evolution and Development* **5**, 394–413.
Stadler, B. M. R., Stadler, P. F., Wagner, G. P., and Fontana, W. (2001). The topology of the possible: Formal spaces underlying patterns of evolutionary change. *Journal of Theoretical Biology* **213**, 241–274.
Steppan, S. J., Storz, B. L., and Hoffmann, R. S. (2004). Nuclear DNA pylogeny of the squirrels (Mammalia: Rodentia) and the evolution of arboreality from c-myc and RAG1. *Molecular Phylogenetics and Evolution* **30**, 703–719.
Thompson, D. W. (1942). *On Growth and Form.* 2nd ed. Cambridge, England: Cambridge University Press.
Thorington, R. W. (1983). Flying squirrels are monophyletic. *Science* **225**, 1048–1050.
Thorington, R. W., and Darrow, K. (1999). Jaw muscles of Old World squirrels. *Journal of Morphology* **230**, 145–165.
Van Valen, L. (1976). Energy and evolution. *Evolutionary Theory* **1**, 179–229.
Van Valkenburgh, B. (1994). Ecomorphological analysis of fossil vertebrates and their paleocommunities. In *Ecological Morphology* (P. C. Wainwright and S. M. Reilly, eds.), pp. 140–166. Chicago: University of. Chicago Press.
Vermeij, G. J. (1970). Adaptive versatility and skeleton construction. *American Naturalist* **104**, 253–260
Vermeij, G. J. (1973). Biological versatility and earth history. *Proceedings of the Natural Academy of Science* **70**, 1936–1938.
Vermeij, G. J. (1974). Adaptation, versatility and evolution. *Systematic Zoology* **22**, 466–477.
VonDassow, G., and Munro, E. (1999). Modularity in animal development and evolution: Elements of a conceptual framework for EvoDevo. *Journal of Experimental Zoology (Molecular Development and Evolution)* **285**, 307–325.
Wagner, G. P., and Altenberg, L. (1996). Perspective: Complex adaptations and the evolution of evolvability. *Evolution* **50**, 967–976.
Wagner, G. P., and Stadler, P. F. (2003). Quasi-independence, homology and the unity of type: A topological theory of characters. *Journal of Theoretical Biology* **220**, 505–527.
Wake, D. B., and Roth, G. (eds.). (1988). *Complex Organismal Functions: Integration and Evolution in Vertebrates: Report of the Dahlem Workshop on Complex Organismal Functions—Integration and Evolution in Vertebrates.* Chichester, England: Wiley.

Webster's New Collegiate Dictionary (1976). Springfield, MA: Merriam.

West-Eberhard, M. J. (2003). *Developmental Plasticity and Evolution.* New York: New York.

Yang, A. S. (2001). Modularity, evolvability, and adaptive radiations: A comparison of the hemi- and holometabolous insects. *Evolution and Development* 3, 59–72.

Zahler, P., and Khan, M. (2003). Evidence for dietary specialization on pine needles by the woolly flying squirrel *Eupetaurus cinereus*. *Journal of Mammalogy* 84, 480–486

Zelditch, M. L., Straney, D. O., Swiderski, D. L., and Carmichael, A. C. (1990). Variation in developmental constraints in *Sigmodon*. *Evolution* 44, 1738–1747.

CHAPTER 20

Variation and Developmental Biology: Prospects for the Future

David M. Parichy
Department of Biology, University of Washington, Seattle, Washington, USA

INTRODUCTION

Recent years have seen great strides in our understanding of organismal development. At the dawn of the twenty-first century, we have identified many key signaling pathways and mechanisms of gene regulation. We have also defined at a molecular level many of the phenomena identified by embryologists during the early and middle years of the twentieth century, including the organizer and limb polarizing activity of vertebrates and the maternal determinants of insect early patterning. It might seem that many of the biggest questions have been answered and that only the relatively minor details need to be elucidated.

In this review and prospectus, I argue that, in fact, many of the biggest questions for developmental biology still remain unanswered. Some of these questions explicitly concern variation. Other questions will be answered only with a deeper appreciation of variation in a developmental context. In both cases, a new incorporation of population thinking into developmental biology will need to be achieved.

Variation

Ernst Mayr introduced the concept of "population thinking" as one of the three major contributions made by Darwin (Mayr, 1976). Although the first two contributions—compelling demonstration of evolution and identification of natural selection as an evolutionary mechanism—have been widely celebrated, the third, replacing typological thinking with population thinking, had not been recognized. Mayr relates typological thinking to Platonic idealism, in which observed variability merely reflects an underlying natural "idea" or type. In this view, variation is simply noise with no inherent meaning; truth lies only in the types themselves and discontinuities between types represent the true order of the natural world.

By contrast, a population-based view of nature holds that variation is the true state of being: the uniqueness of individuals is real and it is the type that is an abstraction. Mayr argues that organisms can be described collectively only in statistical terms: means (and variances) are thus necessary abstractions if we are to summarize the true differences among individuals. Mayr was principally concerned with the ways in which typological and population thinking can influence our views of evolution, with a populational approach providing a clearly superior rationale for understanding organismal differences and how they have come into existence. The success of evolutionary biology following the neo-Darwinian synthesis, and its (implicit or explicit) embrace of population thinking, provides ample evidence for Mayr's assertion.

Developmental biology and one of its foundations, comparative morphology, have a long history of typological thinking (for recent reviews, see Amundson, 1998, 2000; Hall, 1999; Richardson *et al.*, 1999). The typological approach can be seen in Buffon's and Geoffroy Saint-Hilaire's idealistic morphologies, in which all forms were thought to derive from just one or a few types. The typological approach also is evident in the theories of Georges Cuvier's, which defined four "*embranchements*" into which all animals could be classified; variation within an *embranchment* merely reflected deviation from an ideal functional scheme, and life outside of *embranchments* was an impossibility. Although the perspectives of Saint-Hilaire and Cuvier differed dramatically in relating form to function, both held to notions of ideal types. Von Baer, who saw types in terms of embryonic form, rather than adult form, continued this tradition. Likewise, Owen defined the term "archetype" to represent an idealized form, and he proposed the now famous vertebrate archetype that has been reproduced frequently in textbooks. Finally, Haeckel's famous (or infamous) depiction of similarities among vertebrate embryos reflects an inherently typological view that sought to minimize the uniqueness of individuals or species while highlighting an idealized form (Richardson *et al.*, 1997, 1998).

The modern era of developmental biology has its roots in comparative morphology, the experimental embryology of Roux, and the genetics of Morgan, and as such represents a synthesis of several disparate disciplines. In this chapter, I suggest that the synthesis of modern developmental biology has remained essentially typological in outlook, despite operating in parallel with the fundamentally populational neo-Darwinian synthesis. For developmental biology to address the remaining large questions before it, however, will require the adoption of an increasingly populational perspective. In the following sections, I suggest several of these questions, each relating to variation, and how a populational viewpoint will need to be accommodated.

I. MODEL ORGANISMS: EXPANDING THE FOLD

A key factor underlying the last 50 years of progress in developmental biology has been the emergence of a few, key "model organisms." Among these are the venerable fruit fly *Drosophila melanogaster* (aka "the Fly"), the nematode *Caenorhabditis elegans* ("the Worm"), the anuran amphibian *Xenopus laevis* ("the Frog"), the chicken *Gallus gallus*, the mouse *Mus musculus*, as well as more recent additions such as the mustard weed *Arabidopsis thaliana* and the zebrafish *Danio rerio*.

The use of model organisms has greatly facilitated advances through the resulting concentration of resources. Perhaps the most dramatic of resources is the availability of sequence data for the complete or nearly complete genomes of each of these models. Although whole genome sequences and other genomic resources (e.g., expressed sequence tags, gene chips) have appeared only in the last few years, researchers working on the model organisms have long enjoyed the availability of molecular probes, normal tables of development, standardized protocols for molecular biology and histology, methods for embryological and genetic manipulation, and specialized stocks of various sorts, including mutant strains and inbred lines.

Model organisms also have afforded us vast intellectual resources. Although the primary literature on model organism development is vast and continually expanding, less appreciated is the unpublished, communal knowledge, or "lore," about these organisms ("How does one fertilize fish eggs in vitro?" "How much aeration do the frogs really want?" "What is the best way to explant a chicken neural tube?"). Together, these bodies of knowledge have allowed extraordinary progress in dissecting many basic mechanisms of development.

Today's model organisms have been chosen primarily for their convenience. In some cases the organisms were readily (or even too readily) available. In other instances they were chosen because they are easy to breed or to rear. Sometimes optical clarity, invariant cell lineage, or both have been factors, and almost always, speed of development has been important. Model organisms have not been singled out for study because they are likely to be representative of particular groups or modes of development. Nevertheless, it has often been assumed that what occurs during the development of the premier model organisms can be generalized to others of their genus, order, or phylum. In many cases this is true. In some it is not.

As noted by others (Kellogg and Shaffer, 1993; Bolker, 1995; Metscher and Ahlberg, 1999), the very features that make our model organisms attractive to study can also make our inferences suspect when we try to generalize our findings more broadly. For example, rapid development alone likely entails a host of specializations (Hadfield, 2000). More typically, however, model organisms are like any other organisms in exhibiting a host of traits, some of which are relatively derived and some of which are relatively primitive for their group. Thus, even ignoring the specific features that have made some model organisms the favorites of developmental biologists, some unique derived features will be encountered simply by chance. Studies of amphibian gastrulation and *Drosophila* early pattern formation illustrate the sometimes surprising ways in which development of our model organisms can differ from their nonmodel relatives.

The workhorse model amphibian is *X. laevis*, and pioneering studies by Keller and colleagues have elucidated many of the morphogenetic behaviors occurring during gastrulation and mesoderm formation in this species (Keller *et al.*, 2003). Fate mapping (Keller, 1975, 1976) demonstrated that prospective mesoderm resides only within a deep layer of involuting cells at gastrulation. As gastrulation proceeds, the prospective mesodermal cells already are situated within the "correct" tissue layer, between the roof of the archenteron and the overlying ectoderm. In contrast, early studies showed that most frogs, as well as salamanders, have prospective mesoderm on the surface of the early gastrula, in addition to cells in the deeper layer. As gastrulation proceeds in these embryos, the initially superficial cells find themselves lining the roof of the archenteron and must then ingress secondarily to join the rest of the mesoderm (Vogt, 1929; Purcell and Keller, 1993). A potential transitional form has been observed in *Hymenochirus*, the closest relative of *Xenopus* studied to date, with prospective mesodermal cells ingressing from the archenteron as a sheet rather than as individual cells (Minsuk and Keller, 1996). Thus a fundamental aspect of vertebrate development—how the germ layers arise during gastrulation—differs substantially between *Xenopus* and most other amphibians.

Perhaps the best understood developmental patterning system is that of anterior–posterior axis determination in *D. melanogaster*. Whereas early embryological manipulations with other insects pointed to the existence of maternal determinants in the early embryo (Kalthoff and Sander, 1969; Sander, 1975), it was the work of Driever and Nusslein-Volhard that identified the molecular bases for these events (Driever and Nusslein-Volhard, 1988a,b). Of primary importance is *bicoid*, which is deposited as a maternally derived mRNA at the anterior end of the developing *Drosophila* oocyte. Major functions of *bicoid* protein are to repress the translation of *caudal* mRNA, at the posterior end of the embryo, and to upregulate the transcription of the gap gene, *hunchback* (Ephrussi and Johnston, 2004). Embryos that are mutant for *bicoid* lack anterior head and thorax. In the relatively closely related fly, *Megaselia*, a role for *bicoid* is evident not only in head and thorax development, but also in anterior abdomen formation (Stauber *et al.*, 2000).

Despite the pivotal role of *bicoid* in establishing anterior–posterior pattern in *Drosophila*, this is not the case for other insects or even more distantly related flies. The *bicoid* gene arose by duplication of an ancestral *Hox3/zerknüllt* locus in the lineage leading to cyclorrhaphan flies, of which *Drosophila* and *Megaselia* are members. In these flies, *bicoid* is expressed maternally, whereas its paralogue *zerknüllt* is expressed zygotically. In more basal dipterans that lack *bicoid*, the single *Hox3/zerknüllt* locus is expressed both maternally and zygotically and is likely to serve functions now divided between *bicoid* and *zerknüllt* in cyclorrhaphans (Stauber *et al.*, 2002). Beyond the dipterans, the flour beetle *Tribolium castaneum* lacks a *bicoid* gene, but *orthodenticle* and *hunchback* (gap genes downstream of *bicoid* in *Drosophila*) together fulfill an equivalent function (Schroder, 2003; see also Wimmer *et al.*, 2000). Relative contributions of maternal and zygotic genes to anterior–posterior patterning in the wasp *Nasonia* are also likely to differ substantially from *Drosophila* (Pultz *et al.*, 1999). Thus a basic feature of early patterning differs dramatically in its execution across insects.

These examples reveal how a focus on a few model organisms can speed progress toward an understanding of mechanism, but also how a typological view of these mechanisms could lead to erroneous conclusions about their generality. Additional examples of developmental convergence and parallelism (in the following text) provide ample evidence for the importance of studying mechanisms across multiple taxa even within the same order or family. When studies of basic developmental mechanisms are conducted within an explicitly phylogenetic context, fresh insights into the generality of developmental mechanisms seem always to result.

Early embryologists studied a far more diverse range of organisms than most developmental biologists currently employ. Perhaps most dramatically missing from the mainstream are the lophotrochozoans. Modern molecular phylogenies reveal three major clades of animals: Deuterostomia, Ecdysozoa, and Lophotrochozoa (Adoutte *et al.*, 2000; Balavoine *et al.*, 2002; Ruiz-Trillo *et al.*, 2002; Anderson *et al.*, 2004). All of the commonly studied model organisms fall within the deuterostomes (sea urchin, zebrafish, frog, chicken, mouse) or ecdysozoans (*Drosophila*, nematodes). By contrast, the lophotrochozoans (leeches, mollusks, and possibly flatworms) were studied extensively by early workers and then largely fell out of favor. For example, snail embryogenesis was studied by Conklin at the turn of the last century (Conklin, 1897, 1902) and has been studied sporadically since then (Freeman and Lundelius, 1982), but modern advances in molecular and cellular biology typically have not been applied to understanding the development of these organisms (for a notable exception, see: Lambert and Nagy, 2001). Nevertheless, new resources are becoming available for working even with "nonmodel" species (Voss *et al.*, 2001; Newmark and Sanchez Alvarado, 2002; Tessmar-Raible and Arendt, 2003) that should substantially facilitate studies of basic developmental mechanisms across a wider swath of the animal kingdom. A recognition that current biomedical models are not types, or even averages, of their respective groups, but merely sample species drawn for particular purposes, will be essential for a more complete understanding of development.

II. ECOLOGICALLY SIGNIFICANT DIFFERENCES IN FORM BETWEEN SPECIES

A basic goal of developmental biology is to understand how a fertilized egg gives rise to an adult body. This transformation involves increasing complexity through the processes of pattern formation, morphogenesis, differentiation, and growth. To date, we know a great deal about these processes during early embryogenesis, when the body axes and germ layers are established, when the organ rudiments are forming, and when the organism is interacting only minimally with its environment. Yet we still understand relatively little of the mechanisms responsible for the precise forms of larvae, juveniles, and adults that interact extensively with their environments. Whether the shapes of spicules in a sea urchin pluteus, the color pattern of a caterpillar, the skull shape of an owl, or the growth habit of an oak tree, we typically do not know the genes and cell behaviors underlying the expression of these traits. A major remaining goal for developmental biology is to understand the mechanisms responsible for juvenile and adult form, and by extension, how differences in these mechanisms generate variation in form across taxa.

Perhaps of greatest interest are the developmental bases for interspecific differences that are likely to have adaptive significance. Such analyses can be conducted at several phylogenetic levels. For example, the turtle shell represents an evolutionary innovation having clear adaptive significance, and recent analyses are beginning to identify the signaling pathways underlying shell development (Gilbert *et al*., 2001; Rieppel, 2001). Comparisons with squamate reptiles and birds will allow new insights into the developmental and evolutionary origins of this structure.

Besides true novelties, the developmental bases for adaptive modifications to existing form between major groups have started to be dissected with some success. Examples come from studies of arthropod and vertebrate appendages. In most arthropods, thoracic appendages are used for locomotion, but in several groups of crustaceans, anterior thoracic appendages have become specialized for food handling. The evolutionary appearance of these transformed legs, or maxillipeds, is associated with losses in the expression of the Hox genes from anterior thoracic segments (Averof and Patel, 1997). Variation in arthropod Hox gene expression has been associated with other morphological differences as well (Averof and Akam, 1995; Popadic *et al*., 1998; Abzhanov and Kaufman, 2000).

More recently, factors determining appendage number in arthropods have been studied, with a focus on evolutionary changes in the function of Ultrabithorax (Ubx) protein. In *Drosophila*, thoracic limb outgrowth requires *Distal-less* (*Dll*) expression, but in the abdomen where limbs do not form, *Dll* is repressed by Ubx. If Ubx is ectopically expressed in the thorax, it represses *Dll* there as well, and legs do not form (Gonzalez-Reyes and Morata, 1990; Mann and Hogness, 1990; Vachon *et al.*, 1992). Thus *Dll* expression is correlated with limb development in *Drosophila* and a diverse array of other species as well (Panganiban *et al.*, 1994, 1997; Panganiban, 2000). By contrast, crustaceans have abdominal legs and abdominal *Dll* expression—but also abdominal Ubx expression—suggesting a difference in the way Ubx interacts with *Dll* in this group, as compared with *Drosophila.* Ectopic expression in *Drosophila* embryos of native crustacean Ubx and chimeric proteins reveals the presence of a carboxy terminal Ubx domain that allows for the conditional repression of *Dll* in crustaceans, rather than the constitutive repression of *Dll* in *Drosophila* (Ronshaugen *et al.*, 2002). Thus a difference in Ubx function is associated with the absence of abdominal limbs in *Drosophila* and the presence of abdominal limbs in crustaceans. Analyses of phylogenetically more basal Onychophora (which have legs on each body segment, including the abdomen) reveal that an onychophoran Ubx completely fails to repress *Dll* expression because it lacks key amino acid residues (Grenier and Carroll, 2000; Galant and Carroll, 2002) (though this difference is not likely to be functionally significant in generating the many limbs of onychophorans per se, in that *Ubx* and *Dll* expression do not appear to overlap in leg-bearing segments [Grenier *et al.*, 1997]).

Several studies of vertebrates also have been informative in understanding the evolution of adult form across relatively divergent taxa. For example, a recent study of the shark *Scyliorhinus canicula* may reveal a transitional stage in the evolution of limbs from fins (Tanaka *et al.*, 2002). In tetrapods and teleosts, *Tbx5* and *Tbx4* play essential roles in specifying the form of anterior and posterior appendages, respectively. In *Scyliorhinus*, both of these genes are expressed in a manner reminiscent of other vertebrates. A different situation is observed for *sonic hedgehog* (*shh*). In tetrapods, *shh* is expressed in the posterior margin of the limb mesoderm; Shh protein exhibits polarizing activity during limb development and contributes to limb outgrowth by a feedback loop involving bone morphogenetic protein (Bmp) and fibroblast growth factor (Fgf) signals (Riddle *et al.*, 1993; Laufer *et al.*, 1994; Niswander *et al.*, 1994; Khokha *et al.*, 2003). In the *Scyliorhinus* embryo, however, *shh* transcript is conspicuously absent from the developing fin, though it is expressed in other locations (e.g., notochord and floor plate). Thus it is conceivable that a change in *shh* expression is associated with transition from a sharklike, cartilaginous fin to the tetrapod limb (though expression of *shh* in zebrafish fin buds [Krauss *et al.*, 1993; Akimenko and Ekker, 1995] suggests an alternative scenario in which *shh* expression in developing appendages is ancestral and has been lost from *Scyliorhinus*, rather than gained in tetrapods).

Finally, analyses of patterning genes also shed light on the evolution of limblessness in snakes (Cohn and Tickle, 1999). In limbed tetrapods, Hox gene anterior expression boundaries correlate with the morphological thoracic–cervical transition, where the forelimbs will develop (Burke *et al.*, 1995). In python embryos, however, thoracic Hox genes are expressed throughout the trunk, and limbs do not form anteriorly, suggesting that limb axial repatterning to a thoracic identity has reduced or eliminated the anterior limb field. Although vestigial limb buds are present posteriorly, these buds lack *shh* expression in the mesoderm as well as a morphologically identifiable apical ectodermal ridge with *Fgf8* expression, which is required in other tetrapods for limb outgrowth. Thus changes in axial patterning and loss of the Shh–Bmp–Fgf regulatory loop appear to be associated with limb loss in pythons and presumably the ancestors to modern snakes.

The preceding examples analyzed variation in developmental patterning at relatively deep phylogenetic levels. Although such comparisons are necessary for some of the most dramatic evolutionary transformations, such as the loss and gain of limbs, these comparisons also entail several inherent difficulties. For example, distantly related taxa will have accumulated a variety of developmental differences that are independent of the evolutionary transformation in question, making it more difficult to assess what is causally relevant and what is not. For this reason, studies of variation in developmental mechanisms between closely related species sometimes can provide insights that broader comparisons cannot. For instance, studies of pigment pattern formation in salamander larvae identified presumptively ancestral cell–cell interactions required for stripe formation in several species, but novel and redundant mechanisms in one of these species, in which stripes are somewhat more distinctive (Parichy, 1996a,b, 2001). Such developmental elaborations may represent early stages in the evolution of a trait associated with new selective consequences for the expression of that trait. Thus analyses of variation in cellular interactions across closely related species can provide insights into how initially simple characters are molded through selection into more complex and specialized adaptations.

At a molecular level, another problem inherent to comparisons across deep phylogenetic divides is that such comparisons can be limited to known genes and pathways. For example, extensive previous work on limb patterning and morphogenesis made it possible to compare the expression and activity of candidate genes across taxa. In some instances, however, adaptive differences between species are likely to result from changes in genes or pathways that have not yet been studied in model organisms. One approach to identifying "new" genes as well as "old" genes is the genetic analysis of species or populations that are morphologically different but sufficiently closely related to interbreed. Several analyses of teleost fishes highlight these strategies.

One system that has been employed for analyzing variation among closely related species is the adult pigment pattern of *Danio* fishes, including the zebrafish *D. rerio* (Parichy, 2003). Pigment patterns have clear adaptive significance as they are used for mate recognition and mate choice, as well as predation avoidance and schooling (Endler, 1983; Houde, 1997; Couldridge and Alexander, 2002;

Engeszer *et al.*, 2004). *Danio* fishes exhibit a diverse array of patterns, including horizontal stripes, vertical bars, spots, and uniformly dispersed pigment cells. Moreover, pigment patterns exhibited by many *D. rerio* mutants resemble the naturally occurring patterns of other species, suggesting the affected loci as *a priori* candidate genes for contributing to pattern differences across taxa. Since hybrids can be produced between *D. rerio* and other species, and these hybrids typically have horizontal stripes like *D. rerio*, complementation tests can be used to assess whether the same genes isolated as *D. rerio* mutants might contribute to interspecific differences as well. This approach showed that one mutant *panther*, which lacks stripes, fails to complement in crosses with *D. albolineatus*, which also lacks stripes. Molecular cloning and analysis revealed *panther* to be a *Danio* orthologue of the *fms* gene, which encodes a receptor tyrosine kinase expressed by pigment cells in danios, but not previously known to have a role in pigment pattern formation (Parichy *et al.*, 2000; Parichy and Johnson, 2001; Quigley *et al.*, 2005; see also Long *et al.*, 1996; Sucena and Stern, 2000). This example illustrates how forward genetics in a model organism, coupled with interspecific genetic analysis, can be used to find novel genes associated with ecologically significant variation between taxa.

Two other groups of fishes that have been examined in this context are cichlids and sticklebacks. African cichlids have long been the premier model for adaptive radiations, with more than a thousand species arising over the past million years in Lakes Malawi and Victoria (Allender *et al.*, 2003). Despite their recent origins, these species differ dramatically in jaw morphology, color pattern, and behavior. Quantitative trait locus (QTL) mapping has started to define chromosomal regions that contribute to these differences (Albertson *et al.*, 2003a,b; Streelman *et al.*, 2003).

Stickleback fishes also have a long history of evolutionary, ecological, and behavioral research (Peichel and Boughman, 2003), and different populations exhibit a variety of morphological differences, particularly in dermal armor plating and internal osteological characters. One such trait is the pelvis, which is dramatically reduced or absent in some populations. Construction of a genetic map for sticklebacks has allowed the identification of chromosomal regions as well as candidate genes that contribute to this variation (Peichel *et al.*, 2001). In a recent analysis, pelvic reduction in a freshwater stickleback population was shown to segregate primarily as a single Mendelian factor, and QTL mapping associated a large fraction of the segregating variance with a single chromosomal region. In this instance, mapping of candidate genes based on studies of limb development in model organisms placed *Pitx1* within the interval identified by the quantitative trait locus. Subsequent analyses suggest that a site-specific loss of *Pitx1* expression from the prospective pelvic region is causally relayed to pelvic reduction in this population (Shapiro *et al.*, 2004).

Recent studies of danios, cichlids, and sticklebacks highlight the potential for analyses of closely related species to reveal the genetic and developmental bases for adaptive differences in adult form. Moreover, the many QTL and candidate gene studies performed to date are starting to provide insights into the sheer number of genes contributing to morphological differences between species and populations (Voss and Shaffer, 1997; Doebley *et al.*, 1997; Long *et al.*, 1998; Frary *et al.*, 2000; Kopp *et al.*, 2003). Typically, anywhere from one to several major effect loci have been identified. These studies are limited in their power to detect loci of very small effects, and it is difficult to judge how many studies might have failed to detect major genes owing to publication biases. Nevertheless, the frequent identification of loci explaining substantial proportions of phenotypic variance suggests a more important role for fewer genes of large effect than was assumed by standard evolutionary genetic models, which assumed many genes of small effect (Orr and Coyne, 1992). Thus we are starting finally to understand how many genes contribute, at least substantially, to standing morphological differences between species and populations.

At least three major challenges remain, however. A first, largely technical challenge will be the development of additional methods for identifying genes and pathways contributing to variation between species. Although hybrid analyses and QTL mapping can be useful in some cases, many of the most interesting morphological and developmental differences are between species that are closely related, but not so closely related that viable or fertile hybrids can be produced for genetic analysis (e.g., variation in beak morphology [Schneider and Helms, 2003]). For still other species, genetic approaches might be possible in principle but too difficult to accomplish in practice. Development of additional, unbiased approaches to identifying new genes and pathways underlying interspecific differences will make a major contribution to this area. Conceivably, microarray technologies will be useful in this regard, when coupled with appropriate screening strategies and statistical analyses to control for phylogenetic divergence (Rifkin *et al.*, 2003).

A second, largely conceptual challenge will be linking genes to morphological outcomes. To understand how variation in developmental mechanisms generates variation in adult form, identification of genetic differences is only useful to the extent that these differences are placed in the context of defined cellular behaviors, such as proliferation, death, migration, and differentiation. Bridging the morphogenetic gap between genes and phenotypes is a major challenge even for studies of embryogenesis using model organisms. Doing so to understand differences in adult form between species requires a populational approach at multiple levels of biological hierarchy, from genes and their activities, to cells, to tissues, and to organs.

Finally, a third technical and conceptual challenge will be to relate variation between populations or species to variation within populations themselves. Simply because we can detect loci currently associated with moderate to large phenotypic differences, this association need not imply that the differences initially evolved via mutations having moderate or large effects or even by mutations at the identified loci. In *Danio* fishes, for example, a loss of stripes in *D. albolineatus* was associated with variation at the *fms* locus (Parichy and Johnson, 2001). Although this may reflect a saltational loss of stripes owing to a major effect mutation in *fms* itself, it could also reflect more gradual changes at *fms*, evolutionary changes in cellular requirements for *fms*, or changes in other *fms* pathway genes (Quigley *et al.*, 2005). Elucidating the relationships between interspecific and intrapopulational variation will require a deeper understanding of how allelic differences within populations influence trait expression during development.

III. HOW MANY WAYS TO MAKE A PHENOTYPE: DEVELOPMENTAL VARIATION AND MORPHOLOGICAL SIMILARITY

Besides explaining the mechanistic bases for differences in form across species, a major unresolved question is the extent to which the same phenotype in different species arises by different—or the same—underlying mechanisms. Developmental variation sometimes can be identified even for traits that have a common evolutionary origin. Differences in mesoderm formation among anuran amphibians (as noted previously) are one example in which ancestral morphogenetic mechanisms have been altered in the lineages leading to *Hymenochirus* and *Xenopus*. Another classic example is Meckel's cartilage, which is induced through an epithelial–mesenchymal interaction that employs different epithelial tissues in different vertebrates (Hall, 1984). Extensive analyses of nematode vulval development have also identified very different genetic and cellular mechanisms across species (Jungblut and Sommer, 2000; Sommer, 2001), and cellular ablation studies in leeches have shown interpopulation differences in the earliest cellular interactions that specify cell fates (Kuo and Shankland, 2004). Finally, superficially similar adult stripes of zebrafish and its close relative *D. nigrofasciatus* nevertheless arise with very different contributions from temporally distinct populations of larval and adult melanophores (Quigley *et al.*, 2004).

The extent of cryptic developmental variation is especially relevant to cases of homoplasy, in which organisms look the same for reasons other than descent from a common ancestor. Typically, independent origins of a trait have been termed "convergence" if arising by different mechanisms, or "parallelism" if arising by the same mechanisms (Wake, 1991; Hodin, 2000). These alternative modes have profound implications for our understanding of developmental evolution.

If homoplasy consistently arises through different mechanisms (convergence), this would imply there are few limits on how developmental mechanisms can evolve. Alternatively, if homoplastic evolution typically occurs through the same mechanisms (parallelism), this would suggest that some developmental pathways are more likely to be modified than others. Put another way, repeated instances of parallelism argue that some kinds of mechanistic variants are more likely to arise in populations (and so be available for selection) than others. Thus some pathways of evolutionary change would be more likely, even though multiple phenotypic solutions might be equally adaptive in the face of selection. Assessing the mechanistic bases of homoplasy thus provides crucial data for resolving debates over the importance of developmental biases (or "constraints") during morphological evolution (Maynard Smith *et al.*, 1985).

Although many examples of homoplasy have been identified and may hint at underlying mechanism (Wake, 1991; Huber *et al.*, 2000; Parra-Olea and Wake, 2001; Dowling *et al.*, 2002; Wray, 2002; Santos *et al.*, 2003; Smith and Johanson, 2003), we typically do not know the genetic or cellular bases for the traits in question. Recently, however, several studies have provided evidence at the mechanistic level. For example, repeated reductions in the numbers of larval trichomes in *Drosophila* are associated with changes at the *shaven baby* locus (Sucena *et al.*, 2003). Genetic analyses of sticklebacks also have identified common bases for the repeated evolution of pelvic reduction (Shapiro *et al.*, 2004) as well as armor plate reduction (Colosimo *et al.*, 2004; Cresko *et al.*, 2004). Finally, some particularly illuminating examples of mechanisms come from studies of pigmentation. Dark or "melanistic" phenotypes have evolved repeatedly and independently in both birds and mammals, and recent analyses reveal that many of these instances are associated with activating amino acid substitutions in the melanocortin receptor, Mc1r, which functions to promote pigment synthesis (Theron *et al.*, 2001; Eizirik *et al.*, 2003; Mundy and Kelly, 2003; Nachman *et al.*, 2003; Mundy *et al.*, 2004). Together, these examples reveal a striking degree of evolutionary parallelism that would not have been predicted by classical quantitative genetic models of evolutionary change (Barton and Turelli, 1989; Falconer and Mackay, 1996).

By contrast, other studies identify evolutionary convergence in developmental mechanisms. For example, some studies of melanism in amniotes (Nachman *et al.*, 2003), as well as pigment pattern evolution in *Drosophila* (Wittkopp *et al.*, 2003), reveal different genetic bases across species. Likewise, metamorphic failure, or paedomorphosis, has evolved repeatedly and independently in salamanders, and interspecific hybridization studies support a model in which different genetic changes are responsible in different phylogenetic lineages (Voss and Shaffer, 1996).

Thus both evolutionary convergence and parallelism in developmental mechanisms occur, and it remains to be seen what their relative frequencies might be. The answer to this question will dramatically impact our understanding of both developmental mechanisms and their evolution. Such answers will come only from developmental studies of homoplasy in an explicitly phylogenetic context, requiring a populational perspective on the nature of developmental mechanisms and how they can vary.

IV. INTRASPECIFIC DEVELOPMENTAL VARIATION: CANALIZATION AND DEVELOPMENTAL PLASTICITY

The preceding sections have focused on variation in developmental mechanisms across species or populations that may—or may not—be causally related to variation in form. An equally important, remaining problem in developmental biology concerns the nature of developmental variation within populations. Outside of developmental evolutionary biology (or evolutionary developmental biology, depending on one's predilections [Hall, 2000; Gilbert, 2003]), very few developmental biologists concern themselves with variation in cell behaviors and gene activities across species, let alone among individuals within populations. Nevertheless, such variation exists and determining its causes and consequences will be critical for progress both in biomedical and evolutionary research.

Variation in phenotype results from genetic influences, environmental influences, and interactions between the two (Falconer and Mackay, 1996; Lynch and Walsh, 1998). Evolutionary quantitative genetic models have provided a framework for understanding the sources of phenotypic variation, and such models typically have been applied to adult form (or behavior). Nevertheless, the same principles hold for developmental phenotypes, and several studies have examined the statistical bases of morphological variation at different stages of development (Blouin, 1992; Phillips, 1998; Watkins, 2001). Others have sought to provide a framework for partitioning this variance into components associated with genetic and environmental effects on cell behaviors and gene activities (Riska, 1986; Slatkin, 1987; Atchley and Hall, 1991; Cowley and Atchley, 1992). Such statistical approaches have the advantage of being readily incorporated into evolutionary genetic theory. It has not yet been possible to bridge the gap empirically between statistical estimates of quantitative genetics parameters and underlying developmental mechanisms. Nevertheless, such models serve a valuable heuristic purpose in formally identifying the potential sources of variation in developmental mechanisms. In the following discussion, I consider what we know and do not know about genetic and environmental sources of intrapopulational variation in developmental mechanisms.

As is true for interspecific and interpopulation variation in development, intrapopulation variation may or may not have morphological consequences. Understanding why some differences in gene activities and cell behaviors influence morphology while others do not is perhaps the single greatest challenge facing developmental biology in the future. In this regard, an essential piece of information is the extent of functionally significant standing allelic variation for developmentally relevant loci. For model organisms including human, these data are rapidly being compiled in the form of single nucleotide polymorphisms catalogued in the National Center for Biotechnology Information databases: (http://www.ncbi.nlm.nih.gov). Thus there is now the potential for systematically surveying intraspecific variation in coding as well as noncoding regions of the genome. Moreover, these data can sometimes be associated with major phenotypic effects, particularly in the databases of human genetic disease syndromes (e.g., "Online Mendelian Inheritance in Man"). Thorough surveys of these sorts of data will provide new insights into the likelihood that genetic variants in coding or noncoding sequences have phenotypic effects (Rockman and Wray, 2002; Genissel *et al.*, 2004).

A long-standing observation in developmental genetics is the resistance of development to minor genetic and environmental perturbations, a phenomenon known as "canalization" (Waddington, 1942, 1957, 1975). One example of canalization is the insensitivity of most loci to dosage effects, resulting in the recessivity of many mutant alleles. Outside of mutant laboratory stocks, support for the existence of substantial variation in gene product abundance or activity comes from several sources. For example, microarray analyses of gene expression have revealed extensive variation in transcript abundance among individuals within populations (Oleksiak *et al.*, 2002) and between inbred strains (Pavlidis and Noble, 2001). In principle, such differences can arise from both *cis*-acting and *trans*-acting regulatory variation. Demonstration of substantial *cis*-acting regulatory variation comes from recent analyses of allele-specific transcript abundance in heterozygous individuals: 6–18% of surveyed loci exhibited significant differences in transcript abundance between alleles, typically with 1.5–3-fold imbalances (Cowles *et al.*, 2002; Bray *et al.*, 2003; Pastinen *et al.*, 2004). In the face of such variability, understanding how canalization arises—and why sometimes it fails (see the following text)—are fundamental questions if we are to understand how development translates genotypes into phenotypes.

One explanation for canalization in the face of genetic perturbation comes from inherent properties of developmental systems. For example, mathematical analyses of metabolic pathways have been applied to understand the nature of dominance and recessivity (Kacser and Burns, 1981; Dykhuizen *et al.*, 1987). Considering linear pathways of gene products, the sensitivity of the outcome to variations in the activity of any one pathway member diminishes as the total number of pathway members increases. Thus the more complex (or longer) the pathway, the less sensitive it is likely to be to minor perturbations at each step. Analyses of evolutionary change in human and mouse metabolic genes further supports a model in which increased network complexity is associated with increased canalization (Kitami and Nadeau, 2002).

A mathematical approach recently has been applied to understanding networks of interacting developmental genes as well. In two remarkable papers, the networks of genes comprising the *Drosophila* segment polarity and neurogenic networks have been modeled (von Dassow *et al.*, 2000; Meir *et al.*, 2002). These systems were chosen because of the extensive empirical data ordering genes within these pathways. Nonlinear equations were used to describe the interactions among gene products, and variations in their associated parameters were tested for their ability to produce the expected developmental outcome (i.e., simulated cells expressing the correct genes in the correct places). The major finding from these models is the extraordinary robustness of the simulated pathways: Even very large deviations in gene product concentrations and activities were readily accommodated. Empirical analyses of Bmp signaling in *Drosophila* further support the robustness of such developmental genetic networks (Eldar *et al.*, 2002). Thus a classical view of gene regulation in terms of binary switches, with genes either on or off, may be a substantial (and typological) oversimplification. Rather, we may need to view even the most impressive and complex of resolved genetic pathways (Davidson *et al.*, 2002a,b) as merely rarified descriptions of interactive networks having substantial quantitative and stochastic components.

A second explanation for canalization of developmental mechanisms comes from studies of Hsp90 in *Drosophila* and *Arabidopsis* (Rutherford and Lindquist, 1998; Queitsch *et al.*, 2002; Sangster *et al.*, 2004). Hsp90 acts as a chaperone to ensure correct protein folding particularly in response to stress. Studies of Hsp90 mutants reveal dramatically increased frequencies of developmental anomalies in a variety of traits. Thus Hsp90 suppresses the phenotypic expression of underlying genetic variation, presumably in part through its function as a chaperone and apparently also through chromatin modification (Sollars *et al.*, 2003). A recent analysis identifies the epidermal growth factor receptor gene (*Egfr*) as one of the loci for which cryptic variation is buffered by Hsp90 in nature (Dworkin *et al.*, 2003). It will be especially interesting to see what other genes contribute to canalization of phenotypes in a manner analogous to Hsp90, as well as the extent to which variation at such loci contributes to corresponding variation in canalization.

When developmental canalization breaks down, it can do so in response to genetic or environmental influences. Genetically, this can reflect either novel combinations of existing alleles or the introduction of new allelic variants by mutation. That some allelic combinations simply "work" better than others in the face of genetic perturbations is reflected in the common observation of genetic modifiers to laboratory-induced mutations: For the same mutant locus, the severity of phenotypic effects can differ drastically across genetic backgrounds (Rhim *et al.*, 2000; Nadeau, 2001; Taddei *et al.*, 2001; Slavotinek and Biesecker, 2003). Variable penetrance and expressivity of human disease syndromes presumably also reflects the particular combinations of alleles across loci in individual genomes. In the context of mathematical analyses of gene networks (Meir *et al.*, 2002; von Dassow *et al.*, 2000), we can perhaps view such differences in phenotype as a failure of canalization, occurring in individuals having genomes already lying on the fringes of acceptable parameter space. Similarly for newly arising mutations, we can imagine that some alleles will impact the network of interacting genes significantly enough to affect trait expression. Predicting the phenotypic consequences of such perturbations will require a deeper understanding of morphogenetic mechanisms themselves and how variation in these mechanisms depends on the underlying developmental genetic networks. Novel approaches to this problem will allow a fuller characterization of phenotype space that is accessible through short-term evolutionary change (e.g., phenotypes that are one and two mutational steps away from "wild-type," [Dichtel-Danjoy and Felix, 2004]). Such an advance would complement statistical approaches to predicting the effects of standing genetic and phenotypic variation to short-term evolutionary change (Schluter, 1988; Falconer and Mackay, 1996; Lynch and Walsh, 1998; Arnold *et al.*, 2001).

A final important aspect of developmental variation within populations is the breakdown of canalization in response to environmental factors. This responsiveness of development to the environment is typically referred to as "plasticity," and an enormous body of literature has examined such plasticity in an ecological and evolutionary context (Roff, 1992; Stearns, 1992; Callahan *et al.*, 1997; Price *et al.*, 2003; Relyea, 2004). Plasticity has been profoundly understudied by mainstream developmental biology. Nevertheless, there are many examples of plasticity affecting development rate as well as the development of morphological traits that beg for thorough mechanistic analyses. For instance, many amphibians undergo a metamorphosis, and the timing of this larval to adult transformation can be highly dependent on environmental stresses such as temperature and crowding (Wilbur and Collins, 1973; Parichy and Kaplan, 1992). Interactions between stress hormones and thyroid hormones appear to be important, but much more work at the level of molecular mechanisms is required (Denver, 1998). Similarly, individuals of some species of salamanders can choose between metamorphosis to a terrestrial adult form or the acquisition of sexual maturity in an aquatic, otherwise larval form (Ryan and Semlitsch, 1998). Such facultative metamorphic failure, or paedomorphosis, is only beginning to be analyzed at the level of developmental and genetic mechanisms (Voss *et al.*, 2003; Voss and Smith, 2005). Finally, a wide range of discrete alternative morphologies are inducible by environmental stimuli, including predator-induced development of spines in *Daphnia*, large-mouthed cannibalistic morphs in salamander larvae, and facultative wing development in aphids (Westeberhard, 1989; Hoffman and Pfennig, 1999; Nijhout, 1999; Barry, 2000). Instances of developmental plasticity pose outstanding challenges and rewards, for furthering our understanding of how developmental regulatory and morphogenetic mechanisms influence variation within populations and individual life cycles.

V. CONCLUSIONS

In this review I have surveyed developmental variation at several levels of biological organization, from deep phylogenetic divides to variation within individuals. In some instances, variation in developmental processes results in phenotypic variation with clear impacts on individual fitness, and in other instances, no phenotypic effects are discernable. At each of these levels, determining the causes and consequences of this variation will be of tremendous importance for our understanding of basic developmental mechanisms, human health and disease, and organismal evolution. To achieve these goals, developmental biology will need to more fully embrace populational perspectives on biological organization and variation.

REFERENCES

Abzhanov, A., and Kaufman, T. C. (2000). Crustacean (malacostracan) Hox genes and the evolution of the arthropod trunk. *Development* **127**, 2239–2249.

Adoutte, A., Balavoine, G., Lartillot, N., Lespinet, O., Prud'homme, B., and de Rosa, R. (2000). The new animal phylogeny: Reliability and implications. *Proceedings of the National Academy of Sciences USA* **97**, 4453–4456.

Akimenko, M. A., and Ekker, M. (1995). Anterior duplication of the Sonic hedgehog expression pattern in the pectoral fin buds of zebrafish treated with retinoic acid. *Developmental Biology* **170**, 243–247.

Albertson, R. C., Streelman, J. T., and Kocher, T. D. (2003a). Directional selection has shaped the oral jaws of Lake Malawi cichlid fishes. *Proceedings of the National Academy of Sciences USA* **100**, 5252–5257.

Albertson, R. C., Streelman, J. T., and Kocher, T. D. (2003b). Genetic basis of adaptive shape differences in the cichlid head. *Journal of Heredity* **94**, 291–301.

Allender, C. J., Seehausen, O., Knight, M. E., Turner, G. F., and Maclean, N. (2003). Divergent selection during speciation of Lake Malawi cichlid fishes inferred from parallel radiations in nuptial coloration. *Proceedings of the National Academy of Sciences USA* **100**, 14074–14079.

Amundson, R. (1998). Typology reconsidered: Two doctrines on the history of evolutionary biology. *Biology and Philosophy* **13**, 153–177.

Amundson, R. (2000). Embryology and evolution 1920–1960: Worlds apart? *History and Philosophy of the Life Sciences* **22**, 335–352.

Anderson, F. E., Cordoba, A. J., and Thollesson, M. (2004). Bilaterian phylogeny based on analyses of a region of the sodium-potassium ATPase beta-subunit gene. *Journal of Molecular Evolution* **58**, 252–268.

Arnold, S. J., Pfrender, M. E., and Jones, A. G. (2001). The adaptive landscape as a conceptual bridge between micro- and macroevolution. *Genetica* **112–113**, 9–32.

Atchley, W. R., and Hall, B. K. (1991). A model for development and evolution of complex morphological structures. *Biological Reviews of the Cambridge Philosophical Society* **66**, 101–157.

Averof, M., and Akam, M. (1995). Hox genes and the diversification of insect and crustacean body plans. *Nature* **376**, 420–423.

Averof, M., and Patel, N. H. (1997). Crustacean appendage evolution associated with changes in Hox gene expression. *Nature* **388**, 682–686.

Balavoine, G., de Rosa, R., and Adoutte, A. (2002). Hox clusters and bilaterian phylogeny. *Molecular and Phylogenetic Evolution* **24**, 366–373.

Barry, M. J. (2000). Inducible defences in *Daphnia:* Responses to two closely related predator species. *Oecologia* **124**, 396–401.

Barton, N. H., and Turelli, M. (1989). Evolutionary quantitative genetics: How little do we know? *Annual Review of Genetics* **23**, 337–370.

Blouin, M. S. (1992). Genetic correlations among morphometric traits and rates of growth and differentiation in the green tree frog, *Hyla cinerea. Evolution* **46**, 735–744.

Bolker, J. A. (1995). Model systems in developmental biology. *Bioessays* **17**, 451–455.

Bray, N. J., Buckland, P. R., Owen, M. J., and O'Donovan, M. C. (2003). Cis-acting variation in the expression of a high proportion of genes in human brain. *Human Genetics* **113**, 149–153.

Burke, A. C., Nelson, C. E., Morgan, B. A., and Tabin, C. (1995). *Hox* genes and the evolution of vertebrate axial morphology. *Development* **121**, 333–346.

Callahan, H. S., Pigliucci, M., and Schlichting, C. D. (1997). Developmental phenotypic plasticity: Where ecology and evolution meet molecular biology. *Bioessays* **19**, 519–525.

Cohn, M. J., and Tickle, C. (1999). Developmental basis of limblessness and axial patterning in snakes. *Nature* **399**, 474–479.

Colosimo, P. F., Peichel, C. L., Nereng, K., Blackman, B. K., Shapiro, M. D., Schluter, D., and Kingsley, D. M. (2004). The genetic architecture of parallel armor plate reduction in threespine sticklebacks. *PLoS Biology* **2**, E109.

Conklin, E. G. (1897). The embryology of *Crepidula*. *Journal of Morphology* **13**, 3–209.

Conklin, E. G. (1902). Karyokinesus and cytokinesis in the maturation, fertilization and cleavage of *Crepidula* and other gasterpoda. *Journal of the Academy of Natural Sciences of Philadelphia* **12**, 5–116.

Couldridge, V. C. K., and Alexander, G. J. (2002). Color patterns and species recognition in four closely related species of Lake Malawi cichlid. *Behavioral Ecology* **13**, 59–64.

Cowles, C. R., Hirschhorn, J. N., Altshuler, D., and Lander, E. S. (2002). Detection of regulatory variation in mouse genes. *Nature Genetics* **32**, 432–437.

Cowley, D. E., and Atchley, W. R. (1992). Quantitative genetic models for development, epigenetic selection, and phenotypic evolution. *Evolution* **46**, 495–518.

Cresko, W. A., Amores, A., Wilson, C., Murphy, J., Currey, M., Phillips, P., Bell, M. A., Kimmel, C. B., and Postlethwait, J. H. (2004). Parallel genetic basis for repeated evolution of armor loss in Alaskan threespine stickleback populations. *Proceedings of the National Academy of Sciences USA* **101**, 6050–6055.

Davidson, E. H., Rast, J. P., Oliveri, P., Ransick, A., Calestani, C., Yuh, C. H., Minokawa, T., Amore, G., Hinman, V., Arenas-Mena, C., Otim, O., Brown, C. T., Livi, C. B., Lee, P. Y., Revilla, R., Rust, A. G., Pan, Z., Schilstra, M. J., Clarke, P. J., Arnone, M. I., Rowen, L., Cameron, R. A., McClay, D. R., Hood, L., and Bolouri, H. (2002a). A genomic regulatory network for development. *Science* **295**, 1669–1678.

Davidson, E. H., Rast, J. P., Oliveri, P., Ransick, A., Calestani, C., Yuh, C. H., Minokawa, T., Amore, G., Hinman, V., Arenas-Mena, C., Otim, O., Brown, C. T., Livi, C. B., Lee, P. Y., Revilla, R., Schilstra, M. J., Clarke, P. J., Rust, A. G., Pan, Z., Arnone, M. I., Rowen, L., Cameron, R. A., McClay, D. R., Hood, L., and Bolouri, H. (2002b). A provisional regulatory gene network for specification of endomesoderm in the sea urchin embryo. *Developmental Biology* **246**, 162–190.

Denver, R. J. (1998). Hormonal correlates of environmentally induced metamorphosis in the Western spadefoot toad, *Scaphiopus hammondii*. *General and Comparative Endocrinology* **110**, 326–336.

Dichtel-Danjoy, M. L., and Felix, M. A. (2004). Phenotypic neighborhood and micro-evolvability. *Trends in Genetics* **20**, 268–276.

Doebley, J., Stec, A., and Hubbard, L. (1997). The evolution of apical dominance in maize. *Nature* **386**, 485–488.

Dowling, T. E., Martasian, D. P., and Jeffery, W. R. (2002). Evidence for multiple genetic forms with similar eyeless phenotypes in the blind cavefish, *Astyanax mexicanus*. *Molecular Biology and Evolution* **19**, 446–455.

Driever, W., and Nusslein-Volhard, C. (1988a). The bicoid protein determines position in the Drosophila embryo in a concentration-dependent manner. *Cell* **54**, 95–104.

Driever, W., and Nusslein-Volhard, C. (1988b). A gradient of bicoid protein in *Drosophila* embryos. *Cell* **54**, 83–93.

Dworkin, I., Palsson, A., Birdsall, K., and Gibson, G. (2003). Evidence that Egfr contributes to cryptic genetic variation for photoreceptor determination in natural populations of *Drosophila melanogaster*. *Current Biology* **13**, 1888–1893.

Dykhuizen, D. E., Dean, A. M., and Hartl, D. L. (1987). Metabolic flux and fitness. *Genetics* **115**, 25–31.

Eizirik, E., Yuhki, N., Johnson, W. E., Menotti-Raymond, M., Hannah, S. S., and O'Brien, S. J. (2003). Molecular genetics and evolution of melanism in the cat family. *Current Biology* **13**, 448–453.

Eldar, A., Dorfman, R., Weiss, D., Ashe, H., Shilo, B. Z., and Barkai, N. (2002). Robustness of the BMP morphogen gradient in *Drosophila* embryonic patterning. *Nature* **419**, 304–308.

Endler, J. A. (1983). Natural and sexual selection on color patterns in Poeciliid fishes. *Environmental Biology of Fishes* **9**, 173–190.

Engeszer, R. E., Ryan, M. J., and Parichy, D. M. (2004). Learned social preference in zebrafish. *Current Biology* **14**, 881–884.

Ephrussi, A., and Johnston, D. S. (2004). Seeing is believing: The bicoid morphogen gradient matures. *Cell* **116**, 143–152.

Falconer, D. S., and Mackay, T. F. C. (1996). *Introduction to Quantitative Genetics*. Boston: Addison-Wesley.

Frary, A., Nesbitt, T. C., Grandillo, S., Knaap, E., Cong, B., Liu, J., Meller, J., Elber, R., Alpert, K. B., and Tanksley, S. D. (2000). fw2.2: A quantitative trait locus key to the evolution of tomato fruit size. *Science* **289**, 85–88.

Freeman, G., and Lundelius, J. W. (1982). The developmental genetics of dextrality and sinistrality in the gastropod *Lymnaea peregra*. *Roux's Archives of Developmental Biology* **191**, 69–83.

Galant, R., and Carroll, S. B. (2002). Evolution of a transcriptional repression domain in an insect Hox protein. *Nature* **415**, 910–913.

Genissel, A., Pastinen, T., Dowell, A., Mackay, T. F., and Long, A. D. (2004). No evidence for an association between common nonsynonymous polymorphisms in delta and bristle number variation in natural and laboratory populations of *Drosophila melanogaster*. *Genetics* **166**, 291–306.

Gilbert, S. F. (2003). Evo-devo, devo-evo, and devgen-popgen. *Biology and Philosophy* **18**, 347–352.

Gilbert, S. F., Loredo, G. A., Brukman, A., and Burke, A. C. (2001). Morphogenesis of the turtle shell: The development of a novel structure in tetrapod evolution. *Evolution and Development* **3**, 47–58.

Gonzalez-Reyes, A., and Morata, G. (1990). The developmental effect of overexpressing a Ubx product in *Drosophila* embryos is dependent on its interactions with other homeotic products. *Cell* **61**, 515–522.

Grenier, J. K., and Carroll, S. B. (2000). Functional evolution of the Ultrabithorax protein. *Proceedings of the National Academy of Sciences USA* **97**, 704–709.

Grenier, J. K., Garber, T. L., Warren, R., Whitington, P. M., and Carroll, S. (1997). Evolution of the entire arthropod Hox gene set predated the origin and radiation of the onychophoran/arthropod clade. *Current Biology* **7**, 547–553.

Hadfield, M. G. (2000). Why and how marine-invertebrate larvae metamorphose so fast. *Seminars in Cell and Development Biology* **11**, 437–443.

Hall, B. K. (1984). Developmental processes underlying heterochrony as an evolutionary mechanism. *Canadian Journal of Zoology* **62**, 1–7.

Hall, B. K. (1999). *Evolutionary Developmental Biology*. Boston, MA: Kluwer Academic Publishers.

Hall, B. K. (2000). Evo-devo or devo-evo—does it matter? *Evolution and Development* **2**, 177–178.

Hodin, J. (2000). Plasticity and constraints in development and evolution. *Journal of Experimental Zoology* **288**, 1–20.

Hoffman, E. A., and Pfennig, D. W. (1999). Proximate causes of cannibalistic polyphenism in larval tiger salamanders. *Ecology* **80**, 1076–1080.

Houde, A. E. (1997). *Sex, Color, and Mate Choice in Guppies*. Princeton, NJ: Princeton University Press.

Huber, J. L., da Silva, K. B., Bates, W. R., and Swalla, B. J. (2000). The evolution of anural larvae in molgulid ascidians. *Seminars in Cell and Development Biology* **11**, 419–426.

Jungblut, B., and Sommer, R. J. (2000). Novel cell-cell interactions during vulva development in *Pristionchus pacificus*. *Development* **127**, 3295–3303.

Kacser, H., and Burns, J. A. (1981). The molecular basis of dominance. *Genetics* **97**, 639–666.

Kalthoff, K., and Sander, K. (1969). Der Enwicklungsgang der Missbildung "Doppelabdomen" im partiell UV-bestrahlten Ei von *Smittia parthenogenetica* (Diptera, Chironomidae). *Wilhelm Roux's Archiv fur Entwicklungsmech der Organismen* **161**, 129–146.

Keller, R., Davidson, L. A., and Shook, D. R. (2003). How we are shaped: The biomechanics of gastrulation. *Differentiation* **71**, 171–205.

Keller, R. E. (1975). Vital dye mapping of the gastrula and neurula of *Xenopus laevis*. I. Prospective areas and morphogenetic movements of the superficial layer. *Developmental Biology* **42**, 222–241.

Keller, R. E. (1976). Vital dye mapping of the gastrula and neurula of *Xenopus laevis*. II. Prospective areas and morphogenetic movements of the deep layer. *Developmental Biology* **51**, 118–137.

Kellogg, E. A., and Shaffer, H. B. (1993). Model organisms in evolutionary studies. *Systematic Biology* **42**, 409–414.

Khokha, M. K., Hsu, D., Brunet, L. J., Dionne, M. S., and Harland, R. M. (2003). Gremlin is the BMP antagonist required for maintenance of Shh and Fgf signals during limb patterning. *Nature Genetics* **34**, 303–307.

Kitami, T., and Nadeau, J. H. (2002). Biochemical networking contributes more to genetic buffering in human and mouse metabolic pathways than does gene duplication. *Nature Genetics* **32**, 191–194.

Kopp, A., Graze, R. M., Xu, S., Carroll, S. B., and Nuzhdin, S. V. (2003). Quantitative trait loci responsible for variation in sexually dimorphic traits in *Drosophila melanogaster*. *Genetics* **163**, 771–787.

Krauss, S., Concordet, J. P., and Ingham, P. W. (1993). A functionally conserved homolog of the *Drosophila* segment polarity gene hh is expressed in tissues with polarizing activity in zebrafish embryos. *Cell* **75**, 1431–1444.

Kuo, D. H., and Shankland, M. Evolutionary diversification of specification mechanisms within the O/P equivalence group of the leech genus *Helobdella*. *Development* 131, 5859–5869.

Lambert, J. D., and Nagy, L. M. (2001). MAPK signaling by the D quadrant embryonic organizer of the mollusc *Ilyanassa obsoleta*. *Development* **128**, 45–56.

Laufer, E., Nelson, C. E., Johnson, R. L., Morgan, B. A., and Tabin, C. (1994). Sonic hedgehog and Fgf-4 act through a signaling cascade and feedback loop to integrate growth and patterning of the developing limb bud. *Cell* **79**, 993–1003.

Long, A. D., Mullaney, S. L., Mackay, T. F. C., and Langley, C. H. (1996). Genetic interactions between naturally occurring alleles at quantitative trait loci and mutant alleles at candidate loci affecting bristle number in *Drosophila melanogaster*. *Genetics* **144**, 1497–1510.

Long, A. D., Lyman, R. F., Langley, C. H., and Mackay, T. F. (1998). Two sites in the Delta gene region contribute to naturally occurring variation in bristle number in *Drosophila melanogaster*. *Genetics* **149**, 999–1017.

Lynch, M., and Walsh, B. (1998). *Genetics and Analysis of Quantitative Traits*. Sunderland, MA: Sinauer Associates.

Mann, R. S., and Hogness, D. S. (1990). Functional dissection of Ultrabithorax proteins in *D. melanogaster*. *Cell* **60**, 597–610.

Maynard Smith, J., Burian, R., Kauffman, S., Alberch, P., Campbell, J., Goodwin, B., Lande, R., Raup, D., and Wolpert, L. (1985). Developmental constraints and evolution. *Quarterly Review of Biology* **60**, 265–287.

Mayr, E. (1976). Typological versus population thinking. In *Evolution and the Diversity of Life: Selected Essays*, pp. 409–412. Cambridge, MA: Belknap Press of Harvard University Press.

Meir, E., von Dassow, G., Munro, E., Odell, G. M., Eldar, A., Dorfman, R., Weiss, D., Ashe, H., Shilo, B. Z., Barkai, N., Ruden, D. M., Garfinkel, M. D., Sollars, V. E., Lu, X., Sollars, V., Xiao, L., Wang, X., Siegal, M. L., Bergman, A., Milton, C. C., Huynh, B., Batterham, P., Rutherford, S. L., Hoffmann, A. A., Queitsch, C., Sangster, T. A., and Lindquist, S. (2002). Robustness, flexibility, and the role of lateral inhibition in the neurogenic network. *Current Biology* **12**, 778–786.

Metscher, B. D., and Ahlberg, P. E. (1999). Zebrafish in context: Uses of a laboratory model in comparative studies. *Developmental Biology* **210**, 1–14.

Minsuk, S. B., and Keller, R. E. (1996). Dorsal mesoderm has a dual origin and forms by a novel mechanism in *Hymenochirus*, a relative of *Xenopus*. *Developmental Biology* **174**, 92–103.

Mundy, N. I., and Kelly, J. (2003). Evolution of a pigmentation gene, the melanocortin-1 receptor, in primates. *American Journal of Physical Anthropology* **121**, 67–80.

Mundy, N. I., Badcock, N. S., Hart, T., Scribner, K., Janssen, K., and Nadeau, N. J. (2004). Conserved genetic basis of a quantitative plumage trait involved in mate choice. *Science* **303**, 1870–1873.

Nachman, M. W., Hoekstra, H. E., and D'Agostino, S. L. (2003). The genetic basis of adaptive melanism in pocket mice. *Proceedings of the National Academy of Sciences USA* **100**, 5268–5273.

Nadeau, J. H. (2001). Modifier genes in mice and humans. *Nature Reviews Genetics* **2**, 165–174.

Newmark, P. A., and Sanchez Alvarado, A. (2002). Not your father's planarian: A classic model enters the era of functional genomics. *Nature Reviews Genetics* **3**, 210–219.

Nijhout, H. F. (1999). Control mechanisms of polyphenic development in insects: In polyphenic development, environmental factors alter same aspects of development in an orderly and predictable way. *Bioscience* **49**, 181–192.

Niswander, L., Jeffrey, S., Martin, G. R., and Tickle, C. (1994). A positive feedback loop coordinates growth and patterning in the vertebrate limb. *Nature* **371**, 609–612.

Oleksiak, M. F., Churchill, G. A., and Crawford, D. L. (2002). Variation in gene expression within and among natural populations. *Nature Genetics* **32**, 261–266.

Orr, H. A., and Coyne, J. A. (1992). The genetics of adaptation: A reassessment. *American Naturalist* **140**, 725–742.

Panganiban, G. (2000). Distal-less function during *Drosophila* appendage and sense organ development. *Developmental Dynamics* **218**, 554–562.

Panganiban, G., Nagy, L., and Carroll, S. B. (1994). The role of the Distal-less gene in the development and evolution of insect limbs. *Current Biology* **4**, 671–675.

Panganiban, G., Irvine, S. M., Lowe, C., Roehl, H., Corley, L. S., Sherbon, B., Grenier, J. K., Fallon, J. F., Kimble, J., Walker, M., Wray, G. A., Swalla, B. J., Martindale, M. Q., and Carroll, S. B. (1997). The origin and evolution of animal appendages. *Proceedings of the National Academy of Sciences USA* **94**, 5162–5166.

Parichy, D. M. (1996a). Pigment patterns of larval salamanders (Ambystomatidae, Salamandridae): The role of the lateral line sensory system and the evolution of pattern-forming mechanisms. *Developmental Biology* **175**, 265–282.

Parichy, D. M. (1996b). When neural crest and placodes collide: Interactions between melanophores and the lateral lines that generate stripes in the salamander *Ambystoma tigrinum tigrinum* (Ambystomatidae). *Developmental Biology* **175**, 283–300.

Parichy, D. M. (2001). Homology and evolutionary novelty in the deployment of extracellular matrix molecules during pigment pattern formation in the salamanders *Taricha torosa* and *T. rivularis* (Salamandridae). *Journal of Experimental Zoology* **291**, 13–24.

Parichy, D. M. (2003). Pigment patterns: Fish in stripes and spots. *Current Biology* **13**, R947–R950.

Parichy, D. M., and Johnson, S. L. (2001). Zebrafish hybrids suggest genetic mechanisms for pigment pattern diversification in Danio. *Development Genes and Evolution* **211**, 319–328.

Parichy, D. M., and Kaplan, R. H. (1992). Maternal effects on offspring growth and development depend on environmental quality in the frog *Bombina orientalis*. *Oecologia* **91**, 579–586.

Parichy, D. M., Ransom, D. G., Paw, B., Zon, L. I., and Johnson, S. L. (2000). An orthologue of the kit-related gene fms is required for development of neural crest-derived xanthophores and a subpopulation of adult melanocytes in the zebrafish, *Danio rerio*. *Development* **127**, 3031–3044.

Parra-Olea, G., and Wake, D. B. (2001). Extreme morphological and ecological homoplasy in tropical salamanders. *Proceedings of the National Academy of Sciences USA* **98**, 7888–7891.

Pastinen, T., Sladek, R., Gurd, S., Sammak, A., Ge, B., Lepage, P., Lavergne, K., Villeneuve, A., Gaudin, T., Brandstrom, H., Beck, A., Verner, A., Kingsley, J., Harmsen, E., Labuda, D., Morgan, K., Vohl, M. C., Naumova, A. K., Sinnett, D., and Hudson, T. J. (2004). A survey of genetic and epigenetic variation affecting human gene expression. *Physiological Genomics* **16**, 184–193.

Pavlidis, P., and Noble, W. S. (2001). Analysis of strain and regional variation in gene expression in mouse brain. *Genome Biology* **2**, RESEARCH0042.

Peichel, C. L., and Boughman, J. W. (2003). Sticklebacks. *Current Biology* **13**, R942–R943.

Peichel, C. L., Nereng, K. S., Ohgi, K. A., Cole, B. L., Colosimo, P. F., Buerkle, C. A., Schluter, D., and Kingsley, D. M. (2001). The genetic architecture of divergence between threespine stickleback species. *Nature* **414**, 901–905.

Phillips, P. C. (1998). Genetic constraints at the metamorphic boundary: Morphological development in the wood frog, *Rana sylvatica*. *Journal of Evolutionary Biology* **11**, 453–463.

Popadic, A., Abzhanov, A., Rusch, D., and Kaufman, T. C. (1998). Understanding the genetic basis of morphological evolution: The role of homeotic genes in the diversification of the arthropod bauplan. *International Journal of Developmental Biology* **42**, 453–461.

Price, T. D., Qvarnstrom, A., and Irwin, D. E. (2003). The role of phenotypic plasticity in driving genetic evolution. *Proceedings of the Royal Society of London B, Biological Sciences* **270**, 1433–1440.

Pultz, M. A., Pitt, J. N., and Alto, N. M. (1999). Extensive zygotic control of the anteroposterior axis in the wasp *Nasonia vitripennis*. *Development* **126**, 701–710.

Purcell, S. M., and Keller, R. (1993). A different type of amphibian mesoderm morphogenesis in *Ceratophrys ornata*. *Development* **117**, 307–317.

Queitsch, C., Sangster, T. A., and Lindquist, S. (2002). Hsp90 as a capacitor of phenotypic variation. *Nature* **417**, 618–624.

Relyea, R. A. (2004). Fine-tuned phenotypes: Tadpole plasticity under 16 combinations of predators and competitors. *Ecology* **85**, 172–179.

Rhim, H., Dunn, K. J., Aronzon, A., Mac, S., Cheng, M., Lamoreux, M. L., Tilghman, S. M., and Pavan, W. J. (2000). Spatially restricted hypopigmentation associated with an Ednrbs-modifying locus on mouse chromosome 10. *Genome Research* **10**, 17–29.

Quigley, I. K., Turner, J. M., Nuckels, R. J., Manuel, J. L., Budi, E. H., MacDonald, E. L., and Parichy, D. M. (2004). Pigment pattern evolution by differential deployment of neural crest and post-embryonic neural crest lineages in *Danio* fishes. *Development* 131, 6053–6069.

Quigley, I. K., Roberts, R. A., Manuel, J. L., Nuckels, R. J., Herrington, E., MacDonald, E. L., and Parichy, D. M. (2005). Evolutionary diversification of pigment pattern in Danio fishes: Differential *fms* dependence and stripe loss in D. albolineatus. *Development* 132, 89–104.

Richardson, M. K., Hanken, J., Gooneratne, M. L., Pieau, C., Raynaud, A., Selwood, L., and Wright, G. M. (1997). There is no highly conserved embryonic stage in the vertebrates: implications for current theories of evolution and development. *Anatomy and Embryology (Berlin)* **196**, 91–106.

Richardson, M. K., Hanken, J., Selwood, L., Wright, G. M., Richards, R. J., Pieau, C., and Raynaud, A. (1998). Haeckel, embryos, and evolution. *Science* **280**, 983, 985–986.

Richardson, M. K., Minelli, A., and Coates, M. I. (1999). Some problems with typological thinking in evolution and development. *Evolution and Development* **1**, 5–7.

Riddle, R. D., Johnson, R. L., Laufer, E., and Tabin, C. (1993). Sonic hedgehog mediates the polarizing activity of the ZPA. *Cell* **75**, 1401–1416.

Rieppel, O. (2001). Turtles as hopeful monsters. *Bioessays* **23**, 987–991.

Rifkin, S. A., Kim, J., and White, K. P. (2003). Evolution of gene expression in the *Drosophila melanogaster* subgroup. *Nature Genetics* **33**, 138–144.

Riska, B. (1986). Some models for development, growth, and morphometric correlation. *Evolution* **40**, 1303–1311.

Rockman, M. V., and Wray, G. A. (2002). Abundant raw material for cis-regulatory evolution in humans. *Molecular and Biological Evolution* **19**, 1991–2004.

Roff, D. A. (1992). *The Evolution of Life Histories*. New York: Chapman and Hall.

Ronshaugen, M., McGinnis, N., and McGinnis, W. (2002). Hox protein mutation and macroevolution of the insect body plan. *Nature* **415**, 914–917.

Ruiz-Trillo, I., Paps, J., Loukota, M., Ribera, C., Jondelius, U., Baguna, J., and Riutort, M. (2002). A phylogenetic analysis of myosin heavy chain type II sequences corroborates that Acoela and Nemertodermatida are basal bilaterians. *Proceedings of the National Academy of Sciences USA* **99**, 11246–11251.

Rutherford, S. L., and Lindquist, S. (1998). Hsp90 as a capacitor for morphological evolution. *Nature* **396**, 336–342.

Ryan, T. J., and Semlitsch, R. D. (1998). Intraspecific heterochrony and life history evolution: Decoupling somatic and sexual development in a facultatively paedomorphic salamander. *Proceedings of the National Academy of Sciences USA* **95**, 5643–5648.

Sander, K. (1975). Pattern specification in the insect embryo. In *Cell Patterning*, pp. 241–263. Amsterdam, NY: Associated Scientific.

Sangster, T. A., Lindquist, S., and Queitsch, C. (2004). Under cover: Causes, effects and implications of Hsp90-mediated genetic capacitance. *Bioessays* **26**, 348–362.

Santos, J. C., Coloma, L. A., and Cannatella, D. C. (2003). From the cover: Multiple, recurring origins of aposematism and diet specialization in poison frogs. *Proceedings of the National Academy of Sciences USA* **100**, 12792–12797.

Schluter, D. (1988). Adaptive radiation along genetic lines of least resistance. *Evolution* **50**, 1766–1774.

Schneider, R. A., and Helms, J. A. (2003). The cellular and molecular origins of beak morphology. *Science* **299**, 565–568.

Schroder, R. (2003). The genes orthodenticle and hunchback substitute for bicoid in the beetle Tribolium. *Nature* **422**, 621–625.

Shapiro, M. D., Marks, M. E., Peichel, C. L., Blackman, B. K., Nereng, K. S., Jonsson, B., Schluter, D., and Kingsley, D. M. (2004). Genetic and developmental basis of evolutionary pelvic reduction in threespine sticklebacks. *Nature* **428**, 717–723.

Slatkin, M. (1987). Quantitative genetics of heterochrony. *Evolution* **41**, 799–811.

Slavotinek, A., and Biesecker, L. G. (2003). Genetic modifiers in human development and malformation syndromes, including chaperone proteins. *Human Molecular Genetics* **12**(Spec. No. 1), R45–R50.

Smith, M. M., and Johanson, Z. (2003). Separate evolutionary origins of teeth from evidence in fossil jawed vertebrates. *Science* **299**, 1235–1236.

Sollars, V., Lu, X., Xiao, L., Wang, X., Garfinkel, M. D., and Ruden, D. M. (2003). Evidence for an epigenetic mechanism by which Hsp90 acts as a capacitor for morphological evolution. *Nature Genetics* **33**, 70–74.

Sommer, R. J. (2001). As good as they get: Cells in nematode vulva development and evolution. *Current Opinion in Cell Biology* **13**, 715–720.

Stauber, M., Taubert, H., and Schmidt-Ott, U. (2000). Function of bicoid and hunchback homologs in the basal cyclorrhaphan fly *Megaselia* (Phoridae). *Proceedings of the National Academy of Sciences USA* **97**, 10844–10849.

Stauber, M., Prell, A., and Schmidt-Ott, U. (2002). A single Hox3 gene with composite bicoid and zerknullt expression characteristics in non-Cyclorrhaphan flies. *Proceedings of the National Academy of Sciences USA* **99**, 274–279.

Stearns, S. C. (1992). *The Evolution of Life Histories*. New York: Oxford University Press, New York.

Streelman, J. T., Webb, J. F., Albertson, R. C., and Kocher, T. D. (2003). The cusp of evolution and development: A model of cichlid tooth shape diversity. *Evolution and Development* **5**, 600–608.

Sucena, E., and Stern, D. L. (2000). Divergence of larval morphology between *Drosophila sechellia* and its sibling species caused by cis-regulatory evolution of ovo/shaven-baby. *Proceedings of the National Academy of Sciences USA* **97**, 4530–4534.

Sucena, E., Delon, I., Jones, I., Payre, F., and Stern, D. L. (2003). Regulatory evolution of shaven-baby/ovo underlies multiple cases of morphological parallelism. *Nature* **424**, 935–938.

Taddei, I., Morishima, M., Huynh, T., and Lindsay, E. A. (2001). Genetic factors are major determinants of phenotypic variability in a mouse model of the DiGeorge/del22q11 syndromes. *Proceedings of the National Academy of Sciences USA* **98**, 11428–11431.

Tanaka, M., Munsterberg, A., Anderson, W. G., Prescott, A. R., Hazon, N., and Tickle, C. (2002). Fin development in a cartilaginous fish and the origin of vertebrate limbs. *Nature* **416**, 527–531.

Tessmar-Raible, K., and Arendt, D. (2003). Emerging systems: between vertebrates and arthropods, the Lophotrochozoa. *Current Opinion in Genetics and Development* **13**, 331–340.

Theron, E., Hawkins, K., Bermingham, E., Ricklefs, R. E., and Mundy, N. I. (2001). The molecular basis of an avian plumage polymorphism in the wild: A melanocortin-1-receptor point mutation is perfectly associated with the melanic plumage morph of the bananaquit, *Coereba flaveola*. *Current Biology* **11**, 550–557.

Vachon, G., Cohen, B., Pfeifle, C., McGuffin, M. E., Botas, J., and Cohen, S. M. (1992). Homeotic genes of the Bithorax complex repress limb development in the abdomen of the *Drosophila* embryo through the target gene Distal-less. *Cell* **71**, 437–450.

Vogt, W. (1929). Gestaltungsanalyse amamphibienkeimmitortlicher Vitalfarburg. II. Teil gastrulation und mesodermbildung bei urodelen und anuren. *Roux's Archiv fur Entwicklungsmech der Organismen* **120**, 384–706.

von Dassow, G., Meir, E., Munro, E. M., and Odell, G. M. (2000). The segment polarity network is a robust developmental module. *Nature* **406**, 188–192.

Voss, S. R., and Shaffer, H. B. (1996). What insights into the developmental traits of urodeles does the study of interspecific hybrids provide? *International Journal of Developmental Biology* **40**, 885–893.

Voss, S. R., and Shaffer, H. B. (1997). Adaptive evolution via a major gene effect: Paedomorphosis in the Mexican axolotl. *Proceedings of the National Academy of Sciences USA* **94**, 14185–14189.

Voss, S. R., and Smith, J. J. (2005). Evolution of salamander life cycles: A major effect QTL contributes to discrete and continuous variation for metamorphic timing. *Genetics*, in press.

Voss, S. R., Smith, J. J., Gardiner, D. M., and Parichy, D. M. (2001). Conserved vertebrate chromosome segments in the large salamander genome. *Genetics* **158**, 735–746.

Voss, S. R., Prudic, K. L., Oliver, J. C., and Shaffer, H. B. (2003). Candidate gene analysis of metamorphic timing in ambystomatid salamanders. *Molecular Ecology* **12**, 1217–1223.

Waddington, C. H. (1942). Canalization of development and the inheritance of acquired characters. *Nature* **150**, 563–565.

Waddington, C. H. (1957). *The Strategy of the Genes*. London: Allen and Unwin.

Waddington, C. H. (1975). *Canalization of the Development of Quantitative Characters*, pp. 98–103. Ithaca, New York: Cornell University Press.

Wake, D. B. (1991). Homoplasy: The result of natural selection, or evidence of design limitations. *American Naturalist* **138**, 543–567.

Watkins, T. B. (2001). A quantitative genetic test of adaptive decoupling across metamorphosis for locomotor and life-history traits in the Pacific tree frog, *Hyla regilla*. *Evolution* **55**, 1668–1677.

Westeberhard, M. J. (1989). Phenotypic plasticity and the origins of diversity. *Annual Review of Ecology and Systematics* **20**, 249–278.

Wilbur, H. M., and Collins, J. P. (1973). Ecological aspects of amphibian metamorphosis. *Science* **182**, 1305–1314.

Wimmer, E. A., Carleton, A., Harjes, P., Turner, T., and Desplan, C. (2000). Bicoid-independent formation of thoracic segments in *Drosophila*. *Science* **287**, 2476–2479.

Wittkopp, P. J., Williams, B. L., Selegue, J. E., and Carroll, S. B. (2003). *Drosophila* pigmentation evolution: Divergent genotypes underlying convergent phenotypes. *Proceedings of the National Academy of Sciences USA* **100**, 1808–1813.

Wray, G. A. (2002). Do convergent developmental mechanisms underlie convergent phenotypes? *Brain, Behavior, and Evolution* **59**, 327–336.

CHAPTER 21

Phenogenetics: Genotypes, Phenotypes, and Variation

SAMUEL SHOLTIS AND KENNETH M. WEISS
Department of Anthropology, Pennsylvania State University, University Park, Pennsylvania, USA

INTRODUCTION

Mutations in the gene that codes for phenylalanine hydroxylase (PAH) can result in the recessive human disease phenylketonuria (PKU, Online Mendelian Inheritance in Man [OMIM] *261600). PAH converts the amino acid phenylalanine, obtained primarily through diet, into tyrosine. If, as in PKU, PAH does not function properly, the resulting build up of excess phenylalanine can lead to severe mental retardation and a suite of other symptoms. Many mutations are known, and most individuals with PKU are heterozygotes, having a different mutation on each allele. Because nearly all cases of PKU are caused by mutations in PAH, it has long been considered a classic example of a "simple" genetic (or Mendelian) disorder (but see Scriver and Waters, 1999). However, early detection (infants are generally screened) and a diet low in phenylalanine, while reportedly unpleasant, can prevent or at least greatly reduce the severity of most symptoms. In other words, even a PKU genotype does not lead to a PKU phenotype when the *environment*, in this case diet, is appropriately altered.

This classic example illustrates one of the most basic, and often overlooked, problems in biology: what factors determine the relationship between "genotype," an organism's or specific trait's genetic makeup, and "phenotype," the morphological, biochemical, or behavioral character of that organism or trait, including its entire developmental path (Lewontin, 1992; Weiss, 2003). Historically, a one-to-one relationship has dominated views of the genotype–phenotype relationship. Thus much work has been devoted to the study of genes and traits, but little in between. However, even the famous "one gene–one enzyme/protein" hypothesis holds true in only the simplest cases. Alternative splicing, the formation of protein multimers, epigenetics, epistasis, and a host of other complicating factors, including the environment, all act to complicate this relationship (Figure 21-1) or even the very definition of a "gene."

It is crucial to understand that natural selection acts only indirectly on genes (Weiss and Buchanan, 2003). There has been long-standing debate about the

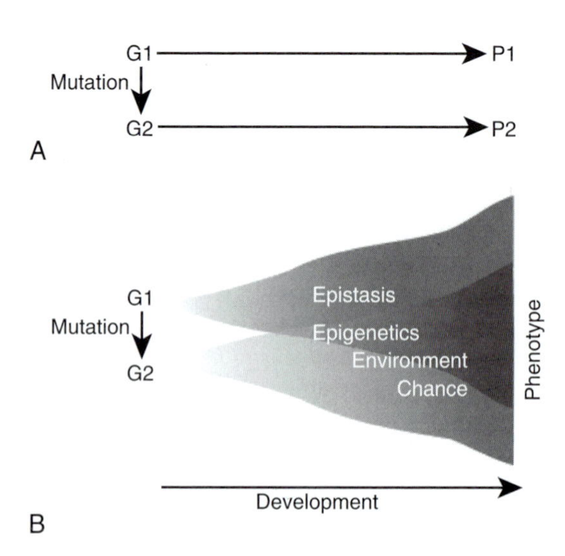

FIGURE 21-1. Old (A) and new (B) views of the relationship between genotype (G) and phenotype (P).

"unit of selection" (e.g., Lloyd, 1992). Although many have postulated the gene as the locus of selection, it is the collection of phenotypes that make up entire organisms that confront selection, not just their genomes. This filtering through phenotypes, which are the result of the interaction between genotype, developmental process, and environment, can lead to variation in the genotype–phenotype relationship. From the point of view of natural selection, as long as an acceptable trait is achieved, the underlying genetic mechanism can vary. Hemoglobin mutations in malarial areas of the world provide a familiar example in which the phenotype of relative malarial resistance can be obtained through several different genetic mechanisms. In fact, in most complex traits in which multiple genes contribute to the phenotype (i.e., stature, blood pressure), there will almost certainly be different genotypes that produce equivalent phenotypes. Where this phenogenetic equivalence exists, there can also be phenogenetic drift for basically the same reasons that we all accept genetic drift (see the following text; Weiss and Fullerton, 2000). We can therefore expect, and selection experiments (Rutledge *et al.*, 1974; Mackay, 1995, 2001) and simulations (Yedid and Bell, 2002) demonstrate, that there will be a many-to-many relationship between genotype and phenotype. Factors other than the genes themselves can be nearly as important, if not equally so. Can PKU really be considered solely, or even mainly, a genetic disorder if it can be treated environmentally? More importantly, can characterizing diseases or traits as genetically determined influence the way we view them, preventing us from seeing environmental factors or solutions (Moore, 2002)?

Ask a geneticist (or any biologist for that matter) about the role of phenogenetics, the relationship between genotype and phenotype, in producing variation during development and over evolutionary time, and you are sure to get the "right" answer. Almost no one in the field maintains the position that the genome is solely responsible for every aspect and nuance of the phenotype of an individual. Nearly all acknowledge the role of epigenetic factors, the environment, gene–environment interactions, and developmental process in the formation of phenotypes. Unfortunately, the extent to which this information is applied and the relative importance it is given is belied by the numerous reports of "genes for" diseases or traits that appear on a regular basis in the news and scientific literature (Kaplan and Pigliucci, 2001). These findings, although not trivial, often only apply to genes with strong effect discovered in isolated families with a particularly severe form of the disease or trait. The deterministic interpretations that these results are often given, the sensationalism with which they are presented, and attempts to generalize the results to all instances of the trait not only overstate the facts but also have the potential to mislead.

The importance of phenogenetic relationships has taken a backseat in biology since the modern synthesis of the 1930s and 1940s and the resulting exclusion of development from studies of evolution, although scientists have argued for much longer than that over the relative importance of genes and traits in evolution (Gilbert *et al.*, 1996; Gilbert 1998, 2003). The successes of population genetics, the theoretical underpinnings of the modern synthesis, further promoted a "gene's eye" view of evolution. Additionally, experimental methods (using inbred plants, flies, or mice reared in homogeneous environments) attempt to remove nuisance factors that may fog the gene to phenotype correlation, cementing the belief that phenotypes can be computed directly from genotypes, but does this accurately model reality (Schlichting and Pigliucci, 1998; Lewontin, 2000; Weiss and Fullerton, 2000)? Even these methods cannot really isolate the effects of genes from the environment. They only demonstrate the effect of the genes in that one specific environment (Sultan, 2003), and we know from numerous examples that genotypes do not always respond linearly to changing environments. This does not minimize the contributions of population genetics or the advances in genetics that have come since the modern synthesis, many of which have been truly groundbreaking. We owe our current advanced state of understanding in genetics to the twentieth century's somewhat deterministic view of nature. Similarly, the use of inbred model organisms is indispensable in many areas of research. However, if the goal of science is to apply lessons learned from the laboratory to nature, the underlying phenogenetic complexities that have thus far been left largely unexplored must become the focus of study, and results from experiments undertaken in these most extreme of unnatural conditions must be interpreted with care. Looking at evolution with a "trait's eye" view is an important first step. Here we present some sobering facts (and only a sampling at that) with the hopes of tempering, not discouraging, research into the genotype–phenotype relationship.

I. MECHANISM VERSUS VARIATION

There are two perspectives with which geneticists have traditionally viewed the relationship between genotype and phenotype (Weiss, 2003). The first view focuses on a causal relationship between genes and traits and can be characterized as "mechanistic." Biologists who take this stance concentrate on the decipherment of the cellular and molecular processes that read and interpret a given DNA sequence for the production of proteins, and therefore it is often presumed phenotypes. Traditionally, the mechanistic view of the phenogenetic relationship invokes an image of genetic determinism and such oft (mis)used analogies as the "genetic blueprint," DNA as the "book of life," or phenotypes "computed" from genotypes (Kay, 2000). However, as will be shown, there can be a great deal of variation and complexity in the genotype–phenotype relationship even when mechanism is the focus of study. Alternative splicing and other mRNA editing, RNA interference, the formation of protein multimers, gene regulation, and a host of other factors all act to complicate the relationship between genes and phenotypes. This relationship is rarely, if ever, one to one.

Genes, in the mechanical or physiological sense, are involved in the production of all traits, but how or whether genes, and their variation, determine traits and the variation in those traits among individuals or populations is a separate question. The many steps between DNA sequence and trait (e.g., regulatory control of expression, mRNA processing, translation, interaction with other genes or proteins and the environment, and chance) add variation to the genotype–phenotype relationship. This "variation" is the focus of study of the second phenogenetic perspective. No two individuals share the same environment, development, the time and space over which genotype and environment interact to create phenotype, or for that matter the same genes (even monozygotic twins accumulate different somatic mutations from the time their individual cells begin to divide). Thus the specific interactions of genes, development, and environment that produce the phenotypes of an individual are unique and can introduce a great deal of variation. The existence of this variation in the genotype–phenotype relationship is not surprising when one views evolution as acting at the level of the phenotype. Comparing the results of gene knock-out experiments between strains of inbred animals (e.g., LeCouter *et al.*, 1998) and the relationship between plants grown in various environments as famously shown by Clausen, Keck, and Hiesey in the 1940s (for review, see Nunez-Farfan and Schlichting, 2001) provide beautiful examples. Strictly Mendelian traits or diseases are rare, and when they are closely studied almost never turn out to be "simple" (Scriver and Waters, 1999). Most biologists know about variation in the expressivity and penetrance of diseases, but few have focused on the evolutionary implications of these and other similar factors, especially in traits that are not disease related. Nongenetic inheritance (i.e., language, learned behaviors, and often important environmental factors such as diet) also plays an important role in the understanding of variation resulting from phenogenetic interactions and not just among humans.

This variation has a huge impact on our ability to map genotype to phenotype, or vice versa. For instance, discovering the "gene for" a particular trait in one individual or family may or may not tell us how that trait develops in others, a fact borne out in the difficulty that is faced in repeating such studies (Long and Langley, 1999; Goring *et al.*, 2001; Ioannidis *et al.*, 2001). In this chapter we will attempt to highlight the importance of phenogenetics in modern biology by exploring the complexity and variation that arises in the relationship between genes and phenotypes whether one focuses on mechanism or variation. We do not suggest that the majority of scientists are unaware of the importance of the phenogenetic relationship but that despite their awareness business continues as usual, and the primacy of genes in evolution and determining phenotypes is taken for granted.

II. FROM GENOTYPE TO PHENOTYPE: MECHANISM

We now know that the "one gene–one enzyme" hypothesis first suggested by Beadle and Tatum (1941) through their work with the bread mold *Neurospora* greatly oversimplifies reality. Many genes play multiple functional roles (pleiotropy) within and among taxa, and because selection works at the phenotypic level, the opposite, that multiple different genes can fulfill equivalent functions, is also true. The updated "one gene–one polypeptide" comes closer to the truth by eliminating the idea of one function, but we will see several examples where even this may not be accurate. Understanding the many-to-many relationships that result from this complexity requires an appreciation of the mechanisms that relate gene structure to protein function. This entails more than transcription and translation, often the only coverage this subject receives in introductory biology courses. The gene-to-protein fidelity that is implied in this necessarily simplistic view is misleading, leaving out the importance of such factors as mRNA editing by splicing or ADAR (see the following text), somatic recombination, and homomultimers and heteromultimers to protein function. Additionally, questions of gene regulation, where and when a particular gene will be expressed and produce a particular protein, are left unasked. Mechanisms that influence the gene to phenotype relationship can be classified into two general categories: those that act prior to and influence transcription itself and those that work during and after transcription at the RNA or protein level.

A. A Quick Digression Concerning DNA Sequence: Arbitrary and Saturated

It is noteworthy to mention that our understanding of these processes do not come from DNA sequence alone. In fact, the sequence of As, Ts, Cs, and Gs that make up a genome would be completely uninterpretable without the vast amount of experimentation that has demonstrated the function of specific DNA elements. There is nothing that we know, for instance, biochemically about the sequence of nucleotides ATG that suggests that it should indicate the start of translation or that it should code for the amino acid methionine, but it has been experimentally shown to do so. In other words, the genetic code is functionally arbitrary. This is also true of the genetic signaling and gene regulation that control cellular differentiation during development (Weiss, 2002). Common signaling pathways (e.g., Wnt, notch/delta, TGF-β) and transcription factors, although often named for the context in which they were discovered, are used generically in many different situations during development and in many different tissues. Gene prediction algorithms that appear to find genes from sequence alone must use experimentally derived information such as the genetic code to look for open reading frames, splice donor and acceptor sites to find exon/intron borders, and regulatory elements like the TATA and CAAT boxes to locate the beginning of genes. An additional layer of complexity is added by the fact that the genome is saturated with such sites. In the extreme case, the two nucleotide canonical splice donor and acceptor sites (GT and AG respectively) would be expected to occur every 16 base

```
TGAAATAAACTGCAATAATTTCTTCGTGAAAACCTGCTGTGA
GCTGAGTGATATAATACTAACATTTATTTTAGCCCTTCTCAT
GTCACTAAGCACTTTATATATCCTATTTAACTTGAATTCTAC
AATAATTCTATGTAATAAATACTAAATCTTGTAAGCAATGGA
GAGAAATATACATGCACTGGCCATCCCTATCTTTAGGAAGTA
TGCAGTCAAACTAAAAAACAAGAGCAATACATCAAACTATTT
AGATTCTTAATAGATAAGGGCATTCCAGAGGAGTATCCATTT
TTCCGGATGAGAGAGCTTTGCTGGAGGCCAGACTGCCGGACG
TGGTGGCTCATGCCTGTAATCCCAGCACTCTGGGAAGCCAAG
GTGGGTGGATCACTTTGAGCTCAGGAGTTGGAGACCAGCCTT
GGCAACACTGCGAAACCTCATCTCTACAAAAAATACAAAAAA
TTAGCCAGCTGCGGTGGTGCACACCTGTAGTTCCAGCTACTA
GGGAAGCTGAGGATCACTTGAGCCTGGGAAGTGGAGGTTGCA
```

FIGURE 21-2. Saturation: splice donor and acceptor sequence elements highlighted in a randomly selected sequence from the human genome.

pairs (bp) by chance alone (Figure 21-2). In a genome with some 3 billion bp (approximately the size of the human genome), that would be more than 180 million sites each! Obviously, just locating these sites alone tells us little about their potential function; the sequence context in which they are found must contain as much information as the sites themselves and must be taken into consideration as well. This saturation and functional arbitrariness makes understanding the mechanisms that lead from gene to functional protein even more difficult.

B. Pretranscriptional Mechanisms

The most universal of the pretranscriptional mechanisms is gene regulation. Basically every cell in every tissue of an organism has an identical set of genetic information (ignoring mutations and some exceptions noted in the following text); what distinguishes these cells and tissue types is the complement of genes active during their development (Freiman and Tjian, 2003). Gene regulation is a hierarchical process that depends on the developmental history of a cell and its interaction with neighboring cells. The mechanism of regulation for any given gene requires several factors. The main components of this process are the actual machinery of transcription, usually RNA polymerase II; promoters, which facilitate the activity of the polymerase; *cis*-acting (on the same chromosome) enhancer, repressor and insulator modules (i.e., transcription-factor-binding sites); and the *trans*-acting transcription factors themselves, often used in concert, which bind to the DNA and influence the ability of RNA polymerase to do its job (Blackwood and

Kadonaga, 1998). Other aspects involve sequence variation related to chromosomal packaging and manipulation in the cell.

The complexity of this system of regulation is best illustrated by the function of the *cis*-regulatory elements. Although regulatory modules are usually found within a few kilobase pairs (kb) 5′ of a gene, they are also found 3′, in introns, and even hundreds of kb from the gene they regulate (Lettice *et al.*, 2002, 2003; Nobrega *et al.*, 2003). These modules, of which there may be many for a particular gene used in different spatial or temporal contexts, contain the binding sites for several transcription factors that may work cooperatively or antagonistically to regulate the expression of a gene. The complement and concentration of transcription factors (TF) present in a cell and the number, location, and specificity of binding sites in a gene's enhancer modules establish its expression domain. However, all of these factors are statistical rather than deterministic. For example, TFs recognize and bind to specific DNA sequences, but these recognition sequences can vary. This variation may influence the affinity of TF binding and therefore the expression levels of a gene (Dermitzakis and Clark, 2002). Because of their short length (as few as 5–10 nucleotides), binding sites evolve quite differently from genes themselves. New genes generally arise through the duplication and divergence of existing genes; new TF-binding sites, however, can easily appear through mutation in raw sequence (Stone and Wray, 2001). Binding sites can also mutate out of existence in the same manner, and this volatility has an important role in phenotypic evolution (see the following text). In addition, binding sites of TFs may overlap and consequently compete with one another. In this case the relative concentration of TFs and the location and number of enhancer and repressor elements present will influence expression (Blackwood and Kadonaga, 1998), which may also be affected by the way different enhancers are arranged in regulatory "cassettes."

There is also an element of chance involved in gene regulation. Individual TFs produced in the cytoplasm of a cell must find their way into the nucleus and locate their specific binding sites in a veritable sea of nucleic acids and proteins. By chance, this doesn't necessarily happen in the same way in every cell of a given tissue, even though we may say that the tissue is expressing the gene. What seems to be important is that "enough" cells (whatever that means) in the tissue behave appropriately (Weiss and Buchanan, 2004). Binding of TF to DNA alone is often not enough to drive expression; also necessary are the protein–protein interaction among several TFs and possibly other proteins. An example is found in tricho-dento-osseous disease (TDO, OMIM #190320; Hart *et al.*, 1997). The disease has been mapped to and mutations found in the homeobox containing transcription factor *Dlx3*. The mutations however do not affect the DNA-binding domain of the gene, but instead truncate the gene, removing a protein interaction domain. The truncated gene can still bind to DNA, but without binding to an as yet unidentified protein partner, its function is impaired (Price *et al.*, 1998; Dodds *et al.*, 2003).

Also important in gene regulation are epigenetic factors, such as DNA methylation or histone acetylation, affecting the structure rather than the sequence of DNA.

For example, high levels of DNA methylation, which generally affects the cytosine in CpG dinucleotides, can make certain regions of the genome inaccessible to the transcriptional machinery. Interestingly, these nonsequence factors can be inherited both cell to cell within an individual's lifetime and across generations. A well-known example is genomic imprinting, in which methylation shuts off the allele of a gene an individual receives from one parent (either maternal or paternal). As many as 70 imprinted genes have been found in mammals, but estimates of the total number are much higher (Murphy and Jirtle, 2003). A different epigenetic mechanism is used in X-inactivation to compensate for the increased gene dosage in XX females compared with XY males. Here, RNA from the gene *Xist* physically covers all but the small pseudo-autosomal region of one randomly chosen copy of the X in each cell (Plath *et al.*, 2002). This RNA coat prevents genes on the inactive X from being expressed.

Phenotypic variation can also arise within an individual's lifetime through somatic mutations that occur during each mitosis and are inherited in descendant mitotic lineages thereafter (as imprinting can be). These can, but need not be, pathological, leading, for example, to cancers. In organisms without sequestered germ lines such as sponges that reproduce by sloughing off groups of cells, offspring genomes can be different from parental even without the usual mechanisms employed in sexual reproduction. In this regard, not all the cells in a sponge need even have an identical genome. In flowering plants, meristems develop individually so that even within a single plant different flowers may accumulate different mutations and thus have different genomes. If these mutations are beneficial and lead that meristem to survive or reproduce more effectively, the "acquired character" can be passed to the next generation and could eventually replace the original in a form of "Lamarckism" that is also completely consistent with our current ideas of genetics and natural selection.

In other systems, several alternative genes, often but not always directly related as members of gene families, are used for a given function, but only one or a subset of these genes is used in a given context. The mammalian adaptive immune system provides an example of this kind of pretranscriptional mechanism that does not fit with the traditional view of the gene to protein relationship. Our immune system must produce a practically limitless amount of variation to cope with an unpredictable array of invaders (Weiss and Buchanan, 2004). It does this in many ways, but of particular interest is the "somatic recombination" of the immunoglobulin genes in the production of antibodies. Antibodies are multimers composed of two heavy and two light immunoglobulin chains. Each of these chains is comprised by several subunits: constant (C), joining (J), and variable (V) regions in the light chain, and heavy chains have these plus an additional diversity (D) region. The genome contains multiple copies of each of these subunits, as many as 51 for the heavy-chain V region (Nossal, 2003). During lymphocyte development, individual copies of each region are chosen, apparently at random, to be transcribed. Recombination brings the chosen regions together into a single transcription unit, and the unused copies are removed from the genome of the cell. This recombination can produce an enormous variety of immunoglobulin chains, but additional variation is added by diversity in the junctions between subunits by including one or several additional nucleotides, often shifting the downstream reading frame and the somatic hypermutability of these regions. A similar mechanism is used in the production of T-cell receptors, another component of the immune system.

This cellular gene choice, expression of a single gene, and often only one allele of that gene, to the exclusion of other closely related genes occurs in several other contexts. For example, olfactory neurons usually only express one of the more than 1000 possible odorant receptor genes (Serizawa *et al.*, 2003; Mombaerts, 2004), and cone cells express either a red or green opsin gene, but not both (Smallwood *et al.*, 2002). In addition many other genes are expressed monoallelically, either through imprinting or other mechanisms (Knight, 2004).

C. Posttranscriptional Mechanisms

Getting transcribed at the right time, place, and level is only the first step in the transition from gene to protein. Posttranscriptional mechanisms that influence the genotype–phenotype relationship include, but are not limited to, alternative splicing, RNA interference, and the joining of individual polypeptides into functional proteins.

Most eukaryotic genes are organized as a series of exons (protein-coding sequence) interrupted by intervening, noncoding sequences known as introns. Following transcription, splice donor and acceptor sites (mentioned in the preceding text) signal cellular machinery to remove the introns forming the mature messenger RNA (mRNA). For many genes this splicing can be done in several different ways, removing or adding coding sequence, leading to altered protein structure and probably altered function. In humans, the prevalence of alternative splicing has been estimated to be as high as nearly 60% of known genes (Lander *et al.*, 2001), but it is difficult to estimate how many of these splice variants, usually identified through ESTs (expressed sequence tags), are functional and how many are merely accidents of cellular processes (Sorek *et al.*, 2004). For the case of alternative splicing, we can update the classic phrase to "one gene–many polypeptides."

Another form of mRNA editing known as ADAR (adenosine deaminase acting on RNA) actively changes specific A nucleotides in mRNA to I (inosine), which acts as G during translation (Reenan, 2001). At least in some cell types and for selected genes, this can result in amino acid changes. For example, the glutamate receptor gene *GluR-B* is edited nearly 100% of the time in neurons changing a glutamine codon to arginine. Reduction in the efficiency of this editing has been associated with human diseases such as amyotrophic lateral sclerosis (ALS) (Kawahara *et al.*, 2004) and malignant gliomas (Maas *et al.*, 2001). What is curious, of course, is why such an elaborate process for editing would evolve rather than just changing the DNA sequence. However, whatever the reason, this system is deeply conserved and has neural function in species from *Drosophila* to mammals. Presumably, it allows spatial and temporal specificity of editing, so that in certain contexts we might see "one gene–a different polypeptide."

Antisense RNA (RNA with sequence complimentary to a functional gene) has long been suspected of playing a role the in the posttranscriptional regulation of gene function (for review, see Wagner and Flardh, 2002), and the recent explosion of interest in RNA interference (RNAi) seems to validate that suspicion. In fact RNA is involved in many other cellular processes, including pretranscriptional gene silencing, RNA editing (both splicing and ADAR), translation, and possibly even DNA editing through reverse transcription (Herbert, 2004). Mattick (2003, 2004) suggests that as much as 98% of transcription in humans is not protein coding (including introns). This means either that our cells produce a large amount of useless RNA or that this noncoding RNA plays an important functional role. Recent work has demonstrated the existence of RNAi in species ranging from bacteria to mammals, suggesting its early evolution and nearly universal application (Wagner and Flardh, 2002; Cerutti, 2003). The mechanism of RNAi is different in different taxa, but the implication is the same: In some cells at some times, genes get transcribed at levels inappropriate for that cell's function. Although nary a day goes by without some new report of the action of RNAi, we are still too early in our understanding of this process to know for sure the size of its role. However, if it is as prevalent as so far seems to be the case, this means that knowing when and how much of a particular gene is being *expressed* in a cell may not mean knowing its importance: One gene–no protein?

Another important general factor to consider in understanding the genotype to phenotype mechanism is that genes only code for strings of amino acids; functional proteins typically are modified and/or require specific proteins to protect them, fold them into functional three-dimensional shape, or transport them in the cell. Many functional proteins are complex forms consisting of multiple copies of the same (homomultimers) or different (heteromultimers) polypeptides (that is, coded by different genes). A functional hemoglobin molecule, for example, consists of two α cluster and two β cluster chains and different members of the α and β clusters are used at different points during development: many genes–one protein.

In this mechanistic sense, because all traits rely on proteins, which with the preceding caveats are coded by the genome, all traits are genetic. However, as we have tried to show, even when one concentrates on this view, there is no simple gene to protein relationship, and as we will show in the following text, producing proteins is only a first step in the series of interactions that lead to phenotype. All of the factors we have seen so far work at the level of the cell, but in multicellular organisms, these complications must be coordinated and choreographed as cells differentiate and like cells congregate into tissues. Amazingly, it almost always works. In the following text, we will discuss the further complication of understanding how variation in the genotype leads to variation in phenotype.

III. FROM GENOTYPE TO PHENOTYPE: VARIATION

Thus far, we have focused on the mechanisms relating genes to phenotypes, but a host of other factors, genetic, developmental, environmental, and stochastic, influence the timing, spatial organization, and precision of these mechanisms and therefore the genotype–phenotype relationship. In certain contexts these factors can lead to additional phenotypic variation from the same genes and in others to stable phenotypes and cryptic genetic variation. Also, there is a commonly held view that evolution takes place by incremental change in end-product genes, those with structural or enzymatic function, for example. However, these genes make up only a small percentage of most genomes. The greater portion of genes is involved in gene regulation, intercellular and intracellular signaling, and protein processing. These genes are used generically and interact during development so that end product or structural genes can be deployed at the right time and place to do their usually narrow task. Change in these regulatory or processing genes, their interactions with other genes or proteins and the environment, or even more likely changes in the regulatory elements (e.g., TF-binding sites) of genes at any stage during development can alter phenotypes and may be more important in phenotypic evolution than changes in the structural genes themselves (e.g., Carroll *et al.*, 2001). It cannot be stressed too greatly that genes only code for polypeptides that are several steps removed from what we usually call a trait. In this section we will discuss some of the ways that traits develop and through that how they can evolve. Important throughout this section is the labile relationship between genotype and phenotype. If we have learned anything in the century and a half since Darwin, it is that the only criterion of evolution is that, if it works, it works. A prescriptive view of the phenogenetic relationship may find that the exceptions outnumber the rules.

A. A Lexicographer's Nightmare: Canalization, Robustness, Plasticity, Polyphenism ...

There is an unfortunately long list of sometimes ungainly terms that pop up in discussions of the relationship between genotype and phenotype. Their meanings overlap. However, there are nuances, and not everyone uses them in exactly the same way. Many of these concepts are discussed in more detail elsewhere in this volume (see, e.g., Gibson, Stearns), so here we will try to reduce them to generalities and give a brief description of the impact that they have on our understanding of the relationship between genes and traits. All have to do with the interaction of the genotype with the environment (here broadly defined to include everything from genetic background to external temperature) to produce phenotype. They can essentially be boiled down to two phenomena, both of which undermine the traditional one-to-one view in phenogenetics: (1) a given phenotype produced reliably despite slight variation in either the underlying genes or the environment (e.g., canalization, robustness, developmental stability, cryptic variation, buffering) or (2) different phenotypes produced by the same genotype in different environments (e.g., phenotypic plasticity, norms of reaction, polyphenism).

Waddington and Schmalhausen, two of the few biologists actually interested in connecting development and evolution during the height of influence of the modern synthesis, introduced the first phenomenon, what Waddington (1942) termed "canalization." The impetus came from the fact that "wild-type" organisms appeared to be much less variable than "mutants." Waddington suggested that this reduced variation resulted from selection; he argued that it would be advantageous for organisms to produce some optimal norm even in the face of minor genetic or environmental perturbation. This makes intuitive sense, but we also tend to think that canalization reduces evolvablity, a potential liability. Recent work has shown that, although this may be true in the short-term, the hidden variation that results can have a large long-term evolutionary impact. The evolution of discrete traits has been a problem in evolutionary theory since its inception, especially because these traits often have a complex genetic architecture so that single mutations are unlikely to produce the necessary change (but see the next section for a situation in which this could occur). Goldschmidt's notion of "hopeful monsters" has been universally panned, but several lines of recent evidence suggest that saltational change may have played a role in evolution (Dietrich, 2003). Probably most important for reviving at least some acceptability for the notion of sudden change was the discovery of the genetic basis of change in the numbers or arrangement of segmental or meristic traits. The classic instance was the finding that Hox gene mutation could produce extra wings or leg–antennal substitutions in flies, examples of traits considered long ago in this context by William Bateson (Bateson, 1894). Change in segment number can be brought about by mutation in a single gene. Though the change is often not completely viable or competitively "fit," it shows that small genetic change can lead to organized morphological change, and some of this resembles evolutionary changes in body plan (Carroll *et al.*, 2001).

A basic idea is that the majority of changes necessary for a discrete shift in phenotype could be phenotypically silent. However, at some point, one more change in this newly created genetic background could generate a qualitatively different phenotype. Lauter and Doebley (2002) suggest that this could be the case in teosinte, the wild ancestor of maize, which harbors genetic variation for invariant traits that distinguish it from maize. This variation manifests by changing the genetic background through maize–teosinte hybrids (Lauter and Doebley, 2002). This phenomenon has also recently been modeled using the secondary structure of RNA (the use of the term "plastogenetic congruence" in this work may take the jargon award! Ancel and Fontana, 2000; Fontana, 2002).

One buffering mechanism that has received much recent attention is the mechanism of heat shock proteins (HSPs). HSPs act as protein chaperons in both normal conditions and under stress, ensuring proper protein folding or degrading damaged proteins (Rutherford, 2000, 2003). Experimental alteration of HSP expression level has demonstrated its role both by ameliorating the harmful effects of mutations when overexpressed (Fares *et al.*, 2002) and by revealing hidden variation when reduced (Rutherford and Lindquist, 1998; Queitsch *et al.*, 2002).

Recently, Siegal and Bergman (2002) have suggested that canalization could evolve even without selection; it could simply be an inevitable result of the interconnectedness of complex developmental systems. Regardless of how it has come about (for a recent discussion see de Visser *et al.*, 2003), one result is that a certain amount of genetic variation has no impact on phenotype under so-called "normal" conditions. To come full circle, Waddington, in a set of famous experiments (Waddington, 1953, 1956), also demonstrated that if hidden genetic variation could be induced into expression by a large enough developmental shock, it could be selected for and eventually expressed in the absence of the shock, a process he called "genetic assimilation."

Since the end of the nineteenth century, we have known examples of the opposite phenomenon, different phenotypes from the same genotype (Gilbert, 2003). Despite this, it is often thought that even if we cannot predict genotype from phenotype, the converse should be relatively easy. Although it may sometimes be true, for many complex traits, the importance of nongenetic factors such as the environment makes even this task daunting. This is relevant to ideas about the potential genetic specificity of natural selection, which also has to screen genetic effects indirectly via phenotypes. One early and famous experimental example mentioned in the preceding text involved growing plants at different elevations, but there are also many natural examples: seasonal morphs in butterflies and caterpillars, predator-induced polyphenism in some fish and other organisms, and environmental sex determination, to name a few. Wing polyphenism in ants (Abouheif and Wray, 2002) provides a particularly instructive example because it also demonstrates the instability of genotype–phenotype relationships. Environmental cues determine whether ants develop into winged queens or wingless nonreproductive castes. All extant ant species share this trait, so it is believed to have evolved only once. However, the genetic mechanism with which wing development is interrupted varies both between and within ant species. Abouheif and Wray (2002) show that wing development is disrupted differently in the two wingless castes (soldiers and workers) of the ant species *Pheidole morrisi* and even between the forewings and hindwings of one caste.

Both of the phenomena discussed in the preceding text have been known for the better part of a century. Although interest grows, the importance of environmental contributions to phenotypes is still underplayed by a majority of biologists.

Also, these are exactly the kinds of things that are minimized by our experimental methods. There has, however, been a call (Gilbert, 2001; Sultan, 2003) to shift biology out of the laboratory and back into the field. This "ecological-developmental biology," as it is called, will be important if we are to understand evolution and development as they happen in nature rather than in the controlled conditions of the laboratory.

B. Developmental Process: Patterning Repeated Traits

Repeated structures abound in nature in both plants and animals. Look in the mirror; your hair, your teeth, your fingers and toes, your ribs, and your vertebrae all represent series of repeated elements, often with some variation on the theme. Stripes on fish, pelage patterns in animals, segments in insects, and branch and leaf patterns in plants all provide additional examples. Is there a gene for each tooth, stripe, or leaf? Almost certainly not. The interaction of many genes and other factors determine the development of these complex traits. Two complimentary models of developmental processes have been proposed to explain how these repeated, serially homologous traits develop, and there is mounting experimental and computer-simulated evidence to suggest that these processes are widespread. The first model relies on gradients of diffusing morphogens (e.g., signaling, transcription, or growth factors) that establish heterogeneous gene expression across an embryo that hierarchically sequester regions with different developmental fates. The identity of these regions may then be determined by a combinatorial code such as the famous Hox code used in anteroposterior (AP) axis determination in both invertebrates and vertebrates. The second model suggests that wavelike patterns, such as those found in teeth and pelage patterns, for example, can emerge through the dynamic interaction of diffusing genetic or chemical factors. Again, combinatorial gene expression is involved, this time involving different combinations in areas of structure growth compared with surrounding inhibition zones.

Complex developmental and regulatory networks underlie most of the traits patterned in this manner and many others. Often these networks are used in several different contexts during development in a single species (e.g., Hox genes also pattern limbs) so that presumably a given network evolved in a single context and was subsequently coopted as novel morphologies evolved (Carroll *et al.*, 2001). Amazingly, many of these networks are highly conserved across deep phylogenetic distance, often having similar roles in vertebrates and invertebrates. In plants similar patterning process are used but with a different set of genes, for example, MADS box rather than homeobox, even though plants do have homeobox genes. Meyerowitz (2002) therefore suggests that the similar processes evolved independently in plants and animals, demonstrating how fundamental these processes are to development and evolution. Although the level of this conservation is sometimes surprisingly high, it fits with our evolutionary view of life and the idea of common ancestry. This characteristic also forms the basis of our ability to use experimental model organisms. In many respects a human is genetically a mouse or even a fly. However, model organisms do not always behave as we expect; for example, knock-out experiments of what seem like important genes in mice often produce an unexpected or "no" phenotype (Thyagarajan *et al.*, 2003), and up to 50% of yeast genes when inactivated have been characterized as nonessential (Thatcher *et al.*, 1998). Of course, it is unknown and possibly unknowable in these cases whether changing the environment would induce effects (Tautz, 2000).

Because the first model is familiar to most and reviews can be easily found (e.g., Carroll *et al.*, 2001), especially concerning axis determination in the fruitfly, we will focus attention on the second. However, one particularly interesting aspect of axis development in flies for our understanding of the genotype–phenotype relationship is that the initial conditions for segmentation are set before the activation of the zygotic genome. So-called "maternal effect" genes form gradients of high to low concentration from both ends (e.g., *bicoid*—anterior; *nanos* and *caudal*—posterior) of the embryo. These gradients establish the heterogeneous environment that allows the developmental cascade of gap, pair-rule, segment polarity, and finally Hox genes to sequester and successively define segments and their identities. Thus the maternal genome plays an important role in determining the phenotypes of its offspring. This type of pattern formation has withstood the test of extensive experimentation and is universally accepted.

Less well known is the second model mentioned in the preceding text, which draws its inspiration from the chemical reaction–diffusion processes first described by Turing (1952). A basic reaction–diffusion process in essence entails the diffusion of two substances in an initially uniform field. One substance, called the "activator," induces its own activity and that of the second substance, the "inhibitor," which reduces the activity of the activator. In such situations periodic wavelike patterns of activator level often emerge (Figure 21-3). The amount of each substance, the level of their effect on one another, and their respective rates of diffusion determine the dynamics of the system. One can easily see how such a process could be at the root of periodically patterned morphological traits. For example, peaks of activator level could cross a threshold initiating a developmental cascade that eventually produces a morphological unit (e.g., tooth, stripe), while valleys correspond to the spaces between units. Importantly, the specific genes used in these processes are not as important as the dynamic properties of the system. In fact the same genes (often fibroblast growth factors or bone morphogenetic proteins) could be used generically in many patterned traits; the process specifies the pattern, not the trait. Computer simulations have shown that reaction–diffusion type processes can accurately model tooth patterns (Salazar-Ciudad and Jernvall, 2002; Streelman *et al.*, 2003), stripes on fish (Kondo, 2002), mollusk shell pigmentation patterns (Meinhardt *et al.*, 1995), feather patterns (Jung *et al.*, 1998), vascular patterns in leaves (Dengler and Kang, 2001), and many others.

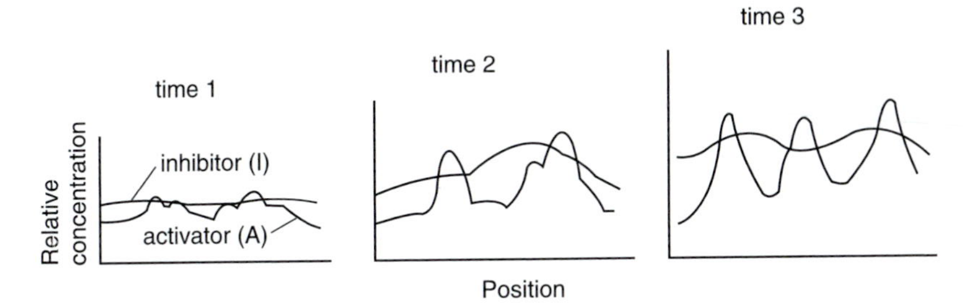

FIGURE 21-3. A basic reaction–diffusion model shown at different sequential times creating a repetitive pattern. The horizontal axis represents a layer of otherwise similar cells. Cells produce an activator (A) that catalyzes its own continued production. As (A) diffuses across the tissue, cells are stimulated to produce (A) and a rapidly diffusing inhibitor (I). Where the concentration of (A) relative to (I) exceeds a threshold level, gene switching occurs leading to activity peaks where units, such as teeth or stripes, form.

One of the most interesting aspects of dynamic patterning processes is that slight tweaking of the parameters (e.g., diffusion rates, the space in which the reaction takes place) genetically or environmentally can lead to large *qualitative* changes in the pattern produced. This could possibly explain evolutionary changes in repeated discrete traits because the genes responsible for the actual construction of the trait would not be affected. Changes that lead to additional peaks of activator could increase the number of units of a repeated trait because the activator simply recruits the existing developmental program. Jernvall (2000) has shown how variation in cusp number in the teeth of seals fits with just these types of changes. Similarly, so-called atavistic traits such as horse's toes could be the result of relatively simple changes in patterning dynamics, rather than requiring back mutations to undo millions of years of evolution (e.g., Weiss and Sholtis, 2003).

These two models are almost certainly not mutually exclusive, and combinations probably are important in the development of many traits. In teeth where the patterning mechanism is still not completely understood, plausible arguments have been made for both types of patterning processes (Thomas *et al.*, 1997; Weiss *et al.*, 1998). What is important for our understanding of phenogenetics is that in both of these processes many of the genes important in the evolution and development of repeated traits determine the pattern rather than the trait.

C. Gene Regulation and the Evolution of Phenotypes

We described the mechanism of gene regulation in the preceding text and discussed how easily new binding sites for transcription factors can evolve. These characteristics of the regulation of gene expression are extremely important in understanding the evolution of phenotypes. Because of the closeness of human and chimpanzee proteins (>98% identity, and we now know that our genomes have similar levels of identity), King and Wilson (1975) suggested early that our morphological differences may be explained by differential gene regulation, rather than changes in end-product genes. In 1975 this was a mere guess, but as our understanding of the processes of gene regulation has grown and examples have been identified, thanks largely to microarray technology, it has proven to be truly prophetic.

Differential gene expression has been demonstrated both within and among several species. Fay and Wu (1999) describe differential gene expression in nine yeast strains correlated with resistance to copper sulfate, an antimicrobial. Oleksiak and others (2002) show variation in gene expression in fish species of the genus *Fundulus*. Enard and colleagues (2002) compared gene expression levels among several primate species, including humans, and found differences. This result received much attention because expression differences were greater in human brain tissue than in liver or blood leukocytes when compared with other primates. However, similar to the case of "genes for" traits, an "expression difference for" interpretation of the evolution of humans is probably far too simplistic. Many other examples can be found (Wray, 2003, reviews several) suggesting that differential gene regulation must play an important evolutionary role.

To complicate things further, we also see conserved expression patterns despite wholesale changes in regulatory machinery. For example, both the *cis*-regulatory elements and the transcription factors that control expression of the gene *Endo16* have changed in two species of sea urchin without changing its expression pattern (Romano and Wray, 2003). Similarly, α-globins and β-globins, which shared a common ancestor approximately 450 million years ago, continue to be expressed coordinately with almost no conserved regulatory elements (Hardison, 1998).

D. Phenogenetic Drift: The Role of Chance in the Evolution of Genotype–Phenotype Relationships

At the crux of all the previously mentioned examples of problems in phenogenetics is the labile relationship between genes and traits. Genetic drift is usually applied to DNA sequences to model the effects of finite population size on the change of allele frequencies over time, when the alleles in question do not affect Darwinian fitness. The safest way to do this is to restrict attention to areas of the genomes in question that are least likely to be affected by selection. This naturally leads to a focus on intergenic or intronic sequence, or third ("wobble") positions in codons and the like.

It is often tacitly assumed that drift does not apply to variation that has function or phenotype, but that is mistaken (a fact well known in population genetics).

The key fact would have to be that phenotypic variation associated with the genetic variation does not affect fitness. This is selectively neutral variation and might include such traits as stature, leaf shape, number of caudal vertebrae, details of facial morphology, and the like. Even these kinds of apparent details are, of course, speculative examples that might not even be affected by genetic variation, and it must be said that some biologists tend to insist that all phenotypic (or, indeed, genetic) variation is likely to affect fitness in some way. Allelic variation associated with such phenotypes can change stochastically as genetic drift, and there can be associated phenotypic drift in traits the alleles affect.

A different kind of stochastic change in phenogenetic relationships can take place even when there is natural selection (Weiss and Fullerton, 2000; Weiss, 2003; Weiss and Buchanan, 2003). The reason is that selection acts on phenotypes, not genotypes, so that alternative genotypes associated with the same phenotype will experience genetic drift relative to each other. This can be referred to as phenogenetic drift and has two major applications. At any given time, what appears to be the same phenotype (or equivalently if it can be known, the same at least with respect to fitness) can be caused by different genotypes. Mapping of quantitative traits, studies of disease or mutational effects, and so on have identified this kind of allelic and locus heterogeneity. A practical implication is that, even supposing the risky proposition to be true that the phenotype can be accurately predicted from knowledge of the genotype, the reverse may not apply. The genotype may not necessarily be reliably predicted from knowledge of the phenotype. This kind of genotypic equivalence has been well known since the earliest models of the genetic basis of quantitative or complex traits.

Another effect of phenogenetic drift can be experienced at the evolutionary macroscale. Over time, selection can favor a trait, such as the expression of a given gene in a given place in an embryo or a particular morphology, while its underlying genetic basis changes. Comparison among taxa that presumably share the trait since they diverged from a common ancestor can reveal that its underlying basis differs (e.g., Abouheif and Wray, 2002; Romano and Wray, 2003).

IV. SUMMARY

The relationship between genotype and phenotype is complex whether one focuses on the mechanisms that relate gene structure to protein function or on variation. Mechanisms such as gene regulation, alternative splicing, ADAR, somatic recombination, and the formation of protein multimers lead to a many-to-many relationship between genes and proteins. The environment can alter phenotype without changing genotype. Genetic mechanisms can lead to stabile phenotypes despite changing genes or environment. Developmental processes often produce patterns rather than traits. In addition, chance further complicates the situation.

Much of modern biology is predicated on the assumption that by knowing the genotype of an individual we can predict the phenotype, or even vice versa, but the examples we show in the preceding text demonstrate that in many situations this may not be possible. As our knowledge grows, these complicating factors become harder to ignore, and biologists need to take them into consideration whether they are trying to decipher the genetic basis of disease or the evolution of species. This may seem a depressing fact (especially to the pharmaceutical industry), but what it in fact means is that we are getting closer in our understanding of genetics, development, and evolution to how they have and continue to work in nature. As in any maturing science, the simple models that formed the foundation of our current knowledge must be reevaluated as new evidence is revealed. We must start by acknowledging that evolution works at the phenotypic level, not directly on genes, and that the resulting relationship between genotype and phenotype is labile and complex.

ACKNOWLEDGMENTS

We thank Anne Buchanan and Jennifer Sholtis for critically reading this manuscript.

REFERENCES

Abouheif, E., and Wray, G. A. (2002). Evolution of the gene network underlying wing polyphenism in ants. *Science* **297**, 249–252.

Ancel, L. W., and Fontana, W. (2000). Plasticity, evolvability, and modularity in RNA. *Journal of Experimental Zoology* **288**, 242–283.

Bateson, W. (1894). *Materials for the Study of Variation Treated with Especial Regard to Discontinuity in the Origin of Species*. London; New York: Macmillan.

Beadle, G. W., and Tatum, E. L. (1941). Genetic control of biochemical reactions in *Neurospora*. *Proceedings of the National Academy of Sciences USA* **27**, 499–506.

Blackwood, E. M., and Kadonaga, J. T. (1998). Going the distance: A current view of enhancer action. *Science* **281**, 61–63.

Carroll, S. B., Grenier, J. K., and Weatherbee, S. D. (2001). *From DNA to Diversity: Molecular Genetics and the Evolution of Animal Design*. Oxford, England: Blackwell.

Cerutti, H. (2003). RNA interference: Traveling in the cell and gaining functions? *Trends in Genetics*, **19**, 39–46.

Dengler, N., and Kang, J. (2001). Vascular patterning and leaf shape. *Current Opinion in Plant Biology* **4**, 50–56.

Dermitzakis, E. T., and Clark, A. G. (2002). Evolution of transcription factor binding sites in Mammalian gene regulatory regions: Conservation and turnover. *Molecular Biology and Evolution* **19**, 1114–1121.

de Visser, J. A. G. M., Hermisson, J., Wagner, G. P., Meyers, L. A., Bagheri-Chaichian, H., Blanchard, J. L., Chao, L., Cheverud, J. M., Elena, S. F., Fontana, W., Gibson, G., Hansen, T. F., Krakauer, D., Lewontin, R. C., Ofria, C., Rice, S. H., von Dassow, G., Wagner, A., and Whitlock, M. C. (2003). Perspective: Evolution and detection of genetic robustness. *Evolution* **57**, 1959–1972.

Dietrich, M. R. (2003). Timeline: Richard Goldschmidt, hopeful monsters and other "heresies." *Nature Reviews Genetics* **4**, 68–74.

Dodds, A. P., Cox, S. A., Suggs, C. A., Boyd, C., Ruiz, R., Hart, T. C., and Wright, J. T. (2003). Characterization and mRNA expression in an unusual odontogenic lesion in a patient with tricho-dento-osseous syndrome. *Histology and Histopathology* **18**, 849–854.

Enard, W., Khaitovich, P., Klose, J., Zollner, S., Heissig, F., Giavalisco, P., Nieselt-Struwe, K., Muchmore, E., Varki, A., Ravid, R., Doxiadis, G. M., Bontrop, R. E., and Paabo, S. (2002). Intra- and interspecific variation in primate gene expression patterns. *Science* **296**, 340–343.

Fares, M. A., Ruiz-Gonzalez, M. X., Moya, A., Elena, S. F., and Barrio, E. (2002). Endosymbiotic bacteria: groEL buffers against deleterious mutations. *Nature* **417**, 398.

Fay, J. C., and Wu, C. I. (1999). A human population bottleneck can account for the discordance between patterns of mitochondrial versus nuclear DNA variation. *Molecular Biology and Evolution* **16**, 1003–1005.

Fontana, W. (2002). Modelling "evo-devo" with RNA. *Bioessays* **24**, 1164–1177.

Freiman, R. N., and Tjian, R. (2003). Regulating the regulators: Lysine modifications make their mark. *Cell* **112**, 11–17.

Gilbert, S. F. (1998). Bearing crosses: A historiography of genetics and embryology. *American Journal of Medical Genetics* **76**, 168–182.

Gilbert, S. F. (2001). Ecological developmental biology: Developmental biology meets the real world. *Developmental Biology* **233**, 1–12.

Gilbert, S. F. (2003). The morphogenesis of evolutionary developmental biology. *International Journal of Developmental Biology* **47**, 467–477.

Gilbert, S. F., Opitz, J. M., Raff, R. A. (1996). Resynthesizing evolutionary and developmental biology. *Developmental Biology* **173**, 357–372.

Goring, H. H., Terwilliger, J. D., and Blangero, J. (2001). Large upward bias in estimation of locus-specific effects from genomewide scans. *American Journal of Human Genetics* **69**, 1357–1369.

Hardison, R. (1998). Hemoglobins from bacteria to man: Evolution of different patterns of gene expression. *Journal of Experimental Biology* **201**, 1099–1117.

Hart, T. C., Bowden, D. W., Bolyard, J., Kula, K., Hall, K., and Wright, J. T. (1997). Genetic linkage of the tricho-dento-osseous syndrome to chromosome 17q21. *Human Molecular Genetics* **6**, 2279–2284.

Herbert, A. (2004). The four Rs of RNA-directed evolution. *Nature Genetics* **36**, 19–25.

Ioannidis, J. P., Ntzani, E. E., Trikalinos, T. A., and Contopoulos-Ioannidis, D. G. (2001). Replication validity of genetic association studies. *Nature Genetics* **29**, 306–309.

Jernvall, J. (2000). Linking development with generation of novelty in mammalian teeth. *Proceedings of the National Academy of Sciences USA* **97**, 2641–2645.

Jung, H. S., Francis-West, P. H., Widelitz, R. B., Jiang, T. X., Ting-Berreth, S., Tickle, C., Wolpert L., and Chuong, C. M. (1998). Local inhibitory action of BMPs and their relationships with activators in feather formation: Implications for periodic patterning. *Developmental Biology* **196**, 11–23.

Kaplan, J. M., and Pigliucci, M. (2001). Genes "for" phenotypes: A modern history view. *Biology and Philosophy* **16**, 189–213.

Kawahara, Y., Ito, K., Sun, H., Aizawa, H., Kanazawa, I., and Kwak, S. (2004). Glutamate receptors: RNA editing and death of motor neurons. *Nature* **427**, 801.

Kay, L. E. (2000). *Who Wrote the Book of Life: A History of the Genetic Code*. Stanford, CA: Stanford University Press.

King, M. C., and Wilson, A. C. (1975). Evolution at two levels in humans and chimpanzees. *Science* **188**, 107–116.

Knight, J. C. (2004). Allele-specific gene expression uncovered. *Trends in Genetics* **20**, 113–116.

Kondo, S. (2002). The reaction-diffusion system: A mechanism for autonomous pattern formation in the animal skin. *Genes Cells* **7**, 535–541.

Lander, E. S., Linton, L. M., Birren, B., Nusbaum, C., Zody, M. C., Baldwin, J., Devon, K., Dewar, K., Doyle, M., FitzHugh, W., Funke, R., Gage, D., Harris, K., Heaford, A., Howland, J., Kann, L., Lehoczky, J., LeVine, R., McEwan, P., McKernan, K., Meldrim, J., Mesirov, J. P., Miranda, C., Morris, W., Naylor, J., Raymond, C., Rosetti, M., Santos, R., Sheridan, A., Sougnez, C., Stange-Thomann, N., Stojanovic, N., Subramanian, A., Wyman, D., Rogers, J., Sulston, J., Ainscough, R., Beck, S., Bentley, D., Burton, J., Clee, C., Carter, N., Coulson, A., Deadman, R., Deloukas, P., Dunham, A., Dunham, I., Durbin, R., French, L., Grafham, D., Gregory, S., Hubbard, T., Humphray, S., Hunt, A., Jones, M., Lloyd, C., McMurray, A., Matthews, L., Mercer, S., Milne, S., Mullikin, J. C., Mungall, A., Plumb, R., Ross, M., Shownkeen, R., Sims, S., Waterston, R. H., Wilson, R. K., Hillier, L. W., McPherson, J. D., Marra, M. A., Mardis, E. R., Fulton, L. A., Chinwalla, A. T., Pepin, K. H., Gish, W. R., Chissoe, S. L., Wendl, M. C., Delehaunty K. D., Miner, T. L., Delehaunty, A., Kramer, J. B., Cook, L. L., Fulton, R. S., Johnson, D. L., Minx, P. J., Clifton, S. W., Hawkins, T., Branscomb, E., Predki, P., Richardson, P., Wenning, S., Slezak, T., Doggett, N., Cheng, J. F., Olsen, A., Lucas, S., Elkin, C., Uberbacher, E., Frazier, M., Gibbs, R. A., Muzny, D. M., Scherer, S. E., Bouck, J. B., Sodergren, E. J., Worley, K. C., Rives, C. M., Gorrell, J. H., Metzker, M. L., Naylor, S. L., Kucherlapati, R. S., Nelson, D. L., Weinstock, G. M., Sakaki, Y., Fujiyama, A., Hattori, M., Yada, T., Toyoda, A., Itoh, T., Kawagoe, C., Watanabe, H., Totoki, Y., Taylor, T., Weissenbach, J., Heilig, R., Saurin, W., Artiguenave, F., Brottier, P., Bruls, T., Pelletier, E., Robert, C., Wincker, P., Smith, D. R., Doucette-Stamm, L., Rubenfield, M., Weinstock, K., Lee, H. M., Dubois, J., Rosenthal, A., Platzer, M., Nyakatura, G., Taudien, S., Rump, A., Yang, H., Yu, J., Wang, J., Huang, G., Gu, J., Hood, L., Rowen, L., Madan, A., Qin, S., Davis, R. W., Federspiel, N. A., Abola, A. P., Proctor, M. J., Myers, R. M., Schmutz, J., Dickson, M., Grimwood, J., Cox, D. R., Olson, M. V., Kaul, R., Raymond, C., Shimizu, N., Kawasaki, K., Minoshima, S., Evans, G. A., Athanasiou, M., Schultz, R., Roe, B. A., Chen, F., Pan, H., Ramser, J., Lehrach, H., Reinhardt, R., McCombie, W. R., de la Bastide, M., Dedhia, N., Blocker, H., Hornischer, K., Nordsiek, G., Agarwala, R., Aravind, L., Bailey, J. A., Bateman, A., Batzoglou, S., Birney, E., Bork, P., Brown, D. G., Burge, C. B., Cerutti, L., Chen, H. C., Church, D., Clamp, M., Copley, R. R., Doerks, T., Eddy, S. R., Eichler, E. E., Furey, T. S., Galagan, J., Gilbert, J. G., Harmon, C., Hayashizaki, Y., Haussler, D., Hermjakob, H., Hokamp, K., Jang, W., Johnson, L. S., Jones, T. A., Kasif, S., Kaspryzk, A., Kennedy, S., Kent, W. J., Kitts, P., Koonin, E. V., Korf, I., Kulp, D., Lancet, D., Lowe, T. M., McLysaght, A., Mikkelsen, T., Moran, J. V., Mulder, N., Pollara, V. J., Ponting, C. P., Schuler, G., Schultz, J., Slater, G., Smit, A. F., Stupka, E., Szustakowski, J., Thierry-Mieg, D., Thierry-Mieg, J., Wagner, L., Wallis, J., Wheeler, R., Williams, A., Wolf, Y. I., Wolfe, K. H., Yang, S. P., Yeh, R. F., Collins, F., Guyer, M. S., Peterson, J., Felsenfeld, A., Wetterstrand, K. A., Patrinos, A., Morgan, M. J., de Jong, P., Catanese, J. J., Osoegawa, K., Shizuya, H., Choi, S., and Chen, Y. J. (2001). Initial sequencing and analysis of the human genome. *Nature* **409**, 860–921.

Lauter, N., and Doebley, J. (2002). Genetic variation for phenotypically invariant traits detected in teosinte: Implications for the evolution of novel forms. *Genetics* **160**, 333–342.

LeCouter, J. E., Kablar, B., Whyte, P. F., Ying, C., and Rudnicki, M. A. (1998). Strain-dependent embryonic lethality in mice lacking the retinoblastoma-related p130 gene. *Development* **125**, 4669–4679.

Lettice, L. A., Horikoshi, T., Heaney, S. J., van Baren, M. J., van der Linde, H. C., Breedveld, G. J., Joosse, M., Akarsu, N., Oostra, B. A., Endo, N., Shibata, M., Suzuki, M., Takahashi, E., Shinka, T., Nakahori, Y., Ayusawa, D., Nakabayashi, K., Scherer, S. W., Heutink, P., Hill, R. E., and Noji, S. (2002). Disruption of a long-range cis-acting regulator for Shh causes preaxial polydactyly. *Proceedings of the National Academy of Science USA* **99**, 7548–7553.

Lettice, L. A., Heaney, S. J., Purdie, L. A., Li, L., de Beer, P., Oostra, B. A., Goode, D., Elgar, G., Hill, R. E., and de Graaff, E. (2003). A long-range Shh enhancer regulates expression in the developing limb and fin and is associated with preaxial polydactyly. *Human Molecular Genetics* **12**, 1725–1735.

Lewontin, R. C. (1992). *Genotype and Phenotype. Keywords in Evolutionary Biology.* (E. F. Keller and E. A. Lloyd., eds.), pp. 137–144. Cambridge, MA: Harvard University Press.

Lewontin, R. C. (2000). *The Triple Helix: Gene, Organism and Environment.* Cambridge, MA: Harvard University Press.

Lloyd, E. A. (1992). *Unit of Selection. Keywords in Evolutionary Biology* (E. F. Keller and E. A. Lloyd, eds.), pp. 334–340. Cambridge, MA: Harvard University Press.

Long, A. D., and Langley, C. H. (1999). The power of association studies to detect the contribution of candidate genetic loci to variation in complex traits. *Genome Research* **9**, 720–731.

Maas, S., Patt, S., Schrey, M., and Rich, A. (2001). Underediting of glutamate receptor GluR-B mRNA in malignant gliomas. *Proceedings of the National Academy of Sciences USA* **98**, 14687–14692.

Mackay, T. F. (1995). The genetic basis of quantitative variation: Numbers of sensory bristles of *Drosophila melanogaster* as a model system. *Trends in Genetics* **11**, 464–470.

Mackay, T. F. (2001). The genetic architecture of quantitative traits. *Annual Review of Genetics* **35**, 303–339.

Mattick, J. S. (2003). Challenging the dogma: The hidden layer of non-protein-coding RNAs in complex organisms. *Bioessays* **25**, 930–939.

Mattick, J. S. (2004). RNA regulation: A new genetics? *Nature Review Genetics* **5**, 316–323.

Meinhardt, H., Prusinkiewicz, P., and Fowler, D. R. (1995). *The Algorithmic Beauty of Sea Shells.* Berlin; New York: Springer-Verlag.

Meyerowitz, E. M. (2002). Plants compared to animals: The broadest comparative study of development. *Science* **295**, 1482–1285.

Mombaerts, P. (2004). Odorant receptor gene choice in olfactory sensory neurons: The one receptor-one neuron hypothesis revisited. *Current Opinion in Neurobiology* **14**, 31–36.

Moore, D. S. (2002). *The Dependent Gene: The Fallacy of "Nature vs. Nurture."* New York: Times Books.

Murphy, S. K., and Jirtle, R. L. (2003). Imprinting evolution and the price of silence. *Bioessays* **25**, 577–588.

Nobrega, M. A., Ovcharenko, I., Afzal, V., and Rubin, E. M. (2003). Scanning human gene deserts for long-range enhancers. *Science* **302**, 413.

Nossal, G. J. (2003). The double helix and immunology. *Nature* **421**, 440–444.

Nunez-Farfan, J., and Schlichting, C. D. (2001). Evolution in changing environments: The "synthetic" work of Clausen, Keck, and Hiesey. *Quarterly Review of Biology* **76**, 433–457.

Oleksiak, M. F., Churchill, G. A., and Crawford, D. L. (2002). Variation in gene expression within and among natural populations. *Nature Genetics* **32**, 261–266.

Plath, K., Mlynarczyk-Evans, S., Nusinow, D. A., and Panning, B. (2002). Xist RNA and the mechanism of x chromosome inactivation. *Annual Review of Genetics* **36**, 233–278.

Price, J. A., Wright, J. T., Kula, K., Bowden, D. W., and Hart, T. C. (1998). A common DLX3 gene mutation is responsible for tricho-dento-osseous syndrome in Virginia and North Carolina families. *Journal of Medical Genetics* **35**, 825–828.

Queitsch, C., T., Sangster, A., and Lindquist, S. (2002). Hsp90 as a capacitor of phenotypic variation. *Nature* **417**, 618–624.

Reenan, R. A. (2001). The RNA world meets behavior: A→I pre-mRNA editing in animals. *Trends in Genetics* **17**, 53–56.

Romano, L. A., and Wray, G. A. (2003). Conservation of Endo16 expression in sea urchins despite evolutionary divergence in both cis and trans-acting components of transcriptional regulation. *Development* **130**, 4187–4199.

Rutherford, S. L. (2000). From genotype to phenotype: Buffering mechanisms and the storage of genetic information. *Bioessays* **22**, 1095–1105.

Rutherford, S. L. (2003). Between genotype and phenotype: Protein chaperones and evolvability. *Nature Reviews Genetics* **4**, 263–274.

Rutherford, S. L., and Lindquist, S. (1998). Hsp90 as a capacitor for morphological evolution. *Nature* **396**, 336–342.

Rutledge, T. T., Eisen, E. J., and Legates, J. E. (1974). Correlated response in skeletal traits and replicate variation in selected lines of mice. *Theoretical Applied Genetics* **45**, 26–31.

Salazar-Ciudad, I., and Jernvall, J. (2002). A gene network model accounting for development and evolution of mammalian teeth. *Proceedings of the National Academy of Sciences USA* **99**, 8116–8120.

Schlichting, C., and Pigliucci, M. (1998). *Phenotypic Evolution: A Reaction Norm Perspective*. Sunderland, MA: Sinauer.

Scriver, C. R., and Waters, P. J. (1999). Monogenic traits are not simple: Lessons from phenylketonuria. *Trends in Genetics* **15**, 267–272.

Serizawa, S., Miyamichi, K., Nakatani, H., Suzuki, M., Saito, M., Yoshihara, Y., and Sakano, H. (2003). Negative feedback regulation ensures the one receptor-one olfactory neuron rule in mouse. *Science* **302**, 2088–2094.

Siegal, M. L., and Bergman, A. (2002). Waddington's canalization revisited: Developmental stability and evolution. *Proceedings of the National Academy of Sciences USA* **99**, 10528–10532.

Smallwood, P. M., Wang, Y., and Nathans, J. (2002). Role of a locus control region in the mutually exclusive expression of human red and green cone pigment genes. *Proceedings of the National Academy of Sciences USA* **99**, 1008–1011.

Sorek, R., Shamir, R., and Ast, G. (2004). How prevalent is functional alternative splicing in the human genome? *Trends in Genetics* **20**, 68–71.

Stone, J. R., and Wray, G. A. (2001). Rapid evolution of cis-regulatory sequences via local point mutations. *Molecular Biology and Evolution* **18**, 1764–1770.

Streelman, J. T., Webb, J. F., Albertson, R. C., and Kocher, T. C. (2003). The cusp of evolution and development: A model of cichlid tooth shape diversity. *Evolutionary Development* **5**, 600–608.

Sultan, S. E. (2003). Commentary: The promise of ecological developmental biology. *Journal of Experimental Zoology Part B (Molecular and Developmental Evolution)* **296**, 1–7.

Tautz, D. (2000). A genetic uncertainty problem. *Trends in Genetics* **16**, 475–477.

Thatcher, J. W., Shaw, J. M., and Dickinson, W. J. (1998). Marginal fitness contributions of nonessential genes in yeast. *Proceedings of the National Academy of Sciences USA* **95**, 253–257.

Thomas, B. L., Tucker, A. S., Qui, M., Ferguson, C. A., Hardcastle, Z., Rubenstein, J. L., and Sharpe, P. T. (1997). Role of Dlx-1 and Dlx-2 genes in patterning of the murine dentition. *Development* **124**, 4811–4818.

Thyagarajan, T., Totey, S., Danton, M. J., and Kulkarni, A. B. (2003). Genetically altered mouse models: The good, the bad, and the ugly. *Critical Reviews in Oral Biology and Medicine* **14**, 154–174.

Turing, A. M. (1952). The chemical basis of morphogenesis. *Philosophical Transactions of the Royal Society of London Series B* **237**, 37–72.

Waddington, C. H. (1942). Canalization of development and the inheritance of acquired characters. *Nature* **150**, 563–565.

Waddington, C. H. (1953). Genetic assimilation of an acquired character. *Evolution* **7**, 118–126.

Waddington, C. H. (1956). Genetic assimilation of the *Bithorax* phenotype. *Evolution* **10**, 1–13.

Wagner, E. G., and Flardh, K. (2002). Antisense RNAs everywhere? *Trends in Genetics* **18**, 223–226.

Weiss, K. (2002). Is the medium the message? Biological traits and their regulation. *Evolutionary Anthropology* **11**, 88–93.

Weiss, K., and Sholtis, S. (2003). Dinner at baby's: Werewolves, dinosaur jaws, hen's teeth, and horse toes. *Evolutionary Anthropology* **12**, 247–251.

Weiss, K. M. (2003). Phenotype and genotype. In *Keywords and Concepts in Evolutionary Developmental Biology* (B. K. Hall and W. M. Olson, eds.), pp. 279–288. Cambridge, MA: Harvard University Press.

Weiss, K. M., and Buchanan, A. V. (2003). Evolution by phenotype: A biomedical perspective. *Perspectives in Biological Medicine* **46**, 159–182.

Weiss, K. M., and Buchanan, A. V. (2004). *Genetics and the Logic of Evolution*. Hoboken, NJ: Wiley.

Weiss, K. M., and Fullerton, S. M. (2000). Phenogenetic drift and the evolution of genotype-phenotype relationships. *Theoretical Population Biology* **57**, 187–195.

Weiss, K. M., Stock, D. W., and Zhao, Z. (1998). Dynamic interactions and the evolutionary genetics of dental patterning. *Critical Reviews in Oral Biology and Medicine* 9, 369–398.

Wray, G. A. (2003). Transcriptional regulation and the evolution of development. *International Journal of Developmental Biology* 47, 675–684.

Yedid, G., and Bell, G. (2002). Macroevolution simulated with autonomously replicating computer programs. *Nature* 420, 810–812.

CHAPTER 22

The Study of Phenotypic Variability: An Emerging Research Agenda for Understanding the Developmental–Genetic Architecture Underlying Phenotypic Variation

BENEDIKT HALLGRÍMSSON,* JEVON JAMES YARDLEY BROWN,† AND BRIAN K. HALL‡

*Department of Cell Biology and Anatomy, Joint Injury and Arthritis Research Group, The Bone and Joint Institute, University of Calgary, Calgary, Alberta, Canada,
†Joint Injury and Arthritis Research Group, University of Calgary, Calgary, Alberta, Canada,
‡Department of Biology, Dalhousie University, Halifax, Nova Scotia, Canada

Variation

INTRODUCTION

Variation exists at all levels of the biological hierarchy, and much biological research is concerned with explaining variation at one level or another. From the determinants of disease to functional morphology, questions in biology tend to focus on explaining specific instances of variation. However, when attention is turned to variation as a topic in itself, as in this volume, the questions that emerge do not address specific instances of variation but rather the tendencies of biological systems to exhibit variation. This general phenomenon—the tendency to vary—is what Wagner and Altenberg (1996) refer to as variability. Variability is an important topic precisely because the transmission of variation from one level of the biological hierarchy to the next is neither straightforward nor intuitive. Instead, biological systems tend to filter the effects of determinants of variation at one level to its expression at the next.

This chapter, as well as much of this book, deals with variability at a specific level, the organismal phenotype. The study of variability at the phenotypic level is concerned with how the architecture of development structures the translation of genetic variation and environmental effects into phenotypic variation. This topic is important because natural selection acts on phenotypes, but the results are interpreted and transmitted from generation to generation at the genetic level. It is interesting precisely because the relationship between variation at the genetic level and variation at the phenotypic level can be very complex. If the relationship between genetic and phenotypic variation were straightforward, the study of development would be largely irrelevant to evolutionary biology (Hall, 1999). Getting at the developmental genetic basis for variability is different from the developmental genetic basis of phenotypic variation in specific developmental contexts. The study of the evolutionary developmental biology of variation is still at a very early stage, but its conceptual framework is well developed and illustrated beautifully through examples such as Shapiro *et al.*'s (2004) study on the developmental genetic basis for pelvic fin reduction in natural populations of sticklebacks (*Gasterosteus aculeatus*) or Abzhanov and colleagues' (2004) study of the developmental basis for variation in beak length in Darwin's finches, studies that are producing some of the most important advances in evolutionary biology today. The study of the developmental basis for phenotypic variability addresses a higher level of abstraction in that it deals with the *tendency* of developmental systems to change, amplify, or reduce the expression of genetic variation at the phenotypic level. This area of study is difficult because variability arises from emergent properties of complex developmental systems. Understanding how variability arises from developmental systems requires approaches that focus on the patterns and nature of the interactions among elements in the system. Such an approach to evolutionary developmental biology is only beginning to be possible now that the maturing of bioinformatics is enabling the modeling of complex developmental systems.

The ambitious goal of this chapter is to outline an emerging research program that addresses the developmental genetic basis of phenotypic variability.

I. VARIABILITY AND THE BIOLOGICAL HIERARCHY

We address the study of variability at a specific level, that of the organismal phenotype. The goal is not to explain particular developmental processes or variation patterns in particular developmental contexts, but to identify the aspects or properties of development that explain how the tendency of developmental systems to exhibit variation is modulated.

Although the mechanistic study of variability at this level is only now becoming possible, questions about variability have been central at other levels of the evolutionary hierarchy. At the molecular genetic level, research on the determinants of mutability is an established area of research with a dedicated journal (*Mutagenesis*). At the population genetic level, the question of how genetic variance is maintained in the face of selection was a central preoccupation of evolutionary biology for much of the last century, and our understanding in this area continues to be refined (Zhang and Hill, 2003; Spichtig, 2004). The study of how genetic variance is modulated by development to produce the organismal phenotype is sandwiched between these two layers of inquiry. As at other levels, questions about variation and variability are related and complementary. In evolutionary developmental biology, the study of variation is concerned with issues such as:

- how the genetic toolkit is recombined in different developmental and phylo genetic contexts to produce different patterns of phenotypic variation;
- which genes out of the tens of thousands present in most genomes are involved in the evolution of morphology and physiology; and
- the roles of gene regulation versus sequence variation in genes themselves in the production of evolutionary change in the phenotype.

These are central concerns in evolutionary developmental biology today (Wilkins, 2002; Carroll *et al.*, 2005). In contrast, questions about variability deal with how variance is structured and modulated by the same developmental processes and systems that are unraveled through questions about variation.

To illustrate the relationship between variation and variability in developmental systems, it is useful to visualize Waddington's (1957) well-worn metaphor, the epigenetic landscape (Figure 22-1). In Waddington's visual metaphor, development of an individual organism—or part of an organism such as an organ, tissue, or individual cell—is like a ball rolling down an undulating, dissected landscape, with the shape of the landscape determined by the underlying genetic architecture. Developmental trajectories are valleys in the landscape. The epigenetic landscape

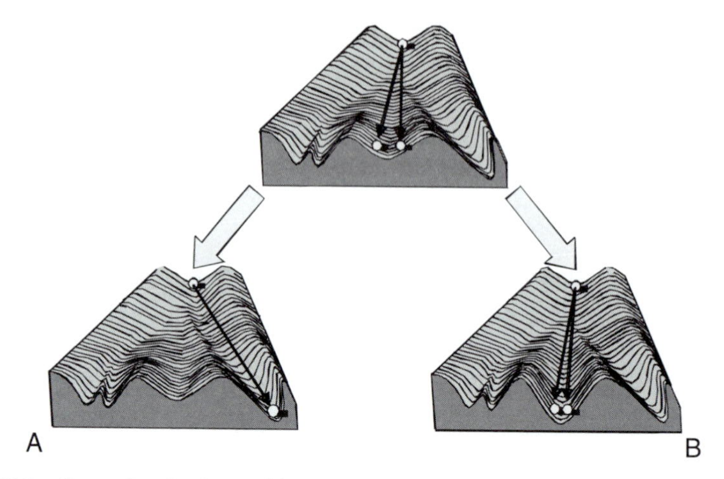

FIGURE 22-1. Two modes of evolution of the epigenetic landscape. In (A) the phenotypic mean is shifted from one valley to another. In (B) the sides of the valley have gotten steeper, depicting a decrease in variance.

represents the probability distribution of developmental outcomes. The ball is more likely to end up at the base of the deepest valley than somewhere on the slope or be shunted to a side valley. As the ball rolls down the landscape, it can be buffeted by external and internal perturbations but tends to return to the base of the valleys in the landscape. The influence of Thom's (1975) catastrophe theory on this conceptualization of development is obvious. The epigenetic landscape captures the idea that changes in the underlying genetic architecture, and thus in the probability landscape of development, can alter both the mean phenotype (the location of the deepest valley) and the variance about that mean phenotype (the steepness of the sides of the valleys) as illustrated in Figure 22-1. Within the metaphor, the study of variation is concerned with explaining how the genetic architecture of development alters locations of valleys in the landscape and determines the valley to be entered, while the study of variability is concerned with the steepness of the sides of the valleys and height of the cusps that separate them. Both modes of evolution depicted in Figure 22-1 are produced by changes in the genetic architecture of development and its interaction with the environment. Particular genetic changes can produce both shifts in mean and changes in variance; there is no reason to assume that there are developmental processes or mechanisms specific to the regulation of variance. The available evidence indicates that many mutations influence both phenotypic variance and mean while others influence the mean and not variance. There are also factors—e.g., Hsp90, a ubiquitous gene product involved in regulating transcription—that influence variance but not the mean (Rutherford and Lindquist, 1998). As we start to build a systems level understanding of development, we will begin to see how phenotypic variation in both mean and variance emerge from variation in the genetic architecture that underlies development. At that point, the study of the developmental genetic architecture of phenotypic variability and variation will converge on a common ground of inquiry.

II. COMPONENTS OF VARIABILITY

The study of phenotypic variability is directed to explaining three related patterns of phenotypic variation (Hallgrímsson *et al.*, 2002):

1. Variation in the sensitivity of developmental systems to genetic or environmental perturbation ("canalization")
2. Variation in how predictable development is under the same environmental and genetic conditions ("developmental stability")
3. The tendency for multivariate variation to be structured by developmental or functional relationships ("morphological or developmental integration")

Canalization, developmental stability, and morphological integration are phenomenological concepts that represent an attempt to impose conceptual order on the complex relationship between genetic and phenotypic variation. The study of these phenomena fall broadly into the area of phenogenetics as defined by Sholtis and Weiss (this volume) and Weiss (2005).

It is important to remember that these components represent a phenomenological classification—simply patterns of variation, not actual processes—a distinction often lost in the literature. This is particularly true when the concepts of developmental stability or canalization are invoked as explanatory factors in behavioral ecology and sociobiology. In the vast majority of cases, the processes that explain these three patterns of variability are not understood, although more is known about integration than about the first and second. When we use the terms "canalization" or "developmental stability," it is important not to fool ourselves into thinking that we are talking about actual developmental processes. The term "canalization," for instance, contains no additional information beyond the observation that developmental systems vary in their sensitivity to perturbation. As argued by Willmore and Hallgrímsson (this volume), confounding pattern and process makes the assumptions that single processes underlie each of canalization, developmental stability, and integration and that these processes are distinct from one another.

Although canalization, developmental stability, and morphological integration represent a phenomenological classification of the complexities of the genotype to phenotype transition, it remains worthwhile to define and measure these phenomena precisely. Canalization refers to the minimization of the effects of genetic or environmental perturbations. Wagner *et al.* (1997) distinguish between canalization of environmental versus genetic perturbations. "Genetic canalization" occurs when phenotypic variation changes without or disproportionately to changes in genetic variation. For instance, when the same mutation expressed on different genetic backgrounds produces different magnitudes of effect, those backgrounds can be said to vary in their sensitivity to the perturbation represented by the mutation and thus in genetic canalization. In this case, some genetic factors that vary among the backgrounds are producing genetic variation in canalization. Similarly, "environmental canalization" occurs when genotypes vary in their sensitivity to the same environmental perturbation. Most experimental situations do not differentiate between genetic and environmental canalization. For instance, when a mutant phenotype exhibits higher variance compared with wild-type littermates, the mutant phenotype is less canalized, but it is not clear whether the additional variation exhibited in the mutant phenotype is of genetic or environmental origin. Developmental stability is measured as the variance left over when genetic and environmental effects are completely accounted for, which is very difficult to do in practice. The most common method is to measure the differences among developmental replicates such as the two sides in bilaterally symmetrical organisms where symmetrical development is expected for developmental reasons. This measure, fluctuating asymmetry (FA) (Van Valen, 1962), confounds environmental effects that affect the two sides differently with hypothetical nonenvironmental sources of variation. This distinction between microenvironmental and larger-scale environmental effects, drawn arbitrarily between effects on the whole organism versus effects on one side and not the other, illustrates the phenomenological as opposed to ontological nature of the distinction between canalization and developmental stability. What is really of interest are the developmental processes that explain the observations that developmental systems vary in their tendency to exhibit FA and phenotypic variance. Morphological integration is assessed through the covariation structure of phenotypic variation. Patterns of integration can be used to make inferences about how the genetic architecture of development is modularized (Wagner, 1996). More importantly, integration is intimately related to the regulation of variance and is, in a sense, a multivariate extension of analyses of variability for single traits. Integration changes whenever changes in variance are not distributed equally across traits. When a particular trait exhibits an increase in variance, the result can be either increased or decreased integration depending on the developmental relationships between that trait and others. For example, if a trait that has direct developmental effects on other traits exhibits increased variance resulting from reduced canalization, the result is increased integration because the increased variation in that trait will be transmitted to others. Conversely, if the increase in variance occurs in a trait that is not tightly connected to other traits, the result is decreased integration because of the increase in the proportion of variance of that trait to its covariance with others. Integration can also vary with changes in the proportions and magnitudes of genetic and environmental variance, although this is less important when genetic and environmental correlations coincide as they appear to do in primate crania (Cheverud, 1982).

III. CURRENT APPROACHES TO UNDERSTANDING THE DEVELOPMENT–GENETIC ARCHITECTURE OF VARIABILITY

The regulation of phenotypic variance is an emergent property of complex developmental systems. Therefore, the study of the developmental–genetic architecture of variability does not lend itself to the standard approaches used to analyze the genetics of complex phenotypes. The reason is that many genes may contribute to variability and do so through many different developmental mechanisms. As Minelli (2003) has argued, there is a strong bias to seek explanations for the developmental basis for evolutionary change at the gene level. Reduction of all explanations to the level of the gene represents an arbitrary choice that may or may not be appropriate given the particular evolutionary transition or developmental process that is being explained. Although determining which genes influence phenotypic variability is important, even completing that huge task for a specific species would not provide satisfactory explanation for patterns of phenotypic variability. Those explanations reside at the developmental process level. The key questions are:

- what properties of developmental processes influence variability; and
- how are those properties affected by different kinds of genetic variation?

Three approaches to investigating the developmental basis for phenotypic variability have emerged in recent years. The next section describes these approaches and provides examples of each.

A. Pattern-Based Approaches

In many areas of science, it is common to approach black-box problems by deriving predictions about the behavior of a system from hypotheses about its internal workings (Hall, 2003a,b). Extrapolating patterns of variability from hypotheses about mechanism and then comparing the predicted with observed patterns is an example of this general approach. Approaches that focus on the analysis and interpretation of patterns of variability have probably yielded most of our current understanding of the mechanisms underlying phenotypic variability. These methods are limited, however, because the observables are very distant from the mechanisms and also because there are often so many possible explanations for the observed patterns that interpretation is ambiguous. Nonetheless, such studies have been and continue to be fruitful avenues of inquiry. Here, we discuss two examples relating to fluctuating asymmetry and canalization. The relationship between ontogeny and developmental stability has been investigated in several systems (Chippindale and Palmer, 1993; Hallgrímsson, 1993, 1998, 1999; Hall, 2003a,b; Hallgrímsson *et al.*, 2003; Kellner and Alford, 2003). The rationale is to determine whether variation accumulates during development or is actively corrected during growth.

We have analyzed the ontogeny of FA for skeletal structures, both in primate postnatal ontogenetic series (Hallgrímsson, 1993, 1999) and for mouse fetal limb development (Hallgrímsson *et al.*, 2003). We reported opposite patterns in these two very different sets of data, with FA variances increasing during postnatal growth in primates and decreasing during mouse fetal limb growth. Decreasing FA with age has also been reported during postnatal skeletal growth in mice (Marquez *et al.*, 2004). The primate postnatal results are hard to interpret because it is difficult to account for the effects of use-related remodeling on skeletal FA. The prenatal development results suggest the existence of mechanisms that reduce the variation captured by FA during ontogeny. An intriguing finding by Marquez *et al.* is that, while FA variances for skull shape decreases, the variance covariation structure for shape is virtually uncorrelated among ages. This is suggestive of the existence of mechanisms that reduce FA since these results can only be obtained if variance is being generated and reduced continually in the cranium throughout ontogeny. A much more thorough discussion of the use of ontogenetic patterning of variance to infer underlying mechanisms is provided in the chapter by Zelditch in this volume.

Another pattern that is an important source of information about the mechanisms underlying variability is how variance components are related to one another. As argued earlier, developmental stability and canalization are defined on the basis of phenotypic variability patterns (FA and among individual variance) and not on an understanding of underlying mechanisms or their relation. This relationship has been investigated in adult mice (Debat *et al.*, 2000), in rhesus macaques (Willmore *et al.*, in press), and during mouse fetal limb development (Hallgrímsson *et al.*, 2002). Debat *et al.* (2000) reported no correlation between FA and phenotypic covariance structure. Their study was based on a natural population of mice and did not control for variation among structures in genetic variance.

We examined the relationship between FA and environmental variance in mouse fetal limb traits (Hallgrímsson *et al.*, 2002) to study test the hypothesis that traits that show higher fluctuating asymmetry variance also show greater sensitivity

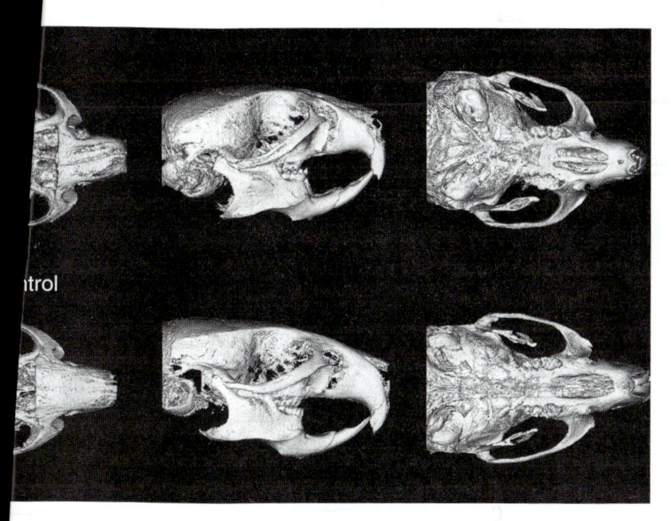

'hree-dimensional reconstructions of computed microtomography scans of a *Bm/Bm*
-type control. In the dorsal view, the neurocranium is cut away to reveal the shortened
achymorph mice. The three views are from left to right dorsal, lateral, and ventral.

rate that a mutation that has a severely deleterious effect and
n skeletal development need not influence FA. This has profound
for the often assumed relationship between FA and fitness in natural

nd example is also a mutation that affects structural proteins in
system, but the result is very different. The *Brachymorph* (*Bm*) pheno-
from an autosomal recessive mutation in the phosphoadenosine-
fate synthase 2 gene (*PAPSS2*). A major skeletal manifestation is
wth of cartilage resulting from undersulfation of glycosaminoglycans
cartilage matrix. This produces a dramatic reduction in the growth of
ocranium, producing an altered craniofacial shape (Figure 22-5). The
t also produces a dramatic decrease in long bone growth in the limbs.
esized that because this mutation affects one component of the cra-
chondrocranium, directly, and the dermatocranium only indirectly, it
uence morphological integration among these craniofacial components.
a set of 34 three-dimensional landmarks obtained through computed
graphy to assess craniofacial size and shape in samples of *Brachymorph*
20) and wild-type controls (N = 20) (Hallgrímsson *et al.*, in preparation a).
gical integration is increased significantly in the *Brachymorph* mutant and
n reduced substantially (Figure 22-6). By contrast, fluctuating asymmetry

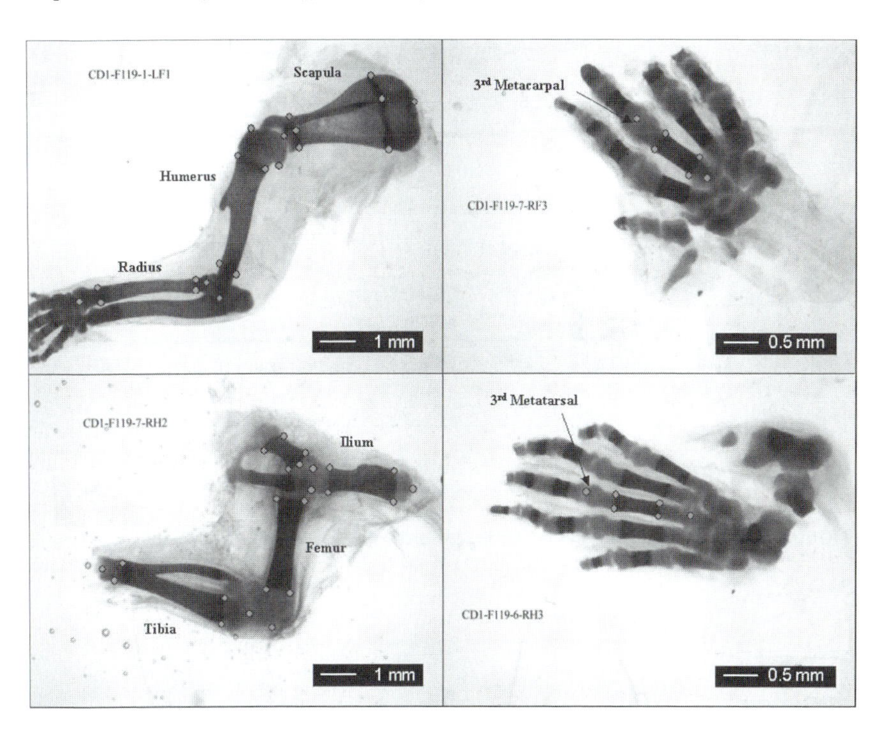

FIGURE 22-2. Landmarks collected for forelimb and hindlimb elements for the CD1 sample. The specimen shown is a neonate (20.5 day sample). Mouse fetuses were cleared and stained with alcian blue for cartilage and alizarin red for bone/osteoid.

to environmental factors. We examined this relationship in a set of 122 interlandmark distances for limb structures in 19-day-old mouse fetuses (Figure 22-2). We found that FA and environmental variances are positively correlated ($r = 0.56$, $p < 0.001$) (Figure 22-3). Willmore *et al.* found a similar but weaker correlation between FA and environmental variance in adult rhesus macaque crania. Together, these results suggest that there is at least some overlap between the mechanisms that regulate FA and canalization. This is not surprising given that the distinction between these two patterns is, to a degree, arbitrary with respect to the range of hypothetical mechanisms that produce them.

B. Perturbation-Based Approaches

Another common approach to understanding a complex system that is hidden from direct view is to subject it to known perturbations and see how it responds. The use of gene knock-outs, knock-downs, or overexpression models are examples of this general approach applied to understanding developmental systems.

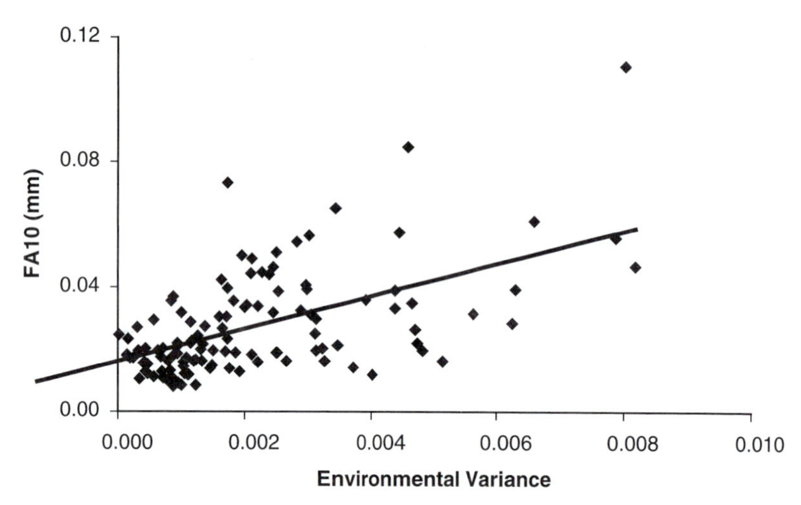

FIGURE 22-3. Size scaled fluctuating asymmetry (FA) plotted against environmental variance for 122 limb traits in gestational day (GD) 19 mouse fetuses ($r = 0.56$, $p < 0.001$). The environmental variance is calculated as $V_E = (1-h^2) \times V_P$.

This approach is useful but cumbersome. By itself, the effect of a particular perturbation does not tell you what a particular gene does. It tells you only that the gene is needed for normal development of the structure in question, although genetic redundancy can complicate even this seemingly simple interpretation. For the study of phenotypic variability, this approach is useful because it can be used to identify developmental pathways or common steps that cross pathways that are particularly relevant to regulating phenotypic variability. In the following text, we present two examples of this approach from work in the Hallgrímsson lab.

To determine whether a major mutation affecting the structure of bone would alter phenotypic variability in the skeletal system, we examined the effect of a null mutation (*oim*) in the pro-α2 chain of collagen I on limb variability in near-neonatal (gestational day [GD] 19) mouse fetuses (Hallgrímsson *et al.*, in preparation a). This mutation produces skeletal effects such as a brittle bone condition that resembles *osteogenesis imperfecta* as well as joint laxity and ligament weakness, deficient myocardial mechanics, and an elevated incidence of craniofacial malformations. We compared samples of 40 individuals per genotype (*oim/oim*, *oim/+*, and +/+), using both geometric morphometrics and interlandmark distances for the same landmarks shown in Figure 22-2. Neither among-individual variance nor fluctuating asymmetry are influenced by this mutation, despite its obvious and severe effects on the mean phenotype for characters such as bone strength and joint function (Figure 22-4). The effects on mean skeletal shape and size at the late fetal stage are also subtle, although other studies have shown reduced growth and

A. Comparisons of FA amor

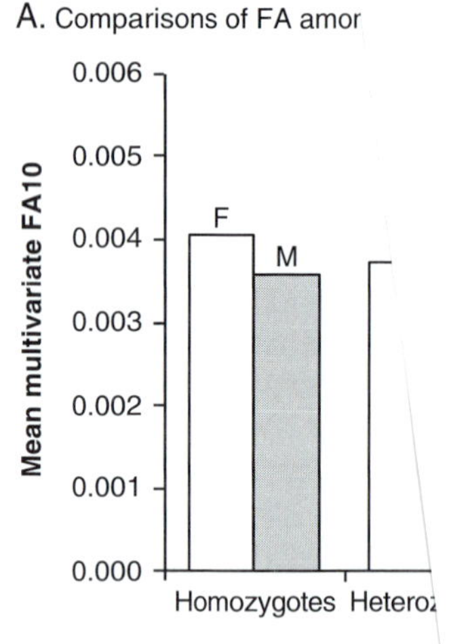

B. Comparisons of individual varian

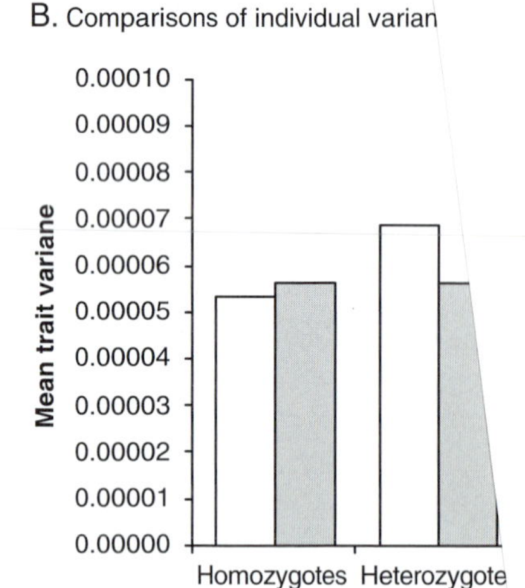

FIGURE 22-4. Multivariate fluctuating asymmetry (FA)
among genotypes for the *oim* mutation and CD1 randomb
are not significant.

shape alteration secondary to fractures during
1998; Phillips *et al.*, 2000). The results of this s
affected by the *oim* mutation is probably not rele
skeletal size and shape or is somehow compens
genetic redundancy if the gene is a member of a r

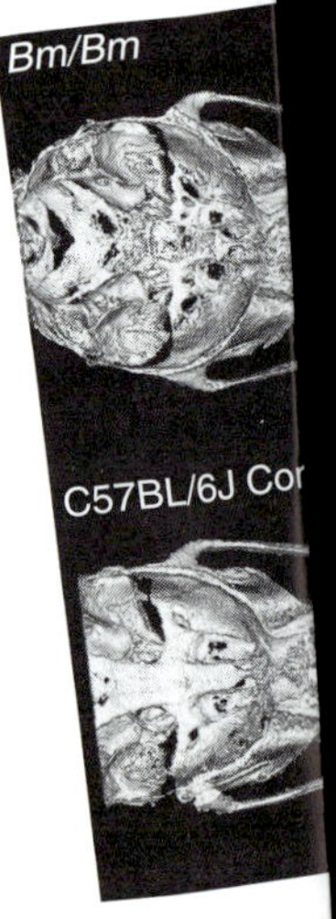

FIGURE 22-5.
mutant and a wil
basicranium in B

results illus
is involved i
implications
populations.
The seco
the skeletal
type results
phosphosul
reduced gr
(GAGs) in
the chond
same defe
We hypot
nium, the
should inf
We used
microtom
mice (N =
Morpholo
canalizati

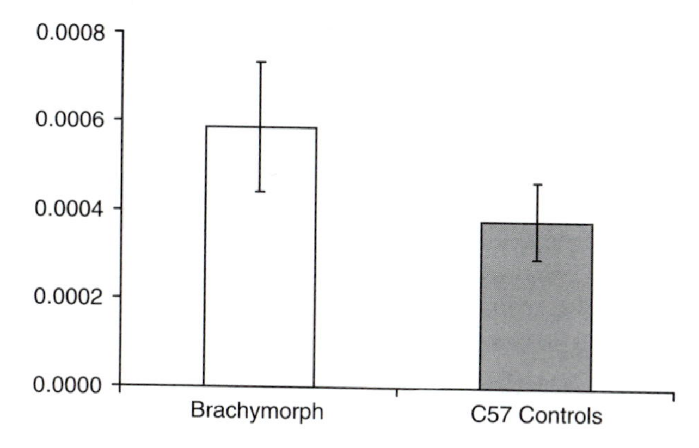

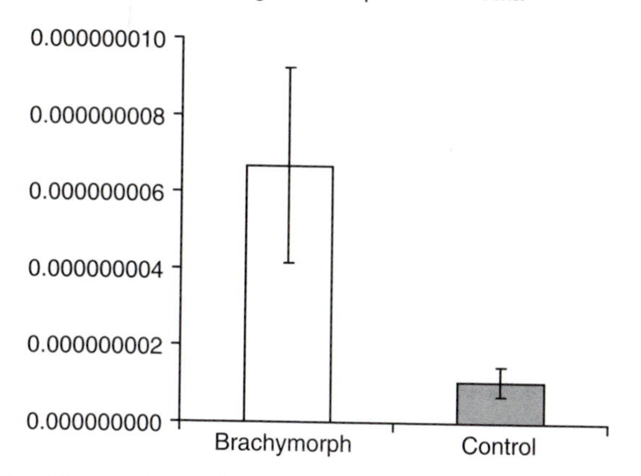

FIGURE 22-6. (A) shows shape variances in *Brachymorph* mice compared with wild-type controls, while (B) shows integration (measured as the variance of the eigenvalues for Procrustes data). Both differences are significant ($p < 0.001$). Statistical methods and experimental details are explained elsewhere (Hallgrímsson *et al.*, in preparation a).

is not significantly affected. The *Bm* mutation thus significantly influences both morphological integration and canalization.

The influence of the *Bm* mutation on canalization and integration probably results from the addition of a developmental influence on shape variation that is not present in the wild type. Since the *Bm* mutation dramatically reduces sulfation of glycosaminoglycans (GAGs), we infer that this influence is variation among individuals in the degree of sulfation—variable expressivity or incomplete penetrance of the mutation. This variation may result from genetic variation at other loci that influence sulfation, environmental effects, or intrinsic effects. We infer that chondrocranial development exhibits greater sensitivity to variation in the sulfation of chondroitin sulfate when the degree of sulfation is low. At normal levels, sulfation probably contributes minimally to phenotypic variation. Because variances are additive (see Van Valen, this volume), the addition of a source of variation not present in the wild type serves to increase overall variance for shape. The effect on integration results from the developmental relationship of the chondrocranium to the dermatocranium. Because the chondrocranium is central and also physically adjacent to dermatocranial elements, variation in the size and shape of this component will produce secondary variation in the neurocranium and the face. Because this variation is correlated with the degree of reduction of the chondrocranium, the result is increased and not decreased integration. What is interesting about the *Brachymorph* example is that, if our explanation is correct, the developmental basis for the increased integration and decreased canalization is quite mundane. In this case, variation that is cryptic in the wild type is "revealed" in the mutant, but the source of that variation is quite straightforward. Similarly, the increase in integration flows obviously from the developmental relationships among craniofacial components. What is also interesting is that this developmental genetic basis for phenotypic variability is particular to a specific developmental context. The gene and developmental processes involved here do not have a more general relationship to variability but rather produce a change in variability as a by-product of the mechanisms by which they influence the phenotypic mean. We suspect that this type of developmental basis for phenotypic variability is typical and not the exception.

C. Model-Driven Approaches

A third approach to understanding phenotypic variability that has produced significant insights is the construction of models that reveal the effects of hypothesized properties of developmental systems on the phenotypic outcome. Of the several examples of this approach, all to date have focused on the implications of abstract properties of developmental systems, such as nonlinear dynamics, thresholds, and network redundancy. Work by Graham, Emlen, and others (Emlen *et al.*, 1993; Graham *et al.*, 1993) focused on the implications of nonlinear dynamics in developmental systems. Their models are highly abstract and difficult to relate to real developmental systems, invoking for instance, interactions among the two sides of developing organisms or oscillatory behavior, which may or may not have a real developmental basis, depending on the context. Equally abstract, but more practical, is the model constructed by Klingenberg and Nijhout (1999) to explore the implications of nonlinear dynamics for the developmental basis of developmental stability. They construct a model consisting of six loci, for which the relationships between developmental parameters and phenotypic outcomes are nonlinear.

Although the developmental noise component of the developmental parameters is kept constant across genotypes, the model produces genetic variation in FA at the phenotypic level as well as a dependency of both phenotypic value and FA on genetic background. Although the model is highly idealized, Leamy *et al.*'s (2002) finding that there is an epistatic genetic basis for FA in the mouse mandible offers some relevant empirical evidence.

The core of the idea that nonlinear dynamics are relevant to the developmental genetic basis for variability is quite simple. Imagine two structures or phenotypic characters that have a nonlinear dependence on some common developmental factor such as the level of transcription of a signaling factor or hormone. Assume that the relationships between the developmental factor and the phenotypic outcome differ between the two structures, and assume also that there is a certain environment/developmental instability variance for the developmental factor that is not altered by the mutation. In this case, a mutation that affects the developmental factor might shift one structure into a region of minimal dependence on the developmental factor and the other into a region of high dependence. In the mutant case, the shift in the developmental factor has moved the phenotypic component represented by the solid line in Figure 22-7 into a region of high sensitivity, while the dashed line remains in an area of low sensitivity. The effect of the mutation is, therefore, to increase the environmental variances of the two structures. Because the increase in variance will differ for the two structures in the case shown

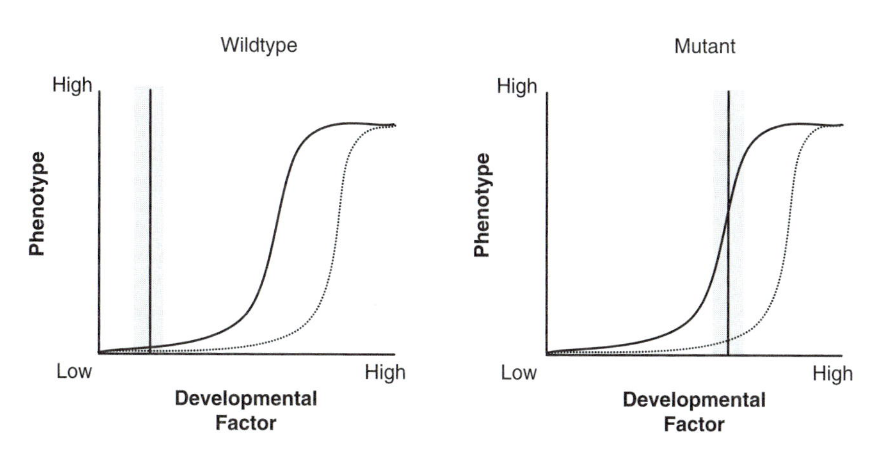

FIGURE 22-7. An illustration of a hypothetical effect of a mutation on the environmental variance and covariation of two phenotypic components. The curves represent the relationships between some developmental factor and the phenotypic outcome for two aspects of the phenotype (solid and dashed lines). The vertical line represents the genotypic value for the wild-type and mutant alleles against some genetic background. The shaded area represents the environmental variation in that factor.

in Figure 22-7, the result will be to reduce the correlation of the two traits even in the absence of differences in the genetic variation for the developmental factor. Thresholds can also produce genetic variation in variability components. Thresholds can be extreme cases of nonlinearity, as in the case of an extreme sigmoid relationship between a developmental determinant and a phenotypic outcome. They can also be real such as when a fusion event either occurs normally or not at all. Real thresholds can also exist when there is a finite capacity to respond (e.g., saturation of receptors or real limit to rate of cell proliferation) to some developmental factor such as a signaling protein so that variation in the amount of that factor above a certain limit produces no phenotypic effect. The *Brachymorph* mutation discussed in the previous section may be an example of a threshold effect. Thresholds are ubiquitous in development, and there are many other examples, of which growth differentiation factor-5 (Gdf-5) and joint formation is one. Ectopic expression of Gdf-5 in mice elicits new joints and suppresses adjacent joints (Storm and Kingsley, 1999; Baur *et al.*, 2000). Another well-known example is segmentation in *Drosophila*. Here, a discrete spatial pattern of concentrations for some gene products (e.g., *Eve* striped 2) is produced by overlaid gradients of expression for others (e.g., bicoid protein, hunchback protein) (Carroll *et al.*, 2005).

If our explanation for the increase in variability in the *Brachymorph* phenotype is correct, this is a case in which variation in the degree of sulfation has no or minimal effect on craniofacial shape and size when it falls within the physiologic range. However, as the sulfation of GAGs is reduced because of the disruption of the functioning of PAPSS2 enzyme, the degree of sulfation begins to affect the growth of cartilage.

The *Brachymorph* example can also be used to illustrate another aspect of developmental organization that is relevant to the developmental genetics of variability. Redundancy and complexity reduce variation by increasing the number of independently varying components that determine the behavior of a system. Redundancy in gene function is a necessary consequence of gene duplication because the initial duplication event is followed by a highly variable period of functional divergence between the paralogous genes (Maslov *et al.*, 2004). As Lande (1977) pointed out, a process composed of a large number of independently varying components will have a lower variance than one with fewer components if the variances of the components remain the same. In the *Brachymorph* example, the null mutation affects an enzyme involved in Paps synthesis, which is the major source of activated sulfate for the production of chondroitin sulfate. Despite the inactive Papss2 protein, some Paps is produced in *Brachymorph* mice. This is probably because of a degree of redundancy between the function of Papss2 and that of a related enzyme, Papss1. If variation in some developmental factor, in this case Paps production, is related to variation in the phenotypic outcome, then the number of independently varying elements that influence that factor will be related to the variance of the phenotypic outcome. In this case, eliminating one important element, Papss2, would be predicted to increase variance purely on the basis of a reduction in the number of regulating components. At a more abstract level of developmental organization, modularity is relevant to the genetic architecture of phenotypic variability. Modularity relates to phenotypic variability because the degree of independence among developmental processes or the development of different structures determines how independently they vary. A structure whose development is influenced by several semi-independent processes or modules will have a lower variance than a structure that is influenced only by one, again assuming that the variances of the outputs of each module are comparable.

IV. A DEVELOPMENTAL SYSTEMS APPROACH TO PHENOTYPIC VARIABILITY

Advances in developmental biology and genetics combined with the development of ever more sophisticated bioinformatics tools and resources is creating the potential for a new and more productive approach to understanding how phenotypic variation and variability is generated in developmental systems. Essentially, this approach combines the construction of increasingly realistic models of development that can be tested by comparing predicted with observed effects of known genetic and environmental perturbations. The construction of models of biological systems that have real predictive power moves the focus of developmental biology from the gene to the larger network or system (Zhu *et al.*, 2004). There are significant theoretical and methodological challenges to taking a developmental systems approach to development and evolution. As Zhu *et al.* (2004) point out, a significant challenge to developmental modeling is that the topology of the systems in development is dynamic and self-generated, whereas current modeling techniques assume a preexisting topology. Significant advances are being made in this field, however. Currently, representative models with real predictive power are being developed for cell cycle regulation in yeast (Stelling and Gilles, 2004). As such models acquire increasing predictive value for more complex developmental systems such as vertebrate limb or craniofacial development, they will become central to understanding how phenotypic variability arises from complex developmental systems.

Models of development will eventually bridge levels of mechanism, containing representations of mechanisms at the gene transcription, posttranscriptional, proteomic, and large-scale developmental process levels. Successful implementation of this approach will reveal which components or properties of developmental systems vary and how variation is translated between levels of developmental mechanisms. As Minelli (2003) has argued, we often lose sight of the relevant level of mechanism for particular developmental or evolutionary explanations by assuming *a priori* that the explanation resides at the level of the gene. In many cases, the relevant variation resides at the level of processes such as cell proliferation, migration, or condensation (Atchley and Hall, 1991; Hall and Miyake, 1995, 2000). Analysis of the genetic networks underlying such processes often reveals that many different genetic changes can produce the same change at the developmental process level. Similarly, there are cases in which variation in different developmental processes can produce very similar phenotypic effects. One example is the regeneration of the lens in urodele amphibians by processes that differ from those responsible for production of the original lens, while other examples include divergent mechanisms of gastrulation or formation of the neural tube (Hall, 1999). The goal of a developmental systems approach is to demystify how variation is translated between levels and how the architecture of development structures the ways in which developmental systems respond to genetic or environmental perturbation.

Currently, there are no experimental models in multicellular organisms for which knowledge of the system is sufficient to fully apply this approach. One of the first issues that arises when we consider the daunting task of modeling complex systems such as the development of the limb or the face is the level of resolution that the models need before yielding useful predictions. The construction of models that completely characterize all molecular level interactions relevant to complex developmental processes is probably very far off and may even be impossible to attain. To be useful, a model has to incorporate relevant known aspects of a system and use those to generate new information. This new information can take the form of predictions about the behavior of the system that can be tested empirically and used to validate and refine the model. Given this view, models need not be built entirely from the bottom or molecular level up (as with cell cycle or lac operon models in yeast), but can be refined simultaneously at multiple levels of organization. In this section, we discuss how a systems approach might be applied to understanding the developmental basis for phenotypic variation and variability in the mouse mandible and the growth of the facial prominences as these systems become more fully understood. Other systems are much better understood in the mouse (e.g., limb development) or in other organisms (*Caenorhabditis elegans* or *Drosophila*). We chose these two systems because they relate to understanding the evolutionary developmental biology of the mammalian craniofacial complex, which is a major area of interest for all three authors.

A. The Regulation of Form in the Mouse Mandible

The mouse mandible is rapidly becoming an important model system for the study of the developmental basis for evolutionary change and phenotypic variation. The driving force behind the development of this model system has been Atchley's and Hall's (1991) model for development and evolution of the mammalian mandible.

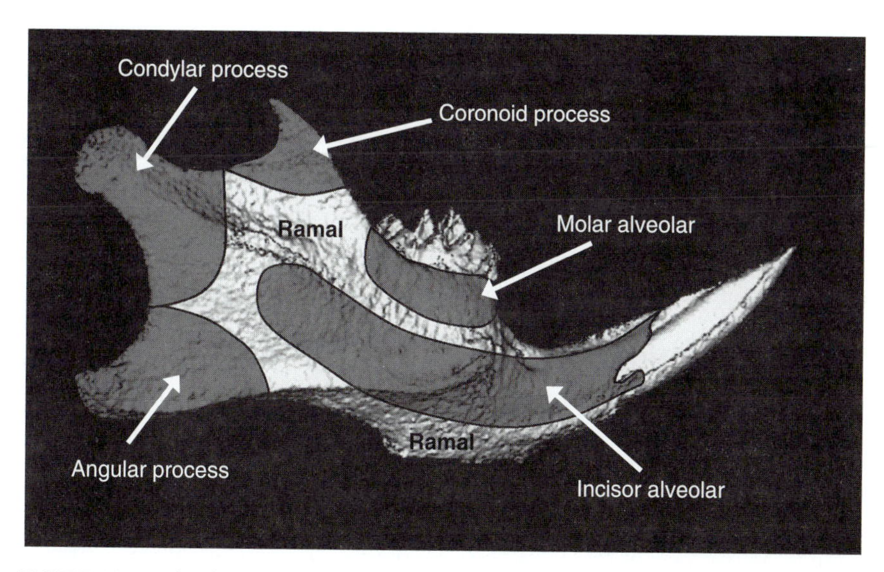

FIGURE 22-8. The division of the mammalian mandible into morphogenetic units according to Atchley and Hall (1991).

Atchley and Hall argued that the mandible is composed of six distinct morphogenetic units, each of which is derived from a separate neural-crest-derived mesenchymal cell lineages and condensation (Figure 22-8). These morphogenetic units have somewhat varying developmental histories. The ramal unit and two alveolar units form by intramembranous ossification, while the three process units form by endochondral ossification. The alveolar units differ, however, in that cells within them differentiate into both fibroblasts and osteoblasts, while those of the ramal unit only form osteoblasts. The three muscle-bearing process units differentiate into hyaline cartilage, which then undergoes endochondral ossification and is replaced by bone (Hall, 2005). While the morphogenetic units of the mandible obviously share developmental mechanisms related to condensation formation, ossification, and growth, it is also clear that they have a degree of independent genetic regulation (Hall, 2003a). This is most evident in the effects of induced mutations that affect some units and not others because of a perturbation to a pathway specific to one or more units within the mandible. This evidence has been reviewed recently by Hall (2003a).

Morphometric analysis of the effects of quantitative trait loci (QTL) has borne out this modular structure of the mandible and also confirmed the view that genetic variation in the mandible can be conceived as being composed of nested subsets that correspond to combinations of the Atchley and Hall morphogenetic units (Klingenberg *et al.*, 2001, 2004). Interestingly, pleiotropic effects of QTLs that affect mandibular shape also show a modular pattern (Ehrich *et al.*, 2003; Cheverud *et al.*, 2004).

The embryological, experimental, and morphometric evidence thus suggests that the development of the mandible can be understood as being composed of sets of developmental processes that are common to the whole structure, processes that are common to some components and not others, and variations on common processes that are under specific genetic control. These common processes are cell division, migration, expansion, and tissue interaction leading to condensation formation and growth. Processes shared by some and not others include the mode of ossification and epigenetic influences produced by muscles that attach to the mandible. The modulation of these common processes among the regions of the mandible are only partially understood, but knock-outs of *msx-1*, *goosecoid*, and *TGFβ-3* perturb development of some units and not others (Hall, 2003a). The developmental genetics of all of the processes involved is far too complex to review here. However, we can use what is known about the formation of mesenchymal condensation and their differentiation into cartilage as an example. Figure 22-9 presents an overview of these processes as compiled by Hall and Miyake (2000).

This model system is still far from the level of resolution needed to generate quantitative predictions from postulated genetic variation. However, there is sufficient

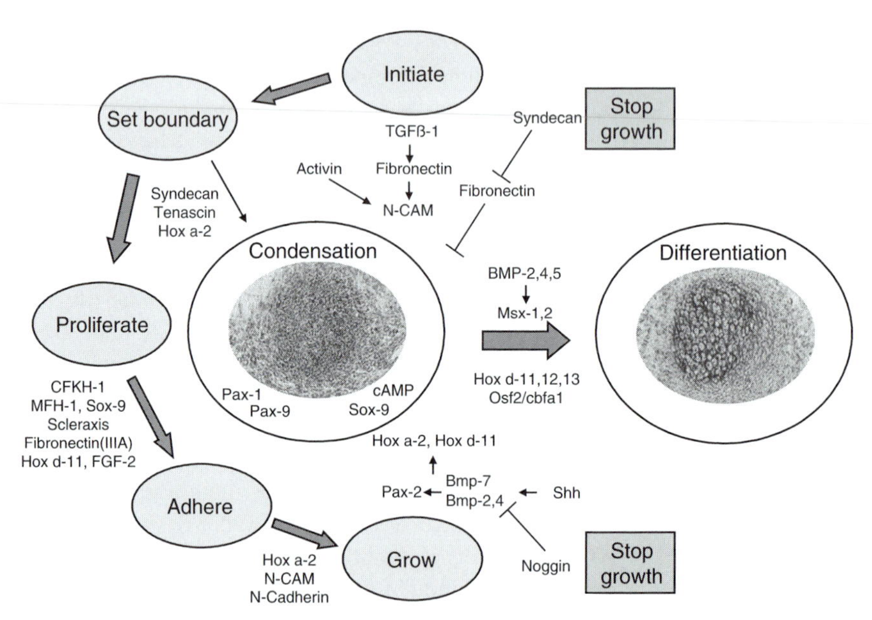

FIGURE 22-9. An overview of the genetic network that regulates mesenchymal condensation formation. After Hall and Miyake (2000). (After Hall, B. and Miyake, T. (2000). All for one and one for all: Condensations and the initiation of skeletal development revisited, Bioessays 22:138–147. Reprinted with permission of Wiley-Liss, Inc., a subsidiary of John Wiley & Sons, Inc.)

detail to generate qualitative predictions about both variation and variability based on the presence of nonlinearities, thresholds, and redundancies at various levels of the developmental hierarchy. Nonlinearity is probably ubiquitous in this system. The boundaries of the condensation are set by a mechanism that involves expression of the cell adhesion glycoproteins syndecan and tenascin and thus depends on the mechanical process of cell adhesion. Although the relationship between the expression levels of these particular extracellular matrix proteins and the mechanics of cell adhesion have not been studied, an analogy can be found in the highly nonlinear relationship between the density of extracellular ligands and integrin expression in the cell membrane (Irvine *et al.*, 2002). It is likely that the expression of particular cell adhesion molecules shows a nonlinear relationship to cell adhesion, with an asymptotic approach to an upper threshold above which additional expression produces no increase in adhesion. If this is true, then a mutation that dramatically reduces the expression of a critical cell adhesion protein will shift variation in cell adhesion to a region of the curve that is more highly dependent on expression of that protein (see Figure 22-7). The condensation process will thus become more variable because of increased sensitivity to the expression level of the cell adhesion protein. If that variation is transmitted to subsequent developmental stages, then a more variable skeletal structure will result. We assume here that the expression of the cell adhesion protein varies among individuals but not that the variation in expression level is increased by the mutation.

Redundancy is also common in this system. For instance, although neural cell adhesion molecule (N-CAM) and N-cadherin are the major cell adhesion molecules involved in cell adhesion within condensations, normal appearing condensations form following knock-outs of either gene (Tavella *et al.*, 1994; Luo *et al.*, 2004). Presumably, this reflects a degree of redundancy both between these proteins and others such as cadherin-11 (Luo *et al.*, 2004). However, if the presence of both proteins as well as other ancillary ones can influence cell adhesion within condensations, we can predict that eliminating either N-CAM or N-cadherin would increase the variance of cell adhesion. This might occur because the overall level of proteins available to produce adhesion is reduced as noted in the preceding text or because the number of independently varying components that influence the eventual outcome (per Lande's argument) has been reduced. Again, if variation in cell adhesion translates to variation in the eventual skeletal structure, then mutations that knock-out or perturb the function of these major cell adhesion proteins should affect the variability of the skeletal structure produced. Alternatively, if cell adhesion within the condensation acts as a threshold in the sense that subsequent developmental events depend only on whether the condensation stage occurs or not, and not on cell adhesion within it, then increased variability will be confined to the condensation stage and not transmitted to subsequent development.

B. The Regulation of Outgrowth of the Facial Prominences in Mice

A model that is less well understood but of great potential importance for understanding the developmental mechanisms related to the evolution of craniofacial morphology is the growth of the prominences that form the face. The mesenchyme that gives rise to the bones, muscles, and connective tissue of the head is derived primarily from neural crest and secondarily from paraxial mesoderm (Couly *et al.*, 1993; JIang *et al.*, 2002). The face forms from paired maxillary and mandibular prominences and a single midline frontonasal prominence that grow anteriorly around the oropharynx and fuse to form the primary and secondary palates and mandibles (Figure 22-10). The major genes involved in regulating the outgrowth of the facial prominences are known. *Bmp-4* is known to stimulate mesenchymal proliferation (Kanzler *et al.*, 2000) and is expressed in the epithelium at the tips of the facial processes (Gong and Guo, 2003). *Fgf-8* and *Shh produce interacting signaling factors that regulate the growth of the maxillary processes. Shh* is expressed in the epithelium, whereas the Ptc-2 protein, the receptor for *Shh*, is found in the mesenchyme (Abzhanov and Tabin, 2004). *Fgf-8* regulates *Barx-1* and *Pax-9*, which are expressed in the growing mesenchyme (Francis-West *et al.*, 2003). Several other signaling and transcription factors such as *Et-1*, *Lhx-6* and *Lhx-7*, *Msx-1* and *Msx-2*, and *Fgf-4* also play important roles in regulating facial prominence growth.

We hypothesize that the growth of the face during its formation influences its eventual size and shape and are testing this hypothesis in ongoing work. This conjecture is supported by the observation that mesenchymal condensation size is a major determinant of adult skeletal element size and that condensation size is influenced by the availability of mesenchyme at the time of formation (Hall and

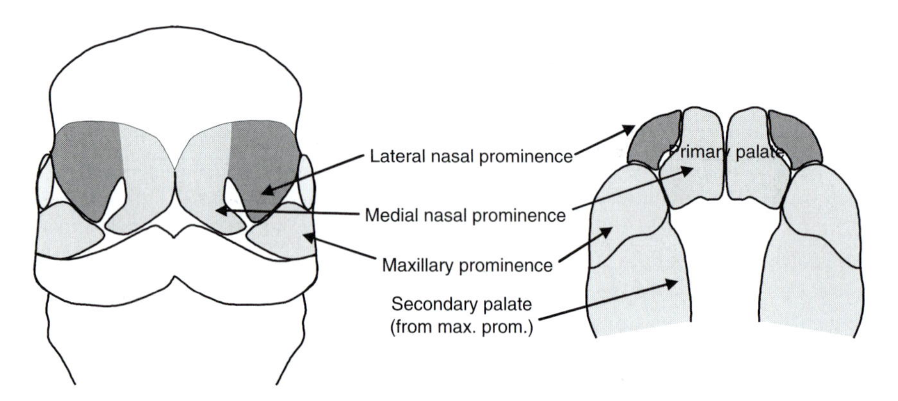

FIGURE 22-10. Schematic diagram of the facial prominences at approximately GD11 in the mouse.

Miyake, 1995). Further, variation in adult beak length in finches is related to variation in *Bmp-4* expression during the formation of the face (Abzhanov *et al.*, 2004). Even in this incompletely understood system, we can make predictions about the developmental-genetics of variability by looking for nonlinearities, thresholds, and redundancy. Several genes are known to have overlapping functions including *Lhx-6* and *Lhx-7, Bmp-5* and *Bmp-7, Msx-1* and *Msx-2,* and *Bmp-2* and *Bmp-4* (Francis-West *et al.*, 2003). Again, we can predict that knocking-out the function of one member of a partially redundant set will increase variability by reducing the number of independently varying components. We know that variability can be increased through perturbations to this system from morphometric analysis of A/WySnJ mice (Hallgrímsson *et al.*, 2004). These mice exhibit delayed growth of the maxillary and nasal prominences and developed cleft lip and palate with incomplete penetrance and variable expressivity (Wang and Diewert, 1992; Wang *et al.*, 1995). Juriloff *et al.* (2004) have shown that nonsyndromic cleft lip in A/WySn mice is produced by two epistatically interacting loci (*clf1* [Chr 11] and *clf2* [Chr 13]). Although the genes at these loci are not yet known, *Msx-1* is misregulated in the facial processes of A/WySnJ mice, with expression persisting past the stage when fusion of the processes occurs (Gong, 2001). Geometric-and-Euclidean distance-matrix-based morphometric analysis of adult A/WySnJ mice shows that morphological integration is very substantially reduced in this strain compared with both C57BL/6J mice and the F1 hybrids of the two strains. Ongoing work is focused on establishing the developmental basis for the reduction in growth of the facial processes, establishing how this particular developmental perturbation increases phenotypic variability and the role that increased variability may play in the etiology of cleft lip and palate.

V. CONCLUSION

The study of the phenotypic variability (canalization, developmental stability, and morphological integration) has largely been conducted outside the mainstream of developmental or evolutionary developmental biology. By focusing on the relationship between variation and variability, this chapter seeks to contextualize the study of the developmental genetics of variability within evolutionary developmental biology. We argue, echoing Hall (1999), that the study of the developmental basis for phenotypic variation or evolutionary change has largely been concerned with explaining changes in mean phenotype. The study of variability, on the other hand, is concerned with the regulation of the variance about that mean or of the tendency of biological systems to exhibit variation. We take the view here that there are not separate and distinct genes responsible for the regulation of the mean phenotype on the one hand and variability on the other. Instead, the same genes that regulate the development of form and whose variation produces changes in mean form also regulate the tendency of the system to vary. The developmental genetic basis for variability most likely resides in emergent properties of the genetic networks that regulate development such as nonlinearity, thresholds, and redundancy. It is our view that understanding the sources of phenotypic variation and the causes of phenotypic variability will converge into a single area of study as we develop a systems-level understanding of how variation is generated and regulated within developmental systems. Then we will start to operationalize the vision of Waddington's epigenetic landscape metaphor.

REFERENCES

Abzhanov, A., and Tabin, C. J. (2004). *Shh* and *Fgf8* act synergistically to drive cartilage outgrowth during cranial development. *Developmental Biology* **273**, 134–148.

Abzhanov, A., Protas, M., Grant, B. R., Grant, P. R., and Tabin, C. J. (2004). Bmp4 and morphological variation of beaks in Darwin's finches. *Science* **305**, 1462–1465.

Atchley, W. R., and Hall, B. K. (1991). A model for development and evolution of complex morphological structures. *Biological Reviews* **66**, 101–157.

Baur, S. T., Mai, J. J., and Dymecki, S. M. (2000). Combinatorial signaling through BMP receptor IB and GDF5: Shaping of the distal mouse limb and the genetics of distal limb diversity. *Development* **127**, 605–619.

Carroll, S. B., Grenier, J. K., and Weatherbee, S. D. (2005). *From DNA to Diversity*. Malden, MA: Blackwell Publishing.

Cheverud, J. M. (1982). Phenotypic, genetic, and environmental integration in the cranium. *Evolution* **36**, 499–516.

Cheverud, J. M., Ehrich, T. H., Vaughn, T. T., Koreishi, S. F., Linsey, R. B., and Pletscher, L. S. (2004). Pleiotropic effects on mandibular morphology II: Differential epistasis and genetic variation in morphological integration. *Journal of Experimental Zoology Part B (Molecular and Developmental Evolution)* **302**, 424–435.

Chippindale, A. K., and Palmer, A. R. (1993). Persistence of subtle departures from symmetry over multiple molts in individual brachyuran crabs: Relevance to developmental stability. *Genetica* **89**, 185–199.

Couly, G. F., Coltey, P. M., and Le Douarin, N. M. (1993). The triple origin of skull in higher vertebrates: A study in quail-chick chimeras. *Development* **117**, 409–429.

Debat, V., Alibert, P., David, P., Paradis, E., and Auffray, J.-C. (2000). Independence between developmental stability and canalisation in the skull of the house mouse. *Proceedings of the Royal Society of London B* **267**, 423–430.

Ehrich, T. H., Vaughn, T. T., Koreishi, S. F., Linsey, R. B., Pletscher, L. S., and Cheverud, J. M. (2003). Pleiotropic effects on mandibular morphology I. Developmental morphological integration and differential dominance. *Journal of Experimental Zoology B (Molecular and Developmental Evolution)* **296**, 58–79.

Emlen, J. M., Freeman, D. C., and Graham, J. H. (1993). Nonlinear growth dynamics and the origin of fluctuating asymmetry. *Genetica* **89**, 77–96.

Francis-West, P. H., Robson, L., and Evans, D. J. (2003). Craniofacial development: The tissue and molecular interactions that control development of the head. *Advances in Anatomy, Embryology, and Cell Biology* **169**, III-VI, 1–138.

Gong, S. G. (2001). Phenotypic and molecular analyses of A/WySn mice. *Cleft Palate–Craniofacial Journal* **38**, 486–491.

Gong, S. G., and Guo, C. (2003). Bmp4 gene is expressed at the putative site of fusion in the midfacial region. *Differentiation* **71**, 228–236.

Graham, J. H., Freeman, D. C., and Emlen, J. M. (1993). Antisymmetry, directional asymmetry, and dynamic morphogenesis. *Genetica* **89**, 121–137.

Hall, B. K. (1999). *Evolutionary Developmental Biology.* Dordrecht, Netherlands: Kluwer.

Hall, B. K. (2003a). Evo-devo: Evolutionary developmental mechanisms. *International Journal of Developmental Biology* **47**, 491–495.

Hall, B. K. (2003b). Unlocking the black box between genotype and phenotype: Cells and cell condensations as morphogenetic (modular) units. *Biology and Philosophy* **18**, 219–247.

Hall, B. K. (2005). *Bones and Cartilage: Developmental and Evolutionary Skeletal Biology.* London: Elsevier Academic Press.

Hall, B. K., and Miyake, T. (1995). Divide, accumulate, differentiate: Cell condensation in skeletal development revisited. *International Journal of Developmental Biology* **39**, 881–893.

Hall, B. K., and Miyake, T. (2000). All for one and one for all: Condensations and the initiation of skeletal development. *Bioessays* **22**, 138–147.

Hallgrímsson, B. (1993). Fluctuating asymmetry in *Macaca fascicularis*: A study of the etiology of developmental noise. *International Journal of Primatology* **14**, 421–443.

Hallgrímsson, B. (1998). Fluctuating asymmetry in the mammalian skeleton: Evolutionary and developmental implications. *Evolutionary Biology* **30**, 187–251.

Hallgrímsson, B. (1999). Ontogenetic patterning of skeletal fluctuating asymmetry in rhesus macaques and humans: Evolutionary and developmental implications. *International Journal of Primatology* **20**, 121–151.

Hallgrímsson, B., Willmore, K., and Hall, B. K. (2002). Canalization, developmental stability, and morphological integration in primate limbs. *American Journal of Physical Anthropology* **35** (Suppl. Yearbook), 131–158.

Hallgrímsson, B., Miyake, T., Wilmore, K., and Hall, B. K. (2003). Embryological origins of developmental stability: Size, shape and fluctuating asymmetry in prenatal random bred mice. *Journal of Experimental Zoology Part B (Molecular and Developmental Evolution)* **296**, 40–57.

Hallgrímsson, B., Dorval, C. J., Zelditch, M. L., and German, R. Z. (2004). Craniofacial variability and morphological integration in mice susceptible to cleft lip and palate. *Journal of Anatomy* **205**, 501–517.

Hallgrímsson, B., Brown, J. J. Y., Ford-Hutchinson, A., Zeldithc M.L., Sheets D.M., and Jirik, F. A. The Brachymorph Mouse and the Developmental-Genetic Basis for Canalization and Morphological Integration, **in preparation a.**

Hallgrímsson, B., Chung, J., and Liu, W. Developmental stability and canalization in mice with deficient type I collagen, **in preparation b.**

Irvine, D. J., Hue, K. A., Mayes, A. M., and Griffith, L. G. (2002). Simulations of cell-surface integrin binding to nanoscale-clustered adhesion ligands. *Biophysiology Journal* **82**, 120–132.

Jiang, X., Iseki, S., Maxson, R. E., Sucov, H. M., and Morris-Kay, G. (2002). Tissue origins and interactions in the mammalian skull vault. *Developmental Biology* **241**, 106–116.

Juriloff, D. M., Harris, M. J., and Dewell, S. L. (2004). A digenic cause of cleft lip in A-strain mice and definition of candidate genes for the two loci. *Birth Defects Research Part A Clinical and Molecular Teratology* **70**, 509–518.

Kanzler, B., Foreman, R. K., Labosky, P. A., and Mallo, M. (2000). BMP signaling is essential for development of skeletogenic and neurogenic cranial neural crest. *Development* **127**, 1095–104.

Kellner, J. R., and Alford, R. A. (2003). The ontogeny of fluctuating asymmetry. *American Naturalist* **161**, 931–947.

Klingenberg, C. P., and Nijhout, H. F. (1999). Genetics of fluctuating asymmetry: A developmental model of developmental instability. *Evolution* **53**, 358–375.

Klingenberg, C., Leamy, L., Routman, E., and Cheverud, J. (2001). Genetic architecture of mandible shape in mice. Effects of quantitative trait loci analyzed by geometric morphometrics. *Genetics* **157**, 785–802.

Klingenberg, C. P., Leamy, L. J., and Cheverud, J. M. (2004). Integration and modularity of quantitative trait locus effects on geometric shape in the mouse mandible. *Genetics* **166**, 1909–1921.

Lande, R. (1977). On comparing coefficients of variation. *Syst Zool* **26**, 214–217.

Leamy, L. J., Routman, E. J., and Cheverud, J. M. (2002). An epistatic genetic basis for fluctuating asymmetry of mandible size in mice. *Evolution* **56**, 642–653.

Luo, Y., Kostetskii, I., and Radice, G. L. (2004). N-cadherin is not essential for limb mesenchymal chondrogenesis. *Developmental Dynamic* **232**(2), 336–344.

Maslov, S., Sneppen, K., Eriksen, K. A., and Yan, K. K. (2004). Upstream plasticity and downstream robustness in evolution of molecular networks. *BMC Evolutionary Biology* **4**, 9.

Marquez, E., and M. Zelditch, (2004). Canalization, developmental stability and integration of rodent skulls (Abstract). *Journal of Morphology* **260**(3), 342.

McBride, D. J., Jr., Shapiro, J. R., and Dunn, M. G. (1998). Bone geometry and strength measurements in aging mice with the oim mutation. *Calcified Tissue International* **62**, 172–176.

Minelli, A. (2003). *The Development of Animal Form: Ontogeny, Morphology, and Evolution*. Cambridge, England: Cambridge University Press.

Phillips, C. L., Bradley, D. A., Schlotzhauer, C. L., Bergfeld, M., Libreros-Minotta, C., Gawenis, L. R., Morris, J. S., Clarke, L. L., and Hillman, L. S. (2000). Oim mice exhibit altered femur and incisor mineral composition and decreased bone mineral density. *Bone* **27**, 219–226.

Rutherford, S. L., and Lindquist, S. (1998). Hsp90 as a capacitor for morphological evolution. *Nature* **396**, 336-342.

Shapiro, M. D., Marks, M. E., Peichel, C. L., Blackman, B. K., Nereng, K. S., Jonsson, B., Schluter, D., and Kingsley, D. M. (2004). Genetic and developmental basis of evolutionary pelvic reduction in threespine sticklebacks. *Nature* **428**, 717–23.

Spichtig, M. (2004). The maintenance (or not) of polygenic variation by soft selection in heterogeneous environments. *American Naturalist* **164**, 70–84.

Stelling, J., and Gilles, E. D. (2004). Mathematical modeling of complex regulatory networks. *IEEE Trans Nanobioscience* **3**, 172–179.

Storm, E. E., and Kingsley, D. M. (1999). GDF5 Coordinates bone and joint formation during digit development. *Developmental Biology* **209**, 11–27.

Tavella, S., Raffo, P., Tacchetti, C., Cancedda, R., and Castagnola, P. (1994). N-CAM and N-cadherin expression during in vitro chondrogenesis. *Experimental Cell Research* **215**, 354–362.

Thom, R. (1975). *Structural Stability and Morphogenesis*. Reading, MA: Benjamin/Cummings.

Van Valen, L. M. (1962). A study of fluctuating asymmetry. *Evolution* **16**, 125–142.

Waddington, C. H. (1957). *The Strategy of the Genes*. New York: MacMillan.

Wagner, G. P. (1996). Homologues, natural kinds and the evolution of modularity. *American Zoology* **36**, 36–43.

Wagner, G. P., and Altenberg, L. (1996). Complex adaptations and the evolution of evolvability. *Evolution* **50**, 967–976.

Wagner, G. P., Booth, G., and Bagheri-Chaichian, H. (1997). A population genetic theory of canalization. *Evolution* **51**, 329–347.

Wang, K. -Y., and Diewert, V. M. (1992). A morphometric analysis of craniofacial growth in cleft lip and noncleft mice. *Journal of Craniofacial Genetics Development Biology* **12**, 141–154.

Wang, K. Y., Juriloff, D. M., and Diewert, V. M. (1995). Deficient and delayed primary palatal fusion and mesenchymal bridge formation in cleft lip-liable strains of mice. *Journal of Craniofacial Genetics and Developmental Biology* **15**, 99–116.

Weiss, K. M. (2005). The phenogenetic logic of life. *Nat Rev Genet* **6**, 36–45.

Wilkins, A. S. (2002). *The Evolution of Developmental Pathways*. Sunderland, MA: Sinauer Associates.

Willmore, K., Klingenberg, C. P., and Hallgrímsson, B. (2005). The relationship between fluctuating asymmetry and environmental variance in rhesus macaque skulls. *Evolution* 54(4).

Zhang, X. S., and Hill, W. G. (2003). Multivariate stabilizing selection and pleiotropy in the maintenance of quantitative genetic variation. *Evolution, International Journal of Organic Evolution* 57, 1761–1775.

Zhu, H., Huang, S., and Dhar, P. (2004). The next step in systems biology: Simulating the temporospatial dynamics of molecular network. *Bioessays* 26, 68–72.

INDEX

Note: *The letter* f *designates illustrations;* t *designates tables.*

H

N

O

P

Q

R

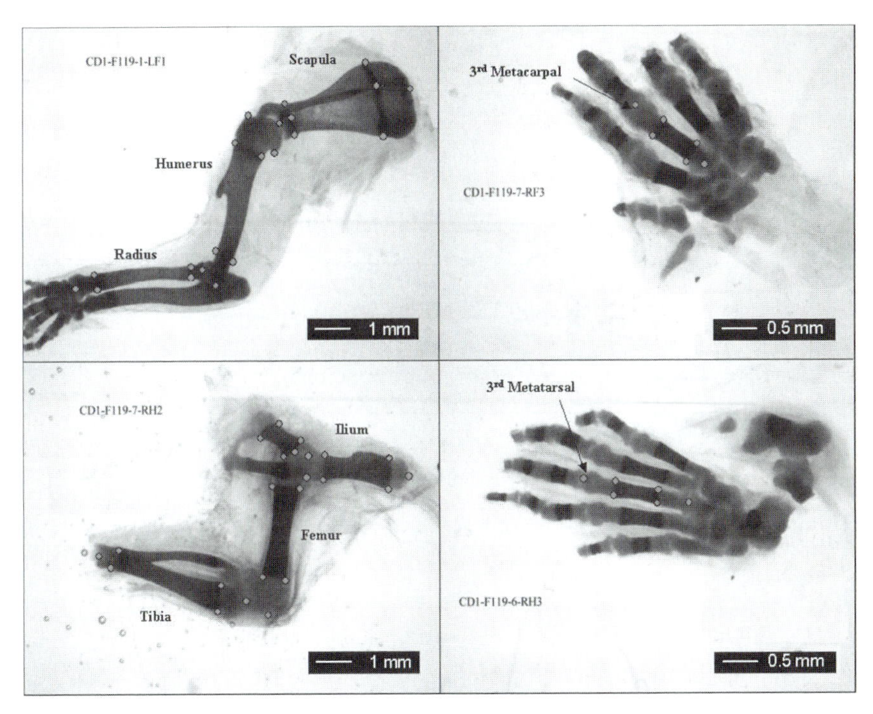

FIGURE 22-2. Landmarks collected for forelimb and hindlimb elements for the CD1 sample. The specimen shown is a neonate (20.5 day sample). Mouse fetuses were cleared and stained with alcian blue for cartilage and alizarin red for bone/osteoid.

to environmental factors. We examined this relationship in a set of 122 interlandmark distances for limb structures in 19-day-old mouse fetuses (Figure 22-2). We found that FA and environmental variances are positively correlated ($r = 0.56$, $p < 0.001$) (Figure 22-3). Willmore *et al.* found a similar but weaker correlation between FA and environmental variance in adult rhesus macaque crania. Together, these results suggest that there is at least some overlap between the mechanisms that regulate FA and canalization. This is not surprising given that the distinction between these two patterns is, to a degree, arbitrary with respect to the range of hypothetical mechanisms that produce them.

B. Perturbation-Based Approaches

Another common approach to understanding a complex system that is hidden from direct view is to subject it to known perturbations and see how it responds. The use of gene knock-outs, knock-downs, or overexpression models are examples of this general approach applied to understanding developmental systems.

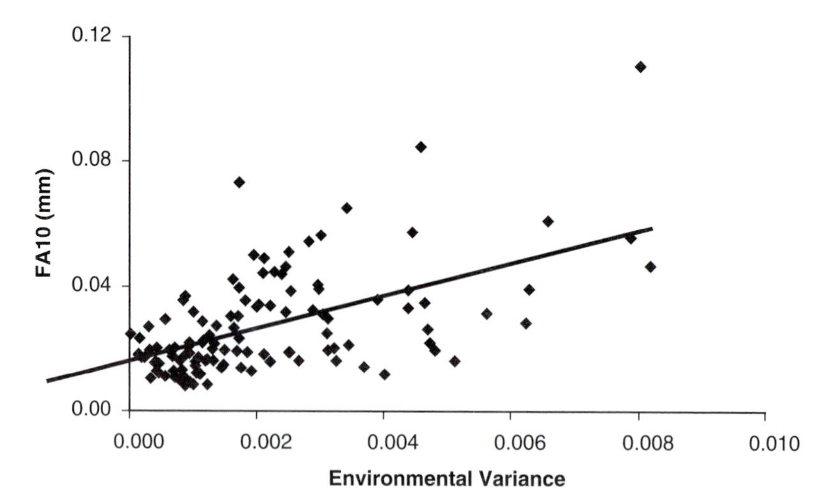

FIGURE 22-3. Size scaled fluctuating asymmetry (FA) plotted against environmental variance for 122 limb traits in gestational day (GD) 19 mouse fetuses ($r = 0.56$, $p < 0.001$). The environmental variance is calculated as $V_E = (1-h^2) \times V_P$.

This approach is useful but cumbersome. By itself, the effect of a particular perturbation does not tell you what a particular gene does. It tells you only that the gene is needed for normal development of the structure in question, although genetic redundancy can complicate even this seemingly simple interpretation. For the study of phenotypic variability, this approach is useful because it can be used to identify developmental pathways or common steps that cross pathways that are particularly relevant to regulating phenotypic variability. In the following text, we present two examples of this approach from work in the Hallgrímsson lab.

To determine whether a major mutation affecting the structure of bone would alter phenotypic variability in the skeletal system, we examined the effect of a null mutation (*oim*) in the pro-α2 chain of collagen I on limb variability in near-neonatal (gestational day [GD] 19) mouse fetuses (Hallgrímsson *et al.*, in preparation a). This mutation produces skeletal effects such as a brittle bone condition that resembles *osteogenesis imperfecta* as well as joint laxity and ligament weakness, deficient myocardial mechanics, and an elevated incidence of craniofacial malformations. We compared samples of 40 individuals per genotype (*oim/oim*, *oim*/+, and +/+), using both geometric morphometrics and interlandmark distances for the same landmarks shown in Figure 22-2. Neither among-individual variance nor fluctuating asymmetry are influenced by this mutation, despite its obvious and severe effects on the mean phenotype for characters such as bone strength and joint function (Figure 22-4). The effects on mean skeletal shape and size at the late fetal stage are also subtle, although other studies have shown reduced growth and

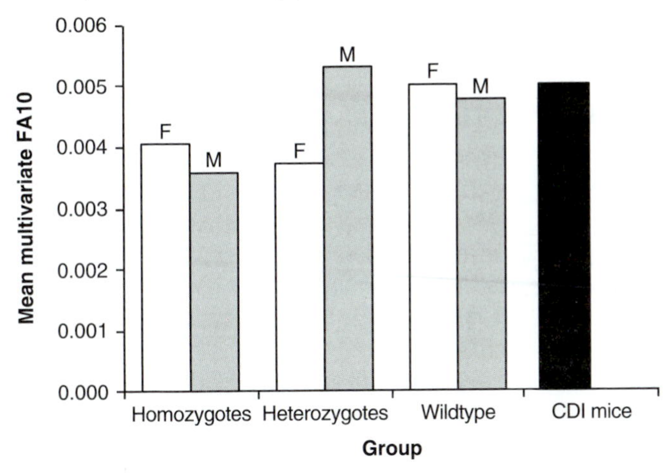

B. Comparisons of individual variance among genotypes

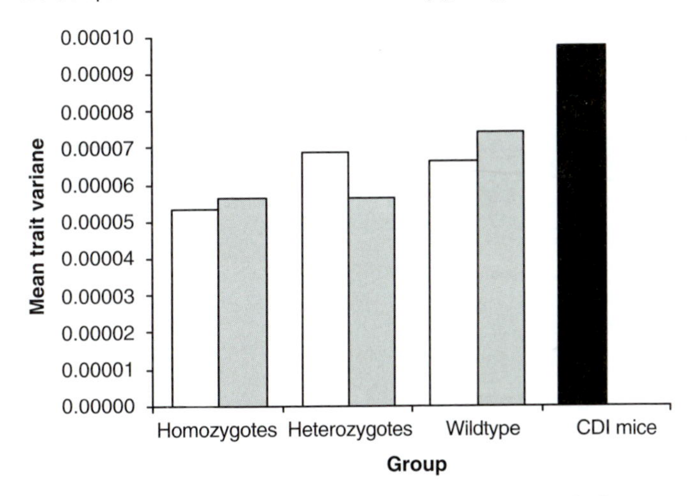

FIGURE 22-4. Multivariate fluctuating asymmetry (FA) and among-individual variance compared among genotypes for the *oim* mutation and CD1 randombred mice. The differences among genotypes are not significant.

shape alteration secondary to fractures during postnatal growth (McBride *et al.*, 1998; Phillips *et al.*, 2000). The results of this study show first that the pathway affected by the *oim* mutation is probably not relevant to phenotypic variability for skeletal size and shape or is somehow compensated for in mutants (perhaps by genetic redundancy if the gene is a member of a multigene family). Second, these

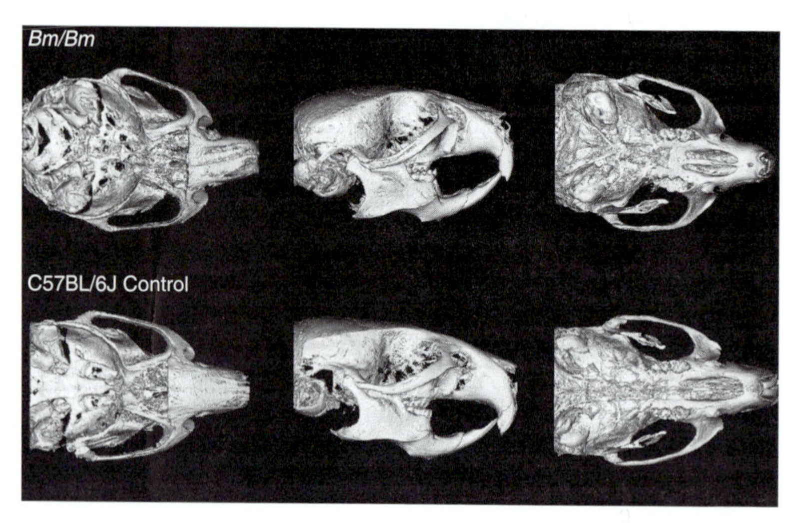

FIGURE 22-5. Three-dimensional reconstructions of computed microtomography scans of a *Bm/Bm* mutant and a wild-type control. In the dorsal view, the neurocranium is cut away to reveal the shortened basicranium in *Brachymorph* mice. The three views are from left to right dorsal, lateral, and ventral.

results illustrate that a mutation that has a severely deleterious effect and is involved in skeletal development need not influence FA. This has profound implications for the often assumed relationship between FA and fitness in natural populations.

The second example is also a mutation that affects structural proteins in the skeletal system, but the result is very different. The *Brachymorph* (*Bm*) phenotype results from an autosomal recessive mutation in the phosphoadenosine-phosphosulfate synthase 2 gene (*PAPSS2*). A major skeletal manifestation is reduced growth of cartilage resulting from undersulfation of glycosaminoglycans (GAGs) in cartilage matrix. This produces a dramatic reduction in the growth of the chondrocranium, producing an altered craniofacial shape (Figure 22-5). The same defect also produces a dramatic decrease in long bone growth in the limbs. We hypothesized that because this mutation affects one component of the cranium, the chondrocranium, directly, and the dermatocranium only indirectly, it should influence morphological integration among these craniofacial components. We used a set of 34 three-dimensional landmarks obtained through computed microtomography to assess craniofacial size and shape in samples of *Brachymorph* mice (N = 20) and wild-type controls (N = 20) (Hallgrímsson *et al.*, in preparation a). Morphological integration is increased significantly in the *Brachymorph* mutant and canalization reduced substantially (Figure 22-6). By contrast, fluctuating asymmetry